ASTRONOMICAL HERITAGE OF THE MIDDLE EAST

COVER ILLUSTRATION:

Top: Zorats Karer (Armenia), an ancient observatory in Armenia. Lower left: explanation of the different phases of the Moon illustrated in astronomical works by al-Biruni (973-1050). Lower right: Maragheh Observatory in northwest Iran.

ASTRONOMICAL SOCIETY OF THE PACIFIC
CONFERENCE SERIES

A SERIES OF BOOKS ON RECENT DEVELOPMENTS IN ASTRONOMY AND ASTROPHYSICS

Volume 520

MS 179, Utah Valley University, 800 W. University Parkway, Orem, Utah 84058-5999
Phone: 801-863-8804 E-mail: aspcs@aspbooks.org
E-book site: http://www.aspbooks.org

ASPCS volumes may be found online with color images at http://www.aspbooks.org.
ASP Monographs may be found online at http://www.aspmonographs.org.

For a complete list of ASPCS Volumes, ASP Monographs, and
other ASP publications see http://www.astrosociety.org/pubs.html.

All book order and subscription inquiries should be directed to the ASP at
800-335-2626 (toll-free within the USA) or 415-337-2126,
or email service@astrosociety.org

ASTRONOMICAL SOCIETY OF THE PACIFIC
CONFERENCE SERIES

Volume 520

ASTRONOMICAL HERITAGE OF THE MIDDLE EAST

Proceedings of a Symposium held at
Yerevan, Armenia
13–17 Nov 2017

Edited by

Sona V. Farmanyan
Byruakan Astrophysical Observatory, NAS RA, Byurakan, Armenia

Areg M. Mickaelian
Byruakan Astrophysical Observatory, NAS RA, Byurakan, Armenia

J. McKim Malville
Department of Astrophysical and Planetary Sciences, University of Colorado, Boulder, Colorado, USA

Mohammad Bagheri
History of Science Department, Encyclopaedia Islamica Foundation, Tehran, Iran

SAN FRANCISCO

ASTRONOMICAL SOCIETY OF THE PACIFIC
390 Ashton Avenue
San Francisco, California, 94112-1722, USA

Phone: 415-337-1100
Fax: 415-337-5205
E-mail: service@astrosociety.org
Web site: www.astrosociety.org
E-books: www.aspbooks.org

First Edition

ASP Conference Series

ISBN: 978-1-58381-927-2
e-book ISBN: 978-1-58381-928-9

Library of Congress (LOC) Cataloging in Publication (CIP) Data:
Main entry under title
Library of Congress Control Number (LCCN): 2019907827

Printed in the United States of America by Sheridan Books, Ann Arbor, Michigan.
This book is printed on acid-free paper.

Contents

Preface ix
S. V. Farmanyan, A. M. Mickaelian, J. M. Malville, and M. Bagheri

Participants xiii

Conference Photograph xvii

Part I. Introduction

Armenian Archaeoastronomy and Astronomy in Culture 3
A. M. Mickaelian and S. V. Farmanyan

Part II. Ancient Astronomy

Environmental and Societal Determinates of Ancient Astronomies 17
J. M. Malville

Astronomical Heritage of Iran 27
S. Nasiri Gheydari

Astronomers and Observatories in Egypt During the Period 969–1914 AD 33
A. A. Hady

Jewish Chronology and Astronomy 45
A. Cohen

Ancient Astronomical Knowledge: the Unity of Diversity 59
I. Simonia

History of Astronomy: Per Aspera Ad Astra 71
A. A. Yeghiazaryan

When Ibn Sīnā Saw Venus on the Disc of the Sun 75
R. Kapoor

Comet Observations in Arab Heritage Books: Did the Arabs Discover the Periodicity of Comets before Edmund Halley? 83
E. K. Konsul

The Expedition of M. R. Štefánik to Observe the Total Solar Eclipse in 1907 from Turkestan 89
P. Hyklová

Part III. Astronomical Heritage for the Future

Development of Astronomy in Turkey 95
S. Alis

Digitized First Byurakan Survey (DFBS) as UNESCO "Memory of the World" Documentary Heritage 105
A. M. Mickaelian, L. A. Sargsyan, G. A. Mikayelyan, L. K. Erastova, K. S. Gigoyan, P. K. Sinamyan and L. R. Hovhannisyan

BAO Plate Archive Heritage 117
A. M. Mickaelian, L. K. Erastova, M. V. Gyulzadyan, G. B. Ohanian, K. S. Gigoyan, G. A. Mikayelyan, P. K. Sinamyan, G. M. Paronyan

Digital Projection of Ancient Astronomical Heritage 127
Ts. Simonia and I. Simonia

Part IV. Astronomical Heritage for Science

The Astronomical Orientations of Sacred Buildings in the Levant During the Bronze and Iron Ages 135
A. Polcaro and V. F. Polcaro

A Pluralistic Vein in the Early Eighteenth Century in Europe 151
M. Capaccioli

Clarification of Unknown Phenomena Using Historical Star Catalogs and Charts . 163
T. Fujiwara and M. Hirai

The Scientific Method: An Approach of Developing Cosmology 171
M. E. Kahil and A. A. Hady

Part V. Astronomical Heritage in Literature and Culture

The Starmus Festival – Our Primary Goal: Science Communication 181
G. Israelian

Kushyar ibn Labban's Scientific Legacy 185
M. Bagheri

Some Astronomical-Cosmic Realities in the Armenian Epic "Sasna Tzrer" 191
G. Broutian

Sunrise as Precondition for the Orientation of Grave-Tombs in Ancient Armenia . 199
H. Y. Simonyan

The First Cosmological Works in the Islamic Period 209
H. Ghalandari

Transylvanian Astronomical Observatories in the 18th Century 213
F. Szenkovits
Vsevolod Viktorovich Stratonov (1869–1938): an Astronomer Driven by Fate from the Middle East to Prague 221
M. Šolc and P. Hyklová
Heavenly Bodies and Phenomena as a Basis for Poetical Images in the Works of Missak Metsarents 229
K. Arakelyan

Part VI. Astronomical Heritage in "Language"

Armenian Names in the Universe 235
A. M. Mickaelian and S. V. Farmanyan

Part VII. Astronomical Heritage for Outreach and Tourism

Astronomical Heritage as a Separate Direction in Scientific Tourism 249
D. Taschyan
Astronomical Tourism Sites in Armenia 253
S. V. Farmanyan and A. M. Mickaelian
Byurakan Astrophysical Observatory as Tourism Center 259
S. V. Farmanyan, V. L. Harutyunyan, G. A. Mikayelyan, and A. M. Mickaelian

Part VIII. Astronomical Heritage for Development

IAU Regional Office of Astronomy for Development: Coordination in Southwest and Central Asia 267
A. M. Mickaelian, S. V. Farmanyan and G. A. Mikayelyan
Astronomy Development in Peru Through Two Generations: the Good and the Bad 279
J. K. Ishitsuka and M. Ishitsuka
Under the Kyrgyz Sky: from Astronomy to Development 285
E. Gradzka
Armenian Astronomical Society (ArAS) Current Activities 295
A. M. Mickaelian, S. V. Farmanyan, and G. A. Mikayelyan

Author Index 305

Preface

The International Conference "Astronomical Heritage of the Middle East" was held on 13–17 November 2017 at the Armenian National Academy of Sciences in Yerevan, Armenia. It was devoted to the role of astronomy in culture and other fields of human activity, and development of these fields due to the knowledge obtained from the Universe. In the modern era, astonomy is probably the field of science which plays a leading role in the formation and development of inter- and multi- disciplinary sciences. For a long time, Astronomy has reached high levels of development, recently new science disciplines have been created, such as Astrochemistry, Astrobiology, Astroinformatics, and Astrolinguistics. Archaeoastronomy plays an important role in culture and in the heritage of nations, chronologies, and calendars created on the basis of astronomical knowledge, etc. Cultural Astronomy also plays an important role in the development of scientific tourism and scientific journalism. The meeting was aimed at the development of problems of astronomy-related interdisciplinary sciences in the countries of the European and Asian regions and preparation of a basis for further possible collaborations by means of presentations of available modern knowledge in various areas of culture by experts from different professions and by joint discussions. A number of outstanding scientists from the mentioned fields, representatives of regional countries (Middle East area) and many Armenian researchers were invited.

The history of Astronomy in Armenia goes back to very old ages. Since ancient times Armenians accumulated astronomical knowledge and have left this heritage in the forms of rock art, ancient observatories, calendars and chronology, records of astronomical events, medieval sky maps, astronomical terms in the language, etc. Nowadays Armenia is one of the developed countries in an astronomical sense as well, though by its economic level Armenia is among the developing countries, and it is situated in a region (Middle East) where efforts are needed to develop and promote cultural astronomy. The famous Byurakan Astrophysical Observatory (BAO) founded in 1946 made a significant contribution in the fields of research of non-stable phenomena and evolutionary processes in the Universe.

The scientific topics of the Conference were presented in the following sessions:

- Astronomical heritage of the Middle East
- Ancient Astronomy
- Astronomy in the Middle Ages
- Astronomy in Ancient Cultures
- Astronomical Heritage along the Silk Road
- Ethnoastronomy
- Astronomical Bases of Philosophy
- Religion and Astronomy
- Planetary Gender
- Astronomy in Folklore and Poetry

- Astronomy in Arts
- Astrolinguistics and Astroheraldry
- Astronomical Observatories as Astronomical Heritage
- Photographic Plates as Astronomical Heritage
- Space World Heritage
- Astro Tourism
- Astronomical Outreach

There was a representative Scientific Organizing Committee (SOC) consisting of 20 members from 13 countries (see below). Altogether, 72 astronomers from 22 countries (Armenia, China, Czech Republic, Ecuador, Egypt, Finland, France, Georgia, Germany, India, Iran, Israel, Italy, Japan, Jordan, Nepal, Peru, Poland, Romania, Spain, Turkey, and USA) attended the meeting. The scientific program consisted of 25 invited and 20 contributed talks. Invited talks by famous scientists were given in the fields of archaeoastronomy, history of astronomy and cultural astronomy followed by a number of contributed ones.

A number of excellent review talks were given on various related topics and many new significant results were presented. Among them, one could mention *Armenian Archaeoastronomy and Astronomy in Culture* by Areg Mickaelian (Armenia), *Development of Astronomy in Turkey* by Sinan Alis (Turkey), *Archeoastronomy in the Middle East (Jordan)* by Awni Moh'd Al-khasawneh (Jordan), *Kushyar ibn Labban's Scientific Legacy* by Mohammad Bagheri (Iran), *Modern cosmology and the "creatio continua" concept by Thomas Aquinas* by Piero Benvenuti (France), *Some astronomical-cosmic realities in the Armenian epic "Sasna Tzrer"* by Grigor Broutian (Armenia), *A pluralistic vein in the early Eighteenth century in Europe* by Massimo Capaccioli (Italy), *Jewish Chronology and Astronomy* by Ariel Cohen (Israel), *Starmus Festival: Musical Astronomical Heritage* by Garik Israelian (Spain), *Environmental and societal determinates of ancient astronomies* by J. McKim Malville (USA), *Digitized First Byurakan Survey (DFBS) as UNESCO "Memory of the World" Documentary Heritage* by Areg Mickaelian (Armenia), *Astronomical Heritage of Iran* by Sadollah Nasiri-Gheydari (Iran), *The Astronomical Orientations of Sacred Buildings in the Levant during the Bronze and Iron Ages* by Vito Francesco Polcaro (Italy), *Iranian National Observatory; building heritage for the future* by Habib Khosroshahi (Iran), *Ancient Astronomical Knowledge—The Unity of Diversity* by Irakli Simonia (Georgia), *Sunrise as Precondition for the Orientation of Grave-Tombs in Ancient Armenia* by Hakob Simonyan (Armenia), *Astronomical Silk Road* by Xiaochun Sun (China), *Ancient Astronomical Knowledge and Rock Art in Armenia* by Karen Tokhatyan (Armenia), *Does "Iron" in Armenian really mean a "Sky Drop"?* by Ararat Yeghikyan (Armenia), and others.

At the end, a general discussion was organized. The Conference provided a good opportunity to discuss further strategy, based on acquired experience, for planning future studies and the developments in the fields of Archaeastronomy and Astronomy in Culture.

This Conference was also rich with social events, including the main touristic attractions of Armenia: Welcome Reception, excursions (BAO tour, including Viktor

Ambartsumian museum, 2.6m and 1m Schmidt telescopes, Zorats Karer—the "Armenian Stonehenge") and the unforgettable Symposium Banquet.

A number of important astronomical meetings have been organized in Armenia, mostly in Byurakan Astrophysical Observatory (BAO): IAU Symposia: #29 in 1966 (Non-Stable Phenomena in Galaxies), #121 in 1986 (Observational Evidence of Activity in Galaxies), #137 in 1989 (Flare Stars in Star Clusters, Associations and Solar Vicinity), #194 in 1998 (Activity in Galaxies and Related Phenomena) and #304 in 2013 (Multiwavelength AGN Surveys and Studies), IAU Colloquium #184 in 2001 (AGN Surveys), as well as the All-European JENAM meeting in 2007. Moreover, Byurakan International Summer Schools (BISS) have been organized since 2006, and six successful schools were held in 2006, 2008, 2010 (combined with the 32nd IAU ISYA), 2012, 2016, and 2018.

The Local Organizing Committee (LOC) consisted mainly of young BAO scientists and other students. We acknowledge all organizations and sponsors supporting the meeting. We would like to thank all participants of the meeting and contributors to this volume, SOC members for their important work on setting up the program of the meeting, and LOC members for their efforts to make the Conference successful.

Sona V. Farmanyan, Areg M. Mickaelian, J. Mckim Malville, and Mohammad Bagheri

September 2018

Organizers and Sponsors

United Nations Educational, Scientific and Cultural Organization (UNESCO)
RA Ministry of Foreign Affairs
RA Ministry of Culture
RA Ministry of Education and Science State Committee of Science (MES SCS)
National Academy of Sciences of the Republic of Armenia (NAS RA)
NAS RA V. Ambartsumian Byurakan Astrophysical Observatory (BAO)
Armenian Astronomical Society (ArAS)
IAU South West and Central Asian Regional Office of Astronomy for Development

Scientific Organizing Committee (SOC)

Piero Benvenuti (Italy, Co-Chair)
Areg Mickaelian (Armenia, Co-Chair)
Juan Antonio Belmonte (Spain)
Grigor Brutean (Armenia)
A. César González-García (Spain)
Habib Khosroshahi (Iran)
Stanislaw Iwaniszewski (Mexico)
Alexander Jones (USA)
Giulio Magli (Italy)
J. Mckim Malville (USA)
Sadollah Nasiri (Iran)
Efthymios Nicolaidis (Greece)
Ray Norris (Australia)
Elma Parsamian (Armenia)
Vito Francesco Polcaro (Italy)
Michael A. Rappenglück (Germany)
Clive Ruggles (UK)
Irakli Simonia (Georgia)
Magda Stavinschi (Romania)
Xiaochun Sun (China)

Local Organizing Committee (LOC)

Areg Mickaelian (Chair)
Sona Farmanyan (Co-chair)
Hayk Abrahamyan
Grigor Brutean
Gor Mikayelyan
Gurgen Paronyan

Participants

Abrahamyan, Hayk, NAS RA Byurakan Astrophysical Observatory (BAO), Armenia

Acharya, Jayanta, Balmikee Campus, Nepal

Aghaee, Alireza, University of Sistan and Baluchestan, Iran

Alis, Sinan, Istanbul University, Department of Astronomy and Space Sciences, Turkey

Altunyan, Ellen, Armenian-Russian Institute, Armenia

Andreasyan, Hasmik, NAS RA Byurakan Astrophysical Observatory (BAO), Armenia

Aramyan, Levon, Technology and Science Dynamics LLC, Armenia

Ataollahi, Hoda, University of Tehran, Iran

Avetisyan, Mane, Yerevan State University, Armenia

Azatyan, Naira, NAS RA Byurakan Astrophysical Observatory (BAO), Armenia

Bagheri, Mohammad, Persian journal Miras-e Elmi (Scientific heritage), Iran

Bailer-Jones, Coryn, Max Planck Institute for Astronomy, Germany

Benvenuti, Piero, International Astronomical Union (IAU), France

Broutian, Grigor, NAS RA Byurakan Astrophysical Observatory (BAO), Armenia

Capaccioli, Massimo, University of Naples Federico II, Italy

Cohen, Ariel, Hebrew University of Jerusalem, Institute of Earth Sciences, Israel

Cristofaro, Ilaria, University of Wales Trinity S. David, MA Cultural Astronomy & Astrology, Italy

Darbinyan, Lilit, Yerevan State University, Armenia

Farmanyan, Sona, NAS RA Byurakan Astrophysical Observatory (BAO), Armenia

Fujiwara, Tomoko, Japan Spaceguard Association, Japan

Gamini, Amir-Mohammad, University of Tehran, Iran

Ghalandari, Hanif, Institute for the History of Science, University of Tehran, Iran

Gigoyan, Kamo, NAS RA Byurakan Astrophysical Observatory (BAO), Armenia

Grandzka, Ewelina, Akademia Ignatianum Cracow, Department of Philosophy, Poland

Green, Daniel W. E., Harvard University, USA

Hady, Ahmed Abdel, Department of Astronomy, Space & Meteorology, Faculty of Science – Cairo University, Egypt

Harutyunian, Haik, NAS RA Byurakan Astrophysical Observatory (BAO), Armenia

Harutyunyan, Varsik, NAS RA Byurakan Astrophysical Observatory (BAO), Armenia

Hovhannisyan, Nona, Yerevan State University, Armenia

Hyklova, Petra, Astronomical Institute of the Charles University, Czech Republic

Ishitsuka, Jose Kaname, Instituto Geofisico del Perú, Perú

Ismaiel, Mohamed, Space Weather Monitoring Center (SWMC), Egypt

Israelian, Garik, Instituto de Astrofisica de Canarias (IAC), Spain

Kahil, Magd Elias, October University Modern Sciences and Arts (MSA), Egypt

Kapoor, Ramesh, Retired, Indian Institute of Astrophysics, India

Khosroshahi, Habib, Institute of Fundamental Research (IPM), Tehran, Iran

Khudanyan, Harutyun, NAS RA Byurakan Astrophysical Observatory (BAO), Armenia

Konsul, Khalil, Arab Union for Astronomy & Space Sciences (AUASS), Jordan

Kostandyan, Gayane, NAS RA Byurakan Astrophysical Observatory (BAO), Armenia

Lopez, Ericson D., Ecuadorian National Observatory, Quito Astronomical Observatory, Ecuador

Magakian, Tigran, NAS RA Byurakan Astrophysical Observatory (BAO), Armenia

Malville, J. McKim, University of Colorado, USA

Martirosyan, Radik, National Academy of Sciences, Armenia

Mickaelian, Areg, NAS RA Byurakan Astrophysical Observatory (BAO), Armenia

Mikayelyan, Gor, NAS RA Byurakan Astrophysical Observatory (BAO), Armenia

Minasyan, Robert, Armenian Institute of Tourism, Armenia

Miqayelyan, Mariam, Armenian-Russian Institute, Armenia

Moh'd Al-khasawneh, Awni, Royal Jordanian Geographic Center, Jordan

Nalbandyan, Naira, Armenian Institute of Tourism, Armenia

Nasiri-Gheydari, Sadollah, Astronomical Society in Iran, Iran

Nikoghosian, Elena, NAS RA Byurakan Astrophysical Observatory (BAO), Armenia

Paronyan, Gurgen, NAS RA Byurakan Astrophysical Observatory (BAO), Armenia

Parsamian, Elma, NAS RA Byurakan Astrophysical Observatory (BAO), Armenia

Pekkola, Marko, Ursa Astronomical Association, Finland

Pikichyan, Hovhannes, NAS RA Byurakan Astrophysical Observatory (BAO), Armenia

Polcaro, Vito Francesco, INAF-IAPS, Italy

Safaei, Iraj, University of Kashan, Iran

Samsonyan, Anahit, NAS RA Byurakan Astrophysical Observatory (BAO), Armenia

Sanayei, Amir Hossein, Sharif University of Technology/ Department of Physics, Iran

Simonia, Irakli, School of Natural Sciences and Engineering, Ilia State University, Georgia

Simonia, Tsitsino, The City Library, Georgia

Simonyan, Hakob, Science Research Center for History and Cultural Heritage, Ministry of Culture, Armenia

Sisoyan, Anna, Yerevan State University, Armenia

Solc, Martin, Charles University of Prague, Astronomical Institute, Faculty of Mathematics and Physics, Czech Republic

Somi, Morteza, University of Tehran, Iran

Sun, Xiaochun, School of Humanities, University of Chinese Academy of Sciences, China

Suvaryan, Yuri, National Academy of Sciences, Armenia

Szenkovits, Ferenc, Babes-Bolyai University, Romania

Tokhatyan, Karen, NAS RA Institute of History, Armenia

Vardumyan, Gohar, NAS RA Institute of History, Armenia

Yeghiazaryan, Anahit, NAS RA Byurakan Astrophysical Observatory (BAO),

Yeghikyan, Ararat, NAS RA Byurakan Astrophysical Observatory (BAO),

Conference Participants

Part I

Introduction

Astronomical Heritage of the Middle East
ASP Conference Series, Vol. 520
Sona V. Farmanyan, Areg M Mickaelian, J. McKim Malville, and Mohammad Bagheri, eds.

Armenian Archaeoastronomy and Astronomy in Culture

Areg M. Mickaelian and Sona V. Farmanyan
NAS RA V. Ambartsumian Byurakan Astrophysical Observatory (BAO), Byurakan 0213, Aragatzotn Province, Armenia; aregmick@yahoo.com

Abstract. We give a review on the Armenian Archaeoastronomy and Astronomy in Culture, including a short overview of archaeoastronomical matters: rock art, zodiacal constellations, ancient observatories, calendars, records of astronomical events by ancient Armenians, and Armenian Medieval astronomy. The current situation with cultural astronomical studies is presented. A comprehensive list of related literature is compiled. Related Armenian institutions involved in archaeoastronomical and cultural astronomical matters are listed and related meetings held in Armenia are described.

1. Introduction

Armenia is extremely rich in archaeoastronomy and culture, including many creations related to astronomical knowledge; ancient calendars, rock art, mythology, etc. On the other hand, this subject is rather poorly developed in Armenia; there are only individual studies on various related issues (especially many studies related to the medieval thinker Anania Shirakatsi, see below) but not coordinated actions to manage this important field of investigation. This article is about Armenian Archaeoastronomy and Astronomy in Culture, a general overview to mention and summarize some activities and related organizations.

2. Armenian Archaeoastronomy

The listing of archaeoastronomical issues include:

- **Zodiac Constellations.** It is believed that the first division of the sky into constellations was made a few thousand years ago, in the 30th-28th centuries B.C., in the Armenian Highland (see *Olcott, W. T. 1911; Flammarion, C. 1880*). According to the American astronomer and historian of science Olcott, the signs of zodiac contain such animals that lived many thousands of years ago in the territory of Armenia and around. It is very probable that ancient people named the constellations after animals living in their countries rather than known from elsewhere. Moreover, many constellations have their own Armenian names which are different from the Greek ones, however, many of them correspond to each other by the meaning.

- **Ancient Observatories.** Two observatories, Zorats Karer,[1] or Karahunge,[2] and Metzamor are especially well known. *Karahunge* is the Armenian twin of the Stonehenge and is considered to be even older. The name derives from kar "stone" and may mean "singing stones." The other famous name is Zorats Kar, and it is the most fascinating historical astronomical building. It is a megalithic assemblage, 200 km from Yerevan, and 3 km from the town Sisian, at an altitude of 1,770 m. The northern latitude is $39°34'$, and eastern longitude is $46°01'$. It is an assemblage of many stones put in a circle and a few arms starting from it. Like many other such buildings, Karahunge was thought to be a religious assemblage. However, only in the middle of the 1980s, Karahunge was first interpreted as an archaeoastronomical monument and was studied by Prof. E. S. Parsamian (1999) and Prof. P. M. Herouni (1998). Estimations give from 7700 to 4000 years ago for the age of Karahunge. There are 222 stones with a total extent exceeding 250 metres, including 84 with holes (with 4-5 cm diameters). Dozens of astronomical stone instruments with an accuracy of 30 arcsec may be found. Forty stones form the central ellipse with 45×36 m sizes, having a ruined stone-cluster in the centre. There is a 8 m wide 8-stone road to the N-E. Some stones were used to find the directions to definite stars. By some estimations (observations of definite stars), the observatory was used during 7700-2200 B.C., for about 5500 years. According to many authors (e.g., Bochkarev & Bochkarev 2005), a comparison of the present state of the monument with its situation a hundred years ago reveals a considerable degradation. Thus, the monument needs urgent protection. The monument is unique of its kind at least in the Trans-Caucasian region and could even be the oldest known observatory in the world. If the estimated age of Karahunge is confirmed by archaeological methods, it clearly should be included in the UNESCO World Heritage list of the most important cultural memorials of our planet.

  Metzamor is the other ancient observatory in Armenia. Metzamor was an ancient town near river Metzamor, 35 km from Yerevan, in Armavir province. There was a settlement since the 5th millennium B.C. It was first interpreted as an archaeoastronomical monument in the middle of the 1960s by Prof. E. S. Parsamian (1985). There is an observatory located outside of the fortress. The most probable estimation of the age is 4600 years. Like Karahunge, Metzamor also needs better study and proper attention both from the Armenian government and world archaeoastronomical community. There are a few other sites in Armenia (Sevsar, Astghaberd, Portablur, Agarak, etc.) that are associated with astronomical activity of our ancient habitants.

- **Armenian Rock Art.**[3] There are numerous petroglyphs of astronomical content in Armenia. Studies of the Armenian rock art present in the territory of modern Armenia (historic Armenia was ten times larger, having a 300,000 square km area) show that the Armenians were interested in heavenly bodies and phenomena. The Earth, the Sun, the Moon, planets, comets, Milky Way, stars, and

---

[1]`http://en.wikipedia.org/wiki/Zorats_Karer`

[2]`http://www.armeniapedia.org/wiki/Karahunj`

[3]`http://www.armrockart.nt.am/`

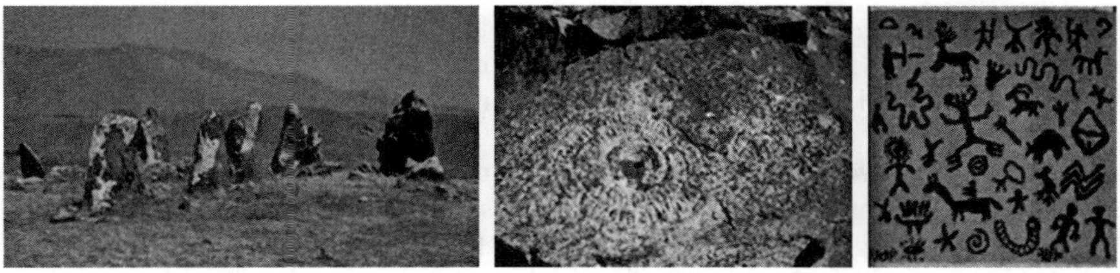

constellations are reflected in these pictures drawn on rocks in mountains around Lake Sevan and elsewhere in Armenia. These pictures and drawings are being studied by a number of historians, archaeologists, and astronomers. However, there is not enough governmental support to organize large-scale studies or at least try to catalog and preserve these ancient treasures.

- **Ancient Armenian Calendar.**[4,5] Ancient (since 2492 B.C.) and medieval (552 and 1085) calendars. According to investigations by H. S. Badalian (1970), B. E. Tumanian (1985), and G. H. Broutian (1997), the Armenian calendar was one of the most ancient in the world, and may even be the most ancient one. Armenians used a Lunar, then Luni-Solar calendar, and since the middle of the 1st millennium B.C. they changed to a Solar calendar, which contained 365 days (12 months by 30 days and an additional month of 5 days). The new year began in Navasard (corresponding to August 11), when the grape harvest was underway and the constellation Orion (Armenian "Haik") became visible in the night sky. Together with the months, all days of any month also had proper names. The year 2492 B.C. was adopted as the beginning. The Armenian Great Calendar was introduced in the VI century, and the difference with the Julian one was re-calculated. It is remarkable that the Mkhitarians from Venice are the oldest publishers of the Armenian and world calendars (since 1775).

- **Astronomical Terms and Names** used in Armenian language since the second and first millennia B.C. (Armenian names of the planets, constellations and some stars).

- **Records of Astronomical Events** by ancient Armenians (e.g., Halley's comet on the 87 B.C. coin by Armenian king Tigranes II the Great, supernovae explosion in 1054 on the location of present Crab nebulae). The coins of Armenian king Tigranes II the Great (95-55 B.C.), silver and copper-bronze tetradrachms and drachms, clearly reveal a star with a tail on the royal tiara which may be associated with the Halley's comet passage of 87 B.C. If so, one has another case when astronomical events can be useful for historical chronological problems; this would be a far earlier record of Halley in Armenia than was previously known from chronicles and also one of the earliest known images of Halley's comet. According to Prof. Pskovskiy, the 1054 Supernova was first seen and recorded in Armenia in May 1054 (and only later in the summer in China). Interestingly, its remnant, the famous Crab nebula, has been studied in detail in

---

[4] http://haytomar.com/

[5] http://en.wikipedia.org/wiki/Armenian_calendar

the Byurakan Astrophysical Observatory and was one of its famous objects of investigation. This nebula has been a natural laboratory for many astrophysical investigations in various multiwavelength ranges.

- **Anania Shirakatsi's** (612-685) **Astronomical Heritage**.[6] One of the most remarkable scientists in the Middle Ages was Anania Shirakatsi (VII century), who had rather progressive astronomical ideas for those times. He was the most important scientist in Armenia, as he was a philosopher, mathematician, geographer, astronomer, chronologist, etc. He has left a few books and writings that survived up to nowadays. Many of them are kept in Matenadaran, the museum of ancient manuscripts. Anania Shirakatsi knew about the spherical shape of the Earth. He also accepted that the Milky Way consisted of numerous faint stars, could correctly interpret Lunar and Solar eclipses, and had other progressive astronomical knowledge for that time. Anania compiled chronological tables, astronomical textbooks, etc. Anania Shirakatsi's works serve as the main source for establishing the ancient Armenian astronomical terminology, including the names of constellations and stars.

- **Medieval Sky Maps and Astronomical Devices** by Ghukas (Luca) Vanandetsi[7] and Mkhitar Sebastatsi[8]. Ghukas Vanandetsi (17th-18th centuries) and Mkhitar Sebastatsi (1676-1749) lived and worked in Europe in the 17th-18th centuries and are known for their detailed charts of the heavens. Lukas Vanandetsi made astronomical instruments, and published the first sky chart with Armenian names of constellations in Amsterdam at the beginning of XVIII century. Mkhitar Sebastatsi was the person who founded the Armenian Catholic Church community in St. Lazar island near Venice, a touristic site for many visitors.

3. Armenian Astronomy in Culture

We are going to open a similar webpage for **Armenian Astronomy in Culture**. Here we list the main topics related to Astronomy in Culture:

---

[6] `http://www.aras.am/Shirakatsi/index.html`

[7] `http://hy.wikipedia.org/wiki/%D5%82%D5%B8%D6%82%D5%AF%D5%A1%D5%BD_%D5%8E%D5%A1%D5%B6%D5%A1%D5%B6%D5%A4%D5%A5%D6%81%D5%AB`

[8] `http://en.wikipedia.org/wiki/Mkhitar_Sebastatsi`

- Astronomy in Ancient Cultures
- Ethnoastronomy
- Astronomy, Religion and Mythology
- Astronomical Bases of Philosophy
- Astronomy and Astrology
- Astronomy in Folklore and Poetry
- Astronomy in Arts
- Astronomy in Fashion
- Astrolinguistics; Astronomical Terms and Names
- Astroheraldry and Astrosymbolism
- Astro Tourism

We have established a research department in the Byurakan Astrophysical Observatory "Archaeo- and Cultural Astronomy," where related studies are being carried out.

4. Publications on Armenian Archaeoastronomy and History of Astronomy in Armenia

Here we give a comprehensive list of 69 references related to Armenian archaeoastronomy and History of astronomy in Armenia:[9]

Abrahamyan, A. G. 1940. Space Science and Chronology, Yerevan

Abrahamyan, A. G. 1944. Anania Shirakatsi's bibliography. A study, Yerevan

Abrahamyan, A. G. 1960. Shirakatsi's new astronomical tables, *History of Natural Sciences and Technology,* No. 1

Abrahamyan, A. G. 1962. Lunar cycles, Yerevan

Abrahamyan, A. G.; Petrosyan G. B. 1970. Anania Shirakatsi, Yerevan

Abrahamyan, A. G.; Petrosyan G. B. 1979. Anania Shirakatsi, Bibliography, Yerevan: "Sovetakan Grogh" publ. house

Abrahamyan, R.; Tumanian B. E. 1956. About Anania Shirakatsi's astronomical works, *Historic-Astronomical investigations,* II issue, M., p. 239-346

Abrahamyan, R.; Tumanian B. E.; Hakobyan T., Melik-Bakhshyan St. 1958. Anania Shirakatsi, Yerevan

Acharyan, H. 1950. Haykakanq, Shirakatsi's month names, *Bulletin NAS RA Soc. Sci.,* No. 12

Anania Shirakatsi 1962. Cosmography, trans. from Ancient Armenian, preface and comments by K.S. Ter-Davtyan and S.S. Arevshatyan, Yerevan

Arakelyan, A. 1955. Anania Shirakatsi, Yerevan

---

[9]see also `http://www.iatp.am/ara/library/index.html`

Arakelyan, A. 1959. Anania Shirakatsi, *History of development of Armenian intellectual culture,* No. 1, Yerevan

Arevshatyan, S. 2014. Anania Shirakatsi's Natural Philosophical Views, *Astronomical Heritage in the National Culture, Proc. Archaeoastronomical Meeting dedicated to Anania Shirakatsi's 1400th anniversary,* held 25-26 Sep 2012 in Byurakan, Armenia. Yerevan, "Gitutyun" Publ. House of NAS RA, p. 68-75

Badalian, H. S. 1940. History of Calendar, Yerevan

Badalian, H. S. 1970. History of Calendar, Publishing House of the Armenian Academy of Sciences, Yerevan

Bochkarev, N. G.; Bochkarev, Yu. N. 2005. Armenian Archaeoastronomical Monuments Carahunge (Zorakarer) and Metsamor: Review and Personal Impressions, *Proceedings of SEAC X Annual Conference: Cosmic Catastrophes,* held in Tartu, Estonia, 2002, eds. Mare Koiva, Izold Pustylnik, & Liisa Vesik, Tartu, p. 27-54

Broutian, G. H. 1997. The Armenian Calendar, Echmiatzin Publishing House, Ejmiatzin, 560p.

Broutian, G. H. 1998. Anania Shirakatsi's "Kharnakhoran," Echmiatzin Publishing House, Ejmiatzin

Broutian, G. H. 2015. The ancient Armenian calendars' connection with the celestial bodies, *Meeting "Relation of Astronomy to other Sciences, Culture and Society,"* 7-10 Oct 2014, Byurakan, Armenia, NAS RA "Gitutyun" Publ. House

Chaloyan, V. 1957. Natural-Scientific views of Anania Shirakatsi, *"Vizantiyskiy vremennik,"* v. XII, M.

Conybeare, F. 1897. Ananias of Shiraq (A.D. 600-650), *Byzantinische Zeitschiff.* Bd. VI

Danielyan, E. 2014. Anania Shirakatsi's Cosmographical and Natural Philosophical Views, *Astronomical Heritage in the National Culture, Proc. Archaeoastronomical Meeting dedicated to Anania Shirakatsi's 1400th anniversary,* held 25-26 Sep 2012 in Byurakan, Armenia. Yerevan, "Gitutyun" Publ. House of NAS RA, p. 76-86

Eynatyan, J. 2014. Anania Shirakatsi's Overheads System, *Astronomical Heritage in the National Culture, Proc. Archaeoastronomical Meeting dedicated to Anania Shirakatsi's 1400th anniversary,* held 25-26 Sep 2012 in Byurakan, Armenia. Yerevan, "Gitutyun" Publ. House of NAS RA, p. 98-106

Farmanyan, S. V.; Mickaelian, A. M. 2014. Sun and Sun Worship in Different Cultures, *Astronomical Heritage in the National Culture, Proc. Archaeoastronomical Meeting dedicated to Anania Shirakatsi's 1400th anniversary,* held 25-26 Sep 2012 in Byurakan, Armenia. Yerevan, "Gitutyun" Publ. House of NAS RA, p. 36-43

Farmanyan, S. V. 2015. Earth and Earth Worship across Cultures, *Meeting "Relation of Astronomy to other Sciences, Culture and Society,"* 7-10 Oct 2014, Byurakan, Armenia, NAS RA "Gitutyun" Publ. House

Farmanyan, S. V.; Mickaelian, A. M. 2015.Cultures across the Universe and the Roots of Armenian Astronomy, *Publ. Astron. Obs. Belgrade,* No. 94

Gurzadyan, V. G.; Vardanyan, R. 2004. Halley's Comet of 87 BC on the coins of Armenian king Tigranes?, *Astronomy & Geophysics,* Vol. 45, No. 4, p. 4.06

Harutyunian, H. A.; Mickaelian, A. M.; Parsamian, E. S. (Eds.) 2014. *Astronomical Heritage in the National Culture, Proc. Archaeoastronomical Meeting dedicated to Anania Shirakatsi's 1400th anniversary* and *XI Annual Meeting of the Armenian Astronomical Society,* held 25-26 Sep 2012 in Byurakan, Armenia, Yerevan, "Gitutyun" Publ. House of NAS RA, p. 220

Harutyunian, H. A. 2014. Armenian Vahagn God as birth of four Cosmic elements, *Astronomical Heritage in the National Culture, Proc. Archaeoastronomical Meeting dedicated to Anania Shirakatsi's 1400th anniversary,* held 25-26 Sep 2012 in Byurakan, Armenia. Yerevan, "Gitutyun" Publ. House of NAS RA, p. 23-28

Harutyunian, H. A.; Mickaelian, A. M. 2014. Anania Shirakatsi's Cosmographical Works and the Methodology of his Natural Science, *Astronomical Heritage in the National Culture, Proc. Archaeoastronomical Meeting dedicated to Anania Shirakatsi's 1400th anniversary,* held 25-26 Sep 2012 in Byurakan, Armenia. Yerevan, "Gitutyun" Publ. House of NAS RA, p. 87-97

Harutyunyan, S. 2000. Armenian Mythology, Hamazgayin Vahe Setyan Publ. House, Beirut, p. 527

Hayrapetian, S. 1941. Anania Shirakatsi's life and activity, *"Collection of scientific materials" of Matenadaran,* No. 1, Yerevan

Herouni, P. M. 1998. Carahunge-Carenish, a Prehistoric Stone Observatory, *Proc. National Academy of Sciences of Armenia,* Vol. 98, 4, p. 307-328

Herouni, P. M. 2006. Armenians and Ancient Armenia: Archaeoastronomy, linguistics, ancient history, *"Giteliq,"* Yerevan, 320 p.

Khrlopyan, G. T. 1959. Anania Shirakatsi's natural-philosophical visions, *Historical - Philological Journal,* No. 2-3

Khrlopyan, G. T. 1964. Anania Shirakatsi's world vision, Yerevan

Malkhasyanc, St. 1961. Anania Shirakatsi (Bibliographic comments)

Martirosyan, H. 2015. Sisian, Zorats Karer, Karahunge, *Meeting "Relation of Astronomy to other Sciences, Culture and Society,"* 7-10 Oct 2014, Byurakan, Armenia, NAS RA "Gitutyun" Publ. House

Mickaelian, A.M. 2008. Preservation and maintenance of the astronomical sites in Armenia, *Proc. International Conference "Starlight: A Common Heritage,"* held in La Palma, Canary Islands, 19-20 April 2007, Eds. Cipriano Marin & Jafar Jafari, IAC, Spain, p. 403-411

Mickaelian, A. M. 2014. Armenian Astronomical Heritage, *Astronomical Heritage in the National Culture, Proc. Archaeoastronomical Meeting dedicated to Anania Shirakatsi's 1400th anniversary,* held 25-26 Sep 2012 in Byurakan, Armenia. Yerevan, "Gitutyun" Publ. House of NAS RA, p. 44-55

Mirumyan, K. 1998. Anania Shirakatsi, Yerevan

Nazaryan, L.S. 2006. Anania Shirakatsi: the great thinker and scientist of Middle Ages, Yerevan, *"Anania Shirakatsi college of Yerevan" educational complex,* "Zangak-97," p. 190

Nazaryan, L. S. 2013. Anania Shirakatsi, Anania Shirakatsi College, Yerevan, p. 242

Nazaryan, L. S. 2014. Anania Shirakatsi's Life and Activities, *Astronomical Heritage in the National Culture, Proc. Archaeoastronomical Meeting dedicated to Anania Shirakatsi's 1400th anniversary,* held 25-26 Sep 2012 in Byurakan, Armenia. Yerevan, "Gitutyun" Publ. House of NAS RA, p. 58-67

Parsamian, E. S. 1985. On Astronomical Meaning of the Small Hill of Metsamor, *Communications of BAO,* Vol. 57, p. 92-100

Parsamian, E. S. 1985. On Possible Astronomical Significance of Megalithic Rings of Angelacot, *Communications of BAO,* Vol. 57, p. 101-103

Parsamian, E. S. 1999. On Ancient Astronomy in Armenia, *Proceedings of the International Conference Oxford VI and SEAC 1999,* ed. J.A. Belmonte, La Laguna, p. 77-81

Parsamian, E. S. 2014. Archaeoastronomy in Armenia, *Astronomical Heritage in the National Culture, Proc. Archaeoastronomical Meeting dedicated to Anania Shirakatsi's 1400th anniversary,* held 25-26 Sep 2012 in Byurakan, Armenia. Yerevan, "Gitutyun" Publ. House of NAS RA, p. 12-22

Petri, W. 1964. Anania Shirakatsi – an Armenian Cosmographer of the 7th century, *Journal of the German Eastern Society,* Vol. 114, Munich

Semyonov, L. 1953. Anania Shirakatsi as an astronomer, *"Ejmiatzin,"* No. 7-8

Ter-Davtyan, K., Arevshatyan S. 1962. Anania Shirakatsi and his cosmographical works, *Anania Shirakatsi. Cosmography.* Yerevan

Ter-Gulanyan, A. 2014. Ancient Astronomical Hieroglyphs of the Armenian Highland and their Echo in Architectural Structures, *Astronomical Heritage in the National Culture, Proc. Archaeoastronomical Meeting dedicated to Anania Shirakatsi's 1400th anniversary,* held 25-26 Sep 2012 in Byurakan, Armenia. Yerevan, "Gitutyun" Publ. House of NAS RA, p. 29-35

Ter-Mkrtchyan, G. 1896. Anania Shirakatsi, *Vagharshapat*

Ter-Mkrtchyan, G. 1979. Anania Shirakatsi, *Armenological studies,* Book A, Yerevan

Tokhatyan, K. 2014. Origins Rock Art and Calendar in Armenia and Anania Shirakatsi, *Astronomical Heritage in the National Culture, Proc. Archaeoastronomical Meeting dedicated to Anania Shirakatsi's 1400th anniversary,* held 25-26 Sep 2012 in Byurakan, Armenia. Yerevan, "Gitutyun" Publ. House of NAS RA, p. 107-133

Tokhatyan, K. 2015. Primitive Art and Petroglyphs of Armenia, *Meeting "Relation of Astronomy to other Sciences, Culture and Society,"* 7-10 Oct 2014, Byurakan, Armenia, NAS RA "Gitutyun" Publ. House

Tumanian, B. E. 1958. Armenian newly revealed astronomical device, *Haypethrat,* Yerevan, p. 31

Tumanian, B. E. 1962. Geocentric and heliocentric systems in Armenia, Yerevan

Tumanian, B. E. 1964. History of Armenian astronomy. Part 1, Yerevan

Tumanian, B. E.; Mnatsakanian, H. H. 1965. Bronze Age belt-calendar, *Mitq,* Yerevan, p. 56

Tumanian, B. E. 1968. History of Armenian astronomy. Part 2, Yerevan

Tumanian, B. E. 1972. History of chronology, *Academy of Sciences of Armenia,* Yerevan

Tumanian, B. E. 1973. Geocentric and heliocentric systems in Armenia, Yerevan

Tumanian, B. E. 1985. History of Armenian astronomy, YSU Publishing House, Yerevan, p. 286

Tumanian, B. E. 1990. Division of the day into parts in ancient and medieval Armenia, YSU Publishing House, Yerevan, p. 37

Tumanian, B. E. 1991. Anania Shirakatsi, Yerevan

Vardumyan, G. 2014. Anania Shirakatsi and "Pagan" Scientists, *Astronomical Heritage in the National Culture, Proc. Archaeoastronomical Meeting dedicated to Anania Shirakatsi's 1400th anniversary,* held 25-26 Sep 2012 in Byurakan, Armenia. Yerevan, "Gitutyun" Publ. House of NAS RA, p. 134-141

Vardumyan, G. 2015. The Universe and Armenian Mythology, *Meeting "Relation of Astronomy to other Sciences, Culture and Society,"* 7-10 Oct 2014, Byurakan, Armenia, NAS RA "Gitutyun" Publ. House

Yerznkyan, N. 2014. Religious-Historical Sublayers in Anania Shirakatsi's Works, *Astronomical Heritage in the National Culture, Proc. Archaeoastronomical Meeting dedicated to Anania Shirakatsi's 1400th anniversary,* held 25-26 Sep 2012 in Byurakan, Armenia. Yerevan, "Gitutyun" Publ. House of NAS RA, p. 142-148

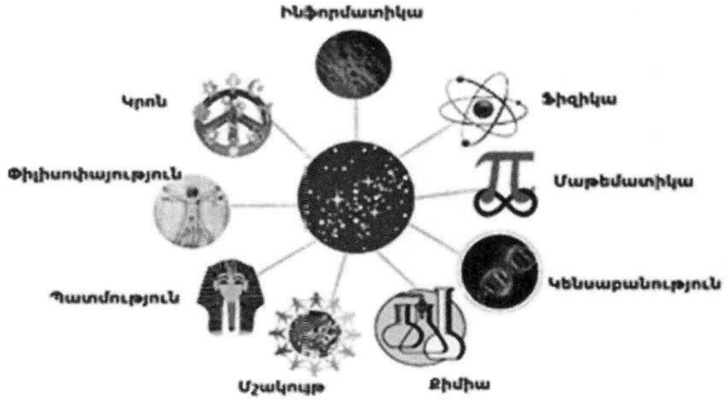

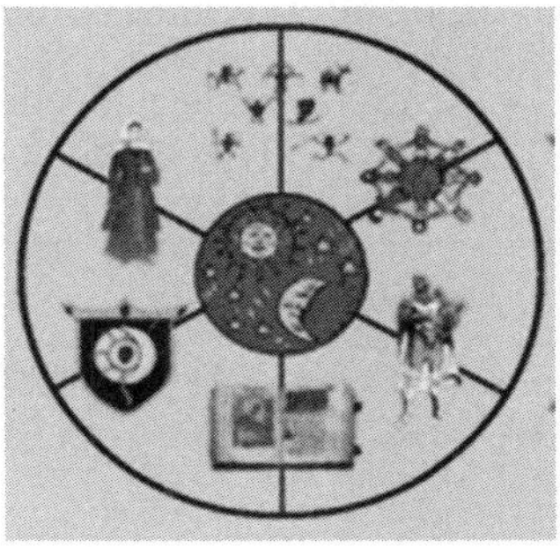

5. Institutions in Armenia involved in AAC

So far, Byurakan Astrophysical Observatory (BAO) coordinates AAC related actions. UNESCO has officially recognized BAO as the coordinator in the field of Astronomy and World Heritage, a sub-project of UNESCO World Heritage project. BAO representatives also are involved in IAU Working Groups *Astronomy and World Heritage (WGAWH)* and *Archaeoastronomy and Astronomy in Culture (WGAAC)* and the European Society for Astronomy in Culture (SEAC, Societé Européenne pour l'Astronomie dans la Culture). Along with BAO, there are several other institutions related to AAC:

- NAS RA V. Ambartsumian Byurakan Astrophysical Observatory (BAO)[10]
- NAS RA Institute of History[11]
- NAS RA Institute of Archaeology and Ethnography[12]
- NAS RA M. Abeghyan Institute of Literature[13]
- NAS RA H. Acharian Institute of Language[14]
- Matenadaran, Mesrop Mashtots Institute of Ancient Manuscripts[15]

6. Meetings in Armenia Related to AAC

During recent years, we have organized a number of meetings, where archaeoastronomy was involved, including those completely devoted to this subject. The last one, held in 2014 was the first meeting on *Relation of Astronomy to other Sciences and Culture* and for the first time it involved the subject *Astronomy in Culture*. The list of these meetings is as follows:

- Joint European and National Astronomy Meeting (JENAM-2007), **Special Session #6: "Archaeoastronomy"**, 21 Aug 2007, Yerevan
- ArAS VIII Annual Meeting "Astronomy and Society," **Session "Archaeoastronomy"**, 7 July 2009, Byurakan

---

[10] `http://www.bao.am/`

[11] `http://www.academhistory.am/en/`

[12] `http://www.archaeology.sci.am/`

[13] `http://litinst.sci.am/en/`

[14] `http://language.sci.am/en`

[15] `http://www.matenadaran.am/`

- **Archaeoastronomy seminar** dedicated to Anania Shirakatsi's 1400th anniversary, 13-14 July 2011, Byurakan
- **Archaeoastronomical meeting** "Astronomical Heritage in the National Culture" dedicated to Anania Shirakatsi's 1400th anniversary combined with ArAS XI Annual Meeting, 25-26 Sep 2012, Byurakan
- **Meeting "Relation of Astronomy to other Sciences, Culture and Society"** (RASCS) combined with ArAS XIII Annual Meeting, Sessions "Archaeoastronomy" and "Astronomy in Culture," 7-9 Oct 2014, Yerevan
- **Young Scientists Conference "Cultural Astronomy in Armenian Highland"**, 20-23 June 2016, Yerevan
- **Meeting "Armenian Astronomical Heritage"**, dedicated to Benik Tumanian's 100th anniversary, 2-3 May 2017, Yerevan
- **International Conference "Astronomical Heritage of the Middle East"**, 13-17 Nov 2017, Yerevan

Abstracts of *JENAM-2007*, the meeting *"Astronomical Heritage in the National Culture"* (2012), *RASCS-2014* and Young Scientists Conference *"Cultural Astronomy in Armenian Highland"* are available in Astrophysical Data System (ADS). Proceedings books have been published for all these meetings except *JENAM-2007* and another one is being published for the International Conference *"Astronomical Heritage of the Middle East"*.

Part II

Ancient Astronomy

Astronomical Heritage of the Middle East
ASP Conference Series, Vol. 520
Sona V. Farmanyan, Areg M. Mickaelian, J. McKim Malville, and Mohammad Bagheri, eds.

Environmental and Societal Determinants of Ancient Astronomies

J. McKim Malville
Department of Astrophysical and Planetary Sciences, University of Colorado, Boulder, Colorado 80309

Abstract. A goal of cultural astronomy is an understanding of the meaning and value of astronomy in ancient cultures and, as a consequence, new insights into the nature of the culture itself. In many cultures there appear to be unique connections between astronomy, cultural traditions and the environment, which include both landscape and skyscape. In some cases, with sufficient knowledge of a culture, one can anticipate its astronomy. Three cases are presented where astronomy appears to be a natural consequence of the culture and its environment.

1. Introduction

The field of archaeoastronomy has undergone a transformation from its earliest roots when goals were largely limited to searches for astronomical alignments. Today, a major concern is the cultural context of the astronomy that is suggested by those alignments. A goal of this paper is to go beyond context and search for meaning. What were people thinking when they designed and built massive construction projects aligned to the heavens? I believe the reason is that certain members of the culture had engaged in what we might identify as mythic thinking, attempting to develop a holistic interpretation of the wider world in which they found themselves. Cassirer (1955) describes mythic thinking as the intuitive understanding of wholeness. Without some sort of narrative involving the interconnectedness between people and the larger world, experiences would remain chaotic and fragmentary, a collection of seemingly unrelated events (Malville 1975). Although myths may often be products of pure imagination and often not falsifiable (the *sina qua non* of modern science), Popper (1961) has suggested that many notable scientific theories have had their origins in mythic thinking. We search for those early myths that might be examples of nascent scientific thinking.

Nabta Playa in southern Egypt contains the oldest known collection of megalithic astronomical alignments, predating by more than a millennium the well-known sites of Stonehenge, New Grange, and the temple culture of Malta. We have no knowledge of the mythmaking of the nomadic pastoralists who built a ceremonial center at Nabta Playa, but we do know something about their stone monuments and can speculate about their mythmaking. In the Late and Terminal Neolithic (7500–5400 BP) nomadic pastoralists built a ceremonial center on the western shore of the seasonal flood lake known as Nabta Playa. The center consists of complex megalithic structures, stone circles, and lines of megaliths crossing the playa. The astronomy of the site appears to be a natural outgrowth of two aspects of the cultures of the nomadic pastoralists: the need to use the brightest stars as navigational aids for annual journeys to the playa for water and forage, and the appearance of monsoon rains around June solstice that brought water to the playa. These nomads were members of a cattle worshiping culture and this combination of June solstice, water, cow worship, and star worship may have been carried by them into the Nile valley and perhaps beyond, continuing for millennia as worship of the cow goddess, Hathor, heliacal rising of Sirius, and the flooding of the Nile.

The second case involves the ceremony of Surya Puja, which is practiced in a number of temples in Tamil Nadu, India. The appearance of naked-eye sunspots at the onset of the Medieval Maximum of solar activity, at the time these were being constructed, apparently produced cognitive dissonance among the Hindu culture, which considered the sun god, Surya, as pure and unblemished.

Finally, in the American Southwest in Chaco Canyon, the appearance of the rising of the December solstice sun over Fajada Butte as viewed from a Great Kiva is the earliest known case of astronomy in Chaco Canyon. It may be the consequence of a horizon calendar and the concern to maintain the cyclic regularity of the seasons.

2. Nabta Playa

Nabta Playa is an internally drained basin located about 100 km due west of Abu Simbel. It is one of the two such basins in the southern Western Desert of Egypt, and was fed by runoff from an enormous area some 1500 square km in size (Malville et al. 1998, 2008). The earliest excavated sites at Nabta have calibrated radiocarbon dates of 10,300–9800 BP. The area began to function as a regional center for nomadic pastoralists during Middle Neolithic (8100–7600 BP), based upon the numerous hearths and the highest frequency of cattle bones of any locality in the Nubian Desert. The Middle Neolithic ended with a short but deep drought, which began around 7600 BP, lasting about 100 years. In the Late Neolithic (7400 to 6600 BP), a new group of nomads began frequenting the playa when it had water. These people, the Ruat El Baquar people (Cattle Herders), were responsible for cattle burials in clay-lined and roofed chambers. The diminished rainfall in the Late Neolithic meant that only a few watered refuges such as Gilf Kebir and Nabta Playa remained (Kuiper and Kropelin 2006). Increasingly skillful navigational techniques, using the stars, would have been required for the nomadic pastoralists to locate these isolated sources of water. Finally, some 30 complex megalithic structures, together with lines of megaliths, were built during the Terminal Neolithic by Bunat El Ansalm people (Megalith Builders). The Terminal Neolithic at Nabta Playa extended from 6600 BP to total abandonment of the area in approximately 5500 BP. These nomads skipped sedentism with its associated agriculture and permanent villages, entering the Late Neolithic with their own repertory of concepts involving astronomy, design of sacred structures, and ritual. Another place where monumental ceremonialism by-passed sedentism appears to have been Göbekli Tepe in southern Turkey.

2.1. The Valley of Sacrifices and the Calendar Circle

Along the western rocky bank of a wadi that entered the playa from the north there are at least ten mounds or tumuli. Built of broken sandstone blocks, the tumuli contained offerings of parts of butchered cattle, goats, and sheep. The largest and perhaps the oldest tumulus contained an entire young cow, the most precious offering that a pastoralist can make. A piece of tamarisk from its roof yielded a calibrated radio carbon date of 7270 years BCE. The animal was lying on its left side, oriented approximately north-south, with its head to the south. This wadi, which is now called the Valley of

Sacrifices, brought water to the playa and would have been an appropriate place to ask the gods for rain by performing cattle sacrifices and other rituals.

The wadi ends with a small sandy knoll containing a circle of stones at its top. A radiocarbon date from a hearth adjacent to the circle yielded a date of 6800 years BP. The circle, approximately four meters in diameter, is composed of individual slabs, ranging in length from 20 to 70 cm set in the sand. Four sets of prominent upright slabs appear to be gates for viewing. When one is lying in the sand, these portals provide sightlines towards the north and toward the position of the rising sun on June solstice. These two alignments may reveal two major astronomical elements in the lives of the nomads. North was important when navigating across the oceans of sand of the Sahara. Even though there was no star at the north celestial pole during this period, the point in the sky around which stars circle would have been a useful navigation tool. June solstice would have been significant for marking a date near the onset of monsoon rains and the flooding of the playa. The association of the stone circle with cattle sacrifices in the Valley of Sacrifices suggests ritual connections between the flooding of the playa, cattle worship, and the solstice sun.

2.2. The Complex Structures

South of the Valley of Sacrifices there are about 30 complex megalithic structures built during the Terminal Neolithic. The largest of the structures is the focus of five radiating megalithic alignments. The builders dug a pit through playa sediments to expose a table rock at a depth of 2.6 m. The rock is a thick lens of hard, quartzitic sandstone that remained after the surrounding softer sediments had been removed by erosion. The northern and western sides of the table rock were shaped and its top was probably worked and smoothed. The oval rock (Figure 1:1) measures 3.3 m by 2.3 m with its long axis aligned slightly west of north. The pit was partially refilled with sand, and a large secondary stone, weighing 2–3 tons, was placed over the center of the table rock (Figure 1:5). This stone was carefully shaped with a large head-like projection giving it the appearance of a cow. It was oriented approximately north-south and held upright by two large slabs set against the structure at its north end.

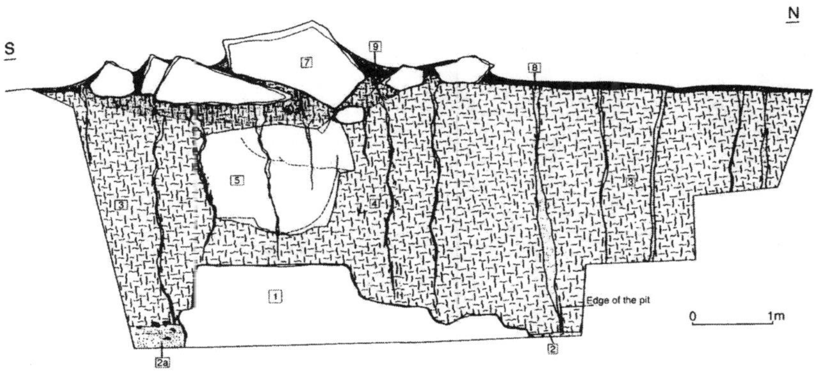

Figure 1. The Central Complex Structure.

2.3. The Alignments

Megaliths are organized along five lines that radiate outward from the largest complex structure (Figure 2). The average weight of the 19 megaliths in the alignments for which we have good measurements, is 2.6 metric tons, ranging upward to the largest megalith, which has a weight of approximately 12 metric tons. The flat sides of the standing megaliths face north. Charcoal from fires in the nearby quarry have given us five radiocarbon dates associated with the quarrying of the megaliths ranging from 6600 BP to 6200 BP.

The brightest star in the region of the sky north of the celestial equator, Arcturus, is a reasonable candidate for the three northern alignments. Each of these three alignments may have been built to account for the changing location of that star due to precession over the period 6600–5700 BP. The two southern alignments may have been associated with stars in the belt of Orion between approximately 6400–6200 BP and toward Sirius and α Centauri, approximately 6600–6300 BP. A dramatic sight would have greeted those who stood at the edge of the flooded playa watching the reflections of stars as they rose above the eastern horizon coming into alignment with the monoliths rising above the dark waters.

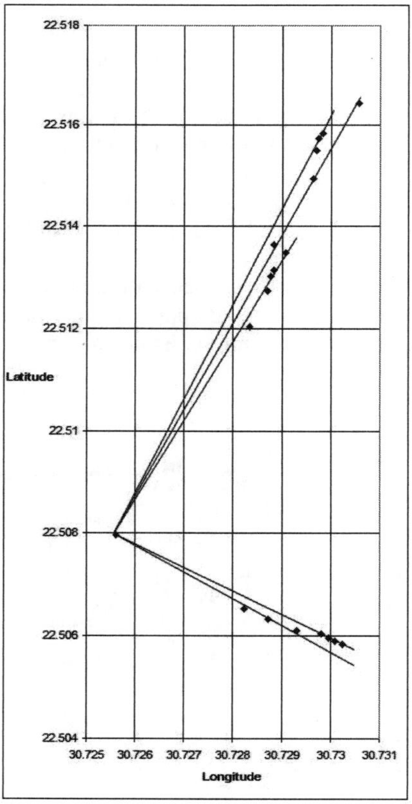

Figure 2. The Five Alignments.

The placing of these megaliths into the playa would have been a significant investment in time and resources in an activity, which transcended survival. It also displays management skills and control over a group of people for extended periods of time, unexpected for Late Neolithic people in the Sahara. That these five alignments converge on the large complex structure, which contains a buried sculpture with the appearance of a cow suggests again the dominance of this animal in the society and, in a fascinating way, the entanglement of cows, nomads, landscape, and skyscape.

2.4. Summary

The evidence for attention to the heavens by these nomads comes in three forms. The repetitive orientation of the faces of the megaliths along a north-south direction, and the cardinal orientations of cattle burials reveals an early symbolic connection to the north, which is what one expects for nomads who need to navigate using the stars. The second bit of evidence for astronomy is found in the stone circle with its two sightlines toward the north and the direction of the rising sun near June solstice; the latter associated with monsoon rains and the flooding of the playa. Finally, the alignments of megaliths, which, if intentionally oriented to the brightest stars of Nabta Playa suggest an even more careful attention to the heavens. Honoring those stars is understandable if the nomads relied upon them for navigation across the Sahara.

Some of the cognitive features of early Dynastic Egypt, such as cattle worship, June solstice, rising stars, and rising of the Nile may have been carried into pre-dynastic cultures in the Nile valley. As the desert dried up, there was perhaps a cyclical movement of nomads between the two ecosystems of the desert and Nile valley in Lower Nubia. Eventually after around 3500 BCE, after the desert became hyper-arid, there was a major migration to the middle Nile Valley in lower Nubia (Gatto 2011). For the pastoralists, cattle may have served as walking larders, providing milk and blood. In contrast, the Predynastic and Old Kingdom residents of the Nile valley had an economy that was based primarily on agriculture and small livestock, such as sheep and goats. The growth of the cult of Hathor is remarkable, especially because it seems to incorporate many of the themes expressed at Nabta Playa such as the onset of flooding, the sun at summer solstice, and the heliacal rising of Sirius, which became a marker for summer solstice and flood. The explicit connection between Hathor, flooding of the Nile, and the heliacal rising of the Nile expressed in the Temple of Hathor at Dendera, built during the Ptolemaic period in the second century CE, appears as a testament to the persistent power of a tradition, which may have had its roots in the nomadic pastoralists of the Western desert.

3. The Surya Puja Temples of South India

There are more than 70 temples in Tamil Nadu in which light of the rising or setting sun, near equinox, enters the sanctuary, and illuminates the image of the god contained therein (Malville and Swaminathan 2005). In many of these temples the event, known as a surya puja, is an elaborate ceremony, organized by the priest of the temple and witnessed by a throng of celebrants. The name of the events suggests the worship (puja) and honoring of the sun god, Surya. The reality is quite different. Since most of these temples are in worship today, it is possible to participate in the ceremony and discuss its meaning with the head priest. The majority of these temples are Shaivite, and the intent of the worshippers is to honor lord Shiva, not the sun. For many of the participants, the sun is seen as a horizontal beam of light, prostrating himself before Shiva.

The best known surya puja in the Tanjor District of Tamil Nadu is held in Nageswarar Temple in Kumbakonam. During the three days of the ceremony some 500 people line up on either side of the horizontal beam of sunlight to watch the illumination of the polished Shiva linga. The priest of the temple has explained that the sun is suffering from leprosy and only by worshipping Shiva could he be cured. That the sun should have leprosy is, indeed, a puzzle. Traditionally light of the sun purifies all that it touches, cleanses laundry, and, indeed, cures one of leprosy. Surya is described as unblemished, pure and white, like a water lily rising out of a muddy pond. Naked-eye sunspots began to appear on the sun in the late eleventh century when the sun entered the period known as the Medieval Maximum of solar activity (1100–1250 BCE). When the sun is dimmed by smoke or fog, large sun spots can be seen, especially at dawn. In the Hindu culture, these sunspots must have produced cognitive dissonance: how could the sun be so afflicted? According to temple traditions at Nageswarar, the sun acquired leprosy due to the curse of a rishi, a Hindu holy man, who was suffering from that disease. When he asked Surya to cure the disease, the sun replied that he was powerless to help. Leprosy was the result of past negative karma, due to misdeeds in a previous life, and the role of the sun was to bestow on human beings both the positive and negative consequence of karmic forces. Upon receiving this response, the rishi cursed the sun for his lack of compassion for the suffering of humans and caused Surya to acquire the disease. Other stories of the leprous sun are found at the temple of Surynarkoil one of the few living temples in which Surya is the primary deity. Elsewhere in India, the ruined sun temple of Khakola Aditya in Varanasi had a ceremony in which dark seeds were thrown into a fire to cleanse spots from the face of the sun (Malville and Singh 1996), which were described as coiled dark snakes. During the same period Samba, the son of Krishna, also suffered from leprosy and was cured in the vicinity of the great sun temple of Konarak (Malville 1989).

4. Fajada Butte in Chaco Canyon, Southwestern United States

Fajada Butte is the single most dramatic topographic feature of Chaco Canyon, the center of a regional system, some 30,000 square miles in size, which flourished from 1000 CE to about 1150 CE in what is now northern New Mexico (Figure 3). At its height it was associated with some 200 outlying communities, the members of which may have travelled to Chaco Canyon around winter solstice for ceremonies and trade (Malville 2008). Before the establishment of the regional system, Fajada Butte may have been a spiritual magnet for early immigrants to the Canyon. They established the Fajada Gap community, a group of 54 small house sites which flourished in the 900s and early 1000s CE. The community contains one of the earliest Great Kivas to be constructed in the canyon. Great Kivas are semi-subterranean structures some 15 to 20 meters in diameter, which may have functioned as community gathering places.

A recent discovery is that the December solstice sun rises over the center of Fajada Butte as viewed from the unexcavated, isolated Great Kiva in Marcia's Rincon (Malville 2014a). This appears to be the earliest example of direct observational astronomy in Chaco Canyon. With a diameter of twenty meters, this Great Kiva is the second largest of twenty-one Great Kivas to have been built in Chaco Canyon. Its location places it within the shadow cast by Fajada Butte at sunrise on December solstice.

Celebrants emerging from the Great Kiva at dawn would have observed the sun rising over the summit of Fajada Butte. A ramp combined with ladders and steps cut into the rock gave access to the summit. Fire pits were constructed at the base of the ramp and at the summit. A fire may have been lit in the two fireboxes at sunrise. Considering the visual power of Fajada Butte combined with winter solstice sunrise, it seems likely that the Great Kiva was intentionally located in order to achieve this astronomical alignment. The surface ceramics of the site as well as the ramp contain pottery, which date from the 900s to early 1000s CE. The ramp is evidence that the small house community was capable of engaging in a cooperative project of substantial proportions. More recent ceramic scatter on the ramp establishes that it was in use up to the late 1200s when the area was abandoned. Even though the residential and ceremonial center of Chaco Canyon was moved further to the west to the vicinity of Pueblo Bonito, winter solstice ceremonies may have taken place at this great kiva and on Fajada Butte until abandonment of the canyon.

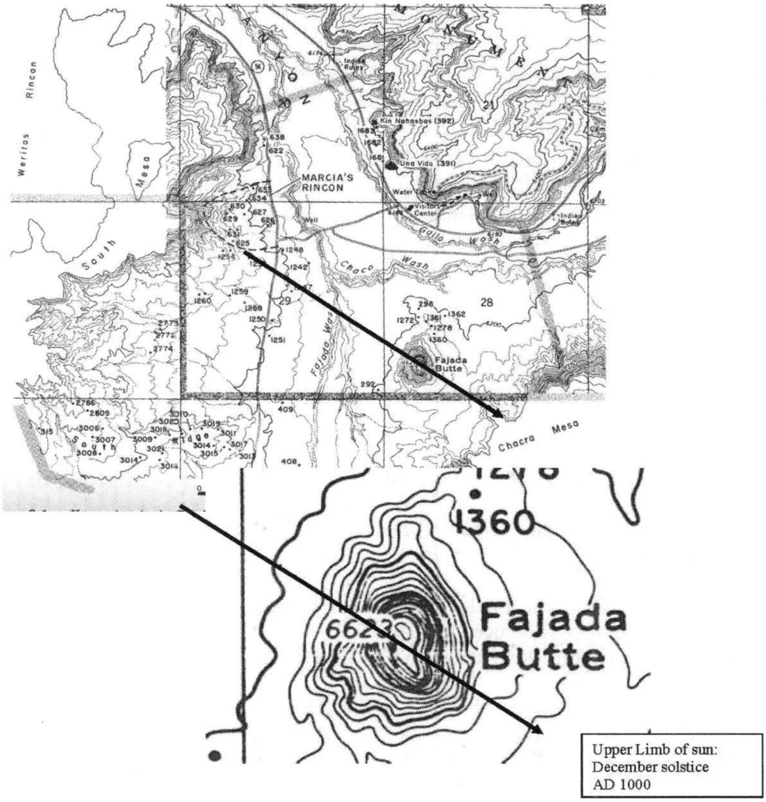

Figure 3. Winter solstice sight-line from the Great Kiva to the summit of Fajada Butte.

Unlike other locations in the Americas where the horizon is smooth or is established by jungle, the horizons of many of the Ancestral Pueblos are irregular, such that a horizon calendar could be established by the changing location of the sun between summer and winter. One tradition among the historic Pueblo is that the Sun has two homes, a northern house in which he stays for a few days around summer and a southern house for winter. There has been some concern that the sun would remain in his southern house too long, when the land was becoming cold. In the Rio Grande pueblos today, near Santa Fe, dances and trade fairs continue to take place near winter solstice, reminding the sun of his responsibility to return to the north again, assuring the continuing cycle of the seasons.

5. Final Remarks

It is possible that the astronomy born out of the harsh conditions of the Western Desert has influenced the long enduring tradition involving Hathor, the rising of the Nile near summer solstice, and the heliacal rising of Sirius. The appearance of naked-eye sunspots on the sun, which, according to Hindu tradition, should be pure and absolutely unblemished, appears to have initiated stories about the leprous sun and ritual celebrations of surya puja. Festivals in Chaco Canyon may have been designed to keep the sun moving between his summer and winter houses, maintaining the cycle of the seasons.

Among the heritage of the Ancestral Pueblos there are other examples of societal and environmental determinants that have been used successfully to predict new astronomical alignments. These include solstice events at a nearly all of the Late Bonito Great Houses in Chaco Canyon (Malville 2008; Munro et al. 2017). In India the outgrowth of these determinants appears as zenith sun alignments at Vijayanagara and orientations of temples at Khajurao to major Hindu festivals (Malville 2014b). Of course, there have been discoveries in archaeoastronomy that were totally unexpected, and from those we can learn something new about the culture and spot potentially exciting new lines of research.

References

Cassirer, E. 1955, The Philosophy of Form, Vol. 2, Mythical Thoughts (New Haven: Yale University Press)

Gatto, M. C. 2011, The Nubian Pastoral Culture as Link between Egypt and Africa: A View from the Archaeological Record, in Egypt in its African Context, ed. K. Exell (Oxford: Archaeopress), 21

Kuper, R. & Kropelin, S. 2006, Climate Controlled Holocene Occupation in the Sahara: Motor of Africa's Evolution, Science, 313, 803

Malville, K. 1975, A Feather for Daedalus—Explorations in Science and Myth, (Palo Alto: Cummings Publishing)

Malville, J. M. 1989, The Rise and Fall of the Sun Temple of Konarak: The Icon Versus the Morning Sun, World Archaeoastronomy, ed. A. L. Aveni (Cambridge: Cambridge University Press), 377

Malville, J. M. 2008, A Guide to Prehistoric Astronomy in the Southwest, (Denver: Johnson Books)

Malville, J. M. 2014a, The Enigmas of Fajada Butte, Astronomy and Ceremony in the Prehistoric Southwest: Revisited, ed. G. E. Munson, T. W. Bostwick, & T. Hull (Albuquerque: Maxwell Museum of Anthropology), 29

Malville, J. M. 2014b, The Astronomy of Indian Cities, Temples, and Pilgrimage Centers, *Handbook of Archaeoastronomy and Ethnoastronomy*, ed. C. L. M. Ruggles (Heidelberg: Springer), 1969

Malville, J. M. & Singh, R. P. B. 1996, Visual Astronomy in the Mythology and Ritual of India: The Sun Temples of Varanasi, Vistas in Astronomy, 39, 431
Malville, J. M. & Swaminathan, R. N. 2005, Surya Puja Temples of South India, in Songs of the Sky, V. D. Chamberlain, J. B. Carlson, M. J. Young (Leicester: Ocarna Books), 310
Malville, J. M., Wendorf. F., Mazar, A. A., & Schild, R. 1998, Megaliths and Neolithic Astronomy in Southern Egypt, Nature, 392, 488
Malville, J. M., Schild R, Wendorf, F., & Brenmer, R. 2008, Astronomy of Nabta Playa, in African Cultural Astronomy, ed. J. Holbrook, R. Medupe, J. Urama, (Hamburg: Springer) 131
Munro, A. M., Hull, T., Mathien, F. J., Malville, J. M., & Cherilynn Morrow, C. 2017, Investigation of Solstice Horizon Interactions at Chacoan Monumental Architecture, Cosmos and Culture, 21
Popper, K. 1961, The Logic of Scientific Discovery, (New York: Basic Books)

Astronomical Heritage of the Middle East
ASP Conference Series, Vol. 520
Sona V. Farmanyan, Areg M. Mickaelian, J. McKim Malville, and Mohammad Bagheri, eds.

Astronomical Heritage of Iran

Sa'dollah Nasiri Gheydari
Department of Physics, College of Science, Shahid Beheshti University, Tehran, Iran

Abstract. There exist various astronomical heritage sites in the Islamic Republic of Iran. It is noteworthy that between the 9th and 15th centuries AD, around 23 observatories were built in Iran, amongst which the Maragheh observatory was the most prestigious observatory in the world. Maragheh observatory, dating back more than 700 years, had been built on a hill near a city of the same name in the East Azerbaijan province of Iran. This observatory was constructed by Fakhruddin Ahmad Maragheie, the prominent architect of that time under the patronage of the Ilkhanid Hulagu and the directorship of Khajeh Nasir al-Din Tusi, a Persian scientist and astronomer. Maragheh observatory is known as the largest historic observatory of Iran, and in 1985, it was inscribed in the Iranian National Heritage List. After the death of Khajeh Nasir in 1250, the reputation of observatory gradually declined, and finally in 1298, it was turned into ruins due to powerful earthquakes and neglect of governors.

1. Introduction

The international conference on "Astronomical Heritage of the Middle East" is devoted to the role of astronomy in culture. Culture is the basis of any promising relationship. As long as nations interact and communicate under the banner of culture, peace will replace conflict, affection will replace violence, and science will replace ignorance in human relations. Culture is a window to a garden where, despite the diversity of its flora, a unique and harmonious odour is emitted from the garden of humanity, revealing a sense of unity in diversity. Among the Asian nations, the culture-driven and culture-spreading nation of Iran has always striven for promoting culture in the world. In this respect, reference can be made to the nomination of personalities on UNESCO's Representative and Anniversary List. 34 renowned Iranian Personalities from a long tentative list are in this celebration list, where 6 of them are astronomers. A few of a long list of astronomers are Abdul-Rahman Sufi, Abdul-Ali Birjandi, Abolfazl Harawi, Abul-Wafa Buzjani, Abu Rayhan Biruni, Abul-Abbas Nairizi, Ahmad Marwazi, Aḥmad Farghani, Ahmad Asturlabi, Avicenna, Fazl ibn Naubakht, Hamed Khojandi, Ghiyath al-Din Jamshid Kashi, Khawaja Nasir al-Din Tusi, and Kushyar Gilani.

More than 23 observatories were constructed between the 9th to 16th centuries in Iran as shown in Fig. 1. Among these, the Maragheh Observatory is one of the most important observatories, dating back to more than 700 years ago. This observatory was constructed by Fakhruddin Maragheie, the prominent architect of that time, by order of Khaje Nasir al-Din al-Tusi, a Persian scientist and astronomer, and its construction took about 13 years. Some historians believe that Maragheh Observatory is the first equipped observatory before the invention of the telescope, and for 300 years, no other

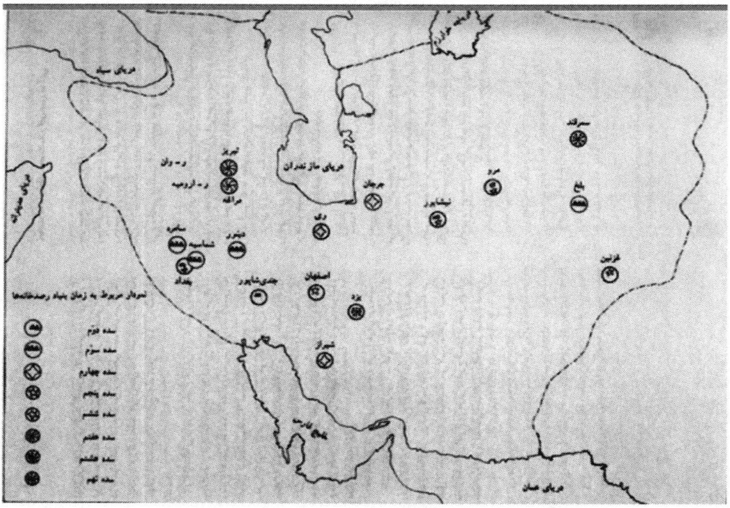

Figure 1. The Observatory sites constructed between 9th to 16th centuries in Iran territory (Vardjavand 1987).

similar observatory was built in the western world. Maragheh Observatory is known as the largest historic observatory of Iran, and in 1985, it was inscribed on the list of National Heritage of Iran. The Ulugh Beg Observatory of Samarkand in Uzbekistan, some ancient observatories such as stone observatories Ujain and Varanasi in India, Uraniborg in Denmark, and the observatory of Shanghai in China, are inspired by Maragheh observatory. It was not just a place for observing the stars; it turned into one of the largest Islamic centers in the world after Gundi-Shapur University and the schools of al-Nizamiyya and al-Mustansiriya in Baghdad.

About 120 scientists, regardless of their nationality and religion, worked in the Observatory. Some of them were: Mu'ayyad al-Din Urzi from Damascus, Najm al-Din Ali Katebi from Qazvin, Fakhr al-Din Ekhlati, engineer from Tbilisi, Najm al-Din Kateb Baghdadi from Iraq, Mohi al-Din Maghrebi from Maghreb, Kamal al-Din Abdul Shaibani, from Iraq, Hussam al-Din Shami from Syria, Fu Mon Ji from China, Abu al-Faraj and Barahbar-aus, prominent christian scientist Qutb al-Din Shirazi, Fakhr al-Din Maraghaeie, Najm al-Din Ostorlabi, Sayyed Rukn al-Din Astarabadi, Sadr al-Din Ali (son of Khajeh), and Asil al-Din al-Hassan (son of Khajeh) from Iran.

It is noteworthy that the wages for different staff members and scientists employed in Maragheh Observatory were as follows:

- Philosophers and mathematicians: 3 Dirhams per day
- Physicians: 2 Dirhams per day
- Jursiprudent: 1 Dirham per day
- Hadith narrator: 1/2 Dirham per day

Maragheh Observatory is the place where Tūsī, with the cooperation of a number of renowned astronomers, mathematicians, and instrument makers, compiled one of the most important astronomical tables, known as the Ilkhani Zīj, in 1272. The astronomers at Maragheh also carried out complicated programs in observational and computational astronomy to update the Ptolemaic parameters. In addition, the Maragheh Observatory represents a turning point in the development of alternatives to Ptolemy's planetary models that, together with innovations developed at the Samarqand Observatory and later at Damascus and Turkey, found their way to Europe and formed a critical part of the mathematical tools that enabled Copernicus to create the heliocentric model of the universe.

Parviz Varjavand, an archaeologist who explored the area, discovered 17 architectural units that together formed the scientific complex including the units that were directly used for astronomical research activities and those that operated as dependent units, such as library, school, foundry, tooling, etc.

The Observatory site before and after excavation is shown in Figs. 2.

Figure 2. The Maragheh Observatory site before and after excavation (Vardjavand 1987).

Figure 3. Left: the remnants of the mural quadrant, the observatory's main instrument. Right: auxiliary observational sites close to the main building of the observatory (Vardjavand 1987).

Different observational instruments were used in Maragheh Obsevatory such as the mural quadrant inside the cylindrical central tower of the observatory, azimuthal quadrant, armillary sphere, etc. The replica of the armillary sphere, Fig. 5, originally made of bronze, is reconstructed of wood in real size by Nasiri Gheydari and Baktash in 1995, maintained in Department of Physics, Zanjan University in Zanjan, Iran. It has the rings that could be adapted to different celestial circles such as meridian, equator, horizon, ecliptic, etc.

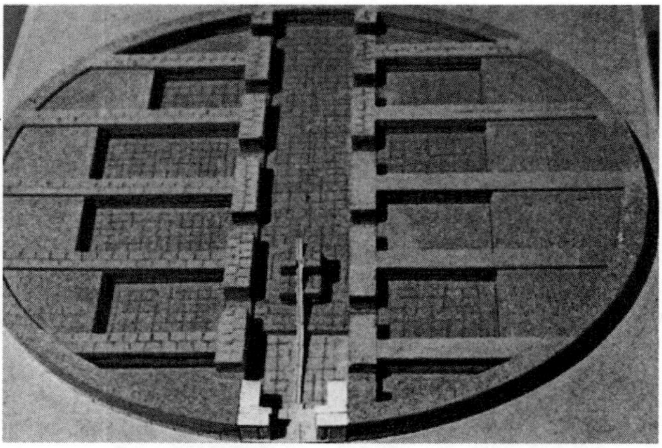

Figure 4. Plan of the central tower with radius of 23.7 m (Vardjavand 1987).

Figure 5. Armillary Sphere reconstructed from wood and maintained in Zanjan University (left). Celestial globe of Margheh (right).

In 1279, a celestial globe of diameter 14.4 cm was constructed by Muhammad al-Urzi, son of the well-known astronomer Mu'ayyad al-Din Urzi. It was expertly made out of bronze, and the celestial constellations as well as the equator, the ecliptic and latitude circles were inlaid in silver and gold. The surrounding rings, representing the horizon, the meridian and the prime vertical, could be adjusted to different geographical

Figure 6. The observatory site covered by a dome-framed tent structure.

latitudes. The globe originally belonged to the famous observatory of Maragheh, but it was transferred to Dresden in 1562 where it has been preserved ever since (Fig. 5).

After Tusi's death, his son was appointed as the Maragheh Observatory Director. It is not known with certainty how long it was active. It turned into ruins as a result of frequent earthquakes and lack of funding by the state and was abandoned later in the middle of the 14th century. The observatory site is currently covered by a dome-framed tent structure and is available for visitors (Fig. 6).

In conclusion, I come back to the issue of culture that today, more than any other time in history, humanity needs to restore, as an invisible and tight string, to promote the solidarity to eliminate the darkness of the world of politics and to eradicate ignorance. This is what we greatly wish for.

I would like to conclude my speech with a few verses by the renowned Iranian poet Ferdowsi:

جز از نیک نامی و فرهنگ و داد   ز رفتار گیتی مگیرید یاد

دلت دار زنده به فرهنگ وهوش   به بد در جهان تا توانی مکوش

جهان ماندگار است و ما رفتنی   به مردم نماند به جز مردمی

The English translation is:

Learn not from worldly behaviours
But prominence, culture and righteousness.

Keep your heart a living with culture and art
Strive not for evil in the world as much as thou art.
The world remains for the world and we depart
Nothing will remain of humankind but human art.

References

Mozaffari, S., Mohammad and Zotti, G. 2012, Ghazan Khan's Astronomical Innovations at Marāgha Observatory, Journal of American Oriental Society, 132, 395
Saliba, G. 1987, "The role of Maragha Observatory in the development of Islamic astronomy: a scientific revolution before the Renaissance," *Revue de Synthase*, 108, 361
Vardjavand, P. 1987, *Kavosh-e Rasadkhaneh-e Maragheh* (Tehran: Amirkabir Publications)

Astronomical Heritage of the Middle East
ASP Conference Series, Vol. 520
Sona V. Farmanyan, Areg M. Mickaelian, J. McKim Malville, and Mohammad Bagheri, eds.

Astronomers and Observatories in Egypt During the Period 969–1914 AD

Ahmed A. Hady
Department of Astronomy & Space and Meteorology, Faculty of Science, Cairo University, Giza, Egypt; aahady@sci.cu.edu.eg

Abstract. The Islamic civilization reached its peak during the period 969–1171 AD prevailing over a vast area that extended from Andalusia in the West to North India in the East. At that time, Egypt achieved remarkable progress in science, particularly in the fields of Astronomy, Medicine, and Engineering. Astronomy was, and still is, of importance from a religious point of view, since it marks the birth of the Moon, and consequently the beginning of every month of the Islamic calendar; accordingly, Islamic feasts are timed. Astronomers also give the exact time of the five prayers of Muslims for each day of the year.

This paper will show the work of astronomers Ibn Yunus al-Misri (952–1009 AD), Hassan Ibn al-Haytham (965–1038 AD), and Ali Ibn Ridwan (998–1061 AD). The astronomical observatories at Cairo during the period 969–1171 AD are included, and the El-Geyoushy and El-Mamoun observatories are described herein.

In this paper, the work of the leading astronomers of the above mentioned period is further reported, with reference to the observatories and astronomers during Ottoman period until 1914 AD. We will review the astronomical observatories erected in this period, especially the observatories in Bulac, Abbassia, and Helwan, all located around Cairo.

1. Introduction

Prehistoric cultures have left astronomical artifacts such as the Egyptian and Nubian monuments, and early civilizations such as the Babylonians, Greeks, Chinese, Indians, Iranians, and Mayas performed methodical observations of the night sky.

In 1244 BCE, the Abu Simbel temple was erected, originally cut into a solid rock cliff, in southern Egypt and located at the second cataract of the Nile River. The axis of the temple was positioned by the ancient Egyptian architects in such a way that every year on October 22 and February 22, the rays of the Sun would penetrate and illuminate the face of the statue, which means that the ancient Egyptians clearly knew the apparent solar motion around the earth.

From naked eye observations, early ideas about the motions of the planets were formed, and the nature of the Sun, the Moon, and the Earth in the universe were explored philosophically. The Earth was believed to be the center of the universe with the Sun, the Moon, and the stars rotating around it. This is known as the geocentric model of the universe, or the Ptolemaic system.

A particularly important early development was the beginning of mathematical and scientific astronomy, which began among the Babylonians, who laid the foundations for the later astronomical traditions that developed in many other civilizations.

The Babylonians discovered that lunar eclipses recurred in a repeating cycle known as Saros.

Astronomy was mostly stagnant in Medieval Europe, at least until the 13th century. However, astronomy flourished in the Islamic world and other parts of the world. This led to the emergence of the first astronomical observatories in the Muslim world by the early 9th century. Muslims in the Middle Ages used astronomy to calculate the direction of Mecca from far-flung corners of the Islamic world. In 964, the Andromeda Galaxy, the largest galaxy in the Local Group, was discovered by the Iranian astronomer al-Sufi (known in the west as Azophi) and first described in his Book of Fixed Stars. The SN 1006 supernova, the brightest apparent magnitude stellar event in recorded history, was observed by the Egyptian Arab astronomer Ali ibn Ridwan and the Chinese astronomers in 1006. Some of the prominent Islamic (mostly Iranian and Arab) astronomers who made significant contributions to the science include al-Battani, Thabit, Azophi, Albumasar, Biruni, Arzachel, al-Birjandi, and the astronomers of the Maragheh and Samarkand observatories. Astronomers during that time introduced many Arabic names now used for individual stars (Shaltout 2014). It is also believed that the ruins at Great Zimbabwe and Timbuktu may have housed an astronomical observatory. Europeans had previously believed that there had been no astronomical observation in pre-colonial Middle Ages sub-Saharan Africa, but modern discoveries show otherwise (Mountser 2001).

However, the invention of the telescope was required before astronomy was able to develop into a modern science. Historically, astronomy included disciplines as diverse as astrometry, celestial navigation, observational astronomy, and the making of calendars, but professional astronomy is nowadays often considered to be synonymous with astrophysics. Astronomy includes the examination, study, and modeling of stars, planets, comets, galaxies, and the cosmos. Most of the information used by astronomers is gathered by remote observation, although some laboratory reproduction of celestial phenomena has been performed (such as the molecular chemistry of the interstellar medium). While the origins of the study of celestial features and phenomena can be traced back to antiquity, the scientific methodology of this field began to develop in the middle of the 17th century. A key factor was Galileo's introduction of the telescope to examine the night sky in more detail.

The mathematical treatment of astronomy began with Newton's development of celestial mechanics and the laws of gravitation, although it was triggered by earlier work of astronomers such as Kepler. By the 19th century, astronomy had developed into a formal science, with the introduction of instruments such as the spectroscope and photography, along with much-improved telescopes and the creation of professional observatories.

In this paper we introduce the astronomers and observatories erected in Egypt during the medieval era up to modern astronomy in Egypt until the year 1903. We show the work of astronomers Ibn Yunus al-Misri, Hassan Ibn al-Haytham, and Ali Ibn Ridwan. The astronomical observatories at Cairo during the period 969–1171 AD are included, as well as the al-Geyoushy and al-Mamoun observatories. Modern Astronomy in Egypt started in 1840, when an astronomical observatory was constructed at Boulac, West Cairo. This observatory was closed in 1860. In 1868 another observatory was built at Abbassia east of Cairo. It continued its function until the end of the century, when observations were not suitable, due to the introduction of the electric tram way in

Cairo. Astronomical observations at Helwan started in 1903 using a 30 inch reflecting telescope.

2. Egyptian Astronomers During the Fatimid Era (969-1171 AD)

2.1. Ibn Yunus al-Misri (952–1009)

Ibn Yunus was born in Cairo at 341 Hegarian (952 AD). The Fatimid kings appreciated his high knowledge in astronomy and mathematics, and they built the El-Hakimi observatory for him on El-Mukattam Mountain near El-Fostat, south of Cairo. He made accurate observations for the Sun, Moon, and planets. He carried out an Almanac and named it El-Hakimi Zeg in acknowledgment of the king, El-Hakem be-Amr Allah. It has four volumes and contains a long introduction and 81 chapters. It is more accurate than the *Almagest* of Ptolemy the astronomer at Alexandria, Egypt (150 AD), and more accurate than similar works done in Baghdad.

Ibn Yunus al-Misri observed the solar and lunar eclipses in Cairo in 978 AD Figure 1 shows an Arabic manuscript, where an explanation of solar eclipses appears. He described this method in his almanac, which was used by the Arabic astronomers in Baghdad during the period of El-Mamoun El-Abasy to determine the radius of the Earth.

He was very successful in spherical trigonometry; he solved many difficult problems by using the vertical project for the celestial sphere on the Horizon and Meridian (Mercer 1994).

Ibn Yunus used a pendulum for time measurement, despite the fact that it has been known for nearly a hundred years that this is based on nothing more than an error made in 1684 by the Savilian Professor of Astronomy at Oxford, Edward Bernard (King 1979).

He found equations and laws to transfer the multiplication process to summation processes before knowing the logarithms.

2.2. Hassan Ibn Al-Haytham (965–1038)

Ibn Al-Haytham was born in El-Basra, Iraq in 965 AD. He was invited by the king, El-Hakem Bamer Allah, to come to Cairo, where he spent most of his life working in the fields of optics and astronomy. He was the founder of the new optics, where he was against the theory of the old optics by Alexandria Bibilotica Astronomers. He explained how solar and lunar eclipses occur. He also made a model for the movements of the six planets in their orbits. He explained the difference between the Sun and stars, and the planets and Moon; where the first are self-illuminated, and the second are reflectors of the radiation coming from the first.

Hassan Ibn Al-Haytham is one of three pearls in the Crown of the Arab-Islamic civilization; the other two are Ibn Sina in Iran, and the most famous astronomer Ibn El-Berony, who lived in Iran and the north of India. Figure 2 shows his photo and eclipse calculations (Ali 2008).

At the same time in Iraq, Alhazen (965–1038) from Basra founded physiological optics and was the first to study the properties of light and convex lenses. He visited Cairo for a short time.

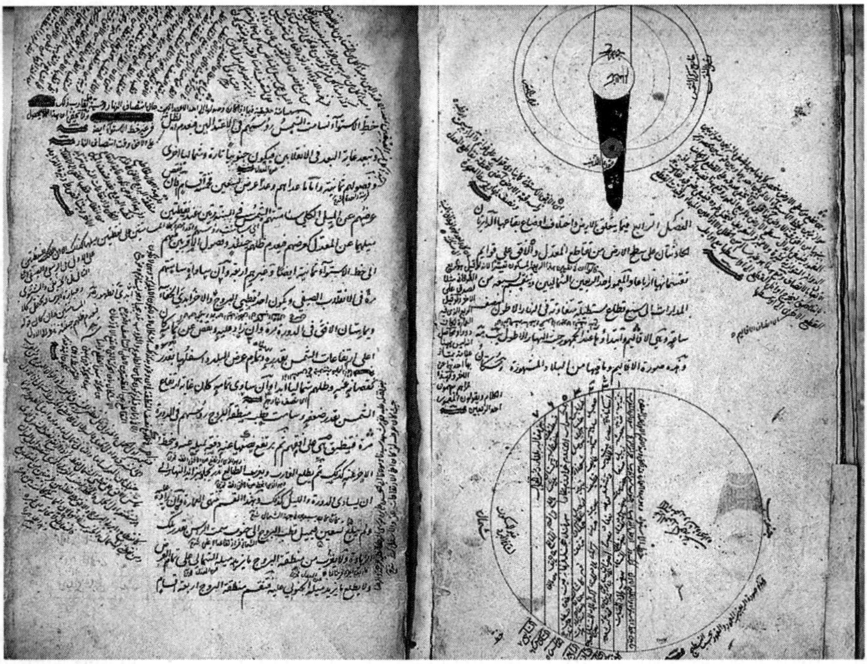

Figure 1. The manuscript El-Hakimi Zeg shows the progress of the Fatimiya in calculating the time of the eclipse.

2.3. Abu'l Hassan Ali ibn Ridwan Al-Misri (988–1061)

Ibn Ridwan was born near Cairo in 998 AD. He was a physicist, but he also had a considerable interest in astronomy. He was a commentator on ancient Greek medicine, but he is better known for providing the most detailed description of the supernova now known as SN 1006, the brightest stellar event in recorded history. Ibn Ridwan states that he himself witnessed the star (supernova of 30 April 1006 AD), though since he was only 18 years old at the time, his description is likely to be largely based on the words of older contemporaries (Stephenson 2002), see Table 1.

Table 1. Historical Supernova

Date (AD)	Type	Magnitude at Max	Discovered by	Remnant
1006	I	-10	Chinese/Arabs	SN1006
1054	II	-5	China/Japan/Armenia	Crab Nebula
1181	II	-1	China/Japan	3C58
1572	I	-4	Tycho Brahe	Tycho
1604	I	-3	Johannes Kepler	Kepler
ca. 1680	II	5?	John Flamsteed	Cas A
1987	II	2.9	Ian Shelton	SN1987A

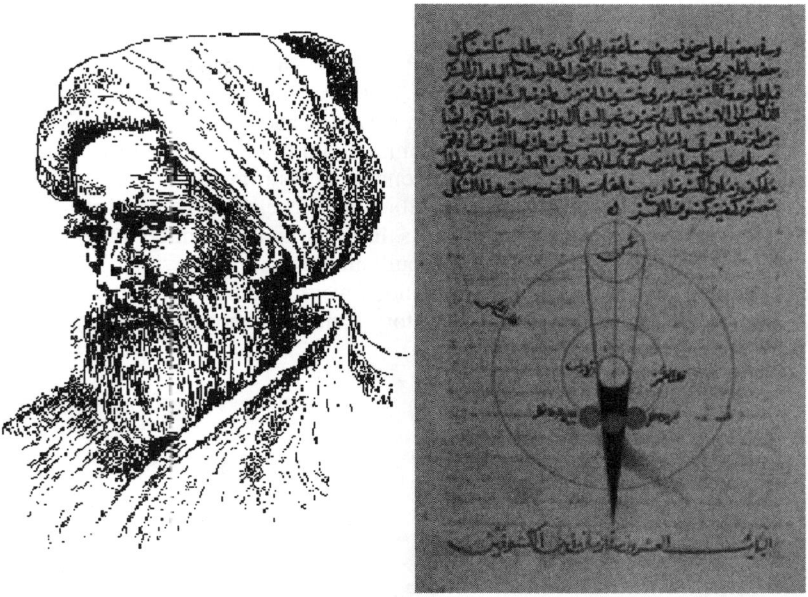

Figure 2. Hassan Ibn Al-Haytham (left), and the total solar eclipse calculations (right).

As an astrological treatise, it had considerable influence in the medical world. The following record in the commentary was discovered by Goldstein. Ali Ibn Ridwan wrote: "I shall now describe a spectacle (Athar) which I saw at the beginning of my studies. This spectacle (Athar) appeared in the Zodiacal sign Scorpio, in opposition to the Sun. The Sun on that day was 15 degrees in Taurus and the spectacle (Nayzak) in the 15th degree of Scorpio. This spectacle (Nayzak) was a large circular body, 2.5 to 3 times as large as Venus. The sky was shining because of its light. The intensity of its light was a little more than a quarter of that of moon light. It remained where it was and it moved daily with its zodical sign until the Sun was in six tile with it (i.e., 60 degrees away) in Virgo, when it disappeared at once (Dufatan Wahidatan)." The positions of the Sun, Moon and planets quoted by Ibn Ridwan clearly indicate a date of AD 1006 April 30.

3. The Astronomical Observatories at Cairo During the Period (969–1171)

3.1. El-Hakimi Observatory

The El-Hakimi Observatory was built on the Moukatam Mountain, south of the Fatimid Cairo by the king Al-Aiziz Be-Allah for Ibn Yunus al-Misri. After the death of the king, his son, Al-Hakim Bamer Allah, supported the Observatory with all the needed instruments and finances. Ibn Yunus al-Misri obtained accurate observations for the Sun, Moon, and planets. He carried out the New Almanac (Zeg) (Khial 2005).

3.2. El-Geyoushy Observatory

The El-Geyoushy Observatory was built by the pre-minister Abo El Kasem El-Geyoushy in 1080 AD on the roof of his mosque (El-Geyoushy Mosque), above the mountain of the Moakatam at Cairo East, to make a new almanac (Zeg) for the festival celebrating 5 centuries since the beginning of Islam. He built an Armillar (astrolabe) with a diameter of 5 meters; it was the best astronomical instrument in the entire Islamic world during that time. The instrument, Armillar, was developed by the Muslim astronomers to increase its accuracy by increasing its diameter and the number of its rings. The pre-minister Abo El-Kasem El-Geyoushy built another astrolabe with a 3.5 meter diameter to observe the Sun. Accurate observations were obtained and a new Almanac (Zeg) was created in 1090 AD. The Observatory contained another astronomical instrument named Sennero Quadrant (see Figure 3).

Figure 3. Photo of El-Geyoushy Observatory.

3.3. El-Mamoun Observatory

In 1093, the pre-minister El-Mamoun Ben El-Batahy built a new Armillar (astrolabe) with a diameter of 2.5 meters and erected it on the roof of Bab El-Naser, an entrance of the Fatimid Cairo (see Figure 4). Many accurate observations were obtained to prepare a new almanac (Zeg). After that, false information reached king Al-amer Be-Ahkam Allah about use of the Observatory for astrological purposes, which is against the Islamic belief. For that, the king killed El-Mamoun (1099). Though the astronomers escaped, the observatory was completely destroyed in 1124 AD.

Figure 4. Bab El-Naser is one of entrances of Cairo from the Fatimid era, and was used as an observatory until 1124 AD.

4. The Ottoman and Khedivate Era in Egypt (1517–1914)

Astronomy in Egypt started strong again during the Ottoman and Khedivate era of Egypt (1517–1914). Hassan al-Jabarti was a Somali astronomer and philosopher who lived in Cairo. Hassan is considered one of the great scholars of the 18th century; he died in 1774.

In 1798 the French scientists erected a small astronomical observatory close to Cairo's center.

In 1825, Khedive Mehmet Ali Pasha started to erect a new astronomical observatory in Cairo's center. In the year 1840, an astronomical observatory was constructed in Boulac, Cairo (Tahrir Square) under the supervision Mahmoud Pasha El-Falaky. This observatory was closed in year 1860 due to light pollution.

Mahmoud Al-Falaki (1815–1885) was an Egyptian astronomer. He was the pioneer of archaeological astronomy, connecting astronomical phenomena and archaeological monuments (Figure 5).

In 1868 another observatory was built in Abbassia, east of Cairo, where Ismail El-Falaki observed the Moon and planets, as well as the daily changes of temperature and rain. It continued its function until the end of the century (see Figure 6). The observations in this observatory were not suitable after that due to the introduction of the electric tram way in Cairo, and it was moved to Helwan city (about 20 km from Cairo).

Astronomical observations at Helwan started in 1903, using a 30-inch reflecting telescope. This 30 inch reflector telescope was provided by British astronomer John Henry Reynolds (1874–1949) (Hady 2008).

Figure 5. Photo of Mahmoud Al-Falaki (1815–1885).

Figure 6. King Fuad and Ismail El-Falaki with his colleagues during the opening day of the Abbassia observatory.

5. Prof. Mahmoud Reda Madwar (1893–1973)

Egyptian Astronomer M. R. Madwar finished his Ph.D. in Astrophysics from University of Edinburgh, Scotland, in the year 1924. He worked at Helwan Observatory beginning in 1926. He established the Astronomy department at Cairo University in 1932. He also supervised the construction of Kattameya observatory starting in 1946, and in 1962, the observatory was erected and started scientific observations (see Figure 7). The obser-

Figure 7. Dome of Kottamia Observatory (left) and Madwar with his equipment (right).

vatory housed a 74 inch (about 1.9 m), reflecting telescope. Positioned at latitude: 29° 55′ 48″ N, longitude: 31° 49′ 30″ E, altitude: 476 meter, it was working until recently, after modification of its software (Hady 2010). The Kattameya mountains lie in the northeastern desert (80 km from Cairo).

6. Conclusion

Astronomy was developed in Cairo, Egypt during the Islamic Fatimid period (969–1171 AD), the peak of the Islamic-Arabic civilization. Three of the most famous Islamic astronomers (Ibn Yunus al-Misri, Hassan Ibn El-Hissam, and Ali Ibn Radwan) lived in Cairo during that period. Kings of the Fatimid Empire supported the building of the observatories in Cairo (El-Hakimi, El-Goushy, and Bab El-Naser observatories), where many accurate observations were obtained using Armillars for the Moon, Sun, and planets. Those observations led to developed Almanacs (Zegs).

The astronomers and observatories in Egypt during the Ottoman and Khedivate era of Egypt (1517–1914) established modern astronomy education and research in Egypt.

References

Ali, Fouad Hassanen 2008, "The Sun of Allah rising on the west, Segred Honika", (in Germany), translated to Arabic, Dar Al-Aalam Al-Arabi, Cairo, Egypt

Clark, D. H. and Stephenson, F. R. 1977, "The historical Supernovae", Oxford, Pergamon Press

Hady, Ahmed A. 2010, "Education in Egypt and Egyptian response to eclipse", Innovation in astronomy educations, Cambridge university press, 282

Hady, Ahmed A. 2008, "Space Science Education in Egypt and the 2006 Solar Eclipse", Advances in Space Research Volume 42, Issue 11, 1800

Khial, Mahmoud 2005, "Islam and Science, Religious Orthodoxy and the Battle for Rationality, Pervez Hoodbhoy", (in English), Translated to Arabic, Supreme Council of Culture, Cairo, Egypt

King, D. A. 1979, "Ibn Yunus and the pendulum: a history of errors", Archives Internationales d'Histoire des Sciences, 29 (104): 35

Mercer, Raymond 1994, "English Orientalists and Mathematical Astronomy," in The "Arabic" Interest of the Natural Philosophers in 17th-Century England, 198

Mountser, Abd El-Halim 2001, "The History of Science and the role of the Arab in its development" Dar El-Maref, tenth Edition, Cairo, Egypt

Shaltout, Mosalam 2014, "Astronomers and Astronomical Observatories at Cairo during the Fatimid Islamic Period," The 2nd International Conference on Arabs' and Muslims' History of Sciences, University of Sharjah, Sharjah, UAE, December 8-11

Stephenson, Richard and Green, David A. 2002, "Historical of Supernovae and their Remnant", Oxford

Astronomical Heritage of the Middle East
ASP Conference Series, Vol. 520
Sona V. Farmanyan, Areg M. Mickaelian, J. McKim Malville, and Mohammad Bagheri, eds.

Jewish Chronology and Astronomy

Ariel Cohen
Institute of Earth Sciences, the Hebrew University of Jerusalem, Jerusalem, Israel

Abstract. Is it possible that the Jewish Biblical Year of Creation (BYC), now 5778 years ago according to the Jewish calendar, or 7526 according to the Greek Orthodox Church, as well as other major biblical events, had been carefully chosen to coincide with what were believed to have been major Astronomical events? We first demonstrate how the Sun-Moon conjunctions occurring at the beginning of the Zodiac were calculated based on an ancient modified detailed astronomical set of data included in al-Khwarizmi's treatise on the Jewish calendar. Then we show that all the main biblical events including the BYC, Exodus, and the building of Solomon's Temple, indeed coincide with such conjunctions. Finally, we demonstrate how the detailed evaluation of the biblical chronological data can lead to the individual sizes of the constellations, thus making the Bible one of the earliest and most important sources for the history and heritage of astronomy originating from the Middle-East, in general, and from the land of Israel, in particular.

1. Introduction

Several basic religious laws such as the ones used in ancient calendars were derived from contemporary scientific findings, and then modified with more precise astronomical data. We chose, as an example, the changing visualization of the formation of the Earth and the Universe. It is clear that in all ancient cultures it had been recognized that there could not be any document other than astronomical descriptions including celestial coordinates of various elements in the sky that could be related to the creation of the universe. In the first millennium BC and the first centuries AC (Anno Christi) the scientific knowledge about the age of the world was very vague, but in several cultures, it was accepted that the age of the present Earth did not exceed a few thousand years. As an illustration we can refer to the Kali-Yuga scale of the age of the present cycle of the world of Hinduism. According to the Hindu tradition, the present cycle started at 3102 BC This was the year that the astronomer Aryabhatta determined based on his calculations in 476 AC to be the starting point of the Kali-Yuga because in that year there had been an extremely rare planetary alignment: all 7 planets including the Moon and the sun had been in the same celestial longitude. This unique arrangement of the planets was believed to have been required for describing the unique cultural event in Hinduism when Krishna left the Earth, returned to his heavenly adobe, and the epoch of their new era began (See Table of the mean motions of the planets p. 17 in Burges 1858). We shall now deal with the epochs of the creation of the world as had been determined based on the Old Testament.

2. The Determination of the Age of the Universe in Astrophysics and in the Jewish Tradition

There are two major versions of the Bible known as the "Septuagint" version which is the Old Testament believed to be the authentic version by the Greek Orthodox Church, mainly in Russia as well as in Armenia, and the "Masoretic" version used currently by practically all other Christians and by the Jews all over the world. One of the major differences between the two versions is the chronology as can be seen in Table 1 where, for example, the ages of several of the Patriarchs when they begot their following generations differ by 100 years each.

Table 1. The ages of the Patriarchs when they begot their following generation. Therefore, following the Septuagint version, Noah gave birth to his son Shem 2144 years after creation (Vaticanus Manuscript).

	Septuagint	Masoretic
Patriarch	Begot	Begot
Adam	230	130
Seth	205	105
Enosh	190	90
Keenan	170	70
Mahalallel	165	65
Jared	162	162
Enoch	165	65
Methuselah	167	187
Lemech	188	182
Noah	502	502

Table 1 and other chronological differences lead to the fact that starting from September 2017 the year 5778 AM (Anno Mundi, the year from creation) used officially in Israel, is the year 7526 AM for the Greek Orthodox Church.

The question raised here is what were the reasons for using first one of the two versions of the chronologies of the Old Testament and then why modify it to the other version? Below we show that the two chronologies are a reflection of the advances in the accuracy of astronomical measurements in ancient times. We show that the ancient sages had visualized the positions of the planets when they were formed and by knowing the daily motions of each planet they calculated backwards and looked for the year in which the stars' celestial longitudes were in agreement with their predetermined conception.

Such an approach reveals that there are profound similarities between this ancient scientific approach and the modern astrophysical methodology used in the present determination of the age of the universe: In astrophysics it is assumed that at first there was the big bang followed by the expansion of the universe. If one can now determine the Hubble expansion rate with a high degree of accuracy, a backward extrapolation of the varying distance between stars can lead to the answer as to when they all started from a practically zero distance. But, with the introduction of increasingly sophisticated methods enabling more accurate determinations of the value of the Hubbles "constant" including the findings of its acceleration, the age of the universe had been changed from 12 to 13.7 billion years, a modification that took place within the past 30 years. We show that the scientific progress is also the reason that made it necessary to change the Anno Mundi scale in the ancient world: a backward extrapolation of the planetary positions, leading to the presumed visualization of their positions when they were formed, using more accurate astronomical input, led to a different Year of Creation (YCR) than the one found using older data. As a consequence, a new version of the biblical chronology had been computed.

But how do we know what was the visualization of the positions of the stars in the sky in the day of creation? We found the answer hidden in a rare manuscript written by one of the best known astronomers of the first millennium AC, Muhammad ibn Musa al-Khwarizmi.

3. The Planets on the Day of Creation

al-Khwarizmi (780–866 AD), the founder of Algebra, also made several contributions in the fields of geography and astronomy. Among his several important treatises in astronomy he included a description of the then newly formed permanent luni-solar Jewish calendar's regulations (Kennedy 1964). In that work he incorporated an old Astronomical Table (Table 2), which included the celestial longitudes of all planets in 3 major epochs: (1) The Biblical day of creation, (2) the beginning of the era of Solomon's Temple, and (3) the epoch, Se, of the Seleucid era, named by al-Khwarizmi as the "Two Horned" epoch. In the original work there is no reference to the origin of the table but as we show below it contains data that can be attributed only to the pre-Talmudic era, namely, before the 3rd century AC, an era in which it was believed that the world was created in the spring and not in the autumn as suggested in the Table. The table contains three sets of data and in our present work, we shall analyze the set of positions of all planets in the first column and just concentrate on the coordinates assigned to the mean Sun and the mean Moon at the epoch of Solomon's Temple.

Table 2. The positions of the planets as reported by Kennedy (1964), based on al-Khwarizmi's treatise. The last value of the first set of data 5 14 had been interpreted by Kennedy to be 0 14.

	The first of the days of Adam; a Friday	The building of the Temple	The beginning of the years of the Two Horned
Mean Sun	5 26	5 26;	6 13; 31, 18
Mean Moon	5 26	5 26;	4 6; 45, 49
Lunar pogee	1 5	9 26; 40, 16	7 26; 17. 19
Saturn	8 15	10 22; 9, 1	8 8; 24, 6
Jupiter	6 5	3 7; 42, 34	3 12; 52, 18, 13
Mars	1 6	1 15; 26, 21	8 12; 14, 46
Venus	4 25	7 12; 11, 47	2 1; 22 3
Mercury	missing in original work	1 13; 19, 19	7 10; 1 18
Lunar node	5 14*	4 26; 34, 11	4 23; 41, 27

In his 1964 article, Kennedy discussed the basic properties of the positions of the planets and made the following remarks which have remained unchallenged except for by us:"al-Biruni remarks that some of the Jews claim that the 'autumnal equinox' took place at the end of the third hour of the Day of Wednesday, the 5th of Tishri after Creation'. Our text has the mean Sun at epoch, four degrees before the autumnal equinox, which is consistent with the above. *But beyond this the numbers for the first two sets of positions seem to make little sense*." (Fig. 1, left). But, as we show below,

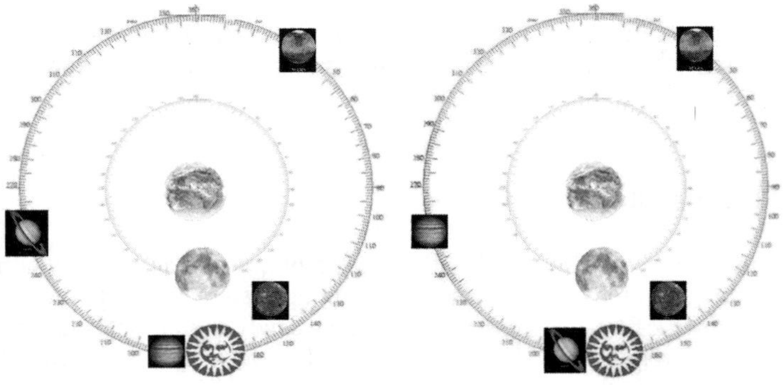

Figure 1. The visualization of the celestial longitudes of the planets as suggested by al-Khwarizmi's Table, claimed by Kennedy (1964), to be meaningless (left), and modified by interchanging Saturn and Jupiter (right).

Kennedy was misled by a misrepresentation of the data for "the first of the days of Adam", and, consequently, was inaccurate in his major statement: the numbers of the first set have a significant meaning, which reveals the basic understanding of the solar system and the first new moon in the Jewish day of creation of the stars and planets. As we show, the table had been modified from a different original table from which we can learn how the sky was visualized by the ancient astronomers who composed the original table in the last centuries BC or the first centuries AC

In fact, when the longitudes of Saturn and Jupiter in the set of positions of the Sun, the Moon and the planets given for "the first of the days of Adam," are interchanged (Fig. 1, right), their assigned values are ingenious: they astonishingly form a unique order of the celestial coordinates of the planets. In our attempts to understand what had been the unseen message which motivated the writers of the original Table to construct the first set of positions as they are, we tried to determine from the set of positions given for the time of the new moon close to the autumn ("Adam's new moon"), what would be the corresponding assumed positions in the preceding spring. At first, when we based our calculations on the motions of the planets, the results at the beginning of spring made also little sense, as stated by Kennedy. But, when we found, as will be elaborated below, that the YCR in the original table made for the spring creation, which is 3 years (37 lunar months) after the newly believed to be the YCR, we assumed that the astronomical positions of Saturn and Jupiter were interchanged in the handwritten table as inserted by al-Khwarizmi into his manuscript. Indeed, just by interchanging the 2 positions, the modified table offered a most interesting result: first, we got the following corrected set of coordinates for the Sun and the planets in "The first days of

Adam": Saturn 6^s 5° = 185°; Jupiter 8^s 15° = 255°; Mars 1^s 6° = 360; Mean Sun 5^s 26° = 176°; Venus 4^s 25° = 145°.

When we used Ptolemy's (2nd century AC) values for the mean daily motions of the planets, the corresponding set of positions for 177.184 days = (29.5 + 793/1080/24) × 6 days earlier, the piositions were found to be as in Table 3.

Table 3. The calculated approximate positions of the outer planets on the first day of the spring (see Fig. 2).

	Ptolemy's values for the Motion of the Planets (°/day)	Mean 177.18356 days earlier	Position of the planet (within 2°)
Saturn	0.033489	179th degree	180
Jupiter	0.083122	239th degree	240
Mars	0.524059	302nd degree	300
Mean sun	0.985635	1st degree	0

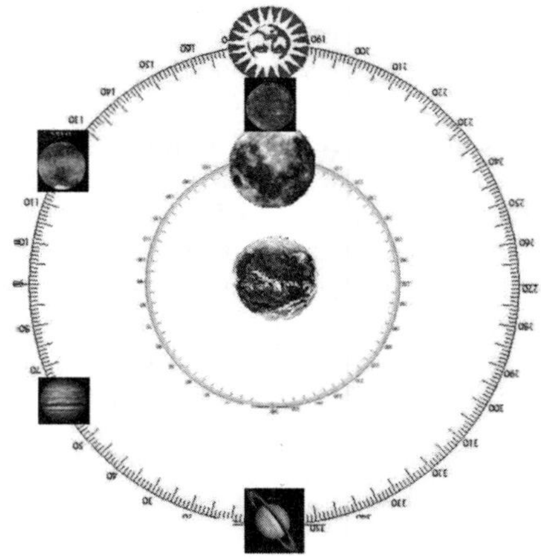

Figure 2. The clear unique order as visualized by the sages for the day of creation.

We now know that we have to go back and look for the year in which we can find such an arrangement of the planets implementing a similar method used by Aryabhatta in the Hindu tradition. In particular, we should look for a new moon occurring at the end of the first day of the spring, the day when all stars including the Sun and the Moon were formed, the weekday of which is Wednesday based on the first chapter of the Old Testament. In Cohen (2009) we describe additional indications that the table refers to the creation of the planets at the specific celestial longitudes being 60 degrees apart including the inferior planets, and we provide the detailed calculation of the position of the inferior planet Venus in day one having the same real celestial longitude as the sun.

4. Calculating the Original YCR Based on the Modified al-Khwarizmi Table

Maimonides devoted two chapters (Maimonides, 12th century AC) to describe the methods used by the ancient Jews in the times of the Sanhedrin (3rd century BC to 3rd century AC) to calculate the time difference in any given year between the new moon of Nisan (the first lunar month of the spring) and the beginning of spring. The first of the two chapters outlines the method used by the ancient sages who applied the value of 365.25 (Y_1) days for the length of the tropical year, in addition to the present value of the length of the mean month in the Hebrew calendar, being 29 days, 12 hours, 44 minutes and 3 1/3 seconds (M_H). This length of the month had been known to the Babylonians as early as the 2nd century BC, and was used by Ptolemy in his *Almagest* (Toomer 1984) in the middle of the 2nd century AC.

In contradiction to the present permanent Hebrew calendar, to be discussed in the next section, in a cycle of 19 years, 235 of the above mentioned mean lunar months, are not equal to 19 years of 365.25 days. The difference after 19 years is close to 87 minutes. Clearly, this difference is increasing by 86.944m after every cycle of 19 years. But other large cycles can be determined to make the difference between integer number of years and a corresponding number of lunar months. Indeed, using the Continued Fraction approach we find that after 1468 years = 536,187 days, the difference between 1468 years and 18,157 months is 3.278 minutes.

The sages who were aware of this large cycle (using a different methodology), used two of such 1468 year cycles, namely 2936 years, as a time difference between the "Building of the Temple" (2nd set of data in al-Khwarizmi's table) and the "First of the Days of Adam." Therefore, the new moon related to the two events took place in practically the same celestial longitude as stated in the Table (λ = 176). We note that al-Khwarizmi thought that the same celestial longitudes are resulting from an integer number of 19 year cycles since the new permanent Hebrew calendar assumed that after 19 years the celestial longitude of the mean Sun remains unchanged. But we claim that the table provides a unique significance to the same celestial longitudes of the new moons of the 2 events when they denote a separation of two major cycles of 1468 years each. This cannot be considered as merely a result of a coincidence. With our conclusion that these large cycles are involved we can proceed and enlighten the methodology used in the determination of the YCR. First, we note that:

2936 years = 1,072,374 days, constitutes of an integer number of weeks + 2 days.

In addition, since 2936 years are a multiple of 4, the spring after 2936 years starts at the same time of the day as in the day of creation. Consequently, all that the astronomer advising the Jewish sages had to do in order to determine the YCR, was to look for a mean new moon that took place on a Friday at 6 PM in Jerusalem (2 days after the creation's weekday of Wednesday), a day after the beginning of spring within the 9th or the 10th centuries BC (in which they assumed that Solomon's Temple was built), which would satisfy the above mentioned constraints. Once such a new moon is found, the YCR would be 2936 years earlier in time.

By using Table VI 6 "of the conjunctions of Sun and Moon" in the *Almagest* (Toomer, p. 278), it can be determined that the new moon close to the beginning of spring on a Friday in the year 821 BC, and, consequently, on a Wednesday in the year 3757 BC, are the only ones that took place in Jerusalem at 6 PM (within 10 minutes). The importance of this result is accentuated in three main conclusions: (1) the visualization of the time of the new moon relative to the spring equinox helped determine the

YCR as well as the year in which Solomon's Temple was built. (2) A major event such as the building of the Temple is believed to have taken place when the sky had the Sun, the Moon, and the zodiac at exactly the same relative coordinates, and (3) it is our claim that the determination of the year related to the building of Solomon's Temple, *used in the biblical chronology, is merely dependent on astronomical calculations*. Indeed, it is off by over 140 years from its later determined, historical accepted value.

5. Changes in the Determination of the YCR in the Hebrew Calendar Resulting from Developments in Astronomy

During the last centuries BC and the first millennium AC it became recognized that the value of Y_1 should be reduced. In his *Almagest*, Ptolemy followed Hipparchus and reduced its value by 1/300 of a day making it equal to 365.24667 (Y_A) days. In his second chapter mentioned above, Maimonides (Maimonides, ch. 10) states that other Jewish sages decided to use a value which was close to the *Almagest* value for a permanent calendar which relied merely on astronomical calculations. But they based their determination of the length of the tropical year, Y_H, on a cycle of 19 years in which there are exactly 235 mean lunar months equalling M_H, making Y_H to be 365.2468222 days.

The YCR can no longer be visualized by the unique order described above in Figure 2, and it was only required that the new moon of the day of creation of Adam will be the first new moon of the autumn and that it should be on a Friday, the weekday when, as written in Genesis, Adam was created. According to Maimonides the new moon 6 months earlier was within 1 day from the beginning of spring. The year that was satisfying such new requirements and was as close as possible to the year 3757 BC, was the year 3760 BC, 3 years earlier, which is the present year of the creation of Adam in the Jewish calendar, a year that starts with "Adam's new moon" (year 2 of the calendar).

But, the original astronomical table had to be modified accordingly before being presented by al-Khwarizmi, and by realizing that in 3 years the celestial longitude of Jupiter varied from being in front of Saturn in 3757 BC to a celestial longitude behind Saturn in the year 3760 BC, they just interchanged the original values between Saturn and Jupiter.

Indeed, even though the length of the tropical year of the *Almagest* is slightly different from the year used in the original table, Figure 3 shows a general agreement for the real celestial longitudes calculated by use of the *Almagest* for the year 3757 BC (upper left of the Figure) whereas there is no resemblance between al-Khwarizmi's modified table and the *Almagest* results for the year 3760 BC (upper right of the figure).

6. Chronological Differences Between the Septuagint and the Masoretic Versions Resulting from Developments in Astronomy.

From the last few centuries BC, major cultures in the Middle East used for the length of the mean lunar month a value of 29.5 days + 44 m (M_A). Such a value has been in use in the Islamic world ever since the 7th century AC as well as by the Jews during the last centuries BC. In addition, this length of the month was used in the Hindu culture to within 0.3 of a second.

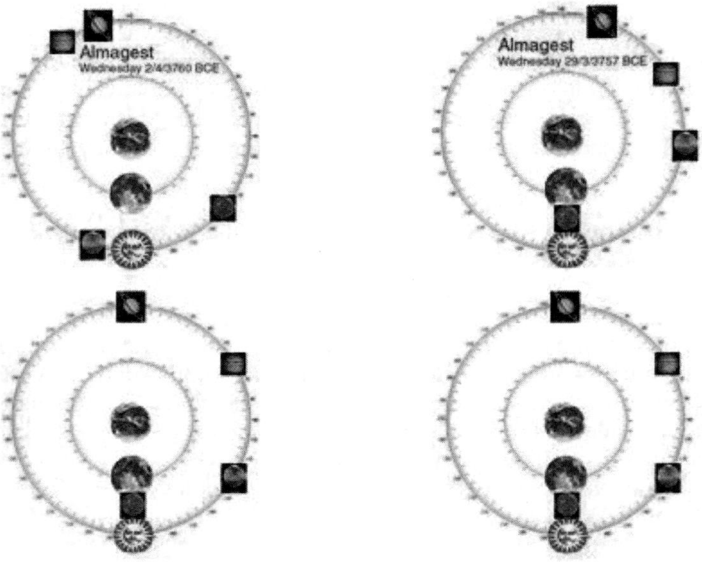

Figure 3. The celestial longitudes of the planets as calculated for 3760 BC (upper left) and for 3757 BC (upper right) using R. H. van Gent's Almagest Ephemeris Calculator `http://www.staff.science.uu.nl/~gent0113/astro/almagestephemeris.htm`

→ what about the lower pair?

We claim that the Septuagint version which was the result of the translation of the contemporary Old Testament by seventy sages from the 12 Tribes of Israel into the Greek language in the 3rd century BC, has a chronology which had been based on astronomical calculations using the length for the tropical year Y_1 and M_A for the length of the month. Therefore, the YCR cannot be the same as in the Masoretic version since going back in time with a different length of the month leads to a different YCR in which the above mentioned constraints apply. In order to reach a different YCR, dissimilar ages of the Patriarchs can contribute in reaching that goal.

When we used the continued fraction approach for the relative values of Y1 and MA, we found that after a grand cycle of 426 years a new moon would occur in practically the same celestial longitude as in the time of creation: 426 x Y_1 - 5269 x M_A = 4 minutes. We, therefore, suggest that the basic large cycle in the Septuagint version is thus 426. According to al-Khwarizmi Table, the building of Solomon's Temple is such an important event in the history of the Hebrews in the Old Testament that it is accompanied with the same new moon at the end of the first day of the spring as in the day of creation. Indeed 10 cycles of 426 years after the day of creation, namely, 4260 years, separate between the two events in the biblical chronology of the Septuagint version (the Vaticanus manuscript), as detailed in the Septuagint chronology.

The fact that there are 4260 years separating the YCR and the Temple's first year has been known in the past to various researchers. For example, J. Hughes states specifically that "the first year of the exodus is 3820 AM, the temple's first year is 4260 AM, and", he adds, "the first year after the destruction of the temple is 4690 AM" (Hughes

2009, p. 239). However, the 4260 years has remained without an explanation, except for by us (Cohen 2005b, in Hebrew). However, as we claim, al-Khwarizmi's manuscript provides the answer: Astronomy had been used by ancient cultures to look for celestial signs in order to determine the years of major events, thus providing the necessary tools to the ancient's understanding of the history of mankind. In fact, the inability to date the past was not only limited to the creation of the world. All the main cultural events described in the Bible, can be shown to have been linked to large astronomical cycles in the Masoretic version of the Bible. The continued fraction approach shows us that the large cycle of 1468 years discussed above is composed of 3 smaller cycles of 483, 483 and 502 years: $483 \times Y_1 = 5974 \times M_H - 27.89$ m, and $502 \times Y_1 = 6209 \times M_H + 59.06$ m.

Consequently, after each such cycle of 483 or 502 years, the new moon and the beginning of spring will be within less than an hour compared to what they were in the day of creation. As can be seen in Table 1 in Cohen 2005a, the major events, namely, the birth of Abraham—the father of the Hebrews, the Exodus, the building of Solomon's Temple, and the building of the second temple, as interpreted from the text, were all in general agreement with such cycles. The agreement with astronomical cycles were considered as the most important clue to the dating of the events. Even though the historians determined dates over 140 and 160 years apart from the corresponding biblical dating of the building of each of the two temples, the biblical chronology of the Masoretic text as interpreted in the 2nd century AC by Joseph ben Halafta, has remained the official biblical chronology of the Masoretic for the Jews.

7. The Ages of the Patriarchs when they Begot their Next Generations

7.1. The Motion of the Mean Sun

As shown above, the main biblical events took place when the mean Sun (MS)–mean Moon (MM) conjunctions took place at the beginning of the Zodiac (the beginning of the sign of Aries). We shall now demonstrate that the ages of the biblical Patriarchs were chosen to correspond to the MS-MM conjunctions that took place at the beginning of the different signs had they been calculated based on the 4th to 3rd centuries BC astronomical input.

Let us assume that the first mean new moon (NM) took place at the beginning of a given sign, and we want to find out what is the number of months and years required for the MS to produce a NM with the MM at the start of the following sign separated by 30 degrees.

By assuming that the lengths of the solar year and the mean synodic month used in the Septuagint Biblical chronology were $Y_1 = 365.25$ days, and $M_A = 29.5$ days and 44 minutes, correspondingly (see above), the MS – MM conjunctions were calculated to occur in steps of 29.106092°:

$$(29 + 12/24 + 44/60/24) \times 360/365.25 = 29.106092\ ^\circ$$

Our goal is, thus, to determine how many steps of 29.105092° are required for a NM to occur on a celestial longitude 30° apart.

It can be seen (Figure 4) that the NM closest to the celestial longitude of 30 degrees occurs after 2215 months = 179 1/12 years: 2215 × 29.106092° = 64469.993° modulo 360 = 29.993°. In 2215 mean months, the MS would, thus, complete 179 + 1/12

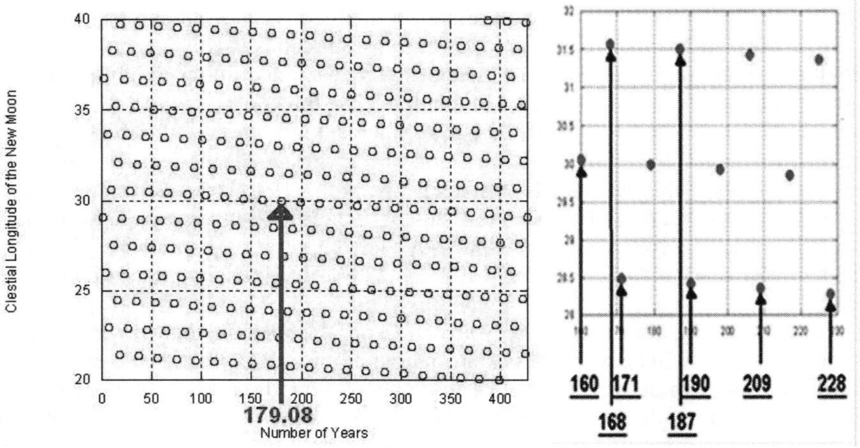

Figure 4. The celestial longitudes of all mean NMs in a cycle of 426 years (left), and enlarged around 179.08 (right).

cycles of 360 degrees. Consequently, if we have a mean NM at the beginning of Libra, then after (179 + 1/12) years, the celestial longitude of a mean NM will be as close as possible to the beginning of Scorpio. Therefore, in order to move the mean NM through all the 12 signs in similar steps, we would need

$$12 \times (179+1/12) = 2149 \text{ years.}$$

This total number of years surprisingly corresponds well with the sum of the years of the first 10 generations of the Patriarchs (12 if we assume that Noah represents three generations, see below) in the Septuagint:

$$230 + 205 + 190 + 170 + 165 + 162 + 165 + 167 + 188 + 502 = 2144.$$

Indeed, the sum is expected to be roughly the same when the mean motions are replaced by the real motions. We also note that the sum does bring the conjunctions close to the celestial longitude of the original conjunctions, as is happening after 2149 years = 5 cycles of 426 + 1 cycle of 19 years.

7.2. The Real Sun (RS)

So why are the ages of all Patriarchs not 179.08 years each, a value that astonishingly represents the mean age of the Patriarchs in the Septuagint version of the bible (assuming, again, that Noah represents 3 generations whose corresponding age when he begot Shem should thus be 3 x 179 1/12 years = 537 1/4 years)? The answer can be derived from the fact that already in the 4th century BC, the astronomers realized that the time required for the Sun to move 30 degrees in the sky is dependent on the time of the year: Hipparchus measured in 175 BC that the aphelion of the Sun's orbit is on the celestial longitude of $L_{ap} = 65.5°$, a value that was believed to be a constant until al-Battani's discoveries in the 9th century AC, and the anomalistic year was, thus, taken by all ancient astronomers to be identical in length to the tropical year; or, that L_{ap} remained constant in the calculations performed by the ancient astronomers including, specifically, those involved in the determination of the chronology of the Septuagint version of the bible.

For the demonstration of our calculations we shall use a model developed by Ptolemy based on Hipparchus' epicycle system for the tropical and the anomalistic years. As illustrated in Figure 5, the RS moves in a counterclockwise motion with a constant speed on the small circle, the center of which represents the celestial longitude of the MS moving clockwise. With a radius R of the MS's orbit (AE in Figure 5), the radius of the RS's circle (BS in Figure 5) was assumed by Ptolemy to be 1/24 R. The celestial longitudes of the 12 signs of the Zodiac are assumed to have an equal extension of 30° each and are all positioned on an outer sphere, the center of which is the Earth.

On the other hand, the RS would be seen from Earth at the beginning of each sign when it moves in unequal time steps from one sign to the next.

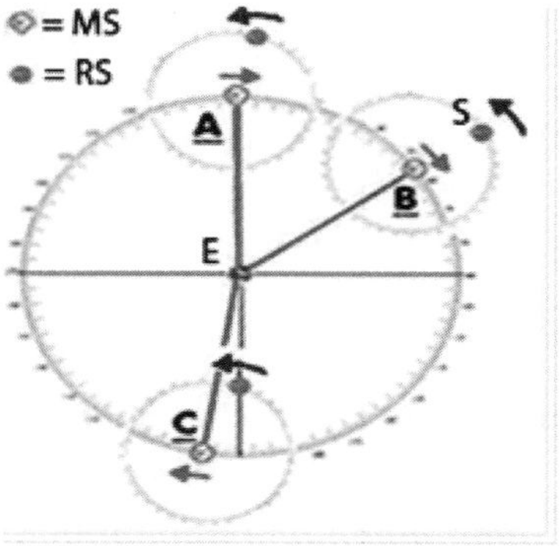

Figure 5. The epicycle model of the motion of the RS. Point A represents the longitude of the MS in the day of creation, whereas point C – at the beginning of the real autumn.

When the aphelion (where the celestial longitudes of the RS and the MS are identical, point B in Figure 5) is assumed to be at the celestial longitude of 65.5 degrees, the calculated celestial longitudes of the MS correspond to the celestial longitudes of the RS at the beginning of each sign as summarized in Table 4.

We note that the *average* time between two consecutive NMs formed by a conjunction of the RS and the real moon (RM) remains M_A. Let us now assume that in the Septuagint year of creation "Adam's NM" (the NM of Tishrei) took place when the RS was at the celestial longitude of 180 degrees corresponding to the beginning of the sign of Libra. At that moment, the MS, representing the time scale mentioned above, is positioned at the celestial longitude of 182.173°. When the RS reaches the celestial longitude of 210° in correspondence with the beginning of the sign of Scorpio, the *MS* would be at the celestial longitude of 211.387°. Consequently, the MS would thus have to complete the required number of full turns plus (211.387-182.173) = 29.214°.

As a result, in order to find the RS at the beginning of the Zodiac signs when a NM takes place, the MS would have to move for the first four signs, starting from Libra through Scorpio, Sagittarius, and Capricorn, in four similar, yet, not identical, steps of a full number of turns plus about 29° from one sign to the following, in accordance with

Table 4. The celestial longitudes of the MS when the RS's longitudes correspond to the start of a sign of the Zodiac.

Sign	Celestial longitude of the RS	Corresponding Celestial Longitude of the MS	RS-MS	Mean Sun steps from one Sign to the next
Libra	180	182.1729	-2.1729	
Scorpio	210	211.3865	-1.3865	29.21357
Sagittarius	240	240.2288	-0.2288	28.84236
Capricorn	270	269.01	0.99	28.78113
Aquarius	300	298.0561	1.9439	29.04612
Pisces	330	327.623	2.377	29.56692
Aries	0	357.8271	2.1729	30.2041
Taurus	30	28.61353	1.38647	30.78643
Gemini	60	59.77119	0.22881	31.15765
Cancer	90	90.99007	0.99007	31.21888
Leo	120	121.9439	-1.9439	30.95386
Virgin	150	152.377	-2.377	30.43309
Libra	180	182.1729	-2.1729	29.79588

the values in Table 4 (calculated for the aphelion at 65.5°) which are between 28.75° and 29.25°.

For the RS to be at the beginning of the signs of Aquarium, Pisces, Aries, Taurus and Gemini when a NM occurs, the first next step of the MS would be a full number of turns plus about 29.5°, followed by a step of a full number of turns plus a little over 30°, a full number of turns plus 30.8°, and 2 steps of a full number of turns plus about 31.2°, correspondingly.

If we now return to Figure 4 in order to find the number of years that are necessary to advance with such required steps surrounding the average value of 179.0833, we get the detailed results in Figure 4 (right). The marked values in Figure 4 (right) are what the editors of the chronology of the Septuagint version had calculated before the determination of the ages of the Patriarchs when they begot their following generations. With the contemporary constraint applied in accordance with the ancient view that Sun-Moon conjunctions are related to historical milestones, it was possible to build up the general structure of the ages of the Patriarchs based on the above calculations to be 228, 209, 190 and 171 for the ages of Adam, Set, Enoch and Keenan. The first sign for which the RS was expected to advance with a step slightly over 30 degrees, was the sixth sign and consequently the first general choice for the age of Jared, the 6th Patriarch, was 160. The ages of 168 and 187 for Methuselah and Lemech were similarly the choice in correspondence with the eighth and the ninth signs.

However, if we assume that they wanted to relate the NMs to real conjunctions occurring at the beginning of each sign, they knew that the number of years should not be exactly those provided by Figure 4. The real conjunctions would differ from the average conjunctions as a result of the difference between the RM against the MM, and as a result of the fact that the first NM was at 1°.

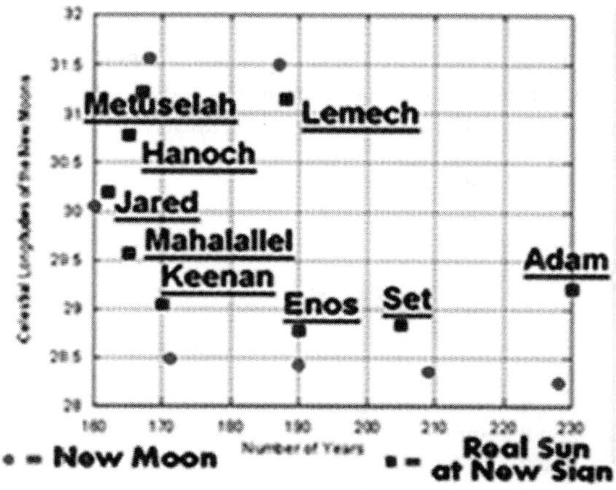

Figure 6. The calculated average NMs (in circles) compared with the ages of 7 Patriarchs and their corresponding steps of the RS (in squares).

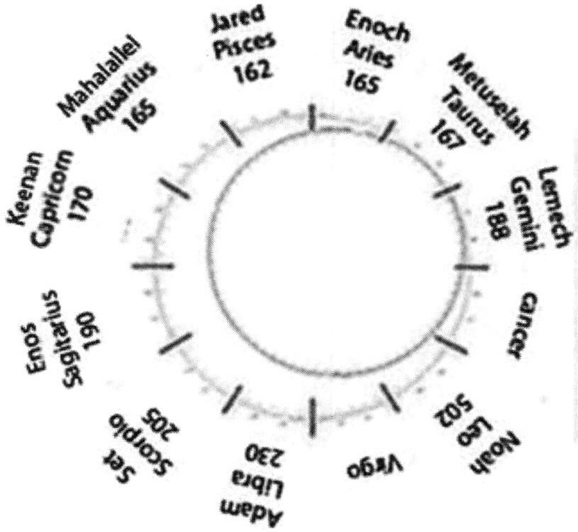

Figure 7. The eccentric visualization of the motion of the Sun, and the Biblical Patriarchs' ages as related to the signs of the Zodiac.

With the real NMs in mind, the ages of Mahalallel and Enoch (Hanoch) could then also be determined at the beginning of the fifth and the seventh signs to be 165 years completing the loop surrounding the average age as calculated above (Figure 6).

Finally, 6175 months after the birth of Noah, the MS changed its longitude by:

$$6175 \times 29.1060917 = 179730.12^\circ = 90.12^\circ \text{ (modulo 360)}$$

Adding 90.12° to the celestial longitude of the MS when Lemech begot Noah would bring it back within 1 degree from its original longitude when Adam was created.

6175 months equal 499 years + 3 months, and according to the Septuagint version of the bible, Noah begot Shem when he was 500 or 502 years old.

Figure 7, thus, summarizes the visualization of the ancient world relating the Patriarchs' ages when they begot their following generations "controlling" each a new sign (with Noah controlling 3 signs) in accordance with the motion of the RS.

References

Burgess, E. 1858, *Surya-Siddhnta, a Text-Book of Hindu Astronomy.*, New Haven, the American Oriental Society

Cohen, A. 2005a, (printed in 2007): "Astronomical Luni-Solar Cycles and the Chronology of the Masoretic Bible", Culture and Cosmos, 9, 2, 47

Cohen, A. 2005b, "The Chronology of the Septuagint Text as Opposed to the Chronologies of the Masoretic and the Samaritan Texts in light of Land of Israel, Greek, and Babylonian Astronomy", Judea and Samaria Res. Studies, Vol. 14, ed. J. Eshel (Ramat-Gan and Ariel), 381 (in Hebrew with an English Abstract).

Cohen, A. 2009, "The Astronomical Coordinates of the Planets in the day of their Creation as Determined by the Jews in Eretz-Israel in the Pre-Mishnah Era", *Judea and Samaria Res. Studies*, Vol. 18, ed. J. Eshel (Ramat-Gan and Ariel), 415 (in Hebrew with an English Abstract).

Hughes, J. 2009, Secrets of the Times: *Myth and History in Biblical Chronology*, Sheffield Academic Press. Kennedy, E. S. (1964), al-Khwarizmi on the Jewish Calendar, *Scripta Mathematica* 27, 55

Maimonides, M. 1956, 12th century: "*Laws of Sanctification of the New Year (LSNY)*" Chaps. 9, 10

Toomer, G. J. 1984, *Ptolemy's Almagest*, London, Duckworth

Astronomical Heritage of the Middle East
ASP Conference Series, Vol. 520
Sona V. Farmanyan, Areg M. Mickaelian, J. McKim Malville, and Mohammad Bagheri, eds.

Ancient Astronomical Knowledge: the Unity of Diversity

Irakli Simonia
The School of Natural Sciences and Engineering of Ilia State University, Tbilisi 0162, Georgia; iraklisimonia@yahoo.com

Abstract. Peculiarities and cultural projection of ancient astronomical knowledge are described. Substances of ancient astronomy including megalithic constructions, symbolism, and manuscripts are considered. Results of investigations of astromegaliths and written materials of Georgia are presented.

1. From Relicts to Enlightenment

Cultures of the ancient nations keep the initial views of the ancients about the universe, the world of stars and planets. These initial imaginations—grains of the primary knowledge, are scattered and irregular, however, they pervade practically all layers and segments of the ancient cultures. Since the pre-historic times, from epoch to epoch, these relict imaginations have been expanding and modernizing steadily. Megaliths and their structures, ancient stony structures spread so widely in different geographical regions throughout the world, are often aligned on the astronomically important points on the horizon, determining constructively both the observational and religious significance of a given complex or monument. The ancients prayed to their gods when watching the rise of bright luminaries: the Sun, the Moon, and stars. Quite often, just the brightest stars in the sky were for an ancient man his gods endowed with almighty power, light, and heat. When appearing above the horizon with ineluctable accuracy and regularity, the bright luminaries-gods were by view of an ancient man determining his daily life, the nearest and distant future. The ancient people gradually learned to divide and use the time for short periods and long seasons.

The megalithic structures transformed from the observational and ceremonial points into the fixators and keepers of time. With accumulating new views and filling the "baggage" of empiric knowledge, the ancient people started to apply more and more widely their constructional complexes for different purposes—to fix the rise of stars, to worship the gods, to define the warm and cold seasons. The time came when these complexes acquired multifunctional significance. The development of civilization, transformation of a human from contemplation and accumulation to construction and creation, as well as invention of new technologies by mankind, have predetermined occurrence and development of new elements of cultures, which included in themselves different crafts, primary architecture and applied arts of the different sub-types and directions. On a stone processed by a man, smelted metal, glass, appeared fragments of the amazing starry sky, wandering luminaries—pictures of the cosmographic nature. Ornaments and engraved images of the scenes of human's "communication" with the heavenly bodies began to decorate the surfaces of dishes of the colored and precious metals, decorations of various types. The obvious progress being accompanied by origination of the scientific fields like mathematics and astronomy, invention of writing, alphabets of different types and forms have further developed mankind's understanding of the surrounding environment and the universe. A man started to reflect the seen and understood on the sheets of parchment together with the empiric data of interpretation of what he has seen, explanations of what he has understood, and the results of what he has analyzed. The sets of the observational data, pictures and drawings, extensive descriptive material have become the basis for the ancient and medieval scientific treatises, dedicated to various astronomical phenomena, the starry sky, wandering luminaries, and of course, the universe in general. The development of different sciences and invention of the exact mechanical and optical instruments in the intellectual environment of the Renaissance and Enlightenment have expanded human knowledge of the same world of stars and planets. Ancient megaliths, the cosmographic ornaments, multilingual manuscripts and books are diverse in their unity, reflecting different paths of uniform "movement" of mankind towards the truth. This unity of diversity is composed of mutually supplementary components that are absolutely unified in spite of differences in their nature. Uniformity of the cosmological imaginations of the ancients and the astronomical knowledge of the periods of Renaissance and Enlightenment have been achieved by different ways and methods. Unity of the worldview in the diversity of its manifestations, as well as the generality of the differences in mankind's understanding of his role in the universe are based on:

1. observation by a man of the near and far orders of the surrounding landscape, relief, and numerous stars in the sky;
2. build-up of the megalithic structures with coordination to the local landscape and alignment with the astronomically significant points of the horizon;
3. observations and worship of the Sun, the Moon, and bright stars;
4. dividing the time into the short periods and long seasons, for daily, religious, and agricultural purposes;
5. reflection of the bright phenomena observed in the sky on stone, glass, metal handicrafts, household items, and decorations;
6. invention of the alphabets and other forms of recording with their further application for description of the celestial phenomena;
7. integration of the visible luminaries and strictly recurring celestial phenomena into the unified world system, with their further description in a form of the schemes, tables, and texts; and
8. the scientific analysis of the earlier collected, preserved, and systematized empirical data by means of the already developed theoretical and experimental research methods.

The principles specified above can be reformulated in detail; however, the unity of diversity of the ancient views and concepts of the universe are visible in the efficiency of the path of development of knowledge from astro-megaliths to orbital telescopes.

2. Solar Stations

The daily life of the ancient people was largely determined by the environment of their habitation. The surrounding natural landscape, having the shape of hills, forest slopes, extensive lowlands, rivers, and lakes) has stimulated the ancient people using these natural features. Their homes and settlements could be located on the lowlands near the water ponds, while the buildings and structures of special purpose could be placed on the heights, peaks of the hills and mountains. These special structures could have defensive, observation, and of course, the religious nature. After completion of a working cycle (a week, as per the modern sense), which had been established by traditions, views, and habits of relevant tribes, the ancient people could devote their free time to various activities, including observation of the celestial luminaries and phenomena.

As a result of the centuries-old accumulation of empirical information, the ancient people learned to "interact" with the sky and align their constructions with the astronomically important points of the horizon. Of course, there existed stricter cycles of vital activity of the ancient people. These cycles were linked to occurrence of certain luminaries (the brightest ones, primarily) in the sky, which, by view of the ancients, were the higher powers—gods. These two aspects were often interpreted uniformly by the ancients: the rising of the bright star as the return of the god of their tribe. Periodic strictly recurring celestial phenomena determined the observational experience and the religious commitment of ancient man. The long-awaited rise of a bright planet or a group of stars once again convinced the ancients in the correctness of their worldview and the necessity of their unwavering submission to the higher powers. Contemplation of the ascending luminaries and the simultaneous carrying out of various ceremonies and sacrifices were an organic part of the actual content of these strict cycles. The phenomena seen once and understood originally passed from generation to generation and acquired new features, details, and elements. Only the widest observational experience could link together a justification for waiting for a celestial phenomenon, correctness of the alignment of the construction, and features of the relevant ceremonies.

Inevitably strict recurrence of the rising and setting of bright celestial bodies allowed the ancient people to understand the intervals and periods of time seeing their virtual ties with climatic conditions, warm and cold seasons, the freezing of ponds, and floods of rivers. The ancients understood time in nature as well, namely, in themselves and the surrounding environment of flora and fauna. The rising of the bright luminaries, as viewed by the ancients, could indicate the approach of the rainy or hot season, or the time of harvesting. Special buildings acquired extremely critical importance. From these stony structures, the ancients observed the starry sky, worshiped the gods, and kept time. In these strange stony complexes built on the peaks of hills, they were storing all their most valuable things: knowledge, faith, and time.

It is not difficult to imagine how valuable these stones were for the ancient people and how they could protect them. Considering that the surrounding landscape determined the location of their houses and their special use buildings, the features of the visual horizon dictated the precise orientation of the elements of these buildings and divided the world of ancient man into the near-range and far-range orders—including the world of the uncountable stars. A particular elevation or depression of the visible horizon could be precisely aligned to the relevant elements of a megalithic structure: entrances, windows, menhirs, and their rows. Just above these hills, at an appropriate time of year, the long-awaited luminaries appeared, permeating with their light the corridors of the megalithic labyrinths. While observing, worshiping, and applying the acquired knowledge, the ancient people moved to the path of understanding their role in the near-range and far-range orders of the universe, as a whole. We are still moving in the initial section of this long path as we expand the knowledge and experience of our ancient predecessors.

The Sun played a special role in the daily life of ancient man, his views on the universe and in the entire progress and development of mankind. For the ancient people, the Sun was the main source of light and heat, indicator of time, and object of special worship. The Sun provided light and heat, transferred life-giving energy to the plants and cultivated crops, cured with its heat, lit the mountainous paths with dazzling light, and marked the approach of the warm and cold seasons with a perfect accuracy. The ancient people noticed regularly recurring movement of the rise and set points of the Sun on the horizon and understood the possibility of using this phenomenon for time keeping, fixing short intervals of time and indicating of the long-range seasons like winter and summer. Regularly recurring solar phenomena and the special power of the Sun's light and heat transformed the brightest celestial luminary into the main deity of the ancients, into the object of their particular worship. They were waiting for its occurrence, worshiping before its rising face and bright golden rays, arranging various ceremonies. The appearance of their god (the Sun) at various points of the horizon (such as the solstices) testified to the ancients about its great power and unequivocally pointed them to the beginning of the new season.

The ancient people devoted to their great deity, the Sun, special complexes of buildings. These megalithic constructions were built on special sites, peaks of mountains and hills, which were too difficult to climb every day even for physically strong men. On specially established days of observations and ceremonies, people climbed (or, were permitted to climb) to their megalithic complexes for observing the sunrise, meeting their god of light and heat, and arranging different ceremonies dedicated to these phenomena, which, in the view of the ancients, established a human's role in the boundless heavenly world. By their understanding, their role was minimal, symbolically expressed in small torches and bonfires, comprising elements of the worship ceremony. Structures of these megalithic complexes were aligned by the ancients according to the sunrise and sunset points in such a manner that the rays of the Sun could penetrate into the particular structural elements: entrances, corridors, and rows of menhirs. Constructional features of these megalithic complexes are uniquely associated with the peculiarities of the movement of the Sun over the sky, its shift along the horizon, and its influence on the consciousness of ancient men.

These ancient megalithic structures have become the first observation sites, fixators and keepers of time, temples for worship to the Sun. The multifunctional nature of these megalithic complexes is obvious. We call these multifunctional complexes "solar stations" (Simonia 2011). At the solar stations the people observed the sunrises and sunsets, fixed the time, and worshiped their god, the Sun. Later on, a level of multifunctionality of these stations increased with expansion of the empirical knowledge of the ancient people, who saw and understood that appearance and movement of the other less bright luminaries (the Moon, stars, and planets) could testify of other facts and phenomena and commemorate the beginning and completion of shorter time intervals. Expressed figuratively, the Sun became the hour hand of the virtual clock, while the stars and planets the minute hand. The increase in the level of multi-functionalism of the solar stations was accompanied by constructional complexity: extension of the new corridors and the menhir systems, mounting of new columns, and creation of new openings in various directions for additional windows. It would not be difficult to imagine how the ancients were observing sunrises through the main entrances to these buildings and observing the rising and setting of other stars from the additional areas of these buildings or through the specially arranged small windows.

The progress of the ancients in their cosmological imagination was positively reflected in their skill in constructing more and more complex buildings. We consider that the invention of solar stations cannot be exclusive to this or that particular ancient culture. Based upon the principle of the unity of diversity as formulated above, we conclude that solar stations could be spread over different regions of the world, at various latitudes, in conditions of different landscapes and climates.

Good examples of solar stations can be found in Georgia. Georgia is an ancient country located at the eastern edge of Europe. High mountains, forests, numerous rivers and lakes, and the coast of the Black Sea are all elements of the landscape of Georgia. During the past millennia, the Georgians formed and developed their own original culture, language, alphabet, literature, applied arts, music, architecture, folklore, and agriculture. One of the most important elements of the Georgian cultural heritage is the development of sciences, especially the exact ones, including mathematics and astronomy. Ancient Georgians' interest in celestial phenomena was manifested during various stages of the pre-historic epochs, including the megalithic era.

Figure 1. Abuli solar station in Georgia.

In Georgia, south of the capital city Tbilisi, in the municipality of Ninotsminda, lies a megalithic complex known as Abuli. The complex is located at 41°22′22″ N, 43°41′06″ E, elevation 2,637 m. The construction of the Abuli complex is characterized by rectangular and spherical forms. The spherical part of the complex is aligned to the southeast. An archaeoastronomical investigation conducted in 2011 and 2012 (Simonia & Jijelava 2012) showed that the eastern part of the horizon is open, and in the northeast direction the visible horizon has negative altitude, that is, it lies below the horizontal. In fact, the Abuli complex is a rounded structure on different levels with a central rectangular-spherical construction. A set of small, stone rooms, symmetrically located, are connected with one another and provide passage from one level to another. The central construction is supported by six columns marking the north, northeast, east, southeast, west, and northwest directions. The narrow corridor divides the building into component parts, north and south. In the northeast direction, there is an architrave entrance into the central edifice. This entrance faces the visible horizon and is oriented with azimuth 59°43′. There are two windows in a spherical part of the central building which, as viewed from the interior, are supplied with small niches. These are oriented with azimuths of 97°50′ and 166°54′. Both windows face the sky; to be more exact, only the mathematical (level) horizon is seen through them. This could undoubtedly have enabled ancient observers to see rising heavenly bodies near the horizon. At the times of the solstices, the rays of the rising sun were seen to penetrate through the architrave entrance. The Abuli complex possesses five or six points (or constructed elements) from which ancient people could have observed astronomical objects and events such as Polaris, the heliacal and acronychal rising of bright stars, and the rising of the Sun and Moon. If people lived around Abuli, then the complex could have been used for religious ceremonies and observations of the heavenly bodies with the aim of determining and preserving time (Fig. 1). If so, then the Abuli complex was indeed is an ancient solar station, or in other words, a multifunctional site for rituals and observation (Simonia & Jijelava 2015).

Near Paravani Lake in the Shaori Mountains of Georgia can be found a large megalithic complex. The Shaori complex consists of two parts: the large Shaori on the highest summit and the small one on the other mountain. The large Shaori building is located at 41°29′2″ N, 43°44′57″ E, elevation 2739 m. The architectural space of the large Shaori comprises an envelope-like shape with the remains of a small column at its center. The inner side of the complex wall contain numerous dolmens. The large Shaori building has a central opening with a door facing southeast. The large Shaori building is connected to the bottom of the mountain by a ceremonial road several meters wide and stepped with recumbent menhirs. These Bronze Age buildings had a dual significance both as a cult center and an observational place (Fig. 2). Religious ceremonies held in Shaori buildings could have included the observations of heavenly bodies. The central opening or door of the large Shaori building faces a true azimuth of 121°40′ as determined by a survey carried out by Ilia State University in 2011. The azimuth of the rising sun at the winter solstice is 121°, so that observers standing near the central column would have seen the rising winter solstice sun through the door. In view of what we know of ancient Georgians' religious beliefs, it is possible that the Shaori complex was not only a ceremonial center with buildings in which diverse rituals were performed, but also an observational site, a type of ancient solar station in which people worshiped their gods and watched the rising of diverse heavenly bodies—the Sun, Moon, and stars—in order to maintain their orientation in time (Simonia et al. 2015). Similar megalithic constructions are located in other parts of Georgia.

3. Ethno-Cosmological Symbolism

The continuous accumulation of new knowledge and practical experience by the antient people led to the development of new technologies, traditions, and methods. Humans

Figure 2. Shaori solar station in Georgia.

learned to smelt metal and glass, and prepare from these materials different items, tools, and ornaments of complex and delicate construction. The daily life and outlook of the ancient people also developed. It was not enough to only remember the objects they saw in the sky, but to reflect them in stone, metal, and glass. The bright, rare phenomena a person could see in the sky, while gradually acquiring more and more systematic understanding, started to be "reflected" on the surfaces of various material objects: dishes, tools, weapons, ornaments, and other items. They could be reflected as a conditionally straight projection (images of stars and planets) or in a form of relative virtual projection (correct geometric shapes, recurring ornaments). Of course, these astronomical projections were formed in line with the religious beliefs of the ancients, reflected in the additional visual and symbolic features of the pictures. Thus, the Sun, Moon, and stars, as well as their symbols, were set in precious stones and with different ornaments and decorations.

In Georgia, archeological artefacts of metal, glass, and ceramic are not infrequently decorated with the images of luminaries, or with pictures and ornaments of a cosmological nature. The ancient Georgians represented on the surfaces of different items, including the items of daily and ritual use, the luminaries they saw in the sky. The religious views of the ancient Georgians also had a certain impact on the straight and symbolic images of luminaries and on the general picture of the world as the ancients understood it. The strictly recurring circular ornaments and closed curves could symbolize rotation of the sky, the steadily recurring return of the luminaries, and a spherical shape of the world in the understanding of the ancients. In the middle of the last century, in Trialeti, in the eastern region of Georgia, a golden goblet decorated with semi-precious stones was discovered (Fig. 3). The goblet dates from the 17th–15th centuries BC and is considered an item of the material culture of Trialeti (National Museum of Georgia, inventory number 9-63/981) (Jorjikashvili & Gogadze 1974). The gold as well as the semi-precious stones of the ornament could symbolize the Sun, especially since the minerals of chalcedony (also known as karneol or sardonyx) are named in number of cultures as the sun-stones, while turquoise minerals are known as the heavenly-stones. Thus, a common, conditionally cosmological picture is formed where the Sun is the basic element of the recurring celestial phenomena and the rotating world of stars. Considering that the cultural traditions of viticulture and winemaking were developing in Georgia during those millennia, we may gain some insight into the common philosophical-cosmological ideas of the ancient Georgians and "find" in the Trialetian golden goblet a material embodiment of the boundless space, of the steady ticking of time, and the extensive harvest of the grapes that depends so strongly on the warm sunbeams.

Another example of the ethno-cosmological symbolism was discovered in the 1970s. A silver bowl with the orderly sequenced pictures of animals and recurring ornaments was found in the Colchis lowland in the western part of Georgia (Fig. 4). The animals depicted on the bown were shown as if running around one center in the counterclockwise direction. The bowl dates to the 2nd century BC (Herzfeld 1947). In the eastern toreutics, the bowls with the images of lotus, a symbol of the Sun, were spread widely. However, in case of the Colchis bowl, the images of animals are placed instead of the lotus, which is undoubtedly associated with the cosmological imaginations of the ancients. The eight-petalled star between the first and the fourth animals can be considered an additional indicator of the cosmological nature of the pictures on the bowl. The Lion, lynx, boar, and deer depicted on the bowl had a special place in the Colchis pantheon. This silver bowl, decorated with the images of animals and regular ornaments, is an example of the ancient Colchis art that demonstrates the views of the ancient Georgians about the world of stars (Simonia & Simonia 2005). The ancient people saw that the large and small groups of stars are easily combined into familiar faces and outlines of animals, which they placed on objects of daily use. Could the pictures of animals on the Colchis bowl be an attempt of the ancient people to see the terrestrial world in the sky, with outlinines of the constellations in the darkness of night? Most likely, the polar region of the sky is reflected on the bowl by its ancient artisan. Is it possible to link the number of animals depicted on the bowl with different seasons of a year, with periods of warm, cold, and moderate weather? Answers to these questions are found in the comparative analysis of the Colchis ornament; one thing is clear: on the silver surface of this sample of ancient art, one sees that space, time, and the rotations stars demonstrated.

The examples of the ethno-cosmological symbolism above testify to the diversity of unity of worldviews of ancient peoples who tried to describe by different methods and symbols the heavenly bodies they were observing in the sky and reflect them as certain referenes in a dynamical world.

4. Letters and Numbers in the Unity of Parchment

As time passed, development of mankind continued as people permanently acquired new knowledge and skills, transforming their outlook and their ways of thinking and acting. New cultural layers were created, technologies were modernized, and instead of

Figure 3. Cosmological symbolism of the Trialeti golden goblet.

passive contemplation of the surrounding world, the more exact analyses of constructivism and representativeness appeared. It was no longer enough just to see, depict, or simply describe. A clear necessity emerged to understand more deeply what was seen, to describe it exactly in detail, and to predict tomorrow's events and circumstances. The ancients recognized the necessity of development of methods for inter-cultural communication, a universal description of the surrounding environment, and to share of knowledge. Invention of scripts of different types was a precondition for a qualitative leap in development of the ancient cultures by enabling a new level of "dialog" between a humanity and the nature. The possibility of a numerical description of the countless stars, and expression of the observed celestial phenomena by the series of letters and numbers, became a unique "instrument" for accumulation, storage, and analysis of the empirical knowledge about the Universe. Written language opened the way to the theoretical description of the world of stars and planets. Thus, in this way the layers of handwritten materials, textual complexes, tables and figures describing movement of planets, constellations, eclipses of the Sun and the Moon were formed.

In different languages, using different alphabets and forms, with their peculiarities, the same celestial phenomena were described, complementing each other and generalizing a common picture of the Universe. Perfection of the systems of accounting and deepening the traditions of written language played important roles in the formation of the exact sciences of mathematics and astronomy. As a result, philosophers and schol-

Figure 4. Projection of the Universe in the Colchis silver bowl.

ars of the ancient era received the precise virtual instruments for studying the universe, a certain initial laboratory of its understanding, which was "based" on the parchment sheet. The heritage in the variety of written materials of different cultures reflects the unity of interest of the ancient scholars to both the process of understanding the creation of the world and the very nature of the celestial phenomena. When considering small details, individual celestial phenomena, or some specific object, the ancient scholars were on the path to the creation of a common model of the Universe and understanding of the world.

Creation of an original alphabet by the ancient Georgians was a precondition to the formation of their written heritage. A key element of this heritage are the manuscripts of astronomical and cosmological nature. More than 300 such manuscripts written using the letters of the ancient Georgian alphabets—Asomtavruli, Nnuskhuri, and later on Mkhedruli—are kept in the libraries and archives of Georgia and other European and Asian countries. Figure 5 shows a fragment of the old Georgian astronomical manuscript (Star Book) describing different interpretations of the celestial phenomena, and mentioning the names of the same luminaries in different languages, reflecting the common Ptolemaic model of the Universe (Simonia 2001). Today, in the epoch of nanotechnology and space missions, representatives of different cultures, scientific schools, and traditions, return again to the common goal of understanding the micro-

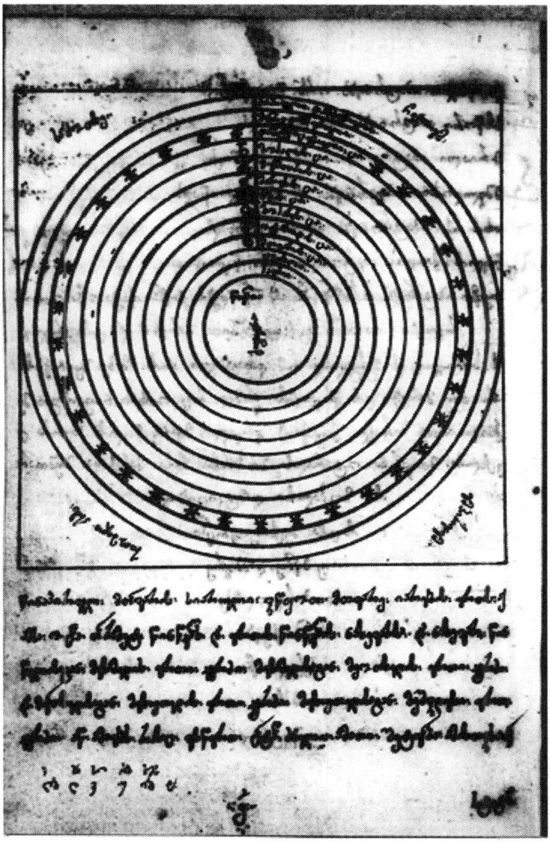

Figure 5. Old Georgian manuscript Star Book.

and macro world. They integrate into the unified complex the cultural variety and erect a "bridge" between the past and the future.

Acknowledgments. The author expresses his gratitude to B. Jijelava, G. Houston, K. Jelia, and Ts. Davlianidze for their participation in the field studies and for valuable discussions.

References

Herzfeld, E. 1947, Zoroaster and his world. Vol.2

Jorjikashvili, L. G. & Gogadze, E. M. 1974, Trialeti monuments of earlier and middle Bronze Age, Tbilisi

Simonia, I. 2001, "Little known aspects of the history of Georgian astronomy," Journal of Astronomical History and Heritage, 4(1), 59

Simonia, I.; Simonia, Ts. 2005, "Metal artifacts as a mirror of ancient Georgian astronomical world view," Current Studies in Archaeoastronomy: Conversation Across time and Space, J. W. Fountain & R. M. Sinclair, eds. (Carolina Academic Press), 435

Simonia, I. 2011, "Astronomical significance of ancient monuments," The Inspiration of Astronomical Phenomena VI, Corsini E.M., ed., ASPCS Vol. 441, 487
Simonia, I. & Jijelava, B. 2012, "Field research of Abuli, Shaori, and Tejisi monuments in mountains of Georgia," Ilia State University, Tbilisi, Private communication
Simonia, I. & Jijelava, B. 2015, "Astronomy in ancient Caucasus," Handbook of Archaeoastronomy and Ethnoastronom, C. L. N. Ruggles, ed. (Springer), 1443
Simonia, I., Jijelava, B., Gigauri, G., & Houston, G. 2015, "Observational and cult sites in prechristian Georgia," Handbook of Archaeoastronomy and Ethnoastronom, C. L. N. Ruggles, ed. (Springer), 1461

Astronomical Heritage of the Middle East
ASP Conference Series, Vol. 520
Sona V. Farmanyan, Areg M. Mickaelian, J. McKim Malville, and Mohammad Bagheri, eds.

History of Astronomy: Per Aspera Ad Astra

Anahit A. Yeghiazaryan
NAS RA V. Ambartsumian Byurakan Astrophysical Observatory (BAO), Byurakan 0213, Aragatzotn province, Armenia; anahit@bao.sci.am

Abstract. This paper contains a brief overview on the development of astronomy and the challenges it faced along the way in various regions around the world.

In Greek, *Astronomia* (*αστρονομια* from *άστρον/ástron*, "star" and *νόμος/nómos*, "law") was the study of the celestial bodies, their interconnection, structure and development. It included the study of cosmic systems and the study of the Universe in general as well.

Astronomy is the oldest of the natural sciences. It dates back to the ancient times and originates from the necessity to navigate through time and space during travels. At the time, even the mere observation of the celestial bodies with the naked eye helped determine a course both on land and at sea. The periodic observation of celestial bodies allowed establishing of certain seasonal patterns. These observations and studies were also the basis for the creation of a number of calendar systems.

There are records of fairly accurate astronomical observations dating back to ancient times. The ancient Egyptians calculated the 365 day year, recognized the sequence of the lunar phases, and predicted the periodicity and the recurrence of solar eclipses with a period 18 years and 10 days as early as 28th century BCE (Before Common Era).

Since BCE the solstice and equinoctial points were calculated in China, the solstice map based on Jupiter's trajectory was created in India, and the spring floods of River Nile were being predicted based on star movements in Egypt. The lunar phase observations were crucial to make field work in the Arabian deserts possible during the hot summer days, and celestial navigation was essential for the development of marine navigation in Greece. Numerous petroglyphs depicting astronomical subjects were found near ancient human settlement sites all over the world, including near Mount Sevsar, Metsamor, and the Vardenyants mountain pass in the Armenian highlands.

Without changing their positions relative to each other, it seemed, the stars were anchored to the heavens, while making their recurring daily journey without fail. The patterns of the stars' everlasting groups were likened to animals, mythological figures or even household items, thus the division of the celestial bodies into constellations was born. The seven days of the week received their names from the Sun, Moon and the five planets of the Solar system, named after the Greek gods Venus, Mercury, Mars, Jupiter and Saturn. The observations helped calculate Sun's movement against the stars, its path and annual period, based on the zodiac constellations.

In Ancient China the sky was divided into 122 constellations, from which 28 were of the Zodiac. Their star catalogue, containing a list of 807 stars, came centuries before the star catalogue of the Greek astronomer Hipparchus. Despite this knowledge, the 12 constellations of the zodiac were the most commonly adopted when it came to the travels of the Sun across the celestial sphere over the course of the year, passing through each one in about one month.

In the 6th century BCE, the Pythagorean school adopted the Geocentric model of the universe; in the 4th century BCE, Eudoxus of Cnidus presented the geometric model of the motions of the Sun, the Moon, and the five planets known in antiquity, which later was further advanced by Aristotle and Ptolemy. During the 6th century BCE, Eratosthenes estimated the radius and the circumference of the Earth in Alexandria, star catalogues were prepared by the Chinese astronomer Shi Shen and Greek astronomers Timocharis and Hipparchus. Later the Christian dogmatism led to the decline of the natural sciences as well as astronomy. Subsequently observations and studies in this field continued mostly in Arabic countries.

In the 7th century CE (Common Era), Anania Shirakatsi, who is considered to be the founder of Armenian natural sciences, discussed the concepts of the spherical Earth, the calendar, and sundials, and presented his calculations of Lunar and Solar eclipses. The works of Shirakatsi, along with a large number of manuscripts by other authors containing information about observations of astronomical events are available at the Mesrop Mashtots Institute of Ancient Manuscripts (commonly referred to as Matenadaran) in Yerevan.

In the 9th century, the Baghdad Caliph Al-Ma'mun initiated the translation of Greek philosophy and scientific works into Arabic, including Ptolemy's famous astronomical study, the *Almagest*. Almost until the end of 15th century CE, astronomy flourished in Central Asia and the Arab world (Al-Battani in the 9th century; Abu al-Wafa in the 10th century; Al-Biruni in the 11th century). Later, starting in the second half of the 13th century, the astronomical lists commissioned by the King Alfonso X of Castile and compiled by European and Mauritian scientists became particularly significant.

In the $12–13^{th}$ centuries, interest in natural sciences in Europe increased. Universities were founded first in Italy and later in a number of other European countries, where natural sciences were studied. The observations of celestial bodies were resumed by German scientists. In 1543 Copernicus published his famous "On the Revolutions of the Heavenly Spheres" in Nuremberg, describing a Sun-centred model; but unfortunately even Danish astronomer Tycho Brahe refused to accept the Copernican system. This system was later confirmed by Italian physicist Galileo Galilei, but during the same 16th century, after 7 years of imprisonment, Giordano Bruno was burned at the stake in Rome on charges of heresy.

In 1609, Galileo built an observatory in Venice, which enabled him to observe the sky, which in turn laid the foundations for miraculous discoveries: mountains were spotted on the Moon, sunspots on the Sun and the diversity of the planetary surfaces was discovered. The Milky Way became known as a multitude of stars, even though invisible to the naked eye. Regrettably the works of Copernicus stayed forbidden for 200 more years.

Kepler's three laws of planetary motion became the basis for the development of celestial mechanics, the cornerstone of which was Newton's law of universal gravitation. Later the orbits and movements of the celestial bodies were clarified based on accurate astronomical observational data and thanks to famous mathematicians (Lagrange, Euler, Laplace, and Gauss). The construction of telescopes and the development of their capabilities contributed to the development of astronomy, astrophysics with its different branches (astrophotography, astrophotometry, astrocolormetrics, astrospectroscopy, etc.), as well as astrochemistry and astrobiology. Ultimately the ground work for humanity's centuries-old dream of space exploration was put into effect. Byurakan Astrophysical Observatory, founded in Armenia in the mid 20th century, has made a significant contribution in astronomy.

Today the scientific studies of astronomers from around the world are accessible to everyone. Through wide spread scientific cooperation there are countless studies, various scientific publications, as well as astronomical companies and regional astronomical centres are established and functioning successfully.

Acknowledgments. The author thanks Arman S. Zaratsyan for his invaluable contribution to the preparation of the above mentioned material.

References

Maunder, E. 1906, Astronomy without Telescopes, London
Olcott, W. 1904, Legends of Stellar Universe, Petersburg
Tumanyan, B. 1964, History of Armenian Astronomy, Yerevan

Astronomical Heritage of the Middle East
ASP Conference Series, Vol. 520
Sona V. Farmanyan, Areg M. Mickaelian, J. McKim Malville, and Mohammad Bagheri, eds.

When Ibn Sīnā Saw Venus on the Disc of the Sun

Ramesh Kapoor
Indian Institute of Astrophysics, Bengaluru, India; rckapoor@outlook.com

Abstract. This paper examines the claim by Abū Alī Ibn Sīnā (980–1037 CE), the renowned Persian natural philosopher, mathematician, physician, and author of works on philosophy, sciences and medical treatises, in one of his books that he saw the planet Venus as a spot on the disc of the Sun.

1. Introduction

Abū `Alī Ibn Sīnā (980–1037 CE), the renowned Persian natural philosopher, mathematician, physician, and author of works on philosophy, sciences, and medical treatises, states in his Compendium of the *Almagest:* "I say that I saw Venus as a spot on the surface of the sun." The date and place of the observation are not given. Yet, Ibn Sīnā was quoted by Naṣīr al-Dīn al-Ṭūsī (1201–1274 CE), and later by the astronomer Cyriacus (ca. 1482 CE). Ibn Sīnā's description amounts to a naked eye observation of the transit of Venus across the disc of the Sun. There indeed was a transit in his lifetime, on May 24, 1032 CE (Julian). However, there are also some questions raised thereabout, the foremost being if it was a sunspot that he may have seen (Goldstein 1969). Goldstein cited Ptolemy who said "even if a planet did transit the Sun, it might not be visible on account of the brightness of the Sun and the small area of it that the planet would cover." However, in Ptolemy's planetary order, a transit by a planet below would happen every conjunction—both inferior and superior.

2. Goldstein's Constraints

Goldstein (1969) looked into earlier claims of the similar kind made by Al Kindi (800–870 CE), Ibn Rushd (1126–98) and Ibn Bajjā (1095–1138/9) but dismissed them on grounds of chronology and feasibility. He examined Ibn Sīnā's Arabic text Compendium of the *Almagest* with the Bibliotheque Nationale, Paris (copy dated 673 AH–1274/5 CE) more closely and quoted in the following manner a crucial comment from the work: "Avicenna here states (in his comments on the beginning of Book IX concerning the order of the planetary spheres): 'I say that I saw Venus as a spot on the surface of the sun,' with the very words cited by Nasir al-Din."

In order to evaluate the claim, Goldstein (1969) listed contact timings of the transit event for a set of latitudes and longitudes where the transit was possible to be seen (Table 1). The contact timings were provided to him by Jean Meeus. From these, he discarded Ibn Sīnā's claim saying "this transit may not have been visible where he lived." Goldstein's conclusion was not location specific. Besides, his contact timings were

Table 1. Contact Timings (UT) for Different Locations

Latitutde	Contact I 15h21m	Contact II 16h07m
30 N	52.8 E	41.3 E
32 N	54.0 E	42.5 E
34 N	55.2 E	43.7 E
36 N	56.5 E	45.0 E
38 N	57.9 E	46.4 E
40 N	59.3 E	47.8 E

based on Meeus' 1958 algorithm that provided timings correct to within about 15 minutes. I have examined the astronomical circumstances with JPL's Horizons Ephemeris system (JPL 2012) in light of the modern transit predictions (Espenak 2012). The circumstances are considered specific to two cities in Iran, namely, Isfahan and Hamadan where Ibn Sīnā may have made his observation.

3. Astronomy in Ibn Sīnā's Times

Ibn Sīnā's were the times when the Islamic astronomy was making great strides. The pre-Islamic Persia was already rich in astronomical literature and bore upon it the influence of Indian works. Ptolemy's monumental work translated into Arabic as *Kitāb al-majistī* in 829–30 CE greatly influenced the Islamic astronomy. Ibn Sīnā lived at Hamadan through the years 1015–23 CE. He moved to Isfahan where he spent the last part of his life during 1023–37 CE. He died and was interred at Hamadan (Afnan 1958). Here below is a list of some of his works:

- Treatise on celestial bodies [*Maqāla fi al-ajrām al-samāwiyya (al-'ulwiyya)*];
- A book of complete astronomical observations *(al-Arṣād al kulliyya)*;
- A commentary on Ptolemy *(Tahrīr al-majistī)*
- A work against astrology [*Maqāla (Risāla) fī ibtāl 'ilm (ahkām) al-nujūm*];
- A treatise on astronomical instruments *(Maqāla fī al-ālāt al-raṣadiyya)*;
- The book of remedy *(Kitāb al-Shifā')*; etc.

Ibn Sīnā's important astronomical observations are summarized as follows:

- The movement of the solar apsides: in the Kitāb al-Shifā´, Ibn Sīnā discussed the movement with respect to the equinoxes regarded fixed since Ptolemy's times.
- A new value for precession: Ibn Sīnā measured it at 1° in 66 years.
- Al Zarquāli of Toledo (1029–87 CE) was first to appreciate the apsidal precession with respect to stars (that he put at $12\overset{''}{.}04$ per year and the apogee at longitude $77°50'$) as distinct from the precession of the equinoxes. Ibn Sīnā's work is an independent recognition of the same but before Al Zarquāli.

- The obliquity of the ecliptic at 23°33′30″ that he cited in his work compared to Ptolemy's value of 23°51′. Quoting the value 23°35′ obtained in the times of Calif Al Māmun (786–833 CE), he commented how the obliquity had decreased further over time (Saliba 2011).

- Ibn Sīnā felt that Ptolemy's observations were deficient and needed to be improved with new observations.

- Ibn Sīnā measured a more accurate value for the longitude of Jurjān (Gurgan) from Baghdad at 9°20′ (conventional value 8.0°; modern 10.3°) from two observations of the culmination of the Moon at the former place where he lived during 1012–14 CE.

4. The 1032 CE Transit Re-constructed

We may now ask how Ibn Sīnā knew that the spot on the disc of the Sun was the planet Venus. It can be only when one is pursuing the target diligently. A transit has to be seen if one describes it in the words Ibn Sīnā did. It is therefore desirable to reconstruct Ibn Sīnā's observation. For that, the question of his location becomes most important. Ibn Sīnā's biographies draw mostly from his disciple Juzjāni's account. In 1023 CE Ibn Sīnā escaped from prison and eventually reached Isfahan to his great welcome and joined the court of Alā-al-Dawā. Most of the accounts concur about his first stay at Hamadan during 1015–23/24 CE and the expedition of 1037 CE. Already sick, his last expedition to Hamadan so deteriorated his condition that he died and was buried there in June or July of 1037. From the biographies we are not able to ascertain if there was an expedition also in 1032 CE, as also whether the observation was from Hamadan or Isfahan. Therefore, we consider astronomical circumstances specific to both Hamadan and Isfahan. The places are 369 km apart. Incidentally, at both, the transit was critically timed because it began close to sunset. However, that is also the best time to view the Sun with the naked eye, subject to right geographical and weather conditions.

4.1. The Circumstances of the Transit at the Two Specific Locations

Contact I of the Venus transit happened at 15:18 UT and Contact II at 15:34 UT (Espenak 2012). At Hamadan (34°47′54″ N, 48°30′53″ E; 1850 m), sunset was at 15:48 UT. At Contact II, the Sun stood at an altitude 2.2°. At Isfahan (32°39'05" N, 51°40′9″ E; 1590 m), sunset was at 15:30 UT. Contact I happened 12 minutes before sunset. At Contact II, the Sun had sunk to the altitude of -1.1°. Seen from an elevation h, the apparent horizon, the point where the line of sight is tangent to the surface of the Earth, is $\cos^{-1} R_{\oplus}/(R_{\oplus} + h)$. In the present case, it is $\sim 1.3° - 1.4°$. The refraction effects at the horizon are approximately equivalent to twice the angular diameter of the Sun. Besides, the contact timings readable from Xavier Jubier's (2012) interactive Google map for the 1032 transit are two minutes earlier. Together these give several additional minutes to an observer. So, an observation of the transit was possible.

4.2. Precision Attained in Measurement of Angles and Time

We need to have an idea of the precision attained by the Islamic astronomers, namely, the computational, instrumental, and the observational precision. In eclipse measurements, typical errors were ~1°, correspondingly the timing error was ~4 minutes. In *al-Zīj al-Hākimī al-kabīr*, Ibn Yūnus (d. 1009 CE) mentions a conjunction of Mercury and Venus in longitude on the evening of May 19, 1000 CE. He noticed Mercury a third of a degree north of Venus. That is in order—the conjunction that happened in the constellation of Gemini was the closest at 17:45 UT with $\Delta\delta = 23\farcm1$.

The next question is if the observation was serendipitous. With reference to a certain epoch, the Islamic astronomers determined dates of greatest elongation and the first conjunction of an inferior planet. Ibn Sīnā would know that the forthcoming conjunction of Venus ($\lambda_{\text{Venus}} = \lambda_{\odot}$) is inferior and, approximately when it is. He may have intended three observations spanning the whole period of invisibility of the planet to determine the exact instant of the conjunction, to the last day or even fraction thereof. One can expect his tables to enable him to compute, more accurately than Ptolemy's, the mean positions of the planet and the Sun spread over several weeks on either side of the day of conjunction. The computed positions would show a peculiar situation of their paths crossing around the time of conjunction.

The orbit of Venus is inclined by 3.39°. The transit can take place depending on where the node is placed with respect to the Sun. We define the path bounded by the points on the ecliptic on either side of the node where grazing incidence would take place as transit zone. Figure 1 is an illustration of the transit event of 24 May 1032 CE computed with the Horizons Ephemeris system, showing the transit zone (Kapoor 2013). It is drawn to the scale using values so derived, including the angular diameters of the objects (57.63″ for Venus and 1888.58″ for the Sun, respectively). The x-axis is the longitude difference $\Delta\lambda = \lambda_{\text{Venus}} - \lambda_{\odot}$ in arcmin of the objects that increases from right to left. The y-axis is ecliptic latitude β (arcmin). The Sun (broken and continuous circle) moves W-to-E (right to left) along the x-axis while Venus (the dot) moves left to right along the sloping line. The first dot (extreme left) corresponds to Venus as at 07:00 UT. As per our computation, the grazing points are ~4° apart, spaced in time by ~58 hours. On May 24, 1032 CE, Venus passed its descending node at 08:38 UT and the grazing latitudes $\beta \sim \pm 0.27°$ on May 23, 05:00 UT and May 25, 12:13 UT respectively. The planet made Contact I with the Sun at 15:18 UT at a position angle PA = 104.445°, and Contact III at 22:32 UT respectively.

4.3. Seeing a Transit Naked Eye

Could the event have been seen with the naked eye? There are claims to seeing Venus with the naked eye within a few degrees of the Sun (Baldwin 1880). Goldstein (1969, 1972) cited a newspaper report of the naked eye sighting of Venus in transit in 1874 from Hawaii, where people kept ready smoked glass to view the Sun at 3:30 in the afternoon. Similarly, Johnson (1883) reported the planet Venus distinctly visible to the unaided eye through a darkened glass during the 6 December 1882 transit he observed from Bridport. That was because one already knew that the planet was in transit. We may bear in mind that:

- The limit of detection (20/20 vision) is about one minute of arc.
- That is also the apparent diameter of Venus while transiting across the face of the Sun.
- A 1′ detection is possible by 100% contrast on a bright extended object;
- It is limited by smearing due to scattering and the resolving ability of the observer.

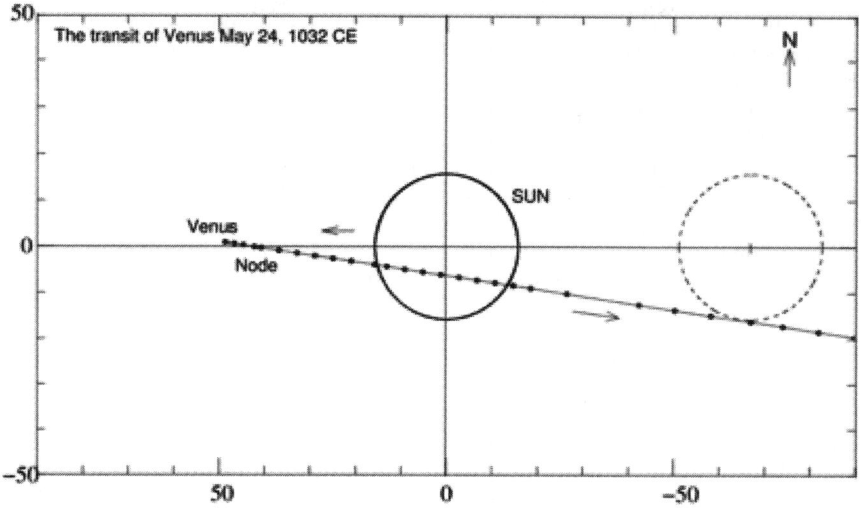

Figure 1. An illustration of the transit event of 24 May 1032 CE (adopted from Kapoor 2013).

Was it the planet or a spot? One is also aware of claims of seeing sunspots with the naked eye, even those with size smaller than one minute of arc. Keller and Friedli (1992) determined the threshold limit of sunspots observable by eye on the basis of observations made by twenty observers. They found that a normal eye can detect a sunspot of 41″ in penumbral diameter and 15″ in umbral diameter. Their measured limit from the tests is 19.″3 that indicates that the visibility is contributed by the umbral and the penumbral diameters both. At 1 AU, a sun spot the size of the Earth will subtend an angle of 17.″6.

Ibn Sīnā does not say where exactly on the body of the Sun he saw the mole. The planet in transit touches the Sun at its eastern limb as in the Figure 1. The position angle (PA) of Venus on the disk of the Sun is measured north to east: $\sim \pi/2 + \sin^{-1}\beta/r$ where β is the ecliptic latitude of the planet at the time of Contact I and r the angular radius of the Sun. For $\beta = -0.0654322°$, the PA is 104.445°.

The PA of Venus is not so large as to discredit the sunspot idea. A large sunspot would normally appear close to the equator. On 03/06 June 1032 CE when Venus rose, the Sun had rotated about its axis so much that the spot, if a sunspot, despite having reached the other limb of the disc, might still be around. Or, did Ibn Sīnā see a spot close to the western limb that vanished as the Sun turned it away a day later?

It is not obvious if the Muslim astronomers had any idea about sunspots or whether transits ever came up for discussion among them. A Sun with spots was philosophically challenging and Ibn Sīnā would have certainly commented upon any claims made in earlier records or works.

Was the Sun active in 1032 CE? For the "mole" to be a sunspot means that the Sun was active in 1032 CE. I have cross-checked with the re-constructed history of the solar cycle (Usoskin 2008). The work provides a detailed analysis of the data on atmospheric

Carbon-14 deposits in tree rings, inversely related to sunspot number and points to periods of grand minima in solar activity in history when the sunspots disappeared almost completely. One such grand minimum may have occurred in the year 1040 CE. A transition from normal high activity to such minima is sudden and can occur within a period of a few years without precursor (Usoskin 2008). That suggests the year 1032 CE could be one of high solar activity. On the other hand, there are pointers to a quiet Sun over a long period of time around the date in a similar study by Clark and Stephenson (1978) based on the ancient records of sunspot observations in China and Korea. Also, Vaquero (2007) contends that the Sun was quiet in this period (see his Fig. 1a on the 50-year moving average of the number of annual naked-eye sunspots recorded over the past 22 centuries). So, the Sun was probably quiet in 1032 CE.

5. The Event in Ibn Sīnā's Words

Muslim astronomers were interested in and made precise measurements of the motions of the planets, particularly around the times of opposition and conjunction, as well as at the ecliptic crossing. It is therefore likely that Ibn Sīnā did not just stumble upon the Venus transit that day. The question of whether the observation was made from Isfahan or Hamadan in the year 1032 CE remains unsettled. The terrain and visibility—the line-of-sight conditions towards the horizon—are also a concern. Pouria Nazemi points out that Isfahan does not have many geographical blockages. The eyesight of our celebrated observer, who at 52 had imperfect health, is also an issue.

The crucial passage in the *Kitāb al-Shifā'* (in Arabic), provided by Nazemi (2012) together with his own translation, is:

أما كرة الزهرة وكرة عطارد فإن الأقدمين رتبوهما تحت كرة الشمس وبعض من تأخر عنهم رتبهما فوق
كرة الشمس أيضا إذ لم يجدوهما تكسفان الشمس وهذا غير واجب فإنه يجوز أن لا يكون مجازهما تحت الشمس
فى السطح الذى يمر بأبصارها ومركز الشمس ومع ذلك تكون تحت كرة الشمس ولا تكسفان الشمس كما يكون
فى أكثر اجتماعات الشمس مع القمر أقول إنى رأيت الزهرة كخال وشامة فى صفحة الشمس.

"But the spheres of Venus and Mercury are under the sphere of the Sun. But some recent scientists said that these spheres are over the sphere of the Sun. Because they couldn't see Venus eclipse the Sun. But this eclipse (transit) is not (always) necessary. Because it is possible they (Venus and Mercury) pass from under the surface of Sun and we can not see them in front of disc of Sun. That is just like the situation that happens during Moon and Sun conjunction. And I say, I saw myself, Venus like a black dot on the disc of Sun."

6. Conclusions

I have considered the astronomical circumstances of the transit episode employing JPL's Horizons Ephemeris system. These suggest that Ibn Sīnā could indeed have obtained a glimpse of Venus on the disc of the Sun just before it set. The location is more likely to be a place north-west of Isfahan, probably Hamadan itself. Ibn Sīnā makes a significant statement in his work *Kitāb al-Shifā'* where he argues that, just as during the Moon and Sun conjunction, a transit does not take place every conjunction because

Venus has a different angle with the ecliptic. In other words, there exist nodes in the path of the planet Venus. The works on the historical sunspot sightings and reconstruction of the solar cycle over the past two millennia suggest that circa 1032 CE, the Sun was rather quiet. That is not to exclude the sunspot possibility altogether, but for a people who meticulously tracked planets to construct planetary paths and constantly improve upon their astronomical tables so that Ptolemy's model could work better, to mistake a sunspot for Venus does not seem a logical proposition.

So, when Ibn Sīnā said he saw Venus on the surface of the sun, he meant it. His observation has in it the first direct proof that the planet is an extended object—with a finite angular (and physical) size. These are fundamental deductions, more significant than even the sighting. These can come only from an actual observation by one fully appreciative of the Ptolemy Model.

Acknowledgments. I am grateful to UNESCO for financial support to attend the conference "Astronomical Heritage of the Middle East" held at RA National Academy of Sciences, Yerevan, Armenia during 13-17 November 2017 where this paper was presented. I thank Prof. Rajesh Kochhar, Pouria Nazemi and Xavier Jubier for useful interactions. John Walker's site was useful in the calculations. Dr. Baba Varghese helped with the scale drawing of Figure 1 and the Library at the Indian Institute of Astrophysics facilitated my search for many research papers and books.

References

Afnan, S. M. 1958, "Avicenna His Life and Works," George Allen & Unwin Ltd, London.

Baldwin, H. L. 1880, "Visibility of Venus in the Daytime," The Observatory, 3, 573

Clark, D. H. and Stephenson, F. R. 1978, "An Interpretation of the Pre-Telescopic Sunspot Records from the Orient," Quarterly J. Royal Astr. Soc., 19, 387

Espenak, F. 2012, "Six Millennium Catalog of Venus Transits: 2000 BCE to 4000 CE," `https://eclipse.gsfc.nasa.gov/transit/catalog/VenusCatalog.html`

Goldstein, B. R. 1969, "Some Medieval Reports of Venus and Mercury Transits," Centaurus, 14, 49

Goldstein, B. R. 1972, "Theory and Observation in Medieval Astronomy," Isis, 63, 39

Johnson, S. J. 1883, "Observation of the Transit of Venus, Dec. 6, 1882, made at Marseilles," Monthly Notices of the Royal Astronomical Society, 43, 75

JPL 2012, Jet Propulsion Laboratory's Horizons system `https://ssd.jpl.nasa.gov/`

Jubier, X. 2012, "The Interactive Google Map for the 1032 Transit," producible through the website `http://xjubier.free.fr/en/index_en.html`

Kapoor, R. C. 2013, "Did Ibn Sīnā Observe the Transit of Venus 1032 AD?" Indian Journal of History of Science, 48, 405

Keller, H. U. and Friedli, T. K. 1992, "Visibility Limit of Naked-eye Sunspots," Quarterly J. Royal Astr. Soc., 33, 83

Saliba, G. 2011, "Avicenna," Encyclopaedia Iranica. viii, Mathematics and Physical sciences

Usoskin, I. G. 2008, "A History of Solar Activity over Millennia," Living Reviews Solar Physics, 5, 3

Vaquero, J. M. 2007, "Historical Sunspot Observations: A Review," astro-ph/0702068

Astronomical Heritage of the Middle East
ASP Conference Series, Vol. 520
Sona V. Farmanyan, Areg M Mickaelian, J. McKim Malville, and Mohammad Bagheri, eds.

Comet Observations in Arab Heritage Books: Did the Arabs Discover the Periodicity of Comets Before Edmund Halley?

Eng. Khalil Konsul
Arab Union for Astronomy & Space Sciences (AUASS), Amman, Jordan

Abstract. Little attention to comets has been paid by Arab historians, and historians of science have mainly focused on observations in Europe and the Far East. Quite recently, three Arab scholsrs have published studies about Middle Eastern cometary observations. This paper discusses the contributions of Dr. Wafeeq Shaker Rida, Dr. Hasan Basurra, and Mr. Mohammad Zahed Abu Ghudda.

1. Introduction

Arab scholars had not paid enough attention to comets and their observation in the books of Arab heritage. Even orientalists who have studied Arab astronomical heritage had not paid any attention to the comets and meteors. Historians of science, on the other hand, focused on the observations of comets made by peoples of the Far East and Europe only. On the other hand, it is generally accepted that Sir Edmund Halley was the first to have realized that at least some comets have periodic orbits.

Quite recently, three contemporary Arab scholars have published studies related to this subject, where they had gathered around 120 texts related to the observation of comets from various books of Arab heritage. This study is based on the following:

1. The Late Dr. Wafeeq Shaker Rida, Dean of Al-Hilla College, Ahlia University at Babylon, Iraq.
2. The Late Dr. Hasan Basurra, Chairman of Astronomy Department, King Abdul Aziz University, Jeddah, Kingdom of Saudi Arabia.
3. The book entitled: "Comet Halley Through Arab History" by Mohammad Zahed Abu Ghudda.

These three sources depend on tens of books and sources in Arab Heritage. So when we say "Observations of Comets in Arab Heritage Books," that means this research has acquired its full legitimacy. In fact, these records of comets observations were not mentioned in astronomical heritage books, but in the books of Arab Historians. The reason for that is as follows: Arabs used to view comets and meteors as if they were ground or atmospheric phenomena. This concept was adopted by Al-Kindi, Al-Biruni and Ikhwan al-Saffa (Brethren of Purity). The same concept was prevailing in other nations as well. So, Arab astronomers in their astronomical books emphasized more important astronomical issues such as the zodiac movement, star locations, planetary movements, eclipses, moon movement, calculations of earth geography, etc. Because the records of comet observations in Arab Heritage books were written by historians rather than by astronomers, the records were of different levels in terms of accuracy and information integration.

2. Historical Texts

2.1. The Concept of Ikhwan al-Saffa (Brethren of Purity) in the 10th century AD

The interpretation of the nature of comets described in the Epistles of Ikhwan al-Saffa (the Brethren of Purity) appeared in the fourth epistle entitled "Physical corporeality in the superior effects/separation of planets with tails," which reads as follows: "And the substance that consists of them, it is gentle smoke and steam that climb upwards, which strongly bond with Saturn and Mercury, and which appear as transparent as crystals, if the sun shines, it can be seen through them. It remains revolving with the heavenly sphere, by rising and setting until it dissolves and vanishes."

2.2. Ibn Malka (died in 1152 AD)

In Volume 2, chapter 3, entitled "On the events of the upper atmosphere such as comets, meteors and aurora," Ibn Malka concluded from his observations of comets the following: "I say that there is a specific conserved effect from celestial forces that keep for comets their shape and governs their decay and times of appearance." This is a marvelous sentence that summarizes all our modern knowledge about the astronomy of comets.

2.3. Abu Maashar Al-Balkhi (787–886 AD)

Recently, Professor Paul Kunitzsch (Germany) and four others researchers published an article about Abu Maashar who wrote in one of his books that he saw a comet behind Venus, and heard from other people that they saw comets behind Jupiter. That was 700 years before Tycho Brahe, who in 1588 AD said that the comet of 1577 AD was farther than the Moon, while the mainstream belief before Brahe used to be that comets were much closer than the Moon.

The fact is that these were serious pioneering studies. In number they are not as extensive as those of the Chinese and Japanese, but they included observations of comets that played a pivotal role in the history of cometary astronomy, such as Newton's comet (1680), the comet of Edmund Haley (1682), the great comet of 1402, daylight comets, anti-tail comets, etc. Here are some of these texts:

1. Antitail Comet (1) From the book "Bright Objects in the Sky of Islamic State," Dr. Hassan Basurah, page 57: "In Safar 1088 Hijri (April 1677 AD), a star with a front ray (anti-tail) extending half Thira'a (yard) appeared in Pleiades at dawn time. It (the anti-tail) stayed there for three days and then disappeared."

2. Antitail Comet (2) From the same book as above, page 61: "In Jumada al-Oula 1261 Hijri (May 1845 AD): in the East, a star with a front ray (anti-tail) extending about one Thira'a (yard) appeared. It (the anti-tail) stayed there for a few days and then disappeared.

   These are very interesting observations, the anti-tail appeared only for a few days, and that is what modern astronomy teaches us.

3. A Comet like a coma that developed a tail: Comet Muharram 278 Hijri (12 May 891 AD): "Two nights before the end of month Muharram, a star with coma appeared. The coma transformed into a tail." Sources: Al-Muntazam, Tariekh Al-Tabari, Albidaieh wa Alnihaieh, Al-Nujum Al-Zaherah, Husn Al-Muhadharah, Rihlat ashitaa wa-assaif.

   This is a text that indicates some great observations of a comet that first develops a coma around its nucleus, and then a tail appears (as it progresses towards the Sun). This is very compatible with modern scientific descriptions of comets.

4. A Daylight Comet (the Great Comet of 1402): In Rajab 804 Hijri (February 1402 AD), a planet as bright as Pleiades appeared with its bright tail rising high up in the sky. It continues rising and setting with its strong brightness seen with that of the Moon, and then was also seen during daytime with the presence of the Sun in the beginning of the month of Sha'ban (the 8th Lunar month). It continued appearing every evening after sunset and staying until the third of the night until the end of Sha'ban, then it completely disappeared. Sources: Badaie Al-Zuhur , Husn Al-Muhadharah, Tareekh Al-Dowlah, Al-Rasuliyyah.

5. Newton's Comet of 1680: In Shawwal 1091 Hijri (November 1680): In late Shawwal, a long tailed star appeared in the sky extending about four Moon's mansions between the South and the North (al-Qibla). The object remained visible for two months before it disappeared. Sources: Tareekh Al-fakhiri, Tareekh Al-amir Al-shihabi, Samt el-nojoom, Tareekh al-yaman fi asr al-istiqlal.

   Therefore, the Arabs did observe the comet of 1680 AD (1091 Hijri). This comet is very important because it was the first one in history whose orbital elements had been calculated. These calculations were made by Sir Isaac Newton in his attempts to apply his laws of mechanics and law of gravity. His method was applied to other comets as well and had been proven to be accurate.

6. Comet brighter than the Sun: In the year 443 Hijri (1051 AD): "A planet with a tail appeared, its brightness was greater than the Sun's. It moved slowly and then disappeared." Source: Rawdh al-Manather.

7. Comet Edmond Halley 1682: In Ramadan 1093 Hijri (September 1682): "In the month of Ramadan, a tailed star appeared from the East at night. After sunset, it looked like the zodiacal light. The object continued like this throughout the month of Ramadan." Sources: Nashr al-mathani, Tareekh Al-yaman fi asr al-istiqlal. This comet of 1682 AD (1093 H) was the first periodic comet to which the astronomer Edmond Halley calculated its orbital elements, and predicted that the comet of 1682 will come back in the late 1758 AD—which happened exactly as predicted. In fact, Edmund Halley had applied Newton's method in calculating the orbit of the comet 1682, being therefore the first person to have applied the laws of mechanics and gravity after Newton. These two great men inaugurated a new era in cometary astronomy.

   In his book *The Greatest Comets in History*, David Seargent mentions that the Arabs had observed Halley's Comet in its apparitions in the years 760, 837, 912, 989, 1066 and 1456 AD. In the books of Arab heritage, I found the following Arab observations of Comet Halley's apparitions in the years 837, 1066, 1145, 1222, 1456, 1682, 1835 and 1910 AD. That means that Seargent mentions three apparitions that are not found in the Arab heritage books that we know about, which are in the years: 760, 912, and 989 AD.

8. The Comet of Al-Mosul (in Iraq): In the book *Kanz-Uddorar & Jame'-Ulghorar* written by Abi Bakr Bin Abdalla Bin Aybak Al-Dawadari, and revised by Ulrich Haarmann, we found the following text about the appearance of a comet in the sky of al-Mosul in Iraq:

   "...He said that the traveling merchant, Shaikh Sharaf-Uddin Al-Sinjari, said the following: I was in Mosul in the year 684 Hijri (1285 AD), and in the night of mid-month of Muharram (the 23rd of March 1285 AD), a very brilliant object had been seen in the sky. It had three tails to the west, people at that time were looking thereat, and one of them was Imad-Uddin Bin Al-Dahhan, Head of Astrologers at that time in Mosul. Notable people asked him while I was hearing: what does the rise of this object mean? He said: this object had appeared in the year 420 H. (1029 AD), and it had had three tails: two tails of them were of the same length as these three tails you are seeing now, and the third tail was much shorter...One more time, this object had appeared in the year 490 H. (1097 AD), when Abdel Mu'men, Caliph of the West, was born...When this object had appeared, it had had three tails; two of them were long and the third one was longer than the tail of Al-Mustanser object, then it had disappeared until the year 553 H. (1158 AD), when Imam Al-Naser Lideen Ellah, Caliph of Baghdad, was born..."

   Here, we note the following:

   (a) This astrological text is optimistic, where all the caliphs that had lived in this period were lucky and have lived for long times, unlike traditional astrological texts, which were pessimistic as far as comets were concerned; the appearance of comets had been linked to natural disasters, wars, or epidemics.

   (b) The comet's appearance are periodic in the following manner: It appeared in 1029 AD, and re-appeared in 1096 AD, 67 years later. It re-appeared in 1158 AD, 62 years later. It re-appeared in 1285 AD, 127 years after its appearance of 1158. If we divide 127 by 2 we get 63.5 years (which means it was missed once. The maximum difference between the periods is: 67 – 62 = 5 years. This might be attributed to orbital perturbations.

   (c) This comet proved the periodicity of comets 420 years before Edmond Halley.

Other records say that comets appeared for a period of time, then they disappeared, then they re-appeared once again. Does that mean that their disappearance was because they orbit the Sun from the backside? No one knows as there was no description to the direction of the tail in the first and second appearance. Some records said after the second appearance "It was the same comet," while other records said "It was another comet."

Acknowledgments. I am deeply indebted to the works of Dr. Wafeeq Shaker Rida, Dr. Hasan Basurra, and Mr. Mohammad Zahed Abu Ghudda. Their works on the historical Arabic observations of comets is pioneering, and it should be translated into English so that non-Arab researchers may benefit from it, in order to correct or complete their knowledge about the history of comets.

References

(1) A Catalogue Of Medieval Arabic and Islamic Observations of Comets During the Period AD 700–1600, Dr. Wafiq Rada in the magazine Zeitschrift Fuer Geschichte Der Arabisch-Islamischen Wissenschaften, Band !3, 1999–2000

معهد تاريخ العلوم العربية و الإسلامية في إطار جامعة فرانكفورت / ألمانيا

2- الدكتور وفيق شاكر رضا / كلية التربية - جامعة بغداد
" المذنبات عند العرب " / من بحوث الندوة القطرية الرابعة لتاريخ العلوم عند العرب 1991) م(

3- الدكتور وفيق شاكر رضا / كلية التربية جامعة البصرة
" أرصاد مذنب هالي عندالعرب " ، 1986 م

4- محمد زاهد أبو غدة / السعودية
" مذنب هالي عبر التاريخ العربي " ، 1986 م

5- الدكتور حسن باصرّة / جامعة الملك عبد العزيز
" أجرام مضيئة في سماء الدولة الإسلامية : وقائع و مشاهدات فلكية للمذنبات و النيازك في كتب التاريخ الإسلامي " ، 2004 م

6- الدكتور حسن باصرّة / جامعة الملك عبد العزيز
" ما دوّن من المذنبات و الشهب و النيازك في التاريخ للإسلامي " ، 2016 م

Astronomical Heritage of the Middle East
ASP Conference Series, Vol. 520
Sona V. Farmanyan, Areg M. Mickaelian, J. McKim Malville, and Mohammad Bagheri, eds.

The Expedition of M. R. Štefánik to Observe the Total Solar Eclipse in 1907 from Turkestan

Petra Hyklová
Astronomical Institute, Charles University in Prague, V Holešovičkách 2, CZ-180 00 Praha 8, Czech Republic; petra.hyklova@gmail.com

Abstract. In 1906-07, the Slovak astronomer Milan Rastislav Štefánik travelled to observe the Turkestan 14 January 1907 solar eclipse at Ura-Tjuba. Due to snowy weather the expedition was unsuccessful, but on the way Štefánik met the expedition from Hamburg-Bergedorf, visited the Pulkovo observatory and was invited to join the Russian expedition. Štefánik later served in the aviation corps of French Army and played a significant role in establishing of the new Czechoslovak Republic of which he became the first Minister of Warfare. His international contacts were essential for success in the diplomacy around the constitution of the independent Czechoslovak Republic in 1918.

1. Life and Work of Milan Rastislav Štefánik (1869-1919)

Štefánik was born on 21 July 1869 Košariska, Slovakia. In 1899–1901, he studied at the Czech Technical University, but soon continued at Charles University in Prague, where he studied in 1901–1904 at the Astronomical Institute. In 1904, Štefánik came to Paris with a recommendation letter for Jules Janssen and started to work at the Meudon Observatory. In the following years he undertook several climbs to Janssen's Observatory at Mt. Blanc, which were physically challenging and took many days. His main task was making observations of the Sun. From 1905 he participated in solar eclipse expeditions. Together with Russian astronomer Alexis Hansky, he improved the spectrohelioscope. In 1905, Štefánik was elected a member of International Union for Cooperation in Solar Research at its first meeting in 1905. Štefánik's papers from his Meudon period were read at sessions of the French *Academie des sciences* before being published in *Comptes Rendus* (Muzeum 2003). Štefánik stayed at the Meudon observatory until Janssen'ss death in 1907. In the same year he was awarded the Prix Jules Janssen, but after losing his job at Meudon, his publication activity declined.

In 1910, when Halley's comet approached Earth, it was calculated that its transit over the solar disc in May 1910 would be observable only from south Pacific. Štefánik was sent to Tahiti by *Bureau des longitudes* and *Bureau central météorologique*, and made meteorological observations as well. However, the transit of a comet nucleus could not be seen due to the small size of the nucleus, which was not known at that time. Štefánik decided to stay at Vavau island until the solar eclipse of 28 April 1911 and therefore started to build an observatory there.

As Tahiti was a French colony and France planned to arrange a system of radio-stations in this area, Štefánik received his first diplomatic assignment to earn the island

Figure 1. Milan Rastislav Štefánik: memorial at Paris-Meudon Observatory (left), in a military airplane (center), and as general of French army in World War I (right).

of Florian from the Equador government for French administration. In 1913 Štefánik came to Equador, helped with improvement of the Quito observatory and built the system of meteorological and télégraphie sans fil stations in Ecuador. The project was not finished due to political changes in France and the outbreak of the First World War. Štefánik, as a French citizen, joined a regiment and his career of astronomer was temporarily suspended. His military career began with training at aviation school in Chartres in 1915 and subsequent reconnaisance flights in Serbia. Since 1916, Stefanik organized the Czechoslovak legions on the French and Italian fronts, among military prisoners in Russia, and among Slovak Americans.

Štefánik rose quickly in rank until he became general of Air Forces and got many military honors: Legion d´honeur, Croix de Guerre and Serbian military medal (Flammarion 1919). In 1915 he joined Czechoslovak resistance in Paris. By 1916, he was one of the most important members. Together with the future first President of Czechoslovakia Tomáš Garrigue Masaryk and the future second President of Czechoslovakia Edvard Beneš, Štefánik co-founded the Czechoslovak National Council that later became the basis of new Czechoslovak government. In 1916-17, Štefánik established the Czechoslovak army in Russia and in the USA (Rajchl 2013).

After the war, Štefánik's political status and future was unclear. He was appointed the Minister of Warfare in the first provisional government of the new republic. His ministerial influence gradually declined, as his agenda was oriented abroad and closely related to Czechoslovak legions, while the Czecho-Slovak army was managed by the Ministry of Defense. In his homeland, Štefánik was virtually unknown, as he had spent all his time abroad since 1913 (Rajchl 2013). Štefánik, at that time engaged to Italian marchioness Giuliana Benzoni, planned to return to Czechoslovakia. Štefánik's decision to travel by plane rather than by train proved unfortunate; on 4 May 1919, the plane crashed on landing at the airport Ivanka pri Dunaji and none aboard survived. Štefánik's unexpected death made him a newly discovered national hero. His memorial statues stand in Bratislava, at Paris-Meudon and Prague-Petřín observatories, a large memorial was erected in Slovakia, and one square in Paris is still named after him.

2. Expedition to Turkestan

Štefánik participated on or organized many eclipse expeditions to Alcosebre, Spain on 30 August 1905, to Ura-Tjuba, Turkestan on 14 January 1907, to Vavau, Tonga

Figure 2. Observation stations at Ura-Tjuba, Turkestan (solar eclipse on 14 January 1907). Left: Pulkovo station, middle: Russian astronomers and Štefánik (on the right side), right: Štefánik's station.

archipelago on 28 April 1911, to Cormeilles-en-Paris on 17 April 1912 and to Passo Quatro, Brasil on 10 October 1912. The 1907 eclipse path crossed Russian Empire and China. Štefánik went on the expedition alone with only few lent astronomical instruments and authorization and recommendation letters from the *Bureau des Longitudes*. He departed from Paris in December 1906 to visit Pulkovo observatory, and to meet his friend Alexis Ganskij from Meudon observatory. Štefánik was warmly welcomed by Ganskij and the Head of the Observatory, Mr. Backlund. They invited him to take part in the Pulkovo expedition (Bepololskij, Vitram, Ganskij), and later temporarily joined the German expedition from observatory in Bergedorf near Hamburg, which travelled to the observation site ad Djizag. Štefánik and Ganskij travelled by train through Moscow and Tashkent to Chernayev, and then in a caravan they continued to the observation site at Ura-Tjuba. It was 1000 m in altitude, near an old fortress 35 meters above the town. Unfortunately, due to continuous snowing for three days, Štefánik could see only a small piece of Sun 12 minutes before the third contact and he was unable perform any part of his scientific program. The expedition departed from Ura-Tjuba on 18 January 1907. Even though Štefánik's expedition was not scientifically succesful, his charming and charismatic character helped him to find friends and invitations to promising projects; on the next expedition in 1910 Štefánik virtually started his political career (Rajchl 1938).

Acknowledgments. Our research was supported by Charles University, project GA UK No. 101-10/250343 and Faculty of Mathematics and Physics of the Charles University.

References

Flammarion, C. 1919, Milan Rastislav Stefanik, presented at General Assembly of La Societe astronomique de France, Paris, 1 June 1919

SNM – Múzeum SRR Myjava Astronomical papers of M. R. Štefánik, 2003

Rajchl, R. 1938, To the Štefánik journey to Turkestan (in Czech), in J. Bartůšek (Ed.), Štefánik. Second volume, Praha – Bratislava, Slovenské nakladatelstvo L. Mazáč, 94

Rajchl, R. 2013, Štefánik astronomer, soldier and diplomat. Pokroky matematiky, fyziky a astronomie Vol. 58, No. 1, 50

Part III

Astronomical Heritage for the Future

Astronomical Heritage of the Middle East
ASP Conference Series, Vol. 520
Sona V. Farmanyan, Areg M. Mickaelian, J. McKim Malville, and Mohammad Bagheri, eds.

Development of Astronomy in Turkey

Sinan Aliş
Istanbul University, Department of Astronomy and Space Sciences, 34119 Istanbul, Turkey; `salis@istanbul.edu.tr`

Abstract. An overview of the historical developments related to astronomy in Turkey is presented. Developments in the Republican Period (since 1923) of Turkey are given; the first institute of astronomy was established in 1933. Recent developments and ongoing projects are also summarized. By 2020, it is expected that the two largest telescopes in Turkey will be 4 m and 2.5 m.

1. Introduction

Archeological records show that some of the oldest civilizations in the world were in ancient Anatolia. Those civilizations, like the Hittites (1600–1178 BCE), created several settlements. Hattusa (6th millennium BCE), Alacahoyuk (2nd millennium BCE) and Gobeklitepe (10th millennium BCE) are amongst them, and each of them have clear evidence that those people were doing astronomical observations. This ancient period is not the focus of this contribution but requires more investigation and deserves more attention from the perspective of archeo-astronomy. On the other hand, the first institutionalized astronomical centre established in the Ottoman Period was the Istanbul Observatory, which was founded in 1575 by the astronomer Takiyuddin (Taqi al-Din). Unfortunately, the observatory did not last long, and with the order of the Sultan Murat the 3rd, it was destroyed in 1580.

After the Istanbul Observatory, there were no astronomical institutions in Turkey until 1911. In 1911, Kandilli Observatory was established and equipped with some telescopes and instruments. The most important developments took place after the establishment of the republic, especially in 1933, when Ataturk initiated the so-called "university reform." The first astronomical institute of the young Turkish Republic was established in 1933 in Istanbul University.

2. Astronomy in the Higher Education

Being the first astronomy department of Turkey, Istanbul University Astronomy Department (1935) enabled establishment of other astronomy departments in the country. Following Istanbul University, Ankara and Ege Universities also established astronomy departments in 1954 and 1962, respectively. For decades, those three astronomy departments were the only astronomy institutes where undergraduate and graduate education was offered. In 1998, a new astronomy department in Erciyes University (Kayseri) was established. Later, in 2012, another department in Inonu University (Malatya) opened.

Recently, in 2014, Ataturk University (Erzurum) opened the latest astronomy department.

Astronomy and astrophysics are not only studied in the astronomy departments. There are several physics departments with an astrophysics branch where research and graduate education were carried out. Amongst those physics departments, the ones in Middle East Technical University (Ankara), Onsekiz Mart University (Canakkale), Sabanci University (Istanbul) and Akdeniz University (Antalya) have the largest number of astrophysics faculty members.

Recently, with developments in satellite communications, some astrophysics groups evolved into a new department called Space Sciences and Technologies. Currently there are two such departments, one in Onsekiz Mart University and one in Akdeniz University.

Table 1 lists astronomy and related departments, and Table 2 lists physics departments with an astrophysics branch or an astrophysics research group.

Table 1. List of astronomy and related departments in Turkey in chronological order.

University	Department	City
Istanbul University	Astronomy and Space Sciences	Istanbul
Ankara University	Astronomy and Space Sciences	Ankara
Ege University	Astronomy and Space Sciences	Izmir
Erciyes University	Astronomy and Space Sciences	Kayseri
Inonu University	Astronomy and Space Sciences	Malatya
Onsekiz Mart University	Space Sciences and Technologies	Canakkale
Akdeniz University	Space Sciences and Technologies	Antalya
Ataturk University	Astronomy and Astrophysics	Erzurum

Table 2. List of physics departments with an astrophysics branch or research group.

University	Department	City
Bosphorus University	Physics	Istanbul
Sabanci University	Physics	Istanbul
Cukurova University	Physics	Adana
Istanbul Technical University	Physics Engineering	Istanbul
Adiyaman University	Physics	Adiyaman
Onsekiz Mart University	Physics	Canakkale
Middle East Technical University	Physics	Ankara
Yildiz Technical University	Physics	Istanbul
Istanbul Kultur University	Physics	Istanbul

The only currently active undergraduate programs are in Istanbul, Ankara, Ege and Erciyes Universities. The number of students are enrolled every year is 70, 40, 30 and 20 for those universities, respectively.

3. Observatories in Turkey

A complete list of observatories in Turkey is given in Table 3. All the observatories in this list are still operational. All of them are devoted to night sky observations except Istanbul University Observatory and Kandilli Observatory, which monitor solar activity on a daily basis.

Figure 1. Istanbul University Observatory and the Astronomy Department at the main campus of Istanbul University in the historical city center.

In a general assembly in 1968, Prof. Dr. Nuzhet Gokdogan from Istanbul University and Prof. Dr. Abdullah Kizilirmak from Ege University emphasized the need for a national observatory rather than the small university observatories. This idea has been elaborated in the forthcoming years and resulted a site testing project for a national observatory (Aslan et al. 1989). After 5 years of site testing, Bakirlitepe (2547 m above sea level) located in the Taurus Mountain Range around the city of Antalya was chosen. There were some years of interruption after site testing due to financial issues which postponed the founding of the observatory until 1995. The inauguration ceremony happened on September 5, 1997. The observatory was financed by the Technical and Scientific Research Council of Turkey (TUBITAK), which is why it is called the TUBITAK National Observatory (TUG). Currently TUG is the most modern observatory in Turkey consisting of 1.5m Russian-Turkish Telescope (RTT150), a 1.0m telescope, and several smaller size robotic telescopes. The Observatory hosts one of the most famous robotic telescopes in the world; ROTSE. Even though the ROTSE project ended, TUG still continues to operate the telescope.

RTT150, the largest telescope in the country, saw first light in 2001. Since then, Turkish astronomers had direct access to spectroscopic observations, which were not avialable before. In 15 years, Turkish astronomers were able to publish radial velocity studies of binary stars, chemical abundances of early-type stars, and recently identified

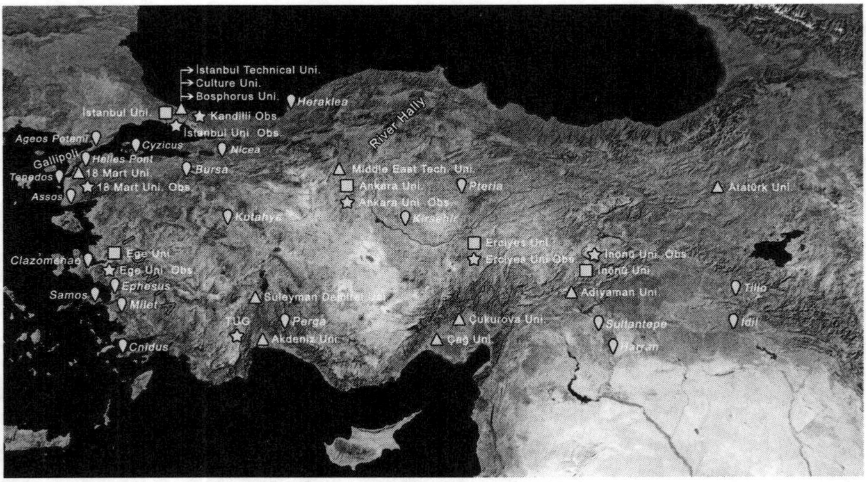

Figure 2. Map of astronomical institutions and observatories in Turkey (Eker et al. 2013). Squares denote astronomy and space sciences departments, triangles denote physics departments with astrophysics branch and space sciences and techonologies departments, stars denote observaories and balloons denote ancient sites with a connection to astronomy.

exoplanets around other stars thanks to the spectroscopic capabilities of the RTT150 (Yilmaz et al. 2017; Caliskan et al. 2015; Ibanoglu et al. 2008).

Table 3. List of currently operational observatories in Turkey. Telescopes in parentheses are being constructed.

Observatory	Founded in	Telescopes
Kandilli Observatory	1911	0.2m
Istanbul University Observatory	1935	0.6m, 0.4m, 0.3m
Ankara University Observatory	1963	0.4m, 0.3m
Ege University Observatory	1965	0.48m, 0.3m
Cukurova University / UZAYMER	1991	0.5m, 0.3m
TUBITAK National Observatory (TUG)	1997	(2.5m), 1.5m, 1.0m, 0.6m
COMU Ulupinar Astrophysical Observatory	2002	1.2m, 0.6m, 0.4m
Inonu University Observatory	2012	0.35m
Adiyaman University Observatory	2014	0.6m
Eastern Anatolia Observatory (DAG)	2020	(4.0m), 0.5m

Since the establishment of the TUBITAK National Observatory, the number and quality of observational astronomy studies have increased. However, the capabilities of the observatory have reached their limit for the community. Lack of larger telescopes prevent studying extragalactic astronomy subjects or generally faint astronomical sources. As a result, a group of astronomers took the initiative and prepared an infrastructure project for a larger telescope (i.e., 4m class). The project called "Eastern Anatolia Observatory" [1] (or DAG) was approved by the Ministry of Development of the Turkish Republic and coordinated by Ataturk University (Erzurum) (Yesilyaprak, Yerli and Keskin, 2016; Yesilyaprak et al., 2016). The DAG telescope will be located on top of a mountain at 3150 m called Karakaya Tepeleri. The telescope will be equipped with active and adaptive optics with two Nasmyth foci. In order to improve stability, the telescope will not have a Cassegrain focus. For the moment, the project is advancing as expected. All necessary infrastructure is in place on the mountain, the building and enclosure are being constructed,and the mirror has been shipped for coating. Recently in 2017, the project made a call for the scientific instruments to be commissioned at the telescope. The DAG project is coordinated by the Astrophysics Research Center of Ataturk University (ATASAM[2]) while Istanbul University, Middle-East Technical University and FMV Isik University strongly contribute to the project with their teams. As the site allows, astronomers in the DAG project prepared for an near-IR telescope. For the moment, it is planned that one Nasmyth focus will be in seeing-limited mode with optical instruments, whereas the other focus will be AO guided for near-IR instruments. As a first light goal, the DAG telescope will be working with a single conjugate adaptive optics (SCAO) system, but for the near future team has plans for an upgrade to a ground-layer adaptive optics (GLAO) system. Adaptive optics system design and manufacturing is being done in Turkey at the FMV Isik University's Optomechatronics Application and Research Center (OPAM).

4. Community and Research Interests

There are nearly 150 researchers in Turkey with a PhD and working on astronomy and astrophysics projects. Institutions of those researchers are given in Tables 2 and 3. Almost the same number of students are enrolled in the graduate programs of the relevant universities. The astronomical community in Turkey gathers regularly every two years in the national astronomy assemblies. This tradition started in 1968 and continues.[3] The group photo of the 20th meeting held in 2016 in Erzurum is shown in Figure 2. In the recent assemblies it was seen that students were more enthusiastic about the national meetings than faculty members. In the 20th national assembly, there were 178 participants, of which 107 were students (75 graduate, 32 undergraduate). Throughout the meeting, which lasted for five days, there were 100 oral and 30 poster contributions presented.

Due to the lack of advanced observational facilities, Turkish astronomers' research interests traditionally focused on stellar astrophysics and solar physics. High-energy astrophysics studies can also be counted in this group as the satellite data become public shortly. The total number of refereed papers by Turkish astronomers is shown in Figure 4 (Eker et al. 2013). There are 838 papers listed in the Science Citation Index database between 1980–2010, where most of those papers are in the fields of stellar astrophysics, galactic structure, galaxies and cosmology, high energy astrophysics, binary

[1] `http://dag.atauni.edu.tr`, `http://dag-tr.org`

[2] http://atasam.atauni.edu.tr

[3] `http://uak.info.tr`

Figure 3. Group photo of the national astronomy assembly in 2016.

stars, stellar evolution, and solar physics. For the recent period of 2001–2010, Turkish astronomers published 53 papers per year (Eker et al. 2013; Bilir et al. 2013).

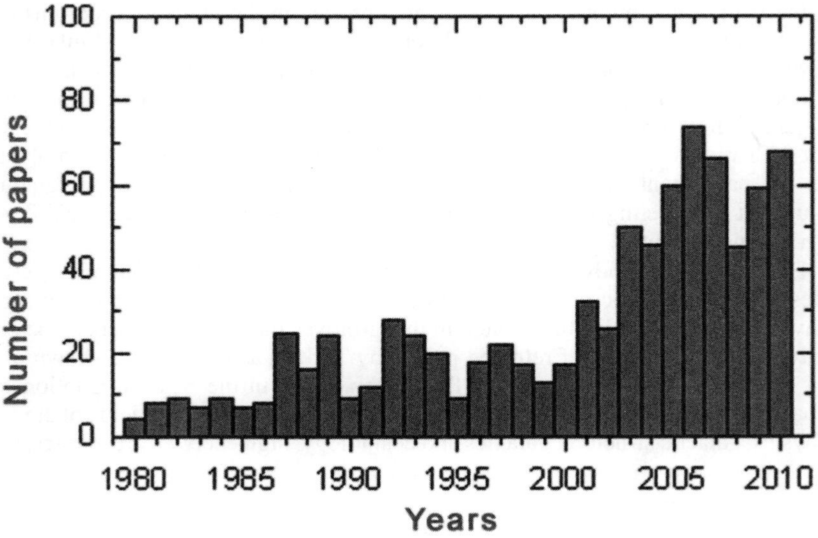

Figure 4. Distribution of the total number of papers of Turkish astronomers (Eker et al. 2013).

5. Public Outreach Activities

Public outreach activities started with university clubs. Among the many clubs today, AAT from Middle East Technical University (Ankara) and AAK from Istanbul University (Istanbul) was founded in 1986 and 1991, respectively.

The Scientific and Technical Research Council (TUBITAK) of Turkey has been publishing a science magazine called "Bilim ve Teknik" (Science and Technology) since 1967. Astronomy related articles and news was always one of the main focus areas of the magazine. Under the leadership of this magazine, the first star party of Turkey was organized in 1998; it has now been 20 years that this nation-wide event has been carried out. This festival-like event was first organized by TUBITAK Science and Technology magazine and is now being organized by the TUBITAK National Observatory. Every year around 400 people participate and camp at the 1900 m site, in the outskirts of the mountain of the national observatory (see Figure 5). This particular event is very important because it led to the organization of many similar star parties all over the country. In 1998, there were no similar events other than this but today several star parties are organized annually.

Figure 5. A series of photos from the 20th star fest in Saklikent in July 2017. Upper left: group photo. Upper right: camping area. Lower left: the author giving a lecture. Lower right: visiting the 1.5m telescope.

Almost all astronomy departments in Turkey organize public events and outreach activities, especially when there is a celestial event. In such events, thousands of people visit observatories in different cities. Ege University Observatory has organized summer schools for high school students and other interested people every year. Istanbul University Observatory recently organized a scientific summer school for science high school students with support from TUBITAK.

A new development in the framework of outreach activities is the opening of a new planetarium. With the sponsorship of the E.C.A. company, there is now a planetarium at the Istanbul University Department of Astronomy and Space Sciences. This is the first planetarium in such an environment, and it is the only 4K resolution planetarium in the country so far (see Figure 6). The capacity of the planetarium is modest as there are only 39 seats, but the opportunity to visit telescopes and have a lecture on astronomy make the planetarium unique compared to other science centers. With the recent infrastructure improvements, the astronomy department at Istanbul University has become a well-equipped complex for public outreach activities.

Figure 6. Recently inaugurated planetarium of Istanbul University. The first and only 4K planetarium in Turkey.

6. International Relations and Collaborations

The Turkish Astronomical Society (TAD) is the representative body of the astronomical community in Turkey. Turkey is a member of the International Astronomical Union (IAU) since 1961. Apart from the national membership, there are also 81 individual IAU members from Turkey. There have been several IAU meetings and symposia organized in Turkey. The last IAU-supported event was the 30th International Summer School for Young Astronomers (ISYA) which happened in July 2008.

Recently, through the Turkish Astronomical Society, Turkey was involved with the Regional Office of Astronomy for Development in Armenia (IAU-SWCA ROAD) to collaborate on outreach and education activities and develop projects together with other regional member countries. It is believed that participation in the IAU SWCA ROAD will help to improve relationships among the neighboring countries, especially between Turkey and Armenia.

Turkey is a part of an agreement with ESA which helps to get some support for researchers and students. Even though there were some efforts for Turkey to be an ESO member state, it was not concluded. This is still a topic of discussion among the community. It is believed that this proposal will be considered again in the near future. One of the largest projects Turkey has participated in is the Spectrum-Roentgen Gamma (SRG) satellite of Russia. As the RTT150 telescope is a joint project between the national observatory and Russian institutions, Turkey contributed 2 million USD to the satellite project. The long awaited project is in the final stage of development now with the help of Germany. Turkey will get some proprietary data from the hard X-ray detector of the SRG (ART-XC) from the Russian side.

7. Conclusion

After a long period of interruption, astronomy again became a field of research in Turkey thanks to the Ataturk's so-called "university reform" where scientists escaping from Nazi Germany were welcomed. Erwin Finlay Freundlich was the founder of the Istanbul University Astronomy Department and the Observatory. Following Freundlich, Thomas Royds, Hans Rosenberg and Wolfgang Gleissberg were the other faculty members who worked in Istanbul University in order to develop and improve scientific activities. Egbert Adriaan Kreiken was the key person to establish Ankara University Observatory which today is called the Kreiken Observatory at Ankara University.[4] The impact of these foreign scientists on Turkish astronomy was crucial and will be always remembered with gratitude, especially Gleissberg and Kreiken, who were very special figures for the Istanbul and Ankara University Observatories, respectively.

Compared to European counterparts, institutionalisation of astronomy in Turkey started late. Since 1933, astronomy in Turkey has become quite developed. In modern Turkey, there are many departments and observatories. Many Turkish students will continue their studies for M.Sc. and Ph.D. degrees abroad in recognized institutions, and some Turkish fellows do Post-Docs all around the world. These people started to return to their home country which helps to improve astronomy in Turkey and also enables research in different scientific topics not studied before.

The most important astronomical project is the DAG telescope which will enable Turkish astronomers to access a 4m class telescope equipped with state-of-the-art instruments as well as adaptive and active optics. Even though the diameter of the telescope is not considered "large" today, the technology and operational principles are quite similar to those of larger telescopes such as Very-Large Telescope (VLT) of ESO. The other important aspect of the DAG telescope is the fact that it will open a new electromagnetic window, the near-infrared, to Turkish astronomers. Another side product of the DAG project will be the start of instrumentation studies, which did not exist before. The DAG telescope will also be accompanied with a coating plant in a couple of years. Thus, astronomers in Turkey will gain more experience on technical issues, and engineers and mechanics will also gain experience with telescopes and instruments.

---

[4] `http://rasathane.ankara.edu.tr`

Acknowledgments. The author would like thank to the organizers for their efforts and hospitality, and UNESCO for travel grants.

References

Aslan, Z. et al., 1989, A&A, 208, 385
Bilir, S. et al., 2013, Scientometrics, 97, 477
Caliskan, S. et al., 2015, New Astronomy, 34, 6
Eker, Z. et al., 2013, Astronomy in Modern Turkey, In Andre Heck (Ed), Organizations, People and Strategies in Astronomy Vol. 2, Venngeist, Duttlenheim, 195
Ibanoglu, C. et al., 2008, MNRAS, 384, 331
Yesilyaprak, C., Yerli, S. K., Keskin, O., 2016, RMxAC, 48, 120
Yesilyaprak, C. et al., 2016, Proceedings of SPIE, 9910, 2U
Yilmaz, M. et al., 2017, A&A, 608, 14

Astronomical Heritage of the Middle East
ASP Conference Series, Vol. 520
Sona V. Farmanyan, Areg M. Mickaelian, J. McKim Malville, and Mohammad Bagheri, eds.

Digitized First Byurakan Survey (DFBS) as UNESCO "Memory of the World" Documentary Heritage

A. M. Mickaelian, L. A. Sargsyan, G. A. Mikayelyan, L. K. Erastova, K. S. Gigoyan, P. K. Sinamyan and L. R. Hovhannisyan
NAS RA V. Ambartsumian Byurakan Astrophysical Observatory (BAO), Byurakan 0213, Aragatzotn province, Armenia; aregmick@yahoo.com

Abstract. The famous Markarian Survey (or the First Byurakan Survey, FBS) was carried out in 1965–1980. Its 2000 plates were digitized in 2002–2007 and the Digitized FBS (DFBS) was created. New science projects have been conducted based on this low-dispersion spectroscopic material. The Armenian Virtual Observatory (ArVO) database accommodates all new data. The project was conducted in collaboration with the Italian, American, and German scientists, as well as the Armenian Institute of Informatics and Automation Problems (IIAP), and continued during 6 years in 2002–2007.

1. Introduction

The Markarian survey (or the First Byurakan Survey, FBS) carried out by B. E. Markarian, V. A. Lipovetski, and J. A. Stepanian in 1965–1980 with the BAO 1m Schmidt telescope and 1.5° prism, was one of the most efficient and important surveys in astronomy. It was the first systematic objective-prism survey, the largest objective-prism survey of the Northern sky (17,000 square deg) and it was a new method for finding Active Galactic Nuclei (AGN). There are a total of $\sim$40,000,000 spectra for $\sim$20,000,000 objects in the entire survey (Markarian et al. 1989). The original aim was the search for galaxies with UV excess (Mazzarella & Balzano 1986; Markarian et al. 1989). It resulted in discovery of 1515 UV-excess (UVX) galaxies, including more than 200 AGN and more than 100 SB galaxies. The Markarian survey led to the classification of Seyferts into Sy1 and Sy2 (Weedman & Khachikian 1968), the definition of Starburst galaxies (Weedman 1977), and several other projects, such as FBS Blue Stellar Objects (BSOs, Mickaelian 2008), late-type stars (Gigoyan & Mickaelian 2012), and optical identifications of IRAS sources: Byurakan-IRAS Galaxies (BIG, Mickaelian & Sargsyan 2004) and Byurakan-IRAS Stars (BIS, Mickaelian & Gigoyan 2006).

The Second Byurakan Survey (SBS) was also carried out with the BAO 1m Schmidt and was a continuation of FBS to fainter magnitudes (Stepanian, 2005). FBS is now digitized and the Digitized First Byurakan Survey[1] (DFBS, Mickaelian et al. 2007; Massaro et al. 2008) is available online. It provides 40,000,000 spectra for 20,000,000 objects at high Galactic latitudes. Detailed descriptions of the FBS, SBS, and DFBS are given in Mickaelian (2016).

---

[1]www.aras.am/Dfbs/dfbs.html

The huge amount of spectral information contained in the FBS plates allowed the development of several other projects based on the FBS, the most important of which is the discovery and investigation of blue stellar objects (Mickaelian 2008), a survey for late-type stars (Gigoyan et al. 2003), and the optical identifications of sources from the IRAS catalogue. The sample of stellar BIS objects is available at CDS (Byurakan-IRAS Stars, Mickaelian & Gigoyan 2006), and a similar catalogue for the BIG galaxies (Byurakan-IRAS Galaxies, Mickaelian & Sargsyan 2004) will soon appear at CDS. The number and classes of new objects discovered within the FBS made clear the need for open access to this information by the entire astronomical community.

2. The Digitized First Byurakan Survey (DFBS)

A number of digitization projects have been accomplished at BAO, including the most important one, the Digitized First Byurakan Survey (DFBS); Mickaelian et al. 2007; Massaro et al. 2008) based on the digitization of the famous Markarian Survey (Markarian et al. 1989). The project consisted of a number of steps:

- development of technical principles of the Project and necessary equipment;
- collection of all photographic plates of the Markarian Survey;
- revision and accounting of the plates and observing journals;
- scanning of a few dozens of plates to set up the necessary parameters for the scanning of all plates;
- input of data from observing journals, and creation of the DFBS Database (Mickaelian et al. 2006);
- creation of the DFBS webpage and user interface;
- scanning of photographic plates;
- archiving on HDDs and DVDs;
- calculating the astrometric plate solution, extracting images and spectra, wavelength calibration, density and flux calibration, performing multiband (UBVR and POSS O/E) photometry;
- making up template low-dispersion spectra;
- numerical classification of low-dispersion spectra;
- visualization of DFBS on the sky map; creation of electronic interactive DFBS map and search system; and
- proposing and discussing new research projects.

The spectral extraction and analysis software is described in Mickaelian et al. (2010) and Knyazyan et al. (2011). the DFBS plate database is available in Vizier, Strasbourg (Mickaelian et al. 2005).

The DFBS project is relevant for preserving a very large database that was a milestone in the history of astronomy and to offer to the scientific community a new tool for investigating the properties of the nearby universe. The participating teams were: Byurakan Astrophysical Observatory (BAO, Armenia), Dipartimento di Fisica, "La Sapienza" Università di Roma (Italy), Cornell University (Ithaca, NY, USA), MIGG Informatica & Ricerca (Italy), Hamburger Sternwarte (HS, Germany), and the Institute for Informatics and Automation Problems (IIAP, Yerevan, Armenia).

Digitization of the FBS plates started in 2002. After several test scans on a set of plates, all of the FBS plates were digitized with an Epson Expression 1680 Pro scanner at the Byurakan observatory. DFBS was the first digitization project conducted by scanner in Armenia, and its results allowed a number of other similar projects. The scanning resolution was 1600 dpi, so the pixel size is about 16 microns. The typical length of an FBS spectrum is ~1.7 mm, which gives 107 pixels along the wavelength direction. An *ad hoc* program SCANFITS written by Stefano Mottola (DLR, Institute of Planetary Research, Berlin) allows the resulting image to be written directly in FITS format with corresponding information about the plate in the header. All 1874 plates in 1139 fields (Mickaelian et al. 2005) were digitized and stored on DVDs.

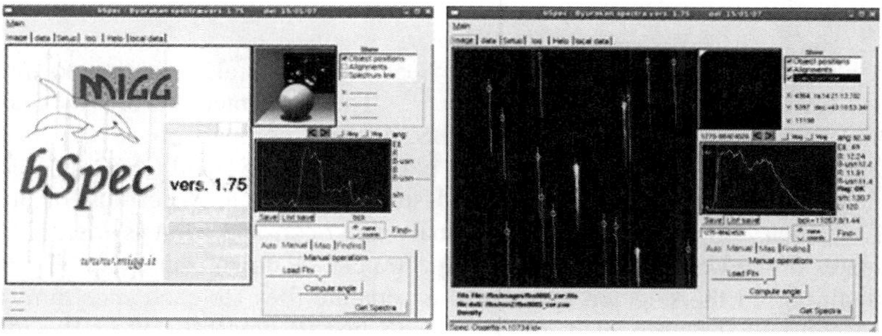

Figure 1. Dedicated software bSpec for the DFBS images and spectra extraction and analysis, an example of spectra extraction.

The next steps were (i) the astrometric (plate) solutions (obtained with an accuracy of $\sim 1''$ rms, or 0.6 pixel), (ii) automatic extraction and classification of the spectral data in a DFBS image, and (iii) wavelength calibration (only an approximate wavelength calibration scale was obtained because the dispersion is strongly non-linear, from about 22 Å/pixel at the blue edge of the spectrum to about 60 Å/pixel at the red edge, with mean dispersion 32.7 Å/pixel and dispersion of about 28.5 Å/pixel at Hγ). The primary scanning parameters and resulting features are given in Table 1.

The DFBS database presently is stored on a dedicated server at the "La Sapienza" University in Rome and can be accessed through a web interface. Mirror sites are available in Yerevan at the IIAP, Trieste (Italy) and Observatoire de Paris (France). The database includes all the digitized plates and the automatically extracted spectra. Only for F-type plates (IIaF, IIAF, IIF, and 103aF) the automatic procedure gives reliable results, and for such plates approximate *B* and *R* magnitudes have also been computed and are available. Visual software named *bSpec* was created and implemented in order to get automated extraction and classification of the spectral data in a DFBS plate (Figure 1). This software performs all the operations necessary to build the DFBS database and was developed by Giuseppe Cirimele and the MIGG Informatica & Ricerca team.

Table 1. Primary scanning parameters and resulting characteristics of the DFBS.

Items	Description
Teams	Byurakan Astroph. Obs., Univ. Roma "La Sapienza," Cornell Univ.
Years	2002–2005
Instrument	Epson Expression 1680 Pro scanner
Scanning options	1600 dpi (15.875 μm pix size), 16 bit, transparency mode, "scanfits"
Plate size	9601×9601 pix, 176 MB file
Spectra	107×5 pix (1700 μm in length)
Dispersion	33 Å/pix average (22–60 Å/pix), 28.5 at Hγ
Spectral resolution	50Å (average)
Astrometric solution	1″ rms accuracy
Scale	1.542″/pix
Photometry	0.3 mag accuracy
Data volume	1874 plates, ~400GB
Number of objects	~20,000,000 (~40,000,000 spectra)

A web page and user interface have been created to allow access to the DFBS database for the astronomical community.[2,3]. The user interface provides access to general information on the FBS and DFBS, an easy comparison of a spectroscopic DFBS plate with the corresponding direct plates from DSS1 and DSS2, and access to the database and to the digitized plates. Users may download portions of the plates in FITS format and perform their own analysis. Each spectrum is identified by its coordinates in the USNO-A2 catalogue. Spectra can be downloaded as a single ASCII file containing all the selected spectra or as separate files for each spectrum (as a tar or zip archive). The user interface (the DFBS portal) presently offers the following operations to the guest user:

- **Sky coverage:** An RA, Dec rectangular sky map shows the position on the sky of each plate. Basic data about each plate are available by clicking on the corresponding sky position (Figure 2, left panel).
- **Plate list:** The list of the plates is shown; it can be sorted by several parameters (e.g., plate numbers, RA or Dec, emulsion types, observers) and downloaded. Basic data about each plate are available by clicking on the plate number.
- **Explore:** It allows the display of a portion of plate around a given central RA, Dec position, comparison with the same portion of the DSS1 or DSS2 (blue, red, or IR), interactive selection of one or more spectra present in the database, their collection (saving in a list) and downloading (ASCII files) to the guest computer.
- **Get Image:** Allows users to select a portion of a plate (presently up to 1024×1024 pixels, i.e., about 26.5′×26.5′) in FITS format and all spectra in this portion

[2]`http://byurakan.phys.uniroma1.it/`

[3]`http://arvo.sci.am/ARVO/DFBS/`

present in the database for downloading (spectra are ASCII files), as well as downloading of the whole selected field.

- **Get Spectra:** Allows downloading all the spectra in the database within a given distance from a selected central position (cone search); the query may be either interactive, by position, or made by uploading an ASCII file containing one or more positions (Figure 2, right panel). Objects may be selected by *B*, *R* or *B*–*R* values. If fewer than 50 objects are selected, the 2D images can also be downloaded by selecting a check box. The table containing the identifier and the main data for each spectrum can also be downloaded as an ASCII file.

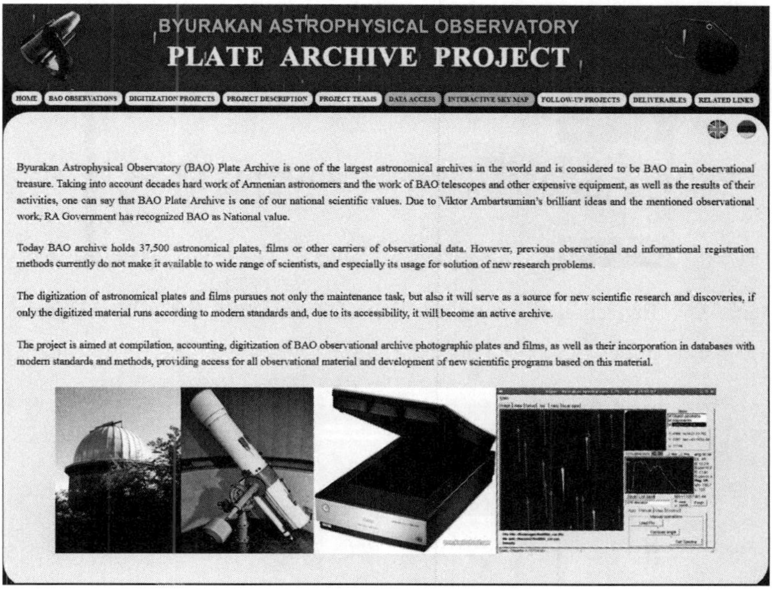

Figure 2. DFBS Sky Coverage (left) and the operation "GETSPECTRA" showing the list of requested spectra and an extraction of a given spectrum (right).

3. DFBS Based Science Projects

In Figure 3 we give an example of extraction of an asteroid spectrum from DFBS using VO software SkyBoT proving how useful the DFBS plates can be for follow-up studies (Thuillot et al. 2007; Berthier et al. 2009; Sarkissian et al. 2012). The search for asteroids in DFBS jointly with IMCEE (Observatoire de Paris, France) colleagues was the most advanced research project. Bright (<15–16 mag) asteroids observed in DFBS are being studied, which are divided into "fast" and "slow" ones depending on their motion during the typical DFBS plate exposure time (20 min), or more or less than 3″. All asteroid spectra are being extracted after they are found by means of SkyBoT. Sample spectra are being modelled similar to Solar spectra. Using these spectra and by means of comparisons with other catalogues, new candidate asteroids are being examined. Spectral analysis of asteroid spectra is being used to obtain definite physical parameters.

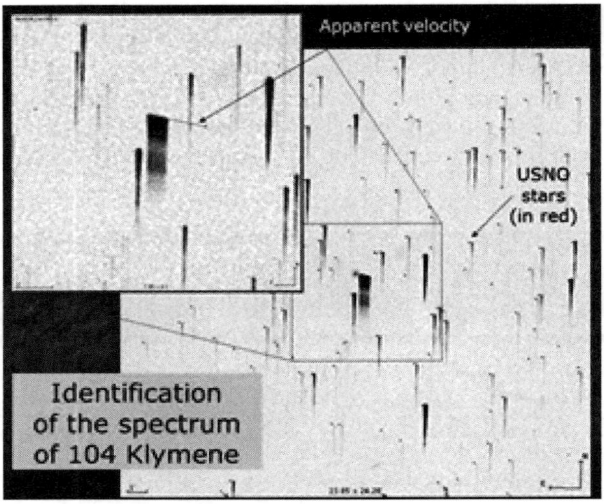

Figure 3. Extraction of an asteroid spectrum from DFBS using VO software SkyBoT

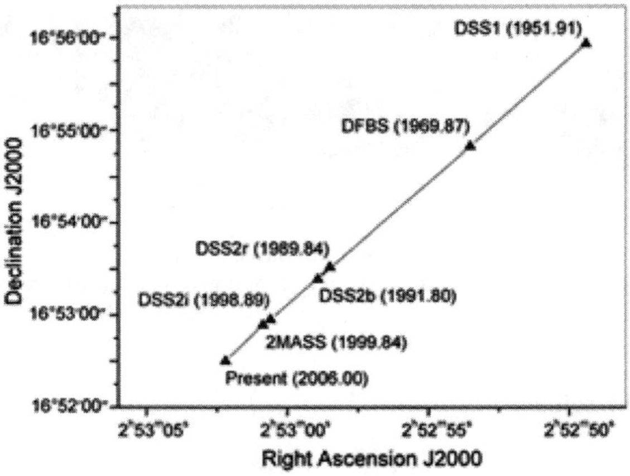

Figure 4. Direction of the motion of the high proper motion (PM=5.050″/yr) star FBS 0250+167, which was found using DFBS plates.

The efficiency of studies of AGB stars based on DFBS low dispersion spectra and follow-up spectroscopy was shown by Nesci et al. (2014). Gigoyan & Mickaelian (2007) have found a very high proper motion (PM) M7 type dwarf star, FBS 0250+167, lying about 3 pc from the Sun. Its PM is 5.050″/yr and it is on the list of the 10 known highest PM stars. DFBS plates from 1969 were used to link measurements between DSS1 and DSS2 and help identify this object and measure its PM (see Figure 4).

The Second Byurakan Survey (SBS; Stepanian 2005) plates are also subject for digitization, as they are hypersensitized and their emulsions are more sensitive to deterioration. 180 plates have been digitized so far. Due to SBS smaller photographic grains, 2400 dpi scanning (10 μm pixel size) is being used and 512 MB files are being obtained for each plate.

Having digitized plates and modern digital observational data, a number of efficient research projects have become possible, such as data discovery, spectral analysis, SED building and fitting, modelling, variability studies, cross-matching, etc. Some examples are variability studies (Samus & Antipin 2012), cross-matching of astronomical catalogs (Malkov 2012), and a search for asteroids and exoplanets using VO tools (Sarkissian et al. 2012).

Our science projects are aimed at discoveries of new interesting objects by searching definite types of low-dispersion spectra in the DFBS, by optical identifications of non-optical sources (X-ray, IR, radio) using the DFBS and DSS/SDSS, by using cross-correlations of large catalogs and selection of objects by definite criteria, etc.

4. DFBS as UNESCO "Memory of the World" Documentary Heritage

The Markarian survey was an outstanding study for all extragalactic (as well as Galactic) astronomy; its main features are:

- The Markarian survey is the first systematic objective-prism survey in the history of astronomy. Objective prism surveys subsequently became the main method for discovery and classification of thousands new AGN, white dwarfs, Hα emission-line objects (both stars and nebulae), and other peculiar objects.
- It is the largest objective-prism survey of the Northern sky (17,000 deg^2), thus providing the largest homogeneous research resource.
- It introduced and applied a new search method for active galaxies, selected by UV-excess.
- Discovery of 1517 UVX galaxies, ~300 AGN, and ~1000 HII galaxies.
- Classification of Seyferts into Sy1 and Sy2 types (Weedman & Khachikian 1968).
- Definition of Starburst (SB) galaxies (Weedman 1977) based on high star formation rate.
- Discovery of many new Blue Compact Dwarf Galaxies (BCDG).
- Discovery of 1103 FBS Blue Stellar Objects (BSOs; Mickaelian 2008) and 1045 Late-type Stars (Gigoyan & Mickaelian 2012; 2nd Part of the FBS).
- Optical identification of 1577 IRAS sources: Byurakan-IRAS Galaxies (BIG; Mickaelian & Sargsyan 2004) and Byurakan-IRAS Stars (BIS; Mickaelian & Gigoyan 2006)); discovery of many new AGN and ULIRGs.

- The Markarian survey led to many other objective prism surveys with better spectral resolution and deeper limiting magnitudes, including the Second Byurakan Survey (SBS, Markarian et al. 1983, Stepanian 2005).

Markarian galaxies are important for various specific extragalactic studies: Mrk 231 is the most luminous infrared galaxy (ULIRG) in the local Universe; Mrk 116 is the most metal-deficient blue compact dwarf galaxy (BCDG; most of the BCDGs are Mrk and SBS galaxies); Mrk 421 and 501 are among the most powerful sources in the Universe. In Table 2 we list some selected Markarian galaxies that are most important for various areas of extragalactic astronomy.

Table 2. Most important Markarian galaxies.

Mrk galaxies	Description
Mrk 231	the closest ULIRG, BAL QSO and most luminous IR galaxy in the Local Universe
Mrk 421, 501	are among the highest known energy sources
Mrk 116 (=IZw18)	the most metal-deficient BCDG (Mrk and SBS)
Mrk 938	the first dynamic merger discovered observationally
Mrk 110	intermediate between NLS1 and BLS1 (FWHM=4900 km/s); understanding BLS1 and NLS1 differences
Mrk 6	shows variations of spectral lines typical of different types of objects (Sy2 & Sy1); very high H column density in X-rays
Mrk 926	one of the rare Sy1 galaxies having LINER properties
Mrk 766	one of the most important NLS1 galaxies
Mrk 273	a wonderful double-double nucleus galaxy
Mrk 266	has multiple structures in the nuclear region
Mrk 231, 507	strongest FeII emitters (FeIIλ4570 / Hα > 2)
Mrk 530, 993, 1018	change their spectra from Sy1.9 to Sy1.0

Many more UVX and emission-line galaxies have been discovered by other surveys and studies similar to the Markarian survey, including Arakelian galaxies having high surface brightness (Arakelian 1975), Kazarian UVX galaxies (Kazarian et al. 2010), the University of Michigan emission-line galaxies (UM; MacAlpine et al. 1982), Case Low-Dispersion Northern Sky Survey galaxies (CG; Pesch et al. 1991), the Montreal blue galaxies (Coziol et al. 1994), SBS UVX and emission-line galaxies (Stepanian 2005), Kiso UV galaxies (KUG; Miyauchi-Isobe et al. 2010), Hamburg/SAO emission-line galaxies (Pustilnik et al. 2005), GALEX UV-luminous galaxies (Hoopes et al. 2007), etc.

Markarian galaxies are reliable objects for studies of active galaxies, as they are bright enough and have been detected in all ranges of electromagnetic radiation; from gamma-ray to radio. The Spectral Energy Distribution (SED) provides the ability to group objects by their shapes and compare to existing physical properties to find various relations and refine the AGN classifications.

The outstanding scientist Viktor Ambartsumian wrote about Markarian galaxies: *"The discovery of Markarian galaxies is a great achievement of our science. Now more and more astronomers both in our country and all over the world seek to study the nature of these objects in more detail on the world's largest telescopes..."*. Ambartsumian also compared Markarian survey plates with the treasures of Matenadaran – collection of ancient manuscripts. In 2011, Markarian Survey and its digitized version, DFBS, entered UNESCO's Documentary Heritage *"Memory of the World"* International Register. It is one of the rarest science treasures that is included in UNESCO lists together with world cultural heritage masterpieces. We show in Figure 5 the related UNESCO Certificate.

The IAU Colloquium #184 in 2001 was dedicated to Markarian and IAU Symposium #304 in 2013 was dedicated to Markarian's 100th anniversary, both related to the Markarian Survey and Multiwavelength AGN Surveys in general. The international symposium on *"Astronomical Surveys and Big Data"* in 2015 was dedicated to the 50th anniversary of Markarian Survey and 10th anniversary of Armenian Virtual Observatory[4] (ArVO), based on DFBS, the digitized Markarian Survey.

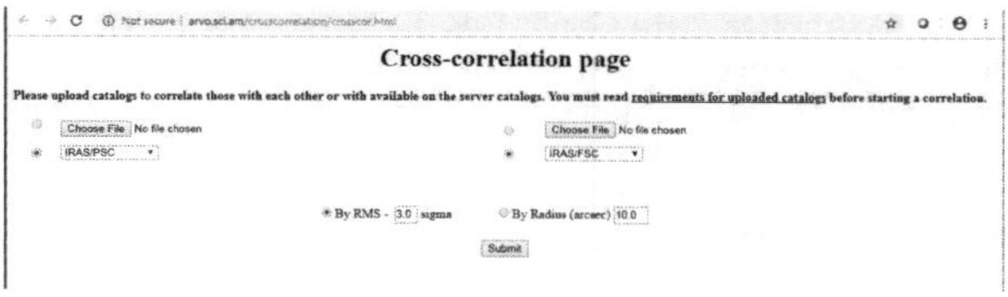

Figure 5. UNESCO certificate awarded for its inscription in the "Memory of the World" documentary heritage International Register.

5. Summary

There are a number of further possible research projects that will be conducted using the digitized plates:

- Correction of ephemerides of known asteroids and search for new asteroids (ex. Thuillot et al. 2007; Berthier et al. 2009).
- Discovery and study of variable stars (e.g., Mickaelian et al. 2011; Nesci et al. 2009).
- Revealing high proper motion stars (e.g., Mickaelian & Sinamyan 2010).
- Study of variability of known blazars and discovery of new blazars.
- Revealing Novae and Supernovae progenitors.
- Discovery of new QSOs.
- Discovery of new white dwarfs (e.g., Sinamyan & Mickaelian 2011).
- Discovery of new late-type stars (e.g., Gigoyan et al. 2010).
- Discovery of optical sources of gamma-ray bursts.

---

[4]`www.aras.am/Arvo/arvo.htm`

- Optical identifications of X-ray, IR, and radio sources (see, for example, Mickaelian & Sargsyan 2004; Mickaelian & Gigoyan 2006; Mickaelian et al. 2006; Hovhannisyan et al. 2009).

References

Berthier, J., Sarkissian A., Mickaelian A., Thuillot W. 2009, Eur. Planetary Sci. Congress, Vol. 4, 526

Erastova, L. K. 2004, A&AT 23, 209

Gigoyan, K. S., Mickaelian A. M. 2007, Astrophysics 50, 54

Gigoyan, K. S., Mickaelian A. M. 2012, MNRAS 419, 3346

Gigoyan, K. S., Sinamyan P. K., Engels D., Mickaelian A. M. 2010, Astrophysics 53, 123

Hovhannisyan, L. R., Weedman D. W., Mickaelian A. M., et al. 2009, AJ 138, 251

Hudec, R. 2014, Proc. Astroplate-2014 Conf., Eds. L. Mišková & S. Vitek. Prague, 1

Kazantseva, L. 2014, Proc. Astroplate-2014 Conf., Eds. L. Mišková & S. Vitek. Prague, 13

Knyazyan, A., Mickaelian A. M., Astsatryan H. 2011, Int. J. Inform. Theories and Appl. 18, 243

Knyazyan, A. V.; Astsatryan, H. V.; Mickaelian, A. M. 2016, Astronomical Surveys and Big Data, Eds. A. M. Mickaelian, A. Lawrence and T. Yu. Magakian. ASP Conf. Series, 505, 236

Markarian, B. E.; Lipovetsky, V. A.; Stepanian, J. A.; et al. 1989, Comm. SAO 62, 5

Massaro, E., Mickaelian A.M., Nesci R.,Weedman D. (Eds.) 2008, The Digitized First Byurakan Survey, ARACNE Editrice, Rome, 78

Mickaelian, A. M. 2006, Rom. Astron. J. 16S, 23

Mickaelian, A. M. 2007, Highlights of Astronomy 14, 594

Mickaelian, A. M. 2008, AJ, 136, 946

Mickaelian, A. M. 2012, Proc. Conf. "50 years of Cosmic Era: Real and Virtual Studies of the Sky." Yerevan, NAS RA, 29

Mickaelian, A. M. 2014, Proc. Astroplate-2014 Conf., Eds. L. Mišková & S. Vitek. Prague, 109

Mickaelian, A. M., Astsatryan, H. V., Sahakyan, V. G., et al. 2009, Proc. CSIT-2009, 420

Mickaelian, A. M., Gigoyan, K. S. 2006, A&A 455, 765

Mickaelian, A. M., Gigoyan, K. S., Nesci, R., Rossi, C. 2006, Mem.S.A.It. 77, 1159

Mickaelian, A. M., Hagen, H.-J., Sargsyan, L. A., Mikayelyan, G. A. 2005, Catalog No. VI/116 at CDS, Strasbourg

Mickaelian, A. M., Hovhannisyan, L. R., Engels, D., et al. 2006, A&A 449, 425

Mickaelian, A. M., Kochiashvili N., Astsatryan H. V., et al. 2009, Proc. CSIT-2009, 424

Mickaelian, A. M., Malkov O. Yu., Samus N. N. (Eds.) 2012, Proc. Conf. 50 years of Cosmic Era: Real and Virtual Studies of the Sky. NAS RA, 251, Yerevan

Mickaelian, A. M., Mikayelyan G.A., Sinamyan P.K. 2011, MNRAS 415, 1061

Mickaelian, A. M., Nesci R., Cirimele G., et al. 2008, Proc. ESAC Workshop on Astronomical Spectroscopy and Virtual Observatory, Madrid, ESA, 29

Mickaelian, A. M., Nesci, R., Rossi C., et al. 2007, A&A 464, 1177

Mickaelian, A. M., Sargsyan L. A., Nesci R., et al. 2010, ASP Conf. Series, Vol. 434: Astronomical Data Analysis Software and Systems XIX, 325

Mickaelian, A. M., Sargsyan, L. A. 2004, Astrophysics 47, 213

Mickaelian, A. M., Sargsyan, L. A., Astsatryan H. V., et al. 2009, Data Science Journal 8, 152

Mickaelian, A. M., Sargsyan, L. A., Gigoyan K. S., et al. 2009, Rom. Astron. J. 18S, 249

Mickaelian, A. M., Sargsyan, L. A., Mikayelyan G. A. 2010, Proc. of Science, 30

Mickaelian, A. M., Sargsyan, L. A., Mikayelyan G. A., et al. 2006, Heron Press Sci. Ser., 82

Mickaelian, A. M., Sargsyan, L. A., Nesci R., et al. 2007, Highlights of Astronomy 14

Mickaelian, A. M.; Abrahamyan, H. V.; Andreasyan, H. R.; et al. 2016b, Astronomical Surveys and Big Data, Eds. A. M. Mickaelian, A. Lawrence & T. Yu. Magakian. ASP Conf. Ser., 505, 262

Mickaelian, A., Sinamyan P. 2010, MNRAS 407, 681

Mickaelian, A. M. 2014, Multiwavelength AGN Surveys and Studies, Proc. IAU S304, 304, 1
Mickaelian, A. M.; Astsatryan, H. V.; Knyazyan, A. V.; et al. 2016a, Astronomical Surveys and Big Data, Eds. A. M. Mickaelian, A. Lawrence and T. Yu. Magakian. ASP Conf. Ser., 505, 16
Nesci, R., Bagaglia M., Nucciarelli G. 2014a, Proc. Astroplate-2014 Conf., Eds. L. Mišková & S. Vitek. Prague, 75
Nesci, R., Gaudenzi S., Rossi C., Pezzotti C., Gigoyan K., Mauron N. 2014b, Proc. Astroplate-2014 Conf., Eds. L. Mišková & S. Vitek. Prague, 91
Nesci, R., Mickaelian A. M., Rossi C. 2009, Astron. Tel. #2338
Osborn, W. 2014, Proc. Astroplate-2014 Conf., Eds. L. Mišková & S. Vitek. Prague, 15
Sarkissian, A., Arzoumanian E., Mickaelian A. M., et al. 2012, Proc. Conf. "50 years of Cosmic Era: Real and Virtual Studies of the Sky." Yerevan, NAS RA, 56
Sinamyan, P. K., Mickaelian A. M. 2009, Astrophysics 52, 76
Sinamyan, P. K., Mickaelian A. M. 2011, Astrophysics 54, 403
Sokolovsky, K. V., Antipin S. V., Zubareva A. M., et al. 2014, ARep 58, 319
Stepanian, J.A. 2005, RMxAA 41, 155
Stupka, J., Benesova E. 2014, Proc. Astroplate-2014 Conf., Eds. L. Mišková & S. Vitek. Prague, 31
Thuillot, W., Berthier J., Sarkissian A., et al. 2007, Highlights of Astronomy 14, 616
Tsvetkov, M.; Tsvetkova K. 2012, Proc. IAU S285: New Horizons in Time-Domain Astronomy, Cambridge Univ. Press, 417
Veron-Cetty, M.-P.; Veron, P. 2010, A&A 518, 10

Astronomical Heritage of the Middle East
ASP Conference Series, Vol. 520
Sona V. Farmanyan, Areg M Mickaelian, J. McKim Malville, and Mohammad Bagheri, eds.

BAO Plate Archive Heritage

Rather similar to previous paper

A. M. Mickaelian, L. K. Erastova, M. V. Gyulzadyan, G. B. Ohanian, K. S. Gigoyan, G. A. Mikayelyan, P. K. Sinamyan, G. M. Paronyan
NAS RA V. Ambartsumian Byurakan Astrophysical Observatory (BAO), Byurakan 0213, Aragatzotn province, Armenia; aregmick@yahoo.com

Abstract. The Byurakan Astrophysical Observatory (BAO) Plate Archive consists of 37,500 photographic plates and films, obtained with the 2.6 m telescope, the 1 m and 0.5 m Schmidt telescopes, and other smaller telescopes during 1947–1991. Its most important part, the famous Markarian Survey (or the First Byurakan Survey, FBS) 2000 plates were digitized in 2002–2007 and the Digitized FBS (DFBS) was created. New science projects have been conducted based on this low-dispersion spectroscopic survey data. Several other smaller digitization projects have been carried out as well, including digitization of some of the Second Byurakan Survey (SBS) plates, photographic chain plates in Coma, where the blazar ON 231 is located, and 2.6 m film spectra of FBS Blue Stellar Objects. However, most of the plates and films are not digitized. In 2015, we started a project to digitize the whole BAO Plate Archive and its scientific products. It is aimed at digitization, extraction and analysis of archival data and building an electronic database and interactive sky map. The Armenian Virtual Observatory (ArVO) database will contain all new data. This project is a collaboration with the Armenian Institute of Informatics and Automation Problems (IIAP) and continued during 4 years in 2015–2018. The final result is an Electronic Database and online Interactive Sky map to be used for further research projects. ArVO will provide all standards and tools for efficient usage of the scientific output and its integration in international databases.

1. Introduction

The astronomical plate archives created during numerous observations at many observatories are the most important part of the astronomical observational heritage. The necessity of digitization of astronomical plates for various national and international projects was emphasized at several Astroplate workshops (Osborn 2014; Hudec 2014; Kazantseva 2014; Nesci et al. 2014a; Stupka & Benesova 2014).

The Byurakan Astrophysical Observatory (BAO) Plate Archive is one of the largest astronomical archives in the world and is considered to be the BAO's main observational treasure. It is the results of decades of hard work by Armenian astronomers using BAO telescopes and other expensive equipment. Today the BAO archive holds some 37,000 astronomical plates, films, or other carriers of observational data. However, these older media do not make the data available to wide range of scientists to address new research problems. Digitization of BAO plates will be a significant contribution to the Wide-Field Plate DataBase (WFPDB) developed in Sofia, Bulgaria (Tsvetkov & Tsvetkova 2012).

Digitization of the BAO Plate Archive and creation of BAO Interactive Astronomical Database (the BAO Plate Archive project, or BAO PAP) has started in February

2015. It is aimed at preservation of BAO's valuable observational material accumulated during 1947–1991, creation of a full Database of all BAO observations, creation of the BAO Interactive Sky Map with visualization of all observations and quick access to the data, development of new research projects based on the existing observational data, and integration of BAO observations into the international databases. A number of young BAO astronomers are involved in this project, which is expected to last 3 years. Project objectives are the preservation of BAO observational archive, preservation of scientific information contained in photographic plates, broad dissemination of observational data, updating observational data to modern standards and usage methods, proposing new science projects, and publicizing BAO activities more widely.

A short description of the BAO Plate Archive was given by Mickaelian (2014a) and more details are given in Mickaelian et al. (2016b).

2. BAO Telescopes and Observing Programs

BAO observers worked with a number of BAO telescopes during 1947–1991 and obtained several tens of thousands of plates, films, and other data products. Table 1 gives a general overview of observations of 10 BAO telescopes that worked on the photographic photometry, electrophotometry, slit and objective prism spectroscopy, and polarimetry of many thousands of astronomical objects. Telescope "Sizes" are given for the mirror and focal length for classical telescopes and for the correcting lens, mirror and focal length for Schmidt type telescopes.

Table 1. Overview of BAO telescopes and observational data.

Telescope	Sizes (cm)	Years	Observing methods	Plates
5″ double-astrograph	13	1947–1950	photometry	3000
6″	15	1947–1950	photometry	3000
8″ Schmidt	20/20/31	1949–1968	photometry	4500
20″ Cassegrain	51/800	1952–1991	electrophotometry	
10″ telescope-spectrograph	25	1953–19??	spectra	
Nebular spectrograph		1954–19??	spectra	
16″ Cassegrain	41/400	1955–1991	electrophotometry	
21″ Schmidt	53/53/183	1955–1991	photometry	12000
40″ Schmidt (AZT-10)	102/132/213	1960–1991	photometry, spectra	7500
ZTA-2.6m	264/1016	1975-1991	photometry, spectra	7000
All Telescopes		1947–1991		37,000

Here we list the main observational projects accomplished on the three most important BAO telescopes (2.6m classical reflector, 1m Schmidt and 0.5m Schmidt; Figure 1).

0.5 m (21″) Schmidt telescope:

- Polarization of variable nebula NGC 2261.

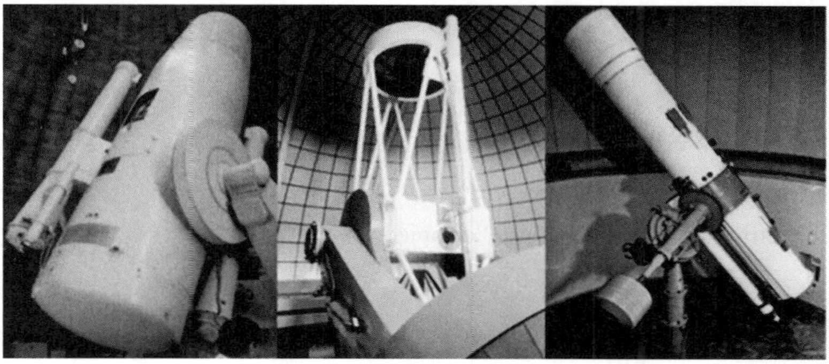

Figure 1. BAO's most important telescopes (from left to right): 1 m Schmidt, 2.6 m classical reflector and 0.5 m Schmidt.

- Nuclei of nearby Sa and Sb galaxies.
- Nuclei of nearby Sc galaxies.
- Search for flare stars in the Pleiades.
- Search for flare stars in Orion.
- Search for flare stars in NGC 7000 (Cygnus).
- Search for flare stars in Praesepe.
- Search for flare stars in Taurus Dark Clouds (TDC).
- Variability of Markarian galaxies.
- Monitoring of extragalactic supernovae.

1 m (40″) Schmidt telescope:

- Detailed colorimetry of bright galaxies.
- First Byurakan Survey (FBS), or Markarian survey (Markarian et al. 1989).
- Search for flare stars in Pleiades.
- Search for flare stars in Orion.
- Search for flare stars in NGC 7000 (Cygnus).
- Search for flare stars in Praesepe.
- Search for flare stars in Taurus Dark Clouds (TDC).
- Second Byurakan Survey (SBS) (Stepanian 2005).
- Extension of the FBS in the Galactic Plane.

ZTA-2.6 m telescope:

- Morphological study of Markarian galaxies.
- Investigation of star clusters.
- Investigation of groups and clusters of galaxies.
- Spectroscopy of FBS blue stellar objects.
- Spectroscopy of FBS late–type stars.
- Spectroscopy of SBS galaxies and stellar objects (BAO/SAO).
- Direct images of the central regions of Markarian galaxies.
- Spectroscopy of T Tauri and flare stars.
- Spectroscopy of Byurakan-IRAS Galaxies (BIG objects).
- Spectroscopy of ROSAT AGN candidates (BAO/HS/OHP/INAOE).

3. BAO Plate Archive Project

The digitization of astronomical plates and films is not only an archive maintenance task, but also serves as a source for new scientific research and discoveries, but only if the data is digitized according to modern standards and made open accessible. The project is aimed at compilation, accounting, digitization of BAO observational archive photographic plates and films, as well as their incorporation in databases with modern standards and methods, providing access to all observational data and development of new scientific programs based on this material.

The Scientific Programs Board (SPB) was created to evaluate the existing observational material, to prioritize plates to be scanned first and to propose new research projects. It consists of BAO Director, DFBS Principal Investigator and ArVO Project Manager Areg Mickaelian (Chair), the Head of BAO Astroinformatics Department Tigran Magakian, and seven other of the most experienced BAO observers, as well as researchers from NAS RA Institute of Informatics and Automation Problems (IIAP) Vladimir Sahakian and Hrachya Astsatryan were involved for their experience in computer science related to databases and computational methods.

The Project Executing Team (PET) consists of 14 members: Kamo Gigoyan (Project Manager, SPB member), Marietta Gyulzadyan (Deputy Manager on Observing Programs), Gurgen Paronyan (Deputy Manager on Technical Issues, Person in charge for BAO Plate Archive), six other PET members involved in scanning and reduction of data (Hayk Abrahamyan, Hasmik Andreasyan, Naira Azatyan, Varsik Harutyunyan, Gayane Kostandyan, and Anahit Samsonyan), Gor Mikayelyan (Database Manager, Web Designer), Sona Farmanyan (Webpage content, dissemination, outreach, and organizational issues) and Aram Knyazyan (NAS RA IIAP, Database Manager). Knarik Khachatryan and Ani Vardanyan also participated in the Project in 2015–2016.
The project consists of the following tasks:

- Development of technical principles of the project, necessary equipment, timeline and budget.

- Collection of all photographic plates (until recently only plates obtained before 1974 had been collected in the BAO Plate Archive).

- Revision and accounting of the plates and observing journals in the BAO Plate Archive.

- Scanning of a few dozen plates for test and educational reasons, and to set up the necessary parameters for the scanning in frame of the main project.

- Input of data from observing journals; creation of the Project Database and development of the principles of organization of data in it.

- Creation of the Project Webpage and User Interface.

- Scanning of photographic plates and films.

- Astrometric solution; extraction of images and spectra; wavelength calibration; density and flux calibration; multiband (UBVR) photometry.

- Making template low-dispersion spectra; numerical classification of low-dispersion spectra.

- Visualization of BAO observations on sky map; creation of electronic interactive sky map and search system.

- Scientific analysis of existing observational material and exploring new research possibilities; proposing and discussing new research projects.

The BAO PAP webpage[1] (Figure 2) contains a wealth of information on BAO observations, previous digitization projects, present project details, teams, follow-up research projects, deliverables and related links (many details will be added during the next few months). The main products will be *Data Access* and *Interactive Sky Map*. The first will contain BAO's Observational Database, the ability to search by any parameter (Dates/Julian dates, Telescope, Observing modes/methods, Instrument, Receiver, Emulsion, Filters, Seeing, Project name, Project PI, Observers, Targets/coordinates, Sky area, Surface, Scale, Spatial resolution, Spectral range, Spectral resolution, Limiting magnitude, Number of nights, Number of exposures, and Links), Data Visualization, and Download of the digitized plates, films, or parts of plates, or spectra or images of individual objects. The Interactive Sky Map will help users visualize the areas observed by BAO telescopes and will give options to check observed areas for a given observation, telescope, observer, observing method, limiting magnitude, etc. There will be a capability to check individual fields for the number of plates and to propose further research projects, including projects on variability, proper motions, and studies of Solar System objects.

[1]`http://www.aras.am/PlateArchive/`

Figure 2. The BAO Plate Archive Project webpage.

4. DFBS and Other Digitization Projects at BAO

A number of digitization projects have been accomplished at BAO, including the most important one, the Digitized First Byurakan Survey (DFBS)[2] (Mickaelian et al. 2007a; Massaro et al. 2008) (see our paper in these Proceedings p. 105) based on the digitization of the famous Markarian Survey (Markarian et al. 1989).

The Second Byurakan Survey (SBS) (Stepanian 2005) plates are also subject for digitization, as they are hypersensitized and their emulsion is more sensitive for deterioration. 180 plates have been digitized so far. Due to the smaller photographic grains in SBF plates, 2400 dpi (10μm pixel size) is being used and 512 MB files are being obtained for each plate.

Photographic spectra of the FBS blue stellar objects (BSOs) have been obtained using 2.6 m telescope and UAGS spectrograph on photographic films. ~700 such spectra have been scanned with 1600 dpi, 16 bit and 650×21 pixel images were obtained (FSB BSOs; Mickaelian 2008) and late-type stars; (Gigoyan & Mickaelian 2012)). All spectra were put in a standard format so that automatic reduction was possible (Figure 3). 101 FBS blue stellar objects were published and a number of planetary nebulae, white dwarfs, hot subdwarfs and HBB stars have been revealed (Sinamyan & Mickaelian 2009).

Another project was the study of long-term variability of ON 231, which appeared in the Coma field, where photographic chains were carried out to discover flare stars. In total 189 plates with a total number of more than 1200 exposures in 1969–1976 with the Byurakan 21″ and 40″ Schmidt telescopes were obtained. This was a valuable material for study of ON231 long-term variability (Figure 4; Erastova 2004).

[2] `http://www.aras.am/Dfbs/dfbs.html`

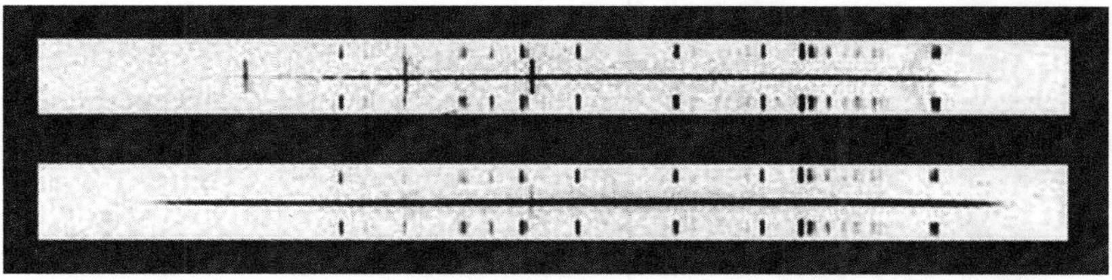

Figure 3. Standard format of FBS spectra with 650×21 pixel images for automatic reduction.

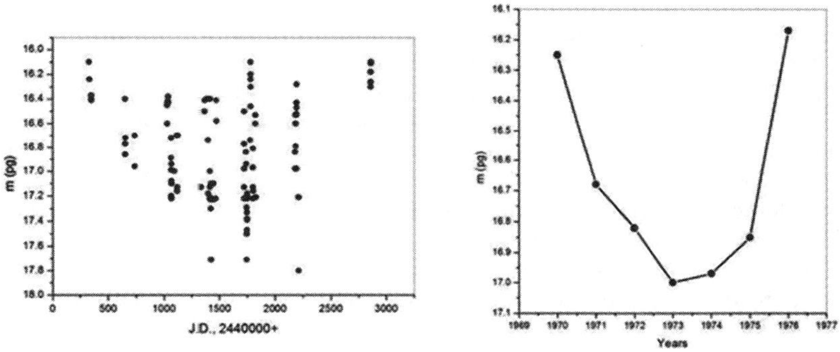

Figure 4. Photometric measurements and variability study of the blazar ON 231 observed in the Coma field.

Having digitized plates and modern digital observational data, a number of efficient research projects have become possible, such as data discovery, spectral analysis, SED building and fitting, modeling, variability studies, cross-matching, etc. Some examples include variability studies (Samus & Antipin 2012), ross-matching of astronomical catalogs (Malkov 2012), and the search for asteroids and exoplanets using VO tools (Sarkissian et al. 2012).

Our science projects are aimed at discoveries of new interesting objects by searching definite types of low-dispersion spectra in the DFBS, by optical identifications of non-optical sources (X-ray, IR, radio) and using the DFBS and DSS/SDSS, by using cross-correlations of large catalogs and selection of objects by definite criteria, etc.

In one of our papers (Abrahamyan et al. 2012), a new tool for cross-correlation of catalogs is presented (Knyazyan et al. 2011), where individual positional error for each object instead of a standard search radius is being taken into account, thus making the result much more accurate and precise (Figure 5).

5. Summary

The BAO Plate Archive is one of the most important astronomical databases available to the astronomical community. A project of its digitization and creation of the electronic database is in progress. At present the main part of the project, the scanning of the

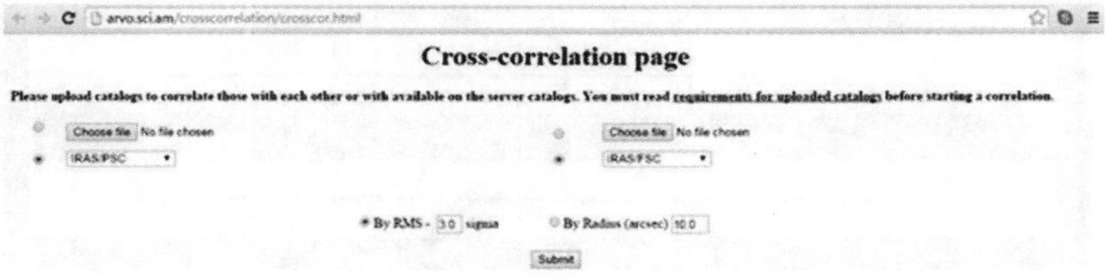

Figure 5. Cross-correlation software developed in frame of ArVO.

plates has produced more than 20,000 digital images of roughly 200 MB each. All they have been stored and double copies are available.

The BAO Plate Archive Project will lead to preservation of BAO's valuable observational data obtained during 1947–1991. However, our goal is not only to create a passive archive of scanned plates and films, but also to make use of those data where more studies are possible. Proper motion and variability studies are most important, as time domain material is contained in historical plates. Such possibilities based on DFBS were shown by Mickaelian et al. (2006a); DFBS as a unique database for proper motion, variability studies, and object classification. New variable stars discovered on digitized plates in the Moscow collection was reported by Sokolovsky et al. (2014).

There are a number of further possible research projects that will be conducted using the plates digitized:

- Correction of ephemerides of known asteroids and search for new asteroids (Thuillot et al. 2007; Berthier et al. 2009).
- Discovery and study of variable stars (Mickaelian et al. 2011; Nesci et al. 2009).
- Revealing high proper motion stars (Mickaelian & Sinamyan 2010).
- Study of variability of known blazars and discovery of new blazars.
- Revealing Novae and Supernovae progenitors.
- Discovery of new QSOs.
- Discovery of new white dwarfs (Sinamyan & Mickaelian 2011).
- Discovery of new late-type stars (Gigoyan et al. 2010).
- Discovery of optical sources of gamma-ray bursts.
- Optical identifications of X-ray, IR, and radio sources (Mickaelian & Sargsyan 2004; Mickaelian & Gigoyan 2006; Mickaelian et al. 2006b; Hovhannisyan et al. 2009).

Acknowledgments. The BAO Plate Archive Project team would like to thank BAO Administration for supporting the project and Scientific Programs Board (SPB) members for useful discussions on possible research programs based on BAO Plate Archive material.

References

Abrahamyan, H. V., Mickaelian, A. M., Knyazyan, A. V., Harutyunyan, G. S. 2012, Proc. Conf. "50 years of Cosmic Era: Real and Virtual Studies of the Sky." Yerevan, NAS RA, 223
Berthier, J., Sarkissian, A., Mickaelian, A., Thuillot, W. 2009, European Planetary Science Congress 2009, held 14-18 September in Potsdam, Germany, 526
Erastova, L. K. 2004, A&AT 23, 209
Gigoyan, K. S., Sinamyan P. K., Engels D., Mickaelian A. M. 2010, Astrophysics 53, 123
Gigoyan, K. S., Mickaelian A. M. 2012, MNRAS 419, 3346
Hovhannisyan, L. R., Mickaelian, A. M., Weedman, D. W. et al. 2009, AJ, 138, 251
Hudec, R. 2014, Proc. Astroplate-2014 Conf., Eds. L. Mišková & S. Vitek. Prague, 1
Kazantseva, L. 2014, Proc. Astroplate-2014 Conf., Eds. L. Mišková & S. Vitek. Prague, 13
Knyazyan, A., Mickaelian A. M., Astsatryan H. 2011, Int. J. "Inform. Theories and Appl." 18, 243
Knyazyan, A. V., Astsatryan, H. V., Mickaelian, A. M. 2016, Astronomical Surveys and Big Data, Eds. A. M. Mickaelian, A. Lawrence and T. Yu. Magakian. ASP Conf. Series, 505, 236
Malkov, O. 2012, Proc. Conf. "50 years of Cosmic Era: Real and Virtual Studies of the Sky." Yerevan, NAS RA, 48
Markarian, B. E., Lipovetsky, V. A., Stepanian, J. A., et al. 1989, Comm. SAO 62, 5
Massaro, E., Mickaelian A.M., Nesci R., Weedman D. (Eds.) 2008, The Digitized First Byurakan Survey, ARACNE Editrice, Rome, 78
Mickaelian, A. M., Sargsyan, L. A. 2004, Astrophysics 47, 213
Mickaelian, A. M., Hagen, H.-J., Sargsyan, L. A., Mikayelyan, G. A. 2005, Catalog No. VI/116 at CDS, Strasbourg
Mickaelian, A. M. 2006, Rom. Astron. J. 16S, 23
Mickaelian, A. M., Gigoyan, K. S. 2006, A&A 455, 765
Mickaelian, A. M., Gigoyan, K. S., Nesci, R., Rossi, C. 2006a, Mem.S.A.It. 77, 1159
Mickaelian, A. M., Hovhannisyan, L. R., Engels, D., et al. 2006b, A&A 449, 425
Mickaelian, A. M., Sargsyan, L. A., Mikayelyan G. A., et al. 2006c, Heron Press Sci. Ser., 82
Mickaelian, A. M. 2007, Highlights of Astronomy 14, 594
Mickaelian, A. M., Nesci, R., Rossi C., et al. 2007a, A&A 464, 1177
Mickaelian, A. M., Sargsyan, L. A., Nesci R., et al. 2007b, Highlights of Astronomy 14
Mickaelian, A. M. 2008, AJ 136, 946
Mickaelian, A. M., Nesci R., Cirimele G., et al. 2008, Proc. ESAC Workshop on Astronomical Spectroscopy and Virtual Observatory, Madrid, ESA, 29
Mickaelian, A. M., Sargsyan, L. A., Astsatryan H. V., et al. 2009a, Data Science Journal 8, 152
Mickaelian, A. M., Astsatryan, H. V., Sahakyan, V. G., et al. 2009b, Proc. CSIT-2009, 420
Mickaelian, A. M., Kochiashvili N., Astsatryan H. V., et al. 2009c, Proc. CSIT-2009, 424
Mickaelian, A. M., Sargsyan, L. A., Gigoyan K. S., et al. 2009d, Rom. Astron. J. 18S, 249
Mickaelian, A., Sinamyan P. 2010, MNRAS 407, 681
Mickaelian, A. M., Sargsyan, L. A., Mikayelyan G. A. 2010, Proc. of Science, 30
Mickaelian, A. M., Mikayelyan G.A., Sinamyan P.K. 2011, MNRAS 415, 1061
Mickaelian, A. M. 2014a, Multiwavelength AGN Surveys and Studies, Proc. IAU S304, 304, 1
Mickaelian, A. M. 2014b, Proc. Astroplate-2014 Conf., Eds. L. Mišková & S. Vitek. Prague, 109
Mickaelian, A. M.; Astsatryan, H. V.; Knyazyan, A. V.; et al. 2016a, Astronomical Surveys and Big Data, Eds. A M. Mickaelian, A. Lawrence and T. Yu. Magakian. ASP Conf. Ser., 505, 16
Mickaelian, A. M., Abrahamyan, H. V., Andreasyan, H. R., et al. 2016b, Astronomical Surveys and Big Data, Eds. A. M. Mickaelian, A. Lawrence & T. Yu. Magakian. ASP Conf. Ser., 505, 262
Nesci, R, Mickaelian, A, Rossi, C. 2009, The Astronomer's Telegram, 2338
Nesci, R., Bagaglia M., Nucciarelli G. 2014a, Proc. Astroplate-2014 Conf., Eds. L. Mišková & S. Vitek. Prague, 75

Nesci, R., Gaudenzi S., Rossi C., Pezzotti C., Gigoyan K., Mauron N. 2014b, Proc. Astroplate-2014 Conf., Eds. L. Mišková & S. Vitek. Prague, 91
Osborn, W. 2014, Proc. Astroplate-2014 Conf., Eds. L. Mišková & S. Vitek. Prague, 15
Sarkissian, A., Arzoumanian E., Mickaelian A. M., et al. 2012, Proc. Conf. "50 years of Cosmic Era: Real and Virtual Studies of the Sky." Yerevan, NAS RA, 56
Samus, N., Antipin, S. 2012, Proc. Conf. "50 years of Cosmic Era: Real and Virtual Studies of the Sky." Yerevan, NAS RA, 39
Sinamyan, P. K., Mickaelian A. M. 2009, Astrophysics 52, 76
Sinamyan, P. K., Mickaelian A. M. 2011, Astrophysics 54, 403
Sokolovsky, K. V., Antipin S. V., Zubareva A. M., et al. 2014, ARep 58, 319
Stepanian, J.A. 2005, RMxAA 41, 155
Stupka, J., Benesova E. 2014, Proc. Astroplate-2014 Conf., Eds. L. Mišková & S. Vitek. Prague, 31
Thuillot, W., Berthier, J., Sarkissian, A., Mickaelian, A. M., Sargsyan, L. A., Iglesias, J., Lainey, V., Birlan, M., Simon, G. 2007, Highlights of Astronomy, 14, 616
Tsvetkov, M., Tsvetkova, K. 2012, Proc. IAU S285: New Horizons in Time-Domain Astronomy, Cambridge Univ. Press, 417
Véron-Cetty, M.-P.; Véron, P. 2010, A&A 518, 10

Astronomical Heritage of the Middle East
ASP Conference Series, Vol. 520
Sona V. Farmanyan, Areg M. Mickaelian, J. McKim Malville, and Mohammad Bagheri, eds.

Digital Projection of Ancient Astronomical Heritage

Tsitsino Simonia[1] and Irakli Simonia[2]
[1]*The City Library, Tbilisi, Georgia;* tsitsino.simonia@yahoo.com

[2]*Ilia State University, Tbilisi, Georgia*

Abstract. This paper discusses the problem of cataloging and preserving old astronomical heritage manuscripts. Examples of old astronomical materials and new initiatives are presented.

1. Particles of Knowledge

Astronomy, as a representative of exact sciences and as a cultural phenomenon, played an important role in development of the mankind. The ancients observed, fixed, remembered, predicted, and, "grain by grain," collected "particles of knowledge" about the Universe. The repetitiveness and regularity of the celestial phenomena had great influence on ancient people, making them return to contemplation of the world of the stars from night to night. Walking on the path of accumulating the knowledge, ancient man tried to understand his role in the Universe. This comprehension was formed over millennia, having passed the distance from primitive megalithic buildings constructed in the context of the heavens to multi-volume star catalogs and atlases. From contemplation to observation, from passive fixation to interactive communication, from watching to measuring—is an evolutionary process on the way to understanding the Universe. "Communication" of ancient man with the world of stars was accompanied by the following basic processes:

1. accumulation of empirical knowledge;
2. formation and development of instrumentation;
3. formation and development of centers of observation, observatories, and research institutes;
4. accumulation of observational experience, relevant skills, and abilities; and
5. invention and development of methods for registration and storage of information.

These synchronous and complementary processes, along with parallel development in other spheres of knowledge and mathematics, led to the regulation of a general process of cognition and formation of the fundamental bases for scientific ideology. Observed phenomena demanded general description, systematization, and formation of fundamental theories. These complicated processes of evolution in scientific ideology demanded great effort and time.

Now, in the 21st century, mankind's path of development is far more evident. However, many problems remain, including the issues of development of relevant fundamental theories on spreading regular knowledge, including astronomical knowledge. Cultural heritage of different kinds and forms, written and printed materials, other relicts of scientific findings can shed new light on the development of the ancient science of astronomy and mathematics. Ancient astronomical treatises, texts, schemes, and tables, are not only carriers of the scientific information proper, but also "reflectors" of the ways for transmission and distribution of knowledge from culture to culture, from continent to continent, from the past to future. Not many of these ways are studied and, sometimes, are even forgotten. Digital data storage technology is also a very important aspect of the astronomical heritage protection problem.

2. Astronomical Manuscripts in Georgia—an Electronic Book

Georgia is an ancient country in the far east of Europe dating from the 7th century BC. The original language and alphabet, literature, music, architecture, applied arts, and oral folklore are just a few of the important parts of the ancient Georgian culture. An organic part of Georgian cultural heritage is ancient philosophical and cosmological theories, material relicts—a witness of the ancient process of comprehending the surrounding world. An important part of ancient Georgian cultural heritage is direct and indirect indicators of the astronomical science, artifacts—material incarnations of the process of evolution in astronomical knowledge. Relying on this conceptual view and gathering together the facts, documents, theories, and hypotheses, we conclude that ancient Georgians followed the traditional path of development of astronomical ideas. This way involved accumulation of empirical knowledge, development of instrumentation, foundation of institutional centers of science, accumulation of observational experience, registration and storage of information, etc. In other words, the development of astronomical ideas in Georgia went in a traditional way from megalithic constructions to optical telescopes.

The territory of Georgia is rich in diverse artifacts that demonstrate the relationship ancient people had with the celestial bodies. The ancient people built and used different stone buildings to observe celestial objects. They reflected the observed phenomena in primitive schemes and ornate pictures cut or carved on stone surfaces. Oral Georgian folklore, legends, and tales reflect ethno-astronomical ideas of ancient Georgians (Simonia et al. 2008). This material is quite scattered, and much of it is lost or poorly studied.

Along with the development of universal human ideas and the formation of ethno-cosmological ideas and accumulation of empirical knowledge, a gradual transformation of the "interaction" of ancient man with the world of stars was taking place. In ancient Georgia, this process was stimulated by the formation of an original alphabet and the appearance of a tradition of written documentation. Ancient Georgians started to write down observed celestial phenomena on skins and later on paper. This became the basis for the formation of a unique layer of ancient Georgian written materials on astronomical and cosmological topics. Original Georgian astronomical treatises, cosmological models, descriptions of instruments and observations, and calendars of different types began to appear. Today all these manuscripts and volumes are in various states of preservation. Classical Persian and Arabic astronomical treatises, works of theoretical and practical significance were translated into the Georgian language (Fig. 2). The earlier Middle Ages were also marked by the appearance of Georgian translations of philosophical-cosmological and calendar-chronological works of Basil the Great, Gregory of Nyssa, John of Damascus and other brilliant representatives of Byzantine culture.

Digital reproduction and storage of ancient astronomical texts and artifacts has high importance. More than 15,000 manuscripts in the Georgian, Persian, Arabic, Greek and other languages are preserved in different universities, libraries, and scientific centers of Georgia. Georgian manuscripts embrace the period of the 5th to 19th centuries and contain original Georgian texts and foreign materials translated into the Georgian language. These manuscripts are quite extensive in their content. The collections preserved in the centers mentioned above contain religious, scientific, literary, and mixed works. The manuscript materials also involve different chronicles, historical notes, descriptions of different events, and instruction material, maps, tables, diagrams and pictures. Among the scientific materials we encounter astronomical, astrological, mathematical, geographical and other works and their fragments. The manuscript materials of astronomical character are in the form of completed treatises and works, or in fragments in the form of separate chapters or parts of chapters. The mixed manuscripts often contain astronomical and astrological fragments. Astronomical material, preserved in the scientific centers, includes from 300 to 400 manuscripts, including original works or translations into Georgian language; approximately 60 Persian and Arabian astronomical materials are in the form of treatises, manuscript books and their fragments. Assessments of the number of manuscripts are approximate and can change with time.

The physical condition of materials is diverse. There are manuscripts preserved in good condition, but a significant fraction of the manuscripts have been damaged by time, and demand restoration and conservation. Taking into consideration the fact that Georgian manuscripts are preserved in the scientific centers and libraries of other countries, including Greece, Egypt, Great Britain, France, Germany, Russia, etc., it should be stressed that: (1) Georgian astronomical manuscripts, scattered in the different centers, were never subject to systematic registration; and (2) the astronomical content of these manuscripts has not been studied in detail; these manuscripts were only subject to general philological and historical analysis. The methods of analysis of ancient astronomy and mathematics have not been used in regard to them yet, with the exception of several astronomical works of the 13th–18th centuries (Abuladze 1990).

With the purposes of virtual organization of scattered manuscript materials, rapid analysis of unknown astronomical works, formation of short descriptions of unknown manuscripts, and simplification of access to the materials for scholars of different countries, we initiated a special project called "Astronomical Manuscripts in Georgia." The main result of this project was formation of the interactive database/electronic book of unknown or insufficiently studied astronomical manuscripts preserved in the scientific centers of Georgia. The key goals of this project were: (1) investigation of unknown or insufficiently studied Georgian, Persian, and Arabic astronomical manuscripts to reveal their scientific significance; and (2) preparation of catalogues of manuscript materials for the electronic book (interactive database) as a reference source, which, in turn, will stimulate further study of these unknown manuscripts. The results of this project are:

1. description and explanation of the content of unstudied and insufficiently studied Georgian, Persian, and Arabic astronomical manuscripts;
2. revealing of scientific data, viewpoints, methods, facts and names important for ancient astronomy in the studied manuscripts;
3. revealing of hitherto unknown astronomical manuscripts; and
4. formation of reference sources for further research (electronic book, website, and printed publications).

The electronic book/database by I. Simonia et al. (2015) "Astronomical Manuscripts in Georgia" was published by Ilia State University Press (Tbilisi, 2015). It was published on DVD discs in the Georgian and English languages. This edition contains descriptions, bulletins, images, fragments, and full texts of little studied and unknown Georgian, Persian, and Arabic astronomical manuscripts of the 9th through 19th centuries. It was the first edition of its kind based on digital technology.

3. Well-known and Little-known Astronomical Heritage Manuscripts

We analyzed more than 400 astronomical, cosmographic, and mixed Georgian, Persian and Arabic manuscripts. More than 300 Georgian and about 60 Persian and Arabian astronomical manuscripts are preserved in the Tbilisi National Center of Manuscripts, in the State Archives, museums, and libraries of Tbilisi, Gori, Kutaisi, Akhaltsikhe, and other cities. These astronomical treatises embrace the period of the 9th through 19th centuries and are written in the ancient Georgian language, late medieval Georgian, and in the Persian and Arabic languages. The material, on the whole, is scattered and involves both theoretical works and complex observational data. These manuscripts in particular are dedicated to descriptions of cosmogony theories, elements of spherical astronomy, various systems of calendars, astronomical instruments, star catalogs, etc. These manuscripts give detailed descriptions of different celestial phenomena, both of regular character and also rare phenomena, including eclipses and appearances of comets. These materials list bright and dim stars, groups of stars, and constellations.

Astronomical manuscripts, surely, are often accompanied by astrological material with different predictions and prognoses. The greater part of the Georgian astronomical manuscripts are anonymous, i.e., authors, translators, or scribes are not specified. However, we also encounter materials by known authors. In the Georgian manuscripts the geocentric views, church calendars, and practical instructions on observation of celestial bodies are common. Many Georgian astronomical manuscripts are included in more general papers and works, primarily of religious nature. Besides these, purely astronomical works, treatises, descriptions, and tables are most common. There are many original Georgian materials and translations. The Persian and Arabic astronomical manuscripts, preserved in the scientific centers of Georgia, are of later period, and describe the issues of cosmogony, practical astronomy, and chronology. All of these have been insufficiently studied, but the centers and archives of Georgia also preserve quite well-known works, including the treatise of 13th century Georgian philosopher and astronomer Abuserisdze Tbileli, and the treatise of 11th century Arabian astronomer Al Sufi, the "Book of Fixed Stars."

Modern digital technologies have great potential for preserving and distributing astronomical heritage records, copies of manuscripts, and images of artifacts. The data

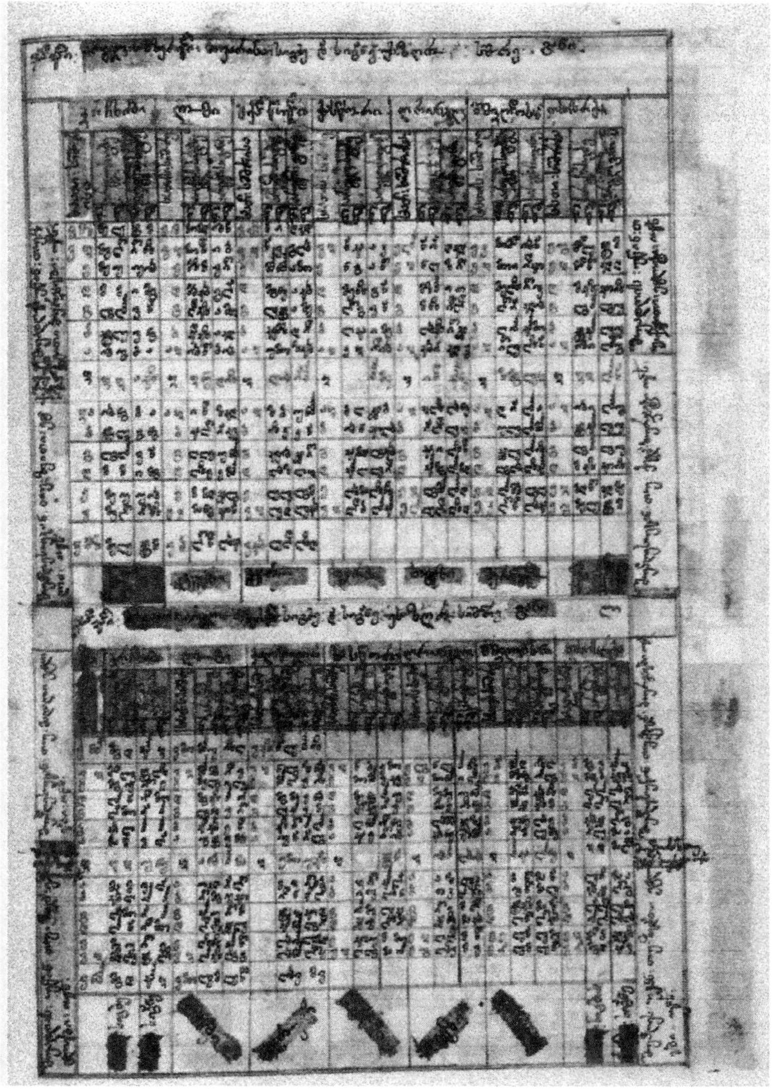

Figure 1. Ulughbeg, Star catalog. It was translated into Georgian by the king of Georgia Vakhtang VI Bagrationi in the 18th century.

storage platforms must be widely accessible, large in capacity, and simple to navigate. Websites, FTP servers, and DVD discs are the best platforms for astronomical heritage data storage. On this basis, we also suggest creating a special electronic visual atlas "On the Meridian of Ancient Astronomy." It must be a large electronic database of unknown or little-studied astronomical heritage materials, including cosmographic images on material artifacts (metal, glass, jewelery), archeological artifacts (stones), and manuscripts with old astronomical drawings and schemes. The main principles of visual data regulation will be: (1) one entry per item—one file with images, brief textual descriptions, and references; and (2) all data ordered by their geographic features (the meridian of data origin). Such an electronic atlas will be a valuable source of data for researchers and students for teaching and research.

References

Abuladze, T. 1990, Tbilisi

Simonia, I., Ruggles, C., & Chagunava, R. 2008, JAHH

Simonia, I., Abuladze, T., Chkhikvadze, N., Samkurashvili, L., Simonia, Ts., Pataridze, K., and Mekadze, G. 2015, ISU Press, Tbilisi

Part IV

Astronomical Heritage for Science

Astronomical Heritage of the Middle East
ASP Conference Series, Vol. 520
Sona V. Farmanyan, Areg M. Mickaelian, J. McKim Malville, and Mohammad Bagheri, eds.

The Astronomical Orientations of Sacred Buildings in the Levant During the Bronze and Iron Ages

Andrea Polcaro[1,2] and Vito Francesco Polcaro[2,3,4]
[1]*Università delgi Studi di Perugia, Perugia, Italy;* andrea.polcaro@unipg.it

[2]*CESAR, Rome, Italy*

[3]*Istituto Nazionale de Astrofisica, IAPS, Rome*

[4]*ACHe, Università di Ferrara*

Abstract. This paper describes the astronomical orientations of sacred buildings in the Levant built during the Bronze and Iron Ages.

1. Introduction

Since prehistory man has tried to represent the reality of the world around him. However, the first clear artistic expression of cult and religion, as well as the first images of the gods, spread between human communities during the Neolithic Period (probably around the 10th to 8th millennium BC) as is shown by the Göbekli Tepe site.[1] Around the 7th millennium BC, in the epoch and the places of the beginning of sedentarization and the farmer life style (the Fertile Crescent), the temple architecture emerges as the fulcrum of life of the new Post-Neolithic agro-pastoral communities, a symbol of the organization of the settlement, the ideological power of the elite, and cultural belonging. In the Near East the more ancient paintings connected with the ritual world, the famous Catal Huyuk pictures on houses in Turkey, mostly referred to funerary aspects.[2] In the 5th to 4th millennium BC, paintings showing not only ritual activities, but also mythological beliefs, are found on houses, tombs, and for the first time, public buildings with evidence of cult purpose. They are found in Turkey (Arslantepe/Malatya), Syria (Tell Huera), Egypt (Naqada tombs), and Jordan (paintings of Tuleylat al-Ghassul). At this last site, on the inner walls of some buildings of area 3, there are the clearest ancient images of the astral world, referred to as divine entities.

[1]The reliefs of the site were discovered carved into enigmatic T-shaped stone pillars. The pillars are disposed in circles and probably represent the ancestors of the different clans, who used the site possibly for funerary and cultic purposes (Schmidt 2007).

[2]Besides bucrania and geometrical designs, the site provided several figurative paintings that are considered one of the most complete data set for the Neolithic religion in the Levant (Hodder 2010).

2. The Calcolithic: Tuleilat al-Ghassul, the Temple and the Star

The site of Tuleilat al-Ghassul is located in Jordan, northeast from the Dead Sea, near the modern city of Amman (northwest from Mount Nebo). Between the end of the 5th millennium and the beginning of the 4th millennium BC, Ghassul appears as a huge settlement of 20-30 ha, with clear attestation of agriculture and permanent domestic architecture. In the Late Chalcolithic Period, Tuleilat al-Ghassul was such a major settlement in the whole Southern Levant that it is now usual to call that period Ghassulian. The excavation of Pontifical Biblical Institute and the following archeological mission to the site, operated in the 1970s by the University of Sidney, brought to light impressive domestic quarters in areas 1 and 3, with houses of different dimension, all centered on a main court with many rooms around it (Mallon et al. 1937, 1940; Hennessy 1977). Some of these houses have particular features, like paintings along all internal walls; most of them also presented many layers of painting, twenty layers one over another, that, as Cameron (1981) had noticed in her work, could indicate a periodical re-drawing of the pictures on particular religious occasions such as festivities. In particular, the well preserved "Star" painting shows in the center a big star with eight rays, four in black and four in red. One ray is marked in a specific way: the circular center of the star has two other white stars, one inside the other (Fig. 1). These smaller stars are inscribed

Figure 1. The Tuleilat al-Ghassul Star Painting.

inside two circles, the external one marked by sixteen red and white triangles, pointing toward the center of the figure. This complicated geometrical representation is easily recognizable as an astral symbol of a god, most probably related to the Sun (Polcaro 2013).

With the discovery in the 1970s of a huge, sacred *themenos* in area E (Seaton 2008), the presence of an important clergy, as a dominant elite of a stratified society and connected with a centralized economy (based on agriculture, as in the following Early Bronze societies), at the site of Tuleilat al-Ghassul during the Late Chalcolithic Period, became clear. This area, later investigated by Stephen J. Burke in 1994–1997, who also excavated two buildings (Temple A and building B), clearly referred to an important shrine (Bourke 2001). In particular, Temple A also presents an external altar in an open area of the *themenos*, connected with the temple main door through a paved road. This important sacred area contains a major temple, a second structure, and the *themenos*. It was in use for one thousand years at least, indicates the maintenance of a cult, was the center of spiritual life of the settlement, and was connected with agricultural activities; this is testified by an adjacent bulk grain storage area.[3] This fact could suggest the presence of a clergy dominating an agricultural stratified society, accumulating the excess food, and organizing the public works, like in the later urbanized societies of the Bronze Age. The main Temple A, the "paved avenue" leading from it to the altar, and the long wall of the *themenos* are more or less parallel and roughly oriented to the east, as well as the long side of the "Sanctuary B," whose entry door faces south. This orientation strongly suggests that the cults celebrated in this area were connected to the rise and culmination of the Sun. In particular, the paved avenue is directed exactly where, at the equinoxes, the Sun appears over Mount Nebo. Furthermore, we notice that the wall surrounding the altar is a semicircle. However, its axis does not coincide with the one of the "paved avenue," instead being aligned with the wall of "Sanctuary B," and with a small structure on the *themenos* wall. This orientation, most probably, is not random.

At the 2009 SEAC Conference in Alexandria, we suggested a connection between the Star painting and the plan of the sanctuary (Polcaro & Polcaro, in press). Overlap the plan of the altar of Temple A on the "Star," put the altar in the centre, and the beginning of the semicircle is in correspondence to the marked ray of the "Star." The end of the larger part of the semicircular wall fits the half red, half pink triangle, the last stone of the semicircle perfectly fits the first pink triangle, and the "paved avenue" one of the red rays. Since we know that the paved avenue is directed east, the marked point thus represents north, and the building painted near the star is Sanctuary B. Orienting the star following our usual rules (north on top, east on right), we could suggest that the first pink triangle denotes the winter solstice period, corresponding on the altar plan to the last southwest stone of the semicircular wall. This gives, with good approximation, the direction of the winter solstice sunset as seen from a point in front of the altar. We could also suggest that the arc denoted by the pink triangles corresponds to the maximum rain period (a period of about 3.5 months, from mid-December to the beginning of March). If this hypothesis is correct, we could imagine that the 16 triangles each correspond to a subdivision of the year, e.g., a 360 day year could be divided into 15 "months" of 24 days plus a shorter month (which may be around the winter solstice), to maintain such a calendar in phase with the rain seasons (Fig. 2). The altar of Temple A could have been the device used to perform this task.[4]

---

[3]See Seaton 2008 for more on the chronology, architecture and finding of Sanctuary A of Tuleilat al-Ghassul.

[4]Recently a comprehensive reassessment study on the Tuleilat al Ghassul painting has been performed, showing several different interpretations of these ancient artworks; for the Star Painting in particular, these interpretations are not always connected to astronomic and calendric meanings (see Drabsch 2015).

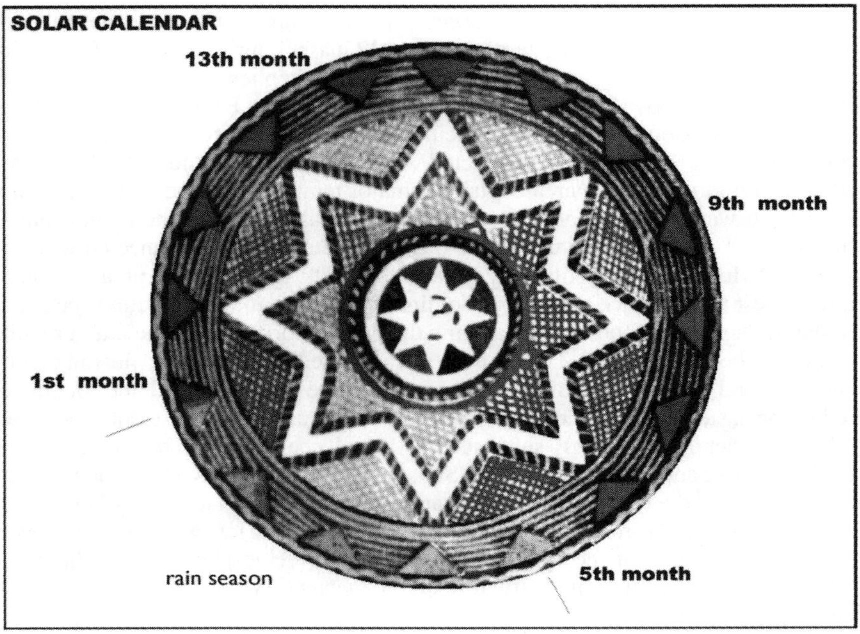

Figure 2. The Hypothesis of the Solar Calendar.

3. The Early Bronze: The HH Temple at Ebla

In the context of the cultures developed since the fourth Millennium in the Near East, the Sumerian in Mesopotamia shows for the first time, through the many oracular texts, the sure attention of a priestly class devoted to the observation of the sky. However, oracles were not the only reason for observing celestial phenomena. Mesopotamian texts show, in fact, particular associations between the erection of temples and the care in choosing the directions of sacred structures. Actually, the temple has always been considered the home of the god, who, as he lived in heaven so dwelt on the Earth through the sacred icon, the divine statue, which represented him within the sancta sanctorum, the cell.

In Syria from the 4th to the 2nd millennium BC, the influence of Mesopotamian culture and the strong veneration for the three main celestial divinities (Shamash the Sun, Sin the Moon, and Ishtar Venus)[5] certainly brought the priestly classes in the North Levant to observe the sky carefully. Considering the textual data of Syria in the 3rd millennium, particularly those of Tell Mardikh/Ebla,[6] there is a strong attention given to the Sun on certain occasions of worship. The clearest reference to this is

[5]For the importance of the goddess Ishtar and the observation of planet Venus in the Sky for the Old-Syrian Ebla of the 2nd millennium BC, see Polcaro (2014a).

[6]The site of Tell Mardikh, ancient Ebla, was discovered in Syria by Paolo Matthiae in 1964 and excavated under his direction by the Sapienza University of Rome for fifty years. In 1975 Ebla became famous for

surely the "Ritual of Regality" (Fronzaroli 1993; Matthiae 2008). The text describes a complicated ritual done during the renewal of royalty; a feast connected to the Eblaite kingship. This ritual is similar to the Egyptian Sed Feast, where the king and queen of Ebla make a pilgrimage to important sanctuaries and mausoleums of the region, bringing gifts to ancestors. The ceremony starts and ends at Ebla. The interesting point for our analysis is the particular attention given, at the beginning of each part of the ritual, to the rising of the Sun, linked to specific places (Polcaro 2014b). In particular, one could assume that one of these places was Temple HH, also called the Temple of the Rock.[7] This building, dated to the Early Bronze Age IVA (around 2500-2400 BCE), was an impressive structure whose cell is built directly on the rock, with 6-meter thick perimeter walls reaching nearly 13 m high (Fig. 3). On the opposite side of the entrance in the western wall of the cell, a natural circular cavity in the rock is present, from which an underground water source probably flowed into the center of the cell (Fig. 4). The entrance between the anteroom and sancta sanctorum looks like a narrow passage about 5 m long, only 1.40 m wide, and had to be at least 3.5 m high. While

Figure 3. The HH Temple of Tell Mardikh/Ebla.

the cavity suggests that there could be a pure water surface working as a mirror on the western side of the cell, the narrow and long door of the sancta sanctorum suggests the

---

the discovery of a large archive of texts in the Royal Palace G, with thousand of cuneiform tablets dated to the 24th century BCE (Matthiae 2010).

[7]Temple HH was discovered in the southeastern sector of the lower city and excavated between 2004 and 2007 (Matthiae 2006, 2007, 2008, 2009).

will to let sunlight through the passage during a particular time of the day for a given, certainly short, period of time.

The Temple of the Rock is very close to the embankment built in the Middle Bronze I, probably following the traces of the oldest Early Bronze walls, just north of the Steppe Gate (also dated to the Middle Bronze). The orientation, 110° ± 1°, of the narrow door of the Temple of the Rock is, however, not in the direction of the gate (which opened more south), but the walls. This wall, in the Early Bronze IV, had to have a thickness of more than 6 m and a height at least equal to that of the great Temple of the Rock. This means that the Sun, before propagating its rays toward the shrine, had to overcome the height of the city walls. Towards the middle of the 3rd millennium BC, the Sun's rays only penetrated the narrow corridor that led to sancta sanctorum between March 25 and July 7 and between August 15 and October 28. The maximum duration of direct lighting in the cell was about 40 minutes and occurred between April 4 and May 22 and between August 31 and October 18. Given the size of the cell, the narrower arc of time in which the Sun's light could reflect on the surface of the water probably present in the sancta sanctorum, was between May 3 and June 4 and between August 20 and September 21. In addition, the days around the spring equinox (on April 11 in this epoch) were those in which the direct light of the sun entered the Temple nearest dawn. During the rest of the year, from November to February, and in the short period between the beginning of July and mid-August, direct sunlight never entered the temple door, leaving the cell in the dark during the day (Polcaro 2014b).

Assuming that the ritual of regality described in Ebla's texts had the Temple of the Rock as a starting point, it would probably have to start when cell lighting conditions occurred, i.e., in a time span between April and June or between September and October, and in particular, shortly after dawn. The ritual moment is often quoted in the text between April and June. If we now accept the analogy between this important North-Syrian regal ritual and the Egyptian Sed Feast, it must be noted that it coincided with the beginning of the New Year. In Egypt, the New Year began with the first known flooding of the Nile, coinciding with the helical rising of the star Sirius. However, in the ancient Near East, according to the widely accepted view of all the scholars, the New Year began in Spring. Thus, it is likely that the months of Gi and itiHa-li-NI mentioned in the texts of the rituals of the Eblaite kingship are to be placed in spring.

The archaeoastronomical analysis of the orientation of the Temple of the Rock of Ebla thus testifies not only the use of astronomical orientation of the temple for calendar purpose, but also as the main ideological connection between the rituals and sky observation. The sunlight's effect inside the temple cells should be connected with the presence of the sacred water pool in the room, giving to ancient Eblaite priests the perception of the holiness of the sacred building.

4. Middle and Late Bronze Age: Pella

The site of Pella, located near the modern Arabic village of Tabaqat Fahl, is mentioned in Egyptian records since the 19th century BC and was named Pella in the Hellenistic Period. Pella is famous for its Roman remains, but in the lower acropolis there are remains of a large sacred building from the 19th century BC. This Middle Bronze Age temple was continuously used and restored until the Iron Age. This sacred area of the Bronze and Iron Ages (Area XXXII) shows impressive continuity as a centre of cultic life of the settlement. The first chronological phase of the main large (32×24 m) "fortress temple" is related to the Middle Bronze Age (1800–1450 BC). It is a typical "temple in antis" with a single wide entrance on the eastern side. Its walls were made by a two-row stone foundation and probable mud-brick elevation (Bourke 2003, 2004). The cella is a simple rectangular space with the door on the shorter eastern side, a paved mud brick floor, and no dividing walls. No ritual objects of any kind were found inside the temple. The massive rectangular shape of the building, with two buttresses flanking the entrance, fits well in the so called "emptybox" temple form found for the Middle Bronze Age in several sites on the western side of the Jordan Valley. Various architectural interventions were made in the Late Bronze Age I (1450–1350 BC). However, the external rectangular shape of the building, and in particular the location and the orientation of the main entrance, were not modified. First, the inner space of the cella was divided in two by a mud brick wall that was oriented north-south, thus obtaining a smaller room on the western side of the temple. Second, the floor inside the new space of the sancta sanctorum was completely redone. The ancient floor was removed to a

Figure 4. The cell of the Temple of the Rock.

depth of 1.5 m, then refilled with medium sized stones, and lastly sealed with a lime plaster floor surface of a yellow-white colour. Third, two large 5×5 m square towers were built on stone buttresses that flanked the Middle Bronze Age entrance of the temple (Bourke 2004). At the end of this period, a strong earthquake probably destroyed the building. It was then rebuilt in the Late Bronze Age II (1350–1150 BC). In this second phase, the temple was reduced in dimension, and the northern wall was moved 5 m to the south. All the walls of the building were rebuilt with a stone foundation and mud brick elevation reducing their dimensions and narrowing the entire structure. The floor of the sancta sanctorum was also re-laid, and two small basalt column bases flanked its entrance, which was narrowed and re-centered to the south. The main entrance of the cella to the east was also provided with a new central colonnade, revealed by three column bases. The wooden remains found near the basaltic stone bases indicate wooden columns, of which the central one was the largest, sustaining the main weight of the structure. The Sidney University archaeologists also hypothesize that in this second phase, the two external square towers of the first temple phase collapsed and were no longer reconstructed (Bourke 2004). Thus, in this second phase, the general shape of the temple, the direction of the main entrance, and the presence of the sancta sanctorum did not change; the architectural innovations were just the reduction of the dimensions and the presence of the two colonnades.

In December 2011, field measurements were made of the main axis of the Middle Bronze Age temple, which had remained the same until the later phase of the Late Bronze (Polcaro, González-García, & Belmonte 2013). The azimuth was 68.75°, and the elevation was 8.5°. This corresponded to a declination of 22.5° ± 0.75°. The orientation towards Jordan valley was 248.75°, and the elevation was −0.5°. However, after inspecting the archaeological data (mainly the consistency in the use of the same orientation pattern throughout the history of the temple), this direction was discarded as a main target, and we assumed that the temple axis was pointing towards the northeast with a declination of 22.5° ± 0.75°. This is consistent with the direction of the sunrise in the summer solstice during the second millennium BC, when the temple was erected (Fig. 5).

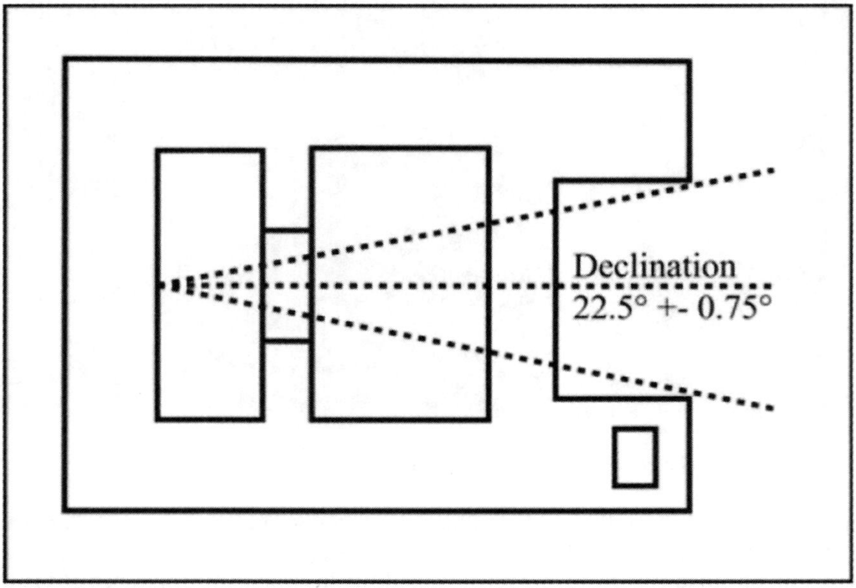

Figure 5. Schematic plan and orientation of the Pella Temple.

The remains found in the temple point towards Baal as the possible deity worshiped there (Polcaro, González-García & Belmonte 2013). Moreover, the constant presence of Egyptian gifts in the Pella temple, since the first construction of the main temple in the Middle Bronze Age, indicates a strong respect for the temple by foreign merchants visiting the settlement. This could be a sign that the deity worshipped in the temple was, since the first installation of his cult in the MB settlement, assimilated as an Egyptian god. In the famous myth of Ugarit, the god Baal was also defeated by Mot, the god of death. He could save himself and be reborn only with the help of his sister Anat and the Sun god Shapash. This huge temple of the Bronze Age was dedicated to a deity that dies and is reborn cyclically and was thus assimilated as an Egyptian god like Osiris. This deity was identified as El and Baal, who were the Lords of the Land in the Southern and Northern Levantine religious traditions. The orientation of the building is consistent with summer solstice sunrise, a moment that, in Near Eastern cultures, was linked with the festivity of dying gods associated with harvesting. It was also a moment of confluence of the underworld and the upper world (Polcaro, González-García & Belmonte 2013). Looking to the calendars known in the Near East during the Bronze Age, the summer solstice festivals occurred in the 4th month, between June and July. This month had associations in Mesopotamia (and in Emar and Mari in the Northern Levant) to cultic festivities dedicated to the dead, ancestors, and dying gods like Dumuzi or Baal (Cohen 1993, p. 454).

After the summer solstice, the longest day of the year, the Sun starts to decrease its presence during the day until the shortest day of the year, the winter solstice. Each day at sunset, the Sun god enters the Netherworld and rules from the particular to the universal. After the summer solstice, he starts to pass a longer and longer time in the Underworld than in the Heavens. Thus, the summer solstice was perceived in the mentality of ancient Near Eastern peoples as the starting point of the confluence period between the worlds above and below; a time when it was possible to communicate with the dead, and when it was necessary to perform rituals in order to protect the living from the dangerous spirits of the Netherworld. In this way, the orientation of the temple of the dying god Baal in Pella towards the summer solstice sunrise could be explained as having a calendrical purpose: identifying the moment of the year when important offerings to the dying god must be dedicated to him, and important festivities linked to his cult must be performed in the temple, i.e., as a way of performing the rites at the correct time looking in the correct direction.

5. The Iron Age: Ain Dara

During the Iron Age the ethnic components of the Levantine population drastically changed. In the Iron Age I the Sea People arrived, while in the Iron Age II-III Assyrians, Babylonians, and finally Persians conquered the Levant. Regarding the Transjordanian kingdoms of Edom, Ammon, and Moab in the Southern Levant, too few archaeoastronomical studies have been done. However, it is possible to state that, at least in the Northern Levantine Aramean culture, astronomical orientation continues to be present in the main temples of the settlements. A case worth mentioning is the temple of Ain Dara (Fig. 6), excavated by Ali Abu Assaf (1990). It is a typical tripartite temple in antis that follows the prevailing sacred architectonic tradition of Bronze Age Syria; the temple, whose associated deity is uncertain, was completely decorated with carved reliefs showing demons and divinities (Akkerman & Schwartz 2003, p. 372).

The main door of the temple (azimuth 140) seems to be roughly oriented in the direction of sunrise around the winter solstice. An interesting feature of the temple is the presence of two large footprints, represented standing in front of the temple entrance, carved on its stone threshold. Similar footprints are also carved on the other two stone thresholds of the antecella and cella doors, as if representing the strides of a

Figure 6. The Temple of Ain Dara, Syria.

giant approaching the sacred statue of the god located on the central back wall of the cella, and on the axis with the entrance; this is clearly a symbol of the god entering his temple. There is also an interesting relationship between the period of illumination of the cella and the large altar of the temple, which is decorated by six slabs with mythological subjects. In particular, it seems that following sunrise during the winter solstice period the sun illuminated each of the six slabs of the altar (except for the sixth one) for more or less 1/12 of the day (Polcaro 2015). When the illumination of the second slab starts, the slab before still remains lit for about 30 min. The process ends at 10:34 a.m., when the fifth slab is no longer illuminated (Fig. 7). It seems that this architecture and conception of the cella illumination functioned as a kind of solar clock, testifying both to the great attention paid by the temple constructors to the sky, and to the maintenance of astronomical knowledge in the Iron Age.

6. The Classical Age: Petra

The Nabateans built several monuments in Petra and elsewhere displaying decoration with a preference for astronomical motifs, possibly as a reflection of their religion (Gawlikowski 1990; Tholbecq 1997). However, due to the lack of direct written accounts and the scarcity of inscriptions, we do not have a clear knowledge of the precise nature of such beliefs, and how they reflected on the calendar or the religious timekeeping system of this ancient society. A statistical analysis of the orientation of their sacred monuments demonstrates that astronomical orientations were often part of an elaborate plan and possibly a trace of the astral nature of Nabataean religion (Belmonte, González-García & Polcaro 2013). Petra and other monuments in the ancient Nabataean kingdom have proven to be marvelous laboratories of the interaction between landscape features and astronomical events, showing impressive hierophanies on particular monuments related to cultic times and worships. Among other findings, the famous Ad Deir has shown a fascinating ensemble of light and shadow effects, perhaps connected with the bulk of Nabataean mythology. However, from the impressive Urn tomb (Fig. 8) eminates a series of suggestive solstitial and equinoctial alignments, which might have helped in its selection as the cathedral of the city (Belmonte, González-García & Polcaro 2013).

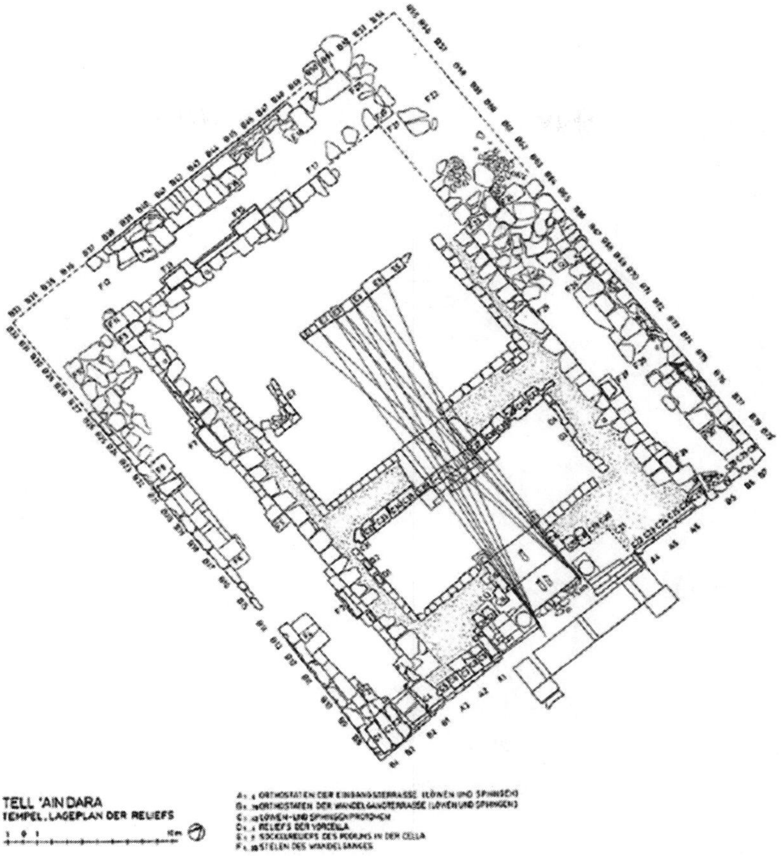

Figure 7. Plan and orientation of the Ain Dara Temple.

7. Conclusions

It is not surprising to find evidences of connection between religion and celestial phenomena in Levant, which has always been the bridge between Egypt and Mesopotamia. For a period of more than 3000 years from the proto-dynastic period, when we have the first evidences of a lunar calendar probably governed by the Nile regime (Kraus 2015), to the Ptolemaic and Roman periods, the Egyptian sky-watchers scrutinized the firmament in a serious attempt to find accurate answers to the previously stated questions. Due to the wandering nature of the Egyptian calendar, the seasons of the tropical year (and the associated climatic phenomena and agricultural activities) did not usually follow the seasons of the civil calendar. Hence, economic activities, although completely under the control of the civil calendar, would be carried out according to the tropical (thus Gregorian) seasons. This is probably the reason why these dates were so important as to be reflected in the orientation of the temples. Indeed, the most interesting example

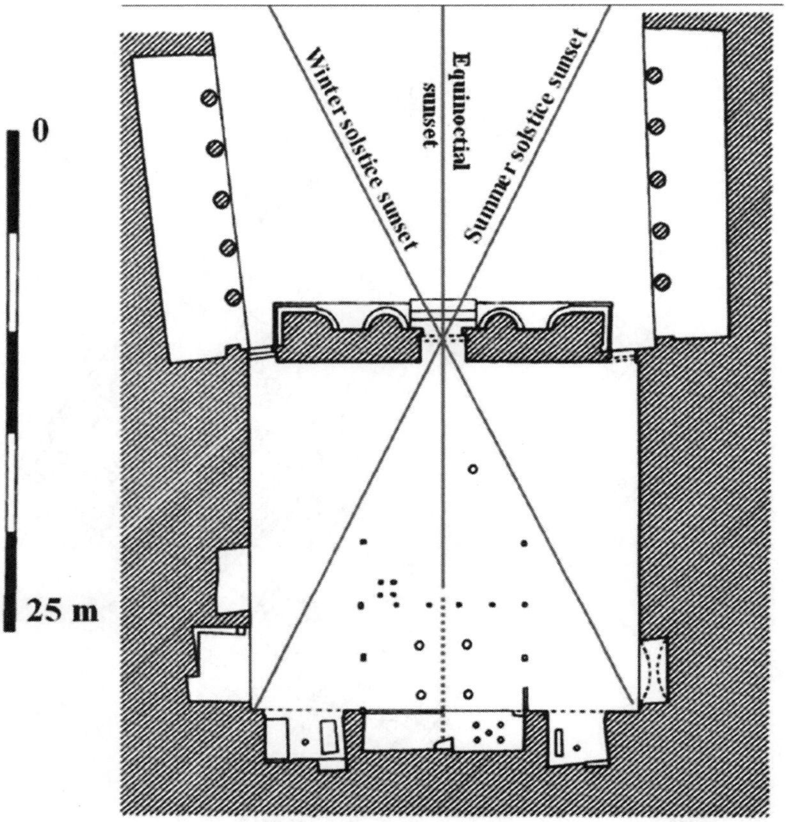

Figure 8. Plan and orientation of the Urn Tomb in Petra, Jordan.

is undoubtedly that represented by the phenomenon famed throughout the world: the illumination of the innermost sanctuary of the main temple of Ramses II at Abu Simbel (Belmonte & Shaltout 2009). Twice in the year, at dawn, the light of the rising sun enters the sancta sanctorum of the main temple of Abu Simbel. The first rays illuminate the figures of Amon-Re, the divinized king, and the shoulder of Re-Horakhty, all of them gods of solar character, while the figure of Ptah, god of the netherworld, stays in darkness. When the light progresses, the figure of Re-Horakhty receives first the full, and afterwards the last, rays of the sun. This marvelous hierophany may have occurred at the beginning of the Peret and Shemu seasons of the ancient Egyptian calendar, during the first decades of the reign of Ramses II, the temple builder.

During the same three millennia, astronomy has been an essential element of cultures that have succeeded in Mesopotamia (Verderame 2015). Sumerian culture in Lower Mesopotamia is one of the first cultures in the world to have a strong interest in sky observation and registration of astronomical phenomena. The strong ideological link between sky and hearth was also present in Mesopotamian religion and cosmology

(Horowitz 2015); the celestial vault also influenced the mind of Sumerians in urban planning. The only features covering part of the horizon were human constructions, especially when the urbanization process reached its mature phase and the Sumerian settlements became real cities with fortification walls and public buildings rising over the houses. The *ziqqurat* is the symbol of sacred Mesopotamian architecture in the Western thought. This monument, standardized at the end of the III millennium BC by the kings of the Third Dynasty of Ur, has changed in size and architecture during the history of Mesopotamia, but remained the highest structure in the city until the end of the Neo-Babylonian Period in the I millennium BC. The *ziqqurat* is the only monument visible over the settlement walls, with a strong visual impact around the urban and the countryside landscape. Despite its simple structural function, a high mud brick platform to sustain an upper temple, the *ziqqurat* appears in Mesopotamian art and literature as a structure of primary importance; a connection between the Earth, domain of the god Enlil, and with the Sky, domain of the god Anu. The ideological function, to connect the Earth and the sky, was also related to the rituals performed in the high temples built above these monuments; usually linked with important seasonal royal rituals.

Preliminary studies on the orientation of the *ziqqurat* of the Moon god Nanna at Ur and the Sun god Utu at Larsa have shown that both buildings are related to particular astronomical orientations that reflect the mythology of these deities (Nadali & Polcaro 2016). The lunar orientation of the *ziqqurat* of Ur can be explained through the mythology of the Sumerian Moon god, his importance at Ur in the period in which the high terrace was built, and finally, the numerous sacred festivities of the city, particularly at the end of the III millennium BC, that are significantly linked to the movements and changes of the Moon in the sky. The high terrace of the city of Larsa, connected to the Ebabbar (one of the most important temples dedicated to Utu, the Sumerian Sun god), shows an orientation to the summer solstice sunrise. In the Old-Babylonian calendars of Mesopotamia, the month of the summer solstice was the time of many festivities linked to the dead and the funerary ideology, but the summer solstice itself is not apparently quoted as a day of a particular solar festivity (Cohen 1993). Therefore, in this case also, we do not have a calendric orientation, but an astronomical one linked to the mythology of the god venerated in that temple. In fact, the day of the summer solstice could easily be considered as the moment of the triumph of Utu, when the daylight prevails over the darkness of the night, i.e., the moment of the maximum power of the Sun god, and his longest daily sky travel. Important aspects of any ancient or modern culture are intrinsically related to how to measure time, how to create a calendar (sacred or profane), how to appropriately orientate sacred structures according to religious regulations, and how to map the sky for proper guidance (either for eschatological reasons or for simpler, more prosaic needs).

All the examples that we have shown demonstrate how, during more than three millennia, these tasks have been performed in Levant, as well as Egypt and Mesopotamia, following an evident dual approach: a cognitive one (calendric function) and a sensory one (ideological/religious function). In fact, from one side, the temple orientations point to identifying a specific moment of the year, when the ritual must be performed. From the other side it is a concrete sign of the presence of the gods, inspiring the sense of the sacred; a tangible connection between the human beings and the house of the divinity, the celestial vault.

References

Akkerman, P. M. G. & Schwartz, G. M. 2003, The archeology of Syria, Cambridge University Press, Cambridge
Abu Assaf, A. 1990, Der Tempel von Ain Dara, Verlag Philipp von Zabern, Darmstadt
Belmonte, J. A., González-García, A. C., Polcaro, A. 2013, Light and Shadows over Petra: astronomy and landscape in Nabataean lands, Nexus Network Journal, 15, 487
Belmonte, J. A. & Shaltout, M. (editors) 2009, In Search of Cosmic Order: Selected Essays on Egyptian Archaeoastronomy, Cairo
Bourke, S. 2001, The Chalcolithic Period, in B. MacDonald et al. (editors), The Archaeology of Jordan, Sheffield, 107
Bourke, S. 2003, Preliminary Report on the University of Sidney's Eighteenth and Nineteenth Seasons of Excavations at Pella (Tabaqat Fahl) in 1996/1997, Annual of the Department of Antiquities of Jordan, 47, 335
Bourke, S. 2004, Cult and Archaeology at Pella in Jordan: excavating the Bronze and Iron Age Temple Precint (1994-2001), Journal and Proceedings of the Royal Society of New South Wales, 137, 1
Cameron, D. 1981, The Ghassulian Wall Paintings, London
Cohen, M.E. 1993, The Cultic Calendars of the Ancient Near East, Bethesda
Drabsch, B. 2015, The Mysterious Wall Paintings of Teleilat Ghassul, Jordan, Monographs of the Sydney University Teleilat Ghassul Project 3, Archaeopress, Oxford
Fronzaroli, P. 1993, Archivi Reali di Ebla, Testi XI. Testi rituali della regalità, Rome
Gawlikowski, M. 1990, Les Dieux des Nabatéens, Aufstieg und Niedergang der römischen Welt II, 18/4, 2659
Hennessy, J. B. 1977, Teleilat Ghassul: An Interim Report, Sydney
Hodder, I. 2010, Religion in the Emergence of Civilization. Catalhöyük as a case of study, Cambridge
Horowitz, W. 2015, Mesopotamian Cosmogony and Cosmology, in C.L.N. Ruggles (editor), Handbook of Archaeoastronomy and Ethnoastronomy, Springer, New York, 1823
Kraus, R. 2015, Astronomy and Chronology: Babylonia, Assyria and Egypt, in C.L.N. Ruggles (editor), Handbook of Archaeoastronomy and Ethnoastronomy, Springer, New York, 31
Mallon, A., Koeppel, R. and Neuville, R. 1934, Teleilat Ghassul I, Pontifical Biblical Institute, Rome
Mallon, A., Koeppel, R. and Neuville, R. 1940, Teleilat Ghassul II, Pontifical Biblical Institute, Rome
Matthiae, P. 2006, Un grand temple de l'époque des Archives dans l'Ebla protosyrienne: Fouilles à Tell Mardikh 2004-2005, CRAI 2006, 447
Matthiae, P. 2007, Nouvelles fouilles à Ebla en 2006: Le Temple du Rocher et ses successeurs protosyriens et paléosyriens, CRAI 2007, 485
Matthiae, P. 2008, Gli Archivi reali di Ebla, Milano
Matthiae, P. 2009 Temples et reines de l'Ébla protosyrienne: résultats des fouilles à Tell Mardikh en 2007 et 2008, CRAI 2009, 747
Matthiae, P. 2010, Ebla la città del trono, Torino.
Nadali, D. & Polcaro, A. 2016, The Sky from the High Terrace: Study on the Orientation of the Ziqqurat in Ancient Mesopotamia, Mediterranean Archaeology and Archaeometry, 16/4, 103
Polcaro, A. 2013, The Tuleilat al-Ghassul Star Painting: A Hypothesis Regarding a Solar Calendar from the Fourth Millennium BC, in L. Feliu et al. (editors), Time and History in the Ancient Near East, Proceedings of the 56th Rencontre Assyriologique Internationale at Barcelona, 26-30 July 2010, Barcelona, 273
Polcaro, A. 2014a, L'ottavo anno da quando Ishtar si manifestò ad Ebla. Venere e il suo simbolismo nella città paleo siriana, in V.F. Polcaro (editor), Atti del XII° Convegno della Società Italiana di Archeoastronomia, Albano Laziale 5-6 ottobre 2012, Napoli, 345
Polcaro, A. 2014b, Architettura templare e orientamenti astronomici: analisi della tipologia nord-siriana dell "Antentempel" nel Periodo Protosiriano, in S. Pizzimenti, L. Romano

(editors), Šime ummiānka Studi in onore di Paolo Matthiae in occasione del suo 75° compleanno offerti dall'ultima generazione di allievi, Contributi e Materiali di Archeologia Orientale XVI, 155

Polcaro, A. 2015, Astronomy in the Levant During the Bronze and the Iron Age, in C.L.N. Ruggles (editor), Handbook of Archaeoastronomy and Ethnoastronomy, Springer, New York, 1789

Polcaro V. F., Polcaro, A. in press, The First Star: An Archaeological and Astronomical Study on the 4th Millennium Ghassulian "Star and Notables" Paintings in Jordan, in M. Rappenglük (editor), Proceedings of the SEAC 17th Annual Meeting, 25-31 October 2009, Alexandria, Egypt, BAR International Series, Oxford

Polcaro, A., Belmonte, J. A., González-García, A. C. 2013, Study on the Orientation of the Bronze Age Temple of Pella, Jordan: the dying god Baal and the Rituals of the Summer Solstice, in I. Sprajc, P. Pehani (editors), Ancient Cosmologies and Modern Prophets, Proceeding of the International Conference of European Society for Astronomy in Culture, Ljubljana, 24-29 September 2012, Supplement of Anthropological Notebooks 19, Ljubljana, 481

Seaton, P. 2008, Chalcolithic Cult and Risk Management at Teleilat Ghassul, BAR International Series 1864, Oxford

Schmidt, K. 2007, Sie bauten die ersten Tempel, München

Tholbecq, L. 1997, Les Sanctuaires des Nabatéens: État de la question à la lumière de recherches archéologiques récentes, Topoi, 7, 1069

Verderame, L. 2015, Mesopotamian Celestial Divination, in C.L.N. Ruggles (editor), Handbook of Archaeoastronomy and Ethnoastronomy, Springer, New York, 1835

Astronomical Heritage of the Middle East
ASP Conference Series, Vol. 520
Sona V. Farmanyan, Areg M. Mickaelian, J. McKim Malville, and Mohammad Bagheri, eds.

A Pluralistic Vein in the Early Eighteenth Century in Europe*

Massimo Capaccioli
Department of Physics, University of Naples Federico II, Via Cinthia, Naples, Italy

Abstract. We review the development of the pluralistic vision in the Western culture and show the existence, in the Mitteleuropa of the 18th century, of an enclave of pluralism in the context of an infinite universe. In 1742, on the frontispiece of his *Atlas novus coelestis*, Johann Gabriel Doppelmayer presented a sort of Brunian "manifesto" in which the Solar System is surrounded by many other planetary systems. Two years later, Leonhard Euler published his *Theoria motuum planetarum et cometarum*, with a clone of the pluralistic iconography of the *Atlas*. In the years following 1751, a fresco was painted in one hall of the Castle of Prague, now the Mathematics Room, under the influence of the Jesuit Joseph Stepling, where the reference to Doppelmayer's and Euler's pictures is evident. While the pluralistic faith of Doppelmayer and Stepling is not ascertained, Euler was a firm pluralist and kept a regular correspondence with Stepling. These facts suggest that the three, while belonging to different cultural and religious environments, shared a common position in favor of the Brunian doctrine and exchanged information, opinions, and even cryptic iconic references to the cosmology of the Nolan philosopher.

1. Introduction

"*The fixed stars are Suns and each illuminates a World.*" This statement, introducing the fifth evening of the *Entretiens sur la pluralité des mondes* by Bernard Le Bovier de Fontenelle, a celebrated booklet published in 1686, although peremptory, rests on no direct evidence. It is rather an act of faith, the expression of an ancient belief that, with alternative fortunes, has been systematically re-proposed throughout the centuries.

The question of the plurality of worlds, that is of the multiplicity of planetary systems similar to ours, has always been associated with the possibility of intelligent life beyond Earth and with the role and future of humanity, in the context of a world model that is an ancient and open problem today. Pluralism—herewith used as a synonym for the plurality of worlds—has its Western roots in the philosophy of atomists. Today, it is again found at the center of a lively scientific debate involving astronomers, physicists, chemists, and biologists. Even the myth of UFOs as alien visitors from deep space keeps inspiring the collective imagination and nourishes science-fiction movie-making and literature, especially after the historic landing of the Apollo 11 mission and the discovery of thousands of extrasolar planets, arousing widespread expectations, hopes and

*For a more detailed discussion, please see the paper presented by M. Capaccioli & I. Cecere at the XXII National Meeting of History of Physics and Astronomy, Genoa and Chiavari, 6-8 June 2002, M. Leone, A. Paoletti & N. Robotti eds., Naples, IISF, pp. 411–425, 2004.

dread, desire for domination and fear of the different. The centuries-old dispute over the closed world and the infinite universe is far from resolved and awaits a scientifically founded answer. Looking back at the evolution of the pluralistic debate starting from the classical thinking, we can see that its revival in the century of the Enlightenment coincides with the appearance of most of the rare visual representations of the multiplicity of the worlds. The basic lack in the use of the iconographic tool has motivated us to look for other examples of pluralistic representations.

In 2001, Imma Cecere (Cecere 2001) conducted a survey of 145 books by 22 pluralist authors of the 18th and early 19th centuries. While confirming a substantial deficiency of this form of non-verbal direct message, this investigation disclosed an intriguing ideological connection that seems to link three large Mitteleuropa capitals of the 18th century culture, Nuremberg, Berlin, and Prague, and crossing the various sects of Christianity in Europe: Catholic, Lutheran, and Calvinist.

2. The Pluralistic Idea Over Time

The hypothesis that the universe is infinite and populated with "*innumerable and infinite globes*" has always generated enthusiastic consent, but also strong skepticism, harsh criticism, and even severe censure. The different positions were mostly dictated by philosophical and religious motivations, and strongly modulated by changes in the social and cultural context.

The earliest and greatest exponents of the pluralistic theory in the ancient world, Democritus, Leucippus, and later Epicurus, who found in Titus Lucretius Carus their chief interpreter in the Latin world, were fiercely opposed, among others, by Plato and Aristotle. The Greek atomists of the 5th century BC came to pluralism by deduction, arguing that, since atoms are infinite and infinite are their possible combinations, there must be an infinite number of worlds that are perpetually created and destroyed. Therefore, the universe has no limit anywhere. It is an empty space populated by an endless number of tiny indivisible particles, the atoms, which move in all directions. The Earth, neither unique, at rest, nor central, would be only one of the many celestial bodies formed by the random aggregation of primitive particles.

The idea of the existence of a multitude of worlds, inaccessible to the senses but accessible to reason, did not meet the favor of Plato. In the *Timaeus*, he argued that his refusal was on the grounds of two precise requests: the uniqueness of the Creator, which implies the uniqueness of creation, and the fact that creation, as composed, should tend towards destruction and decay.

Aristotle, on the other hand, expressed his aversion to pluralism several times; especially in the *De caelo*, where he appealed to the doctrine of the elements tending to their natural places. His cosmology identified the Earth with the physical center of a hierarchy of celestial spheres, those of the Moon and of the planets up to the fixed stars. According to one of the basic principles of his cosmology, everything in the cosmos moves in relation to its center: the earthly element tends spontaneously to the Earth, fire moves away from it, the air and the water find intermediate natural positions. The belief that there could be no more than a single universe arose from the fact that, if it were not so, earth and fire would have more than one natural place to go, in contradiction with physics and logic itself. Finally, in the *Metaphysics*, he condemned the doctrine of the infinite worlds as irreconcilable with the religious and philosophical beliefs of the time.

With the advent and growth of Christianity, given the radical rejection of the idea of a multiplicity of worlds by almost all scholars and theologians, pluralism seemed fated to be extinguished without leaving a trace. Located among the heretical strands of Western culture, it was censured in the same way as any other cosmology alternative to the one endorsed by the Church. In fact, despite the strenuous struggle conducted by Christian scholars since late antiquity (e.g., the Bishops Eusebius of Caesarea in 4th century AC, Theodore of Cyprus in the 5th century, and Augustine of Hippo in the 4th–5th centuries), the debate about the infinity of the universe and the multiplicity of worlds was far from irrelevant. Very often, it managed to evade censorship and vetoes, flowing like an underground river and collecting the consensus of a large group of supporters. In the 13th century, Albertus Magnus described the pluralistic debate as "*one of the most beautiful and noble questions*," adding that, "*if there is only one world or more worlds, it would be desirable for us to investigate it*" (Crowe 1999). Similar thoughts also emerged from Michele Scotus in Spain, William of Auvergne in France, and Roger Bacon in England.

Thomas Aquinas, pupil of Albertus Magnus and Doctor of the Church, was among the greatest detractors of pluralism. To endorse the Christian doctrine of the uniqueness of the world and the finiteness of the universe, he appealed to the passage of the Book of John (1.10) where the Evangelist states that "*the world was created by Him [God].*" This was enough for him to maintain that there is only one world.

At the end of the 13th century, the Bishop of Paris, Etienne Tempier, banned 219 Aristotelian propositions guilty of limiting divine power, among which the thirty-fourth states that "*the First Cause cannot create many worlds.*" It was a new opening towards pluralism that triggered a crisis of the Greek-Christian finite cosmology and prompted some scholars to conduct a critical review of Aristotle's anti-pluralistic arguments (including Jean Buridan, Rector of the University of Paris, William of Ockham, a Franciscan professor in Oxford, and Nicole Oresme, tutor of the future Charles V of France, who authored a commentary to Aristotle's *De Caelo*). Oresme affirmed that no philosophical or scientific doctrine could exclude the possibility of a chain of worlds in time or of innumerable worlds in space. He also refuted the Aristotelian doctrine of motion and natural places, and argued that the restriction of the power of God to the creation of just one universe is equivalent to denying his omnipotence. Nonetheless, he later rejected the pluralist doctrines stating that "*God can and could in his omnipotence create another world besides this, different or similar to it. Not even Aristotle or others are able to prove completely the opposite, but naturally, there has never been nor will there be any other one real world*" (Crowe 1999). Likely, he changed his mind out of fear of triggering a reaction from the ecclesiastical authorities; but he was also motivated by the impossibility, or perhaps the incapacity, of reconciling pluralism with numerous passages of the Holy Books in which the geocentric assumption seems evident, and with one of the main historical beliefs of Christianity, the Atonement.

Later, Nicholas of Cusa sustained the impossibility of conceiving boundaries to space. He imagined the Sun and the Moon populated with inhabitants, and evaluated the idea that life was widespread among stars, but he did not attempt to reconcile the theory of pluralism with the Christian doctrines of the Divine Incarnation and Redemption. According to recent studies, it appears that Cusanus did not really support the infinity of the universe. More than infinite, an attribute that he recognizes only to God, his universe is interminatum. Therefore, "*his world is no longer the medieval cosmos but it is not yet at all the infinite universe of the moderns*" (Koiré 1961).

In the year 1473, the didactic poem *De rerum natura* of Lucretius was published. It had been rediscovered a few decades before. Just 70 years later, Nicholas Copernicus published his major work, *De revolutionibus orbium coelestium*, which, launching a formidable attack on the Aristotelian-Ptolemaic tradition, determined the turning point of scientific thinking and decreed the certificated birth of a new age. In particular, the Copernican lesson of not taking for granted that we occupy a privileged central position in the universe was fraught with consequences, and found in Bruno one of the most brilliant interpreters.

The pluralistic faith tenaciously survived the aversions and interference of the Church, and re-emerged vigorously at the end of the 16th century through the extraordinary intuitions of Giordano Bruno. The Nolan was a passionate pluralist—as documented, in particular, by *La cena delle ceneri* and the *De immenso et innumerabilibus seu de universe et mundo*—and a persuaded supporter of life in the universe. His original contribution placed the heliocentric theory in a perspective of the infinite: the Sun is at the center of an immutable system of fixed stars, and space is infinite, devoid of center, and populated with innumerable worlds. Bruno therefore represents the philosopher of the infinity, of a non-hierarchical universe incompatible with the sky stigmatized by tradition and only partly bound to the Copernican cosmology. He also supports the absolute identity of Heaven and Earth, colliding with Aristotelian dualism. It was precisely this unity, besides the infinity of the universe, which shattered the spheres of Aristotle. His vision of the cosmos, removing more radically than Copernicus the human illusion of being at the center of something, did not threaten only the *ipse dixit*, but also the word of God deposited in the Holy Books. It struck against some historical beliefs of Christianity, in particular against the dogma of all dogmas, the Incarnation and the Redemption. For this, he was sent to the stake with his mouth tightened by a scold's bridle: tragic symbol of eternal violence on freedom of thinking. The extreme punishment inflicted by the Roman Inquisition on the "magician" Bruno was to be exemplary and deeply persuasive, to discourage any form of proselytism. However, the road to pluralism had been definitively established.

The greatest thinkers of the early modern age, Tycho Brahe, Johannes Kepler, and Galileo Galilei, coldly accepted or largely refused the theories of Bruno. Kepler tried to defend his universe from the advance of the new ideas of infinity, believing the Solar System to be something unique in reality and providentially created for man and his needs. He did not oppose, though, the possibility of life on other planets. In 1609, he wrote of an imaginary journey to the Moon, a *Somnium seu opus posthumum de astronomia lunari*, advancing the possibility that the star might be inhabited. This book marked the passage from the fantastic literature on the Moon, inspired by Lucian of Samosata and Ludovico Ariosto, to a fantastic-scientific literature.

Nevertheless, the Epicurean philosophy and the pluralistic question revived in the 1600s, with Pascal, Fontenelle, Huygens, Descartes, and, above all, Isaac Newton, until finally exploding in the following century. Newton, in particular, succeeded in bringing to the debate a more rigorous scientific approach and providing a rigorous and unitary synthesis of the cosmos. The deep meaning of his cosmology lies in the fact that there are no longer a terrestrial and a celestial world, obeying two different physical laws. The universe becomes one, and science asserts its right to study it according to its own rules, without granting any preferential situation.

During the 18th century, the debate on the infinity of the universe and on the plurality of the worlds was enriched by numerous speculative contributions by a large group of convinced supporters, including numerous leading figures in science and philosophy. It was also illustrated by images that, as we will show, tried to visually translate the idea of an infinite space populated with planetary systems similar to ours.

3. Origin and Development of Pluralistic Iconography

The evolution of the pluralistic debate is still at the center of the studies of numerous historians and scientists (cf. Crowe 1999 and references therein). The news concerning the visual representations of this heretical conception of the cosmos is far rarer and less known. The first images of an infinitely populated cosmos of stars can be traced to the period between the 16th and the 17th centuries, to the works of Thomas Digges, William Gilbert, and John Wilkins. In 1576, Digges, a leading exponent of English followers of Copernicus, added to his paternal work, *A Prognostication Everlasting*, an appendix entitled *A perfect description of the caelestial orbes*. There he reported some excerpts from the *De revolutionibus* and, above all, some considerations in favor of pluralistic doctrine. He wrote that one could only imagine what a miserable portion of the divine plan is our corruptible world, since our senses prevent us from perceiving all the rest, owing to the enormous distances. He illustrated his text with a well-known cosmographic diagram in which the Solar System is surrounded by a large number of stars. A little less than a century later, John Wilkins published one of the most important books of 17th century popular science, *The discovery of a new world* (1638), while William Gilbert completed his *De mundo nostro sublunari philosophia nova*, published posthumously in 1651. Both include representations of the universe very close to Digges'.

The three engravings have a common feature: the Solar System is placed at the center of the sky spotted with stars, where the stars are not recognized as planetary systems. The idea of the absolute centrality of our stellar system with respect to the other possible worlds is evident. The stars surrounding it, while at different distances and with different magnitudes and sizes, are still the fixed stars of the Aristotelian cosmology. What emerges is a strongly anthropocentric conception, a sort of compromise solution in which the infinite universe coexists with some legacies of tradition. The innumerable stars are not yet other suns with their planets. The idea that our world can be just one in an infinite multitude, deprived of any privileged position, seems quite far from the Brunian cosmos of countless inhabited worlds. This is no surprise, after all; these were difficult years of transition to a new cosmology freed from the traditional constraints.

Among modern age pluralist authors, Fontenelle was possibly the first to create a new representation of the infinite universe. A picture in his *Entretiens* displays numerous planetary systems surrounding the Solar System, which, while still sitting at the center, is no longer alone. In a splendid London edition,[1] a sort of theatrical curtain creates, through a typically baroque expedient, an illusionistic 17th century dome. Here, in an upward vortex, there are many luminous stars surrounded by planetary orbits, as we will see in the case of Doppelmayer's and Euler's frontispieces to be discussed later.

In an engraved panel of his *Theologie astronomique ou demonstration de l'existence et des attributs de Dieu*, William Derham used the pluralistic iconography already ex-

---

[1]London, aux Lépens de Paul & Isaac Vaillant, Marchands libraires, 1714.

perimented by Fontenelle, but he formulated a comparison between: i) the Aristotelian-Ptolemaic cosmological system, which he rejects, ii) the Copernican cosmological system, accepted but only as a precursor of the new system, and iii) the new system that, starting from the heliocentric universe of Copernicus, develops into a pluralistic model. Finally, in 1742 Thomas Wright published the *Clavis coelestis*. His table XXXII shows stellar shells concentrated around what he calls the Eye of Divine Providence. The picture is historically significant because it explains the content of the IX letter of the *Clavis*: the belief that there are innumerable worlds in infinite universes.

These well-known illustrations are not the only ones found in the scrutiny of the texts of the Enlightenment pluralist authors. Actually, our same survey was motivated by the discovery of an eloquent representation of the infinite worlds on the frontispiece, again well-known but never read in this perspective, of a celestial atlas by Johann Gabriel Doppelmayer. The same iconographic scheme has also been traced in a contemporary work by Euler and in a fresco at the Castle of Prague, testifying to the existence of a pluralistic axis among three great capitals of the 18th century culture, crossing the different souls of Christendom in Europe.

4. A Pluralistic Enclave in the Heart of Europe?

Published in Nuremberg in 1742 by Doppelmayer,[2] the *Atlas coelestis* is a figurate cosmological work accompanied by numerous quantitative diagrams. It includes the planetary systems of Copernicus, Brahe, and Riccioli, the elliptic theories of Kepler, Ward, and Mercator, the lunar theories and those about Halley's comet. A series of ten stellar maps, divided into three groups, illustrate the constellations: one pair of hemispheres centered on the ecliptic poles and another on the equatorial poles, and six maps using gnomonic projections to reproduce the entire celestial vault. The mythological figures corresponding to the constellations are dressed in typically classical costumes.

The *Atlas* is a synthesis of mythology, art, and science. The requirement of a new scientific rigor, characteristic of the Enlightenment, is associated with the taste for the symbolic and artistic representation of the celestial sphere, giving rise to a typographical product of rare beauty. The 30 water-colored tables, engraved by the German painter Johann Justin Preisler, swarm with mythological and allegorical figures ascribed to constellations, planets, and seasons. Child's fleshy figures and cherubs moving impudently amidst several scientific instruments, ornamental scrolls, and planetary orbits, mixed with the heroes of astronomical discovery. The earliest European observatories, such as those of Uraniborg and Paris, are also portrayed in various tables. All is accomplished with an extreme technical skill and inserted in a context still pervaded by frivolous rococo reminiscences.

Doppelmayer's masterwork belongs to the modern cartographic tradition, mostly oriented to a pure description of the celestial sphere and its constellations throughout splendid mythological and allegorical figures (Warner 1979). At the same time, it represents a remarkable encyclopedic "summa" of coeval astronomy and cosmology, motivated by the Enlightenment ideal of a knowledge devoted to the public enjoyment. It is thus perfectly in line with the 17th century cultural parameters which found the

[2]Born in Nuremberg in 1671, Johann Gabriel taught mathematics at the Aegidium Gymnasium of his hometown. Besides teaching, he devoted his life to astronomical and meteorological observations and to physical experiments until he died in 1750 (Wilson 1981).

highest expression and the most effective vehicle of dissemination in the *Encyclopédie* of Diderot and D'Alambert.

The *Atlas* has been analyzed by various scholars of celestial cartography. Two of them, Deborah Jean Warner (Warner 1979) and Patricia Burnham (Burnham 1985), state that, while interesting, it adds nothing significant to the science and to the coeval scientific debate. A careful iconographic and iconological reading of the figured frontispiece (Figure 1) suggests that this might not be the case. The Nuremburg professor appears more like a scholar, with deep interest in the main drives of the contemporary scientific and philosophical debates; and his *Atlas* appears to be an effective vehicle for spreading the coeval scientific thought.

Let us look at the frontispiece in detail (see Fig. 1). The ample hemicycle, opening at the center of the composition and developing in depth into a terrace, secures a unitary system to the different figures of the scene. Among them there are, in the forefront, some of the greatest builders of world systems. Their identification is made simpler by labels, which remove any doubt on their identities. At the left hand, Ptolemy, the Alexandrian astronomer who fathered the longest-lived cosmological tradition, carries a sign with the inscription: *Posteritas bono omnia peracta sunt* ("All things are carried out in the right way by posterity"). Next to him, Copernicus is grasping Ptolemy's left arm, inviting him to look upwards to the Sun, center of the universe. Ptolemy indicates the Earth with his left hand, while holding with the right hand a model of his geocentric system. In turn, Copernicus' right hand indicates the new sky, theorized in the *De revolutionibus*. This representation summarizes the key concepts in the evolution of the cosmological thought, from antiquity to modern age, in the two levels (up and down).[3] It emerges that the two visions are different as well as complementary: the Copernican revolution, in fact, did not improve the methods of astronomy nor did it provide the discovery of new observables, but founded a new system of the world on the ground of the secular Aristotelian-Ptolemaic tradition. The inscription associated to Copernicus, taken from Seneca's *Naturales Questiones*, reads: *Digna res est contemplatione ut sciamus in quo rerum statu simus pigerrimam sortiti an velocissimam sedem circa nos deus omnia an nos agat* (Sen. Nat. Ques. lib. VII cap. II) ("It is a subject worth careful study, in order to know in which physical placing we are, if we have been assigned the laziest home or the fastest one, if God pushes all things around us or He rather pushes us").

On the other side of the hemicycle, Kepler and Brahe stand out. The two astronomers, who gave fundamental contributions to destroy some of the dogmas of the traditional cosmology, are associated with emblematic inscriptions: *Omnia mutantur, stat sol et terra movetur. O curas hominum! O quantum est in rebus inane* (All things are transformed, the sun is still and the earth moves. Or men's concerns! Or how vain are the questions!), for the German, and: *Ne frusta vixisse videar* (Let me not seem to have lived in vain), for the Dane: possibly a recommendation to Kepler to protect his cosmology against Copernicus'.

The ideal nucleus of the composition is in the perspectival focus of the representation: a celestial globe, symbol of the astronomical speculation, placed at the center of a flight of stairs and partly covered by a cloth. The reference to the revealed truth, in this case astronomical truth, is apparent. Two sphinxes, considered since ancient times as

[3]From the ideological and figurative point of view, these gestures refer to the representation of Plato and Aristotle, painted by Raphael in the School of Athens (1509–1511, Rome, Vatican Palace) (Chastel 1982).

Figure 1. Frontispiece of Doppelmayer's *Atlas Coelestis* (Courtesy of the INAF-Capodimonte Astronomical Observatory).

repositories of arcane wisdom, are located at the top of the hemicycle and seem to watch over this small temple of cosmology. All together, a complex of symbols evoking Masonic iconography (Pernety 1758) adorns the work. At the top of the frontispiece, two cherubs unfold a parchment with an unusual representation: embedded in a background of clouds there are innumerable planetary systems, each one with a central luminous star surrounded by planets marked by their own shades. The systems, identical except for the size and the number of orbits, are spread around a dominant Solar System, with the exact number of the known planets and relative satellites, and with the orbit of Halley's comet (not yet reappeared since Halley's guess). The reference to the doctrine of the infinity of the universe and plurality of the worlds is indeed evident. There is also a clear link between the top of the composition, where this Brunian manifesto is unfolded, and the presence, at the bottom, of the four celebrated astronomers. It is worth noticing the absence of equally illustrious scientists such as Galileo, Descartes, and Newton. Quite arbitrarily, Doppelmayer did not consider them as universe makers. Their most important astronomical discoveries are nevertheless symbolized in some tables of the *Atlas*, together with the instrument that made them possible, the telescope.

Doppelmayer's frontispiece is not an *unicum*. We found the same iconographic theme in the frontispiece of the *Theoria motuum planetarum et cometarum* (Figure 2), published in Berlin by Leonard Euler in 1744, whose explicit pluralistic message had seemingly not been noticed before. Here again it appears the motif of a parchment with several planetary systems gathered around what should be the Solar System. The inspiration to the work and the graphic model of Doppelmayer is manifest. The extraordinary similarity between the two leads us to believe that Euler must have known Doppelmayer's *Atlas*. The fact is not surprising. The two books were printed within two years of each other, in places geographically close together. Rather, one wonders why Euler re-used the pluralistic manifesto in the context of a treatise on the motion of planets and comets. It is ascertained that he was a pluralist, but there is no trace of this position in the *Theoria*. Possibly, he wanted to allude to a unitary cosmos, regulated everywhere by the same laws dictated by Newton, which he himself developed the new discipline of celestial mechanics.

But there is never two without three. We serendipitously discovered the existence of a clearly pluralistic fresco, probably coeval of Doppelmayer's and Euler's works, decorating one of the rooms of Prague's castle, now called the New Hall of Mathematics (Figure 3). The dating of the fresco stems from the presence in the fresco of the tower of the Klementinum, developed into an observatory by the Jesuit astronomer Joseph Stepling between 1750 and 1751. The reference to the frontispiece of Doppelmayer's *Atlas* and to Euler's *Theoria* is truly clear: in particular, the detail of the putto bearing the inscription *Sapientissimi Opus*, was certainly known to the client of the painting through an edition of the *Theoria*.

Let us consider the examples of pluralistic iconography in the engravings of Doppelmayer and Euler and in the Prague fresco, and those by Fontenelle, Derham, and Wright. Compared to the 15th–17th century representations, they portray a cosmos no longer limited by a crown of fixed stars, but really infinite and populated with planetary systems similar to ours. On the other hand, the disappearance of the ancient conception of a world enclosed by concentric spheres had opened the way to the vision of a much wider universe, even an infinite one, and with complex relationships, determining at the same time the birth of the problem of the uniformity of the laws of nature postulated by Newton. We may understand in this sense Stepling's intent in placing the heliocentric and infinite universe in the hall dedicated to mathematical studies. This is likely a stratagem similar to that by Osiander, who, in the preface to Copernicus' *De revolutionibus*, proposed to treat the heliocentric model as a reasonable mathematical hypothesis (which explains why it is in the Mathematics Room): a Jesuitical way to smooth out the conflict between the new cosmology and the Christian theology.

Figure 2. Frontispiece of Euler's *De motu planetarum et cometarum* (Courtesy of the INAF-Capodimonte Astronomical Observatory).

In fact, we have no trace of a precise position by the Lutheran Doppelmayer and the Jesuit Stepling in favor of pluralism. On the other hand, the Calvinist Euler gives us what we might call his pluralistic testament in the letter IX of his *Lettres à une princesse d'Allemagne sur divers sujets de physique et de philosophie*. Furthermore, we know of a rich correspondence between Stepling and Euler, and of a common membership of Doppelmayer and Euler to some scientific academies.

Knowledge of these relationships between the three great authors allows us to exclude a priori the simplistic hypothesis of an uncritical use of a common graphical cliché. It is rather possible that Doppelmayer, Euler, and Stepling shared a strong position in favor of the pluralistic doctrine and exchanged information, opinions and even, as in this case, "cryptic" iconographic quotations related to the cosmology of the infinite. Moreover, the geographical and chronological distance between the three almost

Figure 3. Fresco in the Mathematics Room of the Prague Castle.

defines the limits of a sort of pluralistic enclave in the heart of Mitteleuropa in the age of Enlightenment.

In conclusion, the iconographic representations of the pluralistic conception of the world appear to be rather rare and confined in time and space. To date, as we have seen, it turns out that only Fontenelle, Doppelmayer, Euler, Derham, Wright, and Stepling made use of an iconographic model clearly inspired by the pluralist debate, without worrying too much about the risks of direct or indirect sanctions by the Church. This figure is quite small if we consider the proportions of the debate itself and the huge quantity of literary works related to it. Rather, one would expect that pluralistic iconography obtained greater diffusion due to its intrinsic popular/didactic value, i.e., its ability to synthesize the verbal contents of scientific texts and create a non-spoken effective language, a sort of silent instructor, capable of freeing technical literature from semantic or purely linguistic difficulties. It was instead even ignored by a large part of the supporters of the plurality of worlds and the infinity of the universe philosophy, especially during the last decades of the 18th century and early 19th century, everywhere but in Mitteleuropa for a few years at the middle of the 18th century. An explanation could come from the fact that the progress of astronomy, with systematic explorations of the sky and new understanding of the nature of stars, had progressively diminished interest in the pluralistic debate. Only the question of the possibility of other forms of life outside the Earth remained open and gained attention on the iconographic level.

References

Burnham, P. 1985, Celestial Images: an Overviews, in Celestial Images, catalogue of the exhibition edited by D.J. Warner (Boston)

Cecere, I. 2001, Sul fondo antico della Biblioteca dell'Osservatorio Astronomico di Capodimonte: arte e scienza nei frontespizi e nelle antiporte, Laurea thesis (Second University of Naples)

Chastel, A. 1982, Art et Humanisme à Florence au temps de Laurent le Magnifique, Paris, Presses universitaires de France, pp. 494-495 (Second University of Naples)

Crowe, M. 1999, The Extraterrestial Life Debate (Cambridge University Press)

Koiré, A. 1961, From the Closed World to the Infinite Universet (Johns Hopkins Press, Baltimore)

Pernety, A. 1758, Dictionnaire mythohermetique, ChexBauche (Paris: Libraire, à Saint Genevieve e à S. Jean dans le Défert)

Warner, D. 1979, The sky explored. Celestial cartography 1500-1800 (New York-Amsterdam, spec. pp. 9-26: A.R. Liss Publishing)

Wilson, C. 1981, Doppelmayr Johann Gabriel, in the Dictionary of scientific biography, subentry, vol. IV (New York, pp. 166-167: Charles Scribner's Sons)

Astronomical Heritage of the Middle East
ASP Conference Series, Vol. 520
Sona V. Farmanyan, Areg M. Mickaelian, J. McKim Malville, and Mohammad Bagheri, eds.

Clarification of Unknown Phenomena Using Historical Star Catalogs and Charts

Tomoko Fujiwara[1] and Masanori Hirai[2]
[1]*Japan Spaceguard Association, Bisei Spaceguard Center, 1716-3, Okura, Bisei-cho, Ibara, Okayama 714-1411, Japan;* `tomokof@spaceguard.or.jp`

[2]*Gekko Observatory / Fukuoka University of Education, 1-1-6-506, Kasiihama, Higashi-ku, Fukuoka 813-0016, Japan;* `mashirai@opal.famille.ne.jp`

Abstract. In all ages, human beings have tried to understand astronomical phenomena to comprehend what the Universe is, where we are in it, and how we relate to it. With the assistance of large telescopes, satellites, and explorers, we are getting closer to understanding the Universe. However, even with instruments using the most advanced technologies, we cannot observe any phenomena that has already passed or has very long-term variability. Historical records can possibly tell us about stars in the past. In order to determine stellar magnitudes during earlier eras and discover unknown phenomena, we referred to historical star catalogs (Ptolemy's *Almagest*, etc.) and charts kept in libraries all over the world. In this paper, we present historical records of stars and show possible long-term variability and unknown objects found by the analysis of these historical data.

1. Introduction

In order to understand what the Universe is, where we are in it, and how we relate to it, humans have watched celestial bodies with instruments like telescopes and satellites using the most advanced technologies of their times. Through the development of photography during the middle of the 19th century, astronomical phenomena started to be recorded as images and kept as data from that point in time. The benefit of high-performance instruments and high-sensitivity detectors for astronomical observations over a wide range of wavelengths from radio to X-rays and γ-rays enables us to obtain a large amount of information, which can lead to important discoveries for the progress of astronomical science.

However, even if we could obtain the latest observational data, it is not possible to catch a glimpse of the past appearances of objects that have gone or stars that have passed away. The timescale of stellar evolution is much longer than that which we can confirm using the latest observational data. Certainly there still must be other unknown events for us to discover. Only surviving historical records have the potential to tell us unknown information about stars. It is important to analyze such historical data to clarify astronomical phenomena.

So far, 52,011 objects have been recognized as variable stars and registered in General Catalogue of Variable Stars (GCVS). The variation in luminosity occurs through several mechanisms, but are mainly classified into two types: intrinsic variables and extrinsic variable. Intrinsic variability is caused by changes in the physical properties of stars, and the prototypical object is Mira (*o* Cet) as a pulsating variable. It expands and contracts continuously and its magnitude varies between 2.0–10.1 mag with a period of 332 days. Extrinsic variables are caused by external properties. The best-known example is the eclipsing binary Algol (β Per). The bright primary star is regularly eclipsed by the dimmer companion, and the magnitude changes from 2.1 to 3.4 mag regularly every 2.87 days.

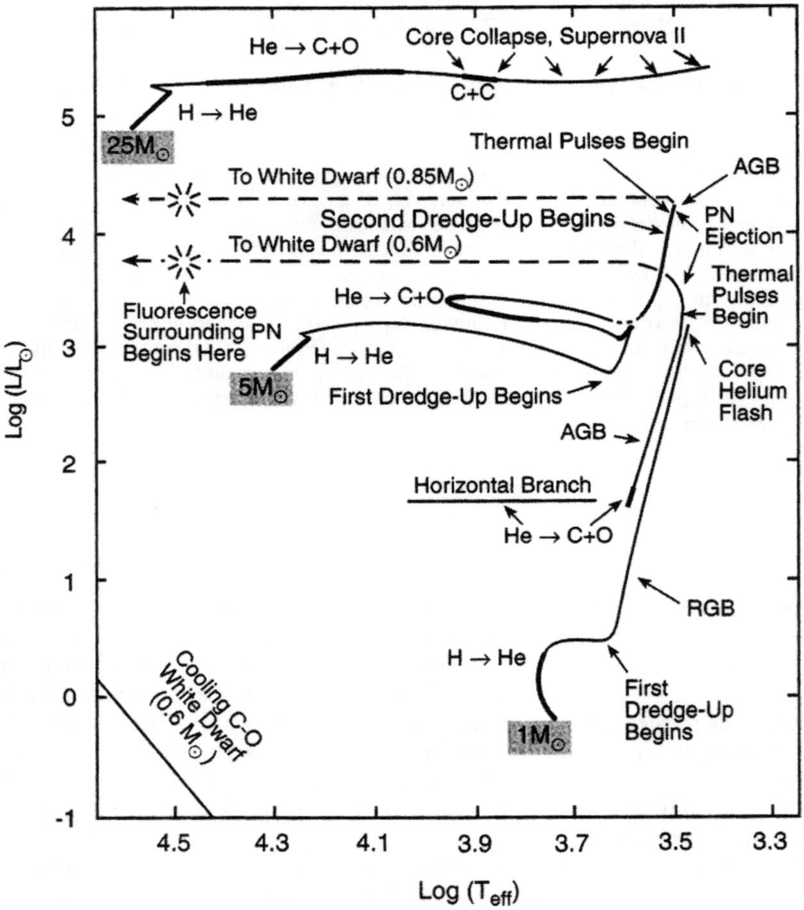

Figure 1. Evolutionary tracks of $1M_\odot$, $5M_\odot$ and $25M_\odot$ stars on the H-R diagram (Prialnik 2000).

Another example of long-term variable stars are Asymptotic Giant Branch (AGB) stars. According to the theory of stellar evolution, after their long stay on the Main Sequence (MS), intermediate and low mass (0.8~8 $M_\odot$) stars transition to the Red Giant Branch (RGB), moving to the right in Hertzsprung-Russel (H-R) diagram. Subsequently they evolve into AGB stars through the Horizontal Branch (HB). AGB stars occupy a strip in the H-R diagram that is almost parallel to, and just above, the RGB. As the star evolves, it rises up along the RGB and brightens.

Recently, photometric and spectroscopic studies of AGB stars have been conducted, but we still do not know their long-term variabilities on timescales longer than several 100~1000 years. Historical data would be helpful and it is the only way to know their past brightnesses. Using historical records, some astronomers have tried to detect variabilities of particular target objects. By analysis of Hevelius' works (e.g., *Prodromus Astronomiae*, 1690), combined with continuous observations obtained in the nineteenth and the twentieth centuries, the long-term period and evolution of the AGB star R Hya was determined by Zijlstra et al. (2002).

2. Data and Analysis

In order to inspect stellar magnitudes of earlier eras, we surveyed surviving historical star catalogs kept in libraries all over the world. Additionally, we used historical charts to compare appearances of the stars at different times. In this section, we describe the characteristics of historical records.

2.1. Historical Star Catalogs

The concept of magnitudes was introduced by Hipparchus in the second century BC. The work of Hipparchus was lost over the years. However, Hipparchus' magnitude system was preserved through subsequent star catalogs. In order to study stellar magnitudes of earlier eras, the following seven catalogs have been carefully studied.

2.1.1. The *Almagest*

The *Almagest* provides the earliest quantitative information on the brightness of the stars. It was written by Ptolemy (or Claudius Ptolemaeus) in ca. 100–170 AD. The *Almagest* is Ptolemy's chief work in astronomy and the book on which his later reputation mainly rests. It includes a star catalog, a list of 1022 stars, arranged in 48 constellations, with the longitude, latitude, and magnitude (from 1 to 6) of each. His observation epoch is about 137 AD. Ptolemy's original catalog has not survived, but there are numerous manuscript copies from the 9th to 16th centuries. We referred to the intensive philological studies conducted by Kunitzsch (1986–1991) and Toomer (1998).

2.1.2. *Kitāb Ṣuwar al-Kawākib*

Kitāb Ṣuwar al-Kawākib, which means Book on the constellations of the fixed stars, was written in Arabic in the 10th century by Abu'l-Ḥusayn ʻAbd al-Raḥmān ibn ʻUmar al-Ṣūfī (903–986). He presented magnitude data based on the results of his own observations. The epoch of his table is 964. We mainly used the Hyderabad edition, which was collated with the oldest extant manuscripts. We also referred the works of al-Bīrūnī (1030) and Schjellerup (1874).

2.1.3. Ulugh Beg's Catalog of Stars

Ulugh Beg (Muḥammad Taragay, 1394–1449) founded a *madrasa*, an institution of higher learning, in Samarkand in 1420. An important result of the scientific work of

Ulugh Beg and his school was the astronomical tables called the '*Zīj*' (epoch 1437 AD). Knobel (1917) revised Ulugh Beg's star catalog made at the Samarkand observatory, using all the contemporary Persian manuscripts kept in Great Britain.

2.1.4. Tycho's *Astronomiae Instauratae Progymnasmata*

Tycho Brahe (1546–1601) observed a supernova in Cassiopeia (Tycho's nova) in 1572. He recorded it in '*Astronomiae Instauratae Progymnasmata*', which was summarized or criticized by others after his death in 1602. In this work, there is a catalog giving the positions and magnitudes of 777 fixed stars. The data in this catalogue are based on Tycho's own observations.

2.1.5. *Uranometria*

Uranometria, written by Johanne Bayer (1572–1625), is composed of a star catalog and star charts. In this catalog, Bayer added 12 southern constellations to Ptolemy's original 48. It depicts magnitudes of 1169 stars.

2.1.6. *Historia Coelestis Britannica*

John Flamsteed (1646–1719) engaged in positional astronomy at Royal Greenwich Observatory and made 20,000 observations of nearly 3,000 stars. His observational data were completely compiled in *Historia Coelestis Britannica* (Flamsteed 1725). It includes stellar equatorial coordinates, ecliptic coordinates, and magnitudes. The observation epoch is 1689.

2.1.7. *Uranometria Nova*

Friedrich Wilhelm August Argelander (1799–1875) observed a few thousand stars with the naked eye. Without measuring instruments, he created *Uranometria Nova* in 1843. The main feature of this work was not the determination of exact positions, but the recording of all stars visible to the naked eye. His *Uranometria Nova* records 3256 stars with equatorial coordinates and magnitudes.

3. Historical Star Charts

In order to compensate for the lack of historical magnitude data, we also use star charts. The following star charts are drawn as images and it is difficult to sample magnitude data recorded in some works. However the appearance of a star informs us that it would be brighter than 6th mag that could be observed with the naked eye.

3.0.1. *Uranometria*

In Section 2.1.5 above, we mentioned *Uranometria*, compiled by Bayer in 1603. This work also includes star charts composed of 49 plates, 48 figures of the Ptolematic constellations, and one figure displaying 12 new southern constellations. Additionally, two charts of the northern and southern hemispheres are shown.

3.0.2. *Prodromus Astronomiae*

Johannes Hevelius (1611–1687) made various observations and results were published as *Prodromus Astronomiae* in 1690. His star charts are composed of 56 figures (54 star charts and 2 additional charts of the northern and southern hemispheres). On the star charts, his constellations are drawn in reverse, as they would be seen on a celestial globe.

3.0.3. *Atlas Novus Coelestis*

The major work of Johann Gabriel Doppelmayr (1671?–1750) is *Atlas Novus Coelestis*, a collection of diagrams with explanations intended as an introduction to the fundamentals of astronomy. There are 36 double-page size engravings in this work, and two charts of the northern and southern hemispheres drawn using the azimuthal equidistant projection.

3.0.4. *Atlas Celéste*

Another work of Flamsteed, *Atlas Coelestis*, was published in 1729. It contains 28 large format charts. In 1776, the 2nd edition of this work was edited by French astronomer Jean Fortin (1750–1831?) and published in Paris as *Atlas Céleste de Flamstéed.* For common use, Fortin reduced Flamsteed's original charts to one-third size and updated the positions of stars to the epoch 1780 instead of 1690, the epoch of Flamsteed's own observations.

3.1. Analysis

In all of the star catalogs mentioned in Sect. 2.1, stellar magnitudes were estimated with the naked eye and were classified from 1st to 6th magnitude based on the Hipparchus system. Before we could use the historical catalogs and charts listed above, we first had to check the reliability of these star catalogs as scientific data. We verified the independence of catalogs on the basis of a statistical test for the distribution of magnitude differences taken from each pair of these works, and the consistency of magnitude data by comparison of the data compiled in them with modern data (Fujiwara et al. 2004), and with magnitude systems and light ratios (Fujiwara & Yamaoka 2005).

4. Results and Discussion

4.1. AGB stars

We investigated the historical data for some bright AGB stars (see Table 1). BL Ori, UU Aur, and Y CVn had not been recorded in any catalogs or charts. These three objects would be too dim to be observed with the naked eye. For the other objects, we compare historical data to results of modern observations.

Historical magnitudes of $o1$ Ori, U Hya, χ Cyg, and 19 Psc are within the range of magnitude variability known from modern observations. The magnitude records of 57 Peg suggest that the star has been brightening for several hundred years (see Fig. 2). The absence of records before the sixteenth century may indicate that this star was dimmer and may not have been observeable with the naked eye. It is possible that the variability of 57 Peg can be explained by stellar evolution.

4.2. Discovery of Unidentified Objects in Bayer's *Uranometria*

Referring to the coordinates of stars recorded in Bayer's star charts in *Uranometria*, we matched each object with fixed stars visible to the naked eye. As for transient objects,

Table 1. Characteristics of bright AGB stars.

Name	Variability Type	$V_{max}-V_{min}$	Period (d)	Sp. Type	Distance (pc)
o1 Ori	SRb	4.65–4.88	–	M3S	166
BL Ori	Lb	8.5–9.7	–	C6,3	595
UU Aur	SRb	8.20–10.00	235	C5,3	415
U Hya	SRb	7.00–9.20	450	C7,3	175
Y CVn	SRb	8.20–10.00	158	C5,4	260
χ Cyg	Mira	3.30–14.20	407	Se	106
57 Peg (GZ Peg)	SRa	4.95–5.23	93	M4S	234
19 Psc (TX Psc)	Lb	6.90–7.70	–	C7,2	315

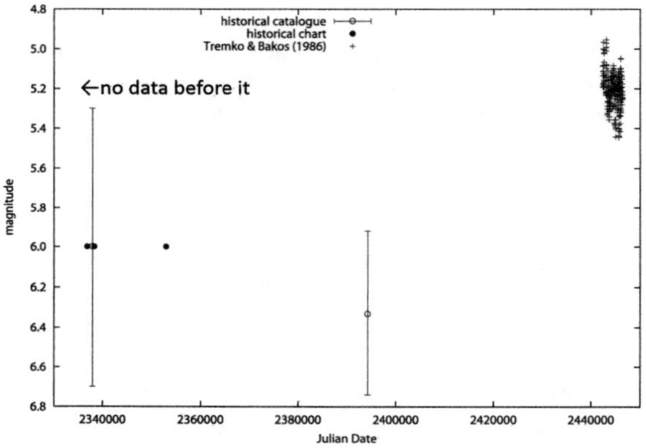

Figure 2. Apparent brightnesses of 57 Peg taken from modern observations and historical records.

a supernova (SN1572) and a peculiar variable (P Cyg) were found. Through the 48 figures of Ptolematic constellations and one figure of the 12 new southern constellations, 13 unknown objects were found (see Fig. 3). These objects could not be identified with stars (including dim ones) within a radius of 2 arcmin. The investigation of these objects is still in progress and will be discussed in another paper.

5. Conclusions

- By comparison of historical records to modern observations, the long-term variability of the AGB star 57 Peg has been measured and appears consistent with the theory of stellar evolution.
- On the charts of Bayer's *Uranometria*, 13 unidentified historical objects were discovered.
- The historical records mentioned in this study are useful to better understand astronomical data from earlier eras and to discover unknown phenomena.

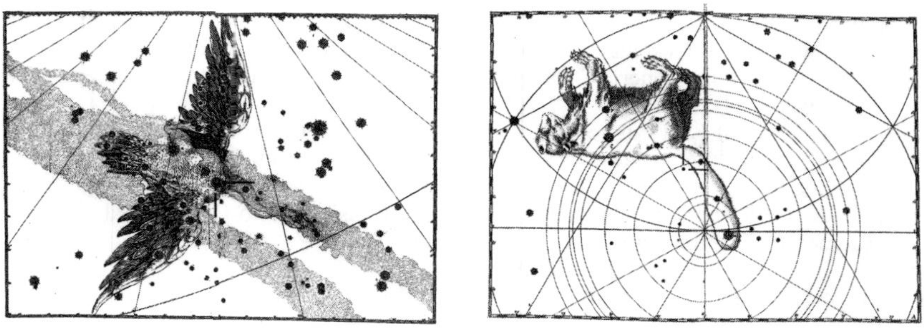

Figure 3. Historical star catalogs. *Left:* P Cyg. *Right:* Unknown object in the constellation Ursa Minor.

Acknowledgments. This research made use of the Simbad database operated at CDS, Strasbourg, France.

References

Alksnis, A., et al., 2001, General Catalog of Galactic Carbon Stars by C. B. Stephenson, Third edition, Baltic Astronomy, 10, 1A.

Argelander, W. A., 1843, Uranometria Nova, Berlin, Berlag von Simon Schropp und Comp.

Bayer, J., 1603, Uranometria, reprinted in 1987, London, British Library.

al-Bīrūnī, 1030, Al-Qānūn al-Mas'ūdī, vol.3, reprinted in 1956, Hyderabad, Dāiratu'l-Ma'ārif-il-'Oṣmānia.

Brahe, T., 1602, Astronomiae Instauratae Progymnasmata, reprinted in 1969, Bruxelles, Impression Anastaltique.

Doppelmayr, J. G., 1742. Atlas novus coelestis, Nuremberg.

ESA, The Hipparcos and Tycho Catalogues, 1997.

Flamsteed, J., 1725, Historia Coelestis Britannica, Tome III, London.

Flamsteed, J., 1795, Atlas céleste de Flamstéed, 3ème édition, edited by de Lalande, J. J. & Méchain, P., Paris.

Flamsteed, J., 1989, Atlas céleste de Flamstéed (1776), 2ème édition, edited by Fortin, M. J., Paris, rebind edition by Koseisha, Tokyo.

Fujiwara, T., Yamaoka, H., Miyoshi, S. J., 2004, A&A, 416, 641–646.

Fujiwara, T., Yamaoka, H., 2005, Journal of Astronomical History and Heritage, 8(1), 39–47.

Gillispie, C. C.,1970–, Dictionary of Scientific Biography, New York, Scribner.

Habing, H. J., Olofsson, H., 2003, Asymptotic Giant Branch Stars, Springer.

Hevelius, J., 1993, Prodromus Astronomiae (1690), rebind edited by Yabuuchi, K., Chijinshokan, Tokyo.

Hirshfeld, A., Sinnott, R. W., Ochsenbein, F., 1991, Sky Catalogue 2000.0, Vol. 1, 2nd ed., Cambridge, Cambridge U. P. & Sky Publishing Corporation.

Knobel, E. B., 1917, Ulugh Beg's Catalogue of stars (1437), Washington, The Carnegie Institution of Washington.

Prialnik, D., 2000, An Introduction to the Theory of Stellar Structure and Evolution, Cambridge, Cambridge University Press.

Ptolemäus, C., 1986–1991, Der Sternkatalog des Almagest Die arabisch-mittelalterliche Tradition Teil I-III (AD 2c.), translated into German and edited by Kunitzsch, P., Wiesbaden, Otto Herrassowitz.

Samus N.N., Kazarovets E.V., Durlevich O.V., Kireeva N.N., Pastukhova E.N., General Catalogue of Variable Stars: Version GCVS 5.1, Astronomy Reports, 2017, vol. 61, No. 1, pp. 80-88. See http://www.sai.msu.su/gcvs/gcvs/

Schjellerup, H. C. F. C., 1874, Description des étoiles fixes, reprinted in 1987, Frankfurt am Main, Institut für Geschichte der Arabisch-Islamischen Wissenschaften.
Stephenson, C. B., 1984, General Catalog of S Stars, second edition, Publ. Warner & Swasey Observatory, 3, 1, 1.
al-Ṣūfī, 986a, Ṣuwar al-Kawākib, edited in 1954, Hyderabad, Dāiratu'l-Ma'ārif-il-'Oṣmānia.
al-Ṣūfī, 986b, Ṣuwar al-Kawākib, facs. ed. in 1986, Frankfurt am Main, Publications of the Institute for the History of Arabic-Islamic Science.
Toomer, G. J., 1998, Ptolemy's Almagest (AD 2 c.), translated into English and annotated by Toomer, G. J., Princeton, Princeton University Press.
Tremko, J., Bakos, G. A., 1986, JRASC, 80, 230–239.
Zijlstra, A. A., Bedding, T. R., Mattei, J. A., 2002, MNRAS, 334, 498–510.

Astronomical Heritage of the Middle East
ASP Conference Series, Vol. 520
Sona V. Farmanyan, Areg M. Mickaelian, J. McKim Malville, and Mohammad Bagheri, eds.

The Scientific Method: An Approach of Developing Cosmology

Magd E. Kahil[1,2] and Ahmed A. Hady[3]
[1]*October University for Modern Sciences and Arts, Giza, Egypt;*
mkahil@msa.eun.eg

[2]*Nile University, Giza, Egypt*

[3]*Astronomy Department, Cairo University, Giza, Egypt*

Abstract. An pedagogical approach to help students understand the development of current theories of cosmology via applications of the scientific method is described in this paper. This method is essentially used to critique ideas and concepts such as science and pseudo-science, soft science and hard science, as well as deterministic and non-deterministic lines of thought. From this perspective, evaluating cosmological theories using the scientific method enables a researcher to suggest new concepts based on the apparent anomalies that are found through meticulous observations. This makes a researcher skeptical of all observed phenomena and motivated to update the methods of experimentation. Accordingly, the need to introduce different mathematical ideas is essential. A good example to illustrate this idea is presented in this paper, which starts by introducing the geocentric model and its defects and progresses all the way to the Big Bang model, which requires a theory that combines elementary particles with large-scale cosmology. One of the most challenging ideas is implementing the concept of geometrization of physics.

The work assigned to students, especially at the freshman level, is to appreciate the importance of studying cosmology. One of the advantages of introducing this line of thinking is to connect biology with cosmology in the new field of astrobiology in order to pave the way for comprehending the origin of life.

1. Introduction

The aim of our present work is to revisit teaching astronomy by applying the scientific way of thinking. This approach is running successfully at MSA-University in Giza (2007–present) and at the American University in Cairo (2002–2016). This program is producing very promising results for helping students appreciate the importance of studying astronomy, and cosmology in particular, and their impact on developing education in different areas. It also opens students' minds to the necessity of multidisciplinary studies for a better awareness of contemporary problems in the natural sciences.

The assessment required is developing the chronological cosmological theories using the scientific method. This is a direct application of the process of implementing the concept of skepticism, assessing the current theories or principles, and trying to find some that seemingly have been overlooked. This is the starting point for skepticism, obtaining another theory/principle, a new concept, able to be falsified in time, in order to keep science renewable and creative. Accordingly, in this paper we are going to give an applicable example that implements the scientific method, and show how it is applied for obtaining accurate observations (Carey 2011). Moreover, such an approach may be implemented in an instructor's plan for exploring novel trends in multidisciplinary studies, e.g., mathematical biology and econophysics, quantum biology, astrobiology, etc.

2. Skepticism and the Laws of Cosmology: a Method for Developing its Theories

A healthy way for checking the validity of theories emerges from turning beliefs in observation into skepticism for every single item. In our study, we show that a gradual development of astronomy arises from revisiting old concepts and checking them against the common laws of nature.

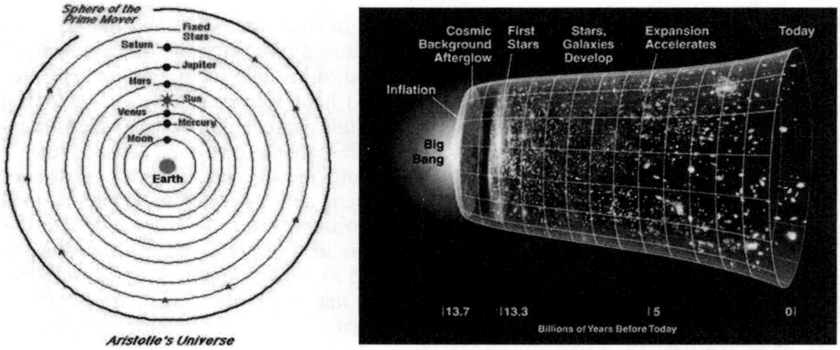

Figure 1. The geocentric model was used historically to explain the nature of celestial objects.

The basic idea of the scientific method may be expressed in terms of the following steps (Carey 2011):

- Asking why?
- Offering an explanation.
- Testing the explanation.

Before offering an explanation and testing it, one must understand the proper meaning of the word "why." The question cannot be answered unless the proper type of observation to answer the question is clear. Being skeptical gives two main questions able to falsify the geocentric model: the retrograde motion of outer planets, and the motion with acceleration in circular orbits. These two arguments lead to a better heliocentric model, which succeeded in revealing the origin of retrograde motion by interchanging the position of the Sun and Earth. Yet this argument is not a satisfactory explanation of the motion with acceleration in circular orbits, which led Kepler to replace circular orbits with elliptical orbits.

In the process of following the concept of skepticism, one must introduce a speculation centered on "Why" objects revolve in elliptic orbits, which gives us a motivation to study Newtonian laws of motion and gravity, which is an application of the second part of the Scientific method listed above. The concept of moving in an elliptical orbit is a result of equating centripetal force with the attraction due to gravity and equating the forces.

Thus, imposing the concept of skepticism on both models and observations may give rise to the new question: why is the law of gravity useful at the solar system level, while no apparent motion is seen at stellar level? This is an application of testing a hypothesis described in the third part of the scientific method above. This is clearly realized by observing that there is no apparent effect of the force of gravity between the fixed stars, which seem motionless on the celestial sphere. Accordingly, one can conclude that the appearance of two stars near each other in the sky is in fact just the projection of those objects. Although the appearance shows that the stars are near each other, one may be in the background, in a different space and time.

Introducing students to the concept of space-time is a good time to discuss Olber's paradox and extend the first law of inertia into 4-dimensions. This is the beginning of explaining both Special and General Relativity. Then one may raise a subsequent question: what about moving in an accelerating frame, i.e., writing Newton's second law in 4-dimensions. This may help to introduce the equivalence principle that led to General Relativity. Students at that level are unable to comprehend the finer details of General Relativity using tensor calculus; to overcome this obstacle we present a remedy in the next section.

The concept of skepticism can be applied to the following statement of special relativity: the speed of light in a vacuum is constant in any inertial frame. A question about the possibility of a variable speed of light in vacuum along space-time could be used to help student think critically about relativity.

An intriguing question supports the need to extend relativity to become a bi-metric theory of gravity. The importance of this lesson is helping students understand that scientific theories can be falsified, and they are not dogmas admitting no exceptions. n a subsequent step, one could question the following statement that describes the equivalence principle: "an inertial frame in a uniform gravitational field is equivalent to a reference frame that has a constant acceleration with respect to the inertial frame." What about non-uniform gravitational fields?

On the other hand, meticulous observations shed light on the scientific model being considered, such as the investigation of the bending of light (gravitational lensing) and the advance of the perihelion of Mercury's orbit, which strongly supported General Relativity. Nevertheless, when explaining the Cosmos using General Relativity, we encounter the problem of the cosmological constant, evidence of the skepticism that led Einstein to revisit his field equations before the discovery of the expansion of the universe. This was a turning point, because the discovery of the expansion of the universe led people change their ideas from the steady-state universe to the Big Bang model. At the student level, one may observe the history of the expanding universe (see Fig. 1), leading to an important conclusion that seeing is not always believing. This gives rise to the study of the Big Bang theory from the skeptical point of view, which leads us to ask about how reliable the assumption of a homogeneous and isotropic distribution of matter is. This question has been supported by recent observations of WMAP, which gives some indication that the Big Bang theory should be revised, opening the door to thinking about string theory and the consequences of brane-world models, leading to the multiverse paradigm.

Such a result is not enough due to a pending question: owing to Newton's third law, for every action there is an equal and opposite reaction, the question is based on the scheme of generalization: what about the inequality of the two actions?

This question motivates us to investigate new horizons of physical biology when expressing biology as physics. Such a tendency led Grandpirre (2008) to consider biology as a multiple quantum system. At the end of this continuous application of skepticism, one may indeed be able to connect quantum conscious with the humanities, leading to the possibility of using physics in understanding non-linear cases such as econophysics. Consequently, teaching cosmology using the scientific method has not only opened the minds of students to appreciate science, but also to yearn for solving real and important problems in society. This may satisfy the strategy of IAU 2010-2020 as shown in Fig 2.

3. Practical Steps to Implement the Syllabus

In this section, we suggest a practical syllabus that could be used to apply the scientific method with developing theories of cosmology and other relevant contemporary theories in nature and society.
Selected Questions: the questions below evaluate the students' level of understanding of the scientific method for developing cosmology and may be subdivided into two categories. Short questions based on imagination centered on the statement "What would happen if..."

- the earth were orbiting a star with a mass half that of the Sun?
- the astronomical unit were equal to one light year?
- a biological body were made of silicon chains?
- the speed of light were 10 m/sec?
- the earth orbited a red giant star?
- the light year were equal to the distance between Earth and Jupiter?
- the Planck constant approached zero?

These questions allow for more than one answer. The purpose of this type of question is not to stick to a unique type of explanation or prevent students from exploring a falsifiable explanation.

The second type of question is centered on writing about some interesting cosmological exploration or changing theories in a brief report about a multidisciplinary topic. An advantage of this approach is implementing creativity in thinking, as in the famous dictum of Carl Sagan: "Science is a way of thinking much more than it is a body of knowledge." Such questions can be based on assignments to watch different scientific videos about alien planets, the microscopic uuniverse, time travel, or multiverse and parallel universes (Driessen 2014). Each of the above mentioned topics is sufficient to keep students fascinated with the wonders of science. After watching the videos, students can be asked to research independently or in groups and prepare an oral presentation as well as submitting a mini-research paper about their abilities to

Figure 2. The IAU Strategic Plan for Astronomy in the Developing World.

comprehend such an issue. Consequently, such an assignment may enhance freshman students' curiosity and help them to be interested in studying the actual mathematical formulations of cosmological theories.

4. Selected Readings for Facilitating the Syllabus

To explain the current theories of cosmology, we recommend beginning by illustrating the significance of theories using the textbook *Mr. Tompkins in Wonder Land* and *Mr. Tompkins Explores the Atom* by George Gamow, rather than the popular book *A Brief History of Time* by Steven Hawking. Such a preference may help non-science students understand scientific theories as a fairy tale. Later, the Hawking book canb be used to simplify the physics without any mathematical formulas or equations.

4.1. The Role of New Suggested Multidisciplinary Studies

One of the advantages of presenting this type of syllabus is offering a glimpse of astrobiology. This concept is vital to viewing astronomy as the cornerstone of all other subjects. This structure gives students a way of thinking about the possibility of extraterrestrial life and the true significance of this notion. Such an approach starts by displaying scientific videos e.g., NOVA, series about planets like Earth, Darwinian Planets, etc. This type of work may shed some light on the meaning and context of extraterrestrial life. Indeed, such an introductory approach may focus students' attention to help them understand that the building blocks of biological life come from the formation of elements in the cores of stars.

In order to keep them interested, there are some simple questions after watching the videos, such as: (1) What is being explained? (2) Is there a vague concept in this video? (3) What are the obstacles to comprehending such an idea? Through the questions above, a student may be able to revisit the current subjects at the freshman level, owing to the necessity to revisit astronomy as an inspiring force in the development of different subjects, e.g., quantum biology and its implication on theory of cognitive psychology.

5. Concluding Remarks

Our strategic plan was to introduce astronomy as an aid to different fields of science. We introduced a practical study to show how studying cosmology may open new questions in multidisciplinary studies. We present the concept of studying astronomy and cosmology, not for curiosity's sake alone, nor for fun, but for developing scientific thinking in students (Burda, Jurkiewicz, & Nowak 2003).

From this perspective, an old study was found in which studying the distribution of stars in galaxies influenced economists to introduce the common laws of insurance (Asano et al. 2015). Recently, the flavor of Riemannian geometry as a tool to express gravity was extended to introduce information geometry in curved spaces to solve risk management problems (Kahil 2014).

Acknowledgments. Magd Kahil would like to thank the organizing committee, especially Prof. Areg Mickaelian and Dr. Sona Farmanyan, for their great support to achieve this task.

References

Asano, M., Basieva, I., Khernnikov, A., & Yamato, I. 2015, "Quantum Information Biology: from Infromation Interpretation of Quantum Mechanics to Applications in Molecular Biology and Cognitive Psychology," arXiv:1503.02515

Table 1. Suggested syllabus for teaching the scientific method using cosmology as essential to multidisciplinary studies

Learning Outcomes	Assessments (graded)	Activities (non-graded)	Resources/ Contents
Recognizing the meaning of scientific thinking	(1) Introduction (2) Midterm exam (3) Concluding remarks, Final exam	Week 1, 6, 11: Discussion Week 5, 14: Review sessions	Scientific method: lecture notes + PowerPoint presentation Cosmological theory Darwinian theory
Applying ways of thinking in our daily lives.	Assignment 1: How to test an advertisement? Project (phase 1: Assigning duties) Assignment 4 Astrobiology; A new Approach Activity (short questions)	Week 2, 6, 12 (2), (6) Discussion to facilitate the assigned task (12) A short video + Discussion	Ways of explaining; lecture notes + PowerPoint presentation Testing Cosmological theories; exercise: discussion debate Testing Darwinian theory: discussion debate
Evaluating scientific method	Falsification vs. verification Time traveling Project (phase 2: Materials feedback) Assignment 3: meaning of time Activity: Beyond Darwin's theory (short graded questions)	Week 3, 8, 13 (3) Briefing an article (8), (13) Video for Time traveling	Fallacies in the name of science Lecture notes+PowerPoint presentation Falsifying current cosmological theories Problems in Big Bang theory: Discussing a scientific paper Falsifying evolution: A probable question: A discussion debate
Evaluating the credibility and validity of information relating to science	(4) Assignment 2: Testing the credibility of a news network Spontaneous human combustion. Activity (Short graded questions) Problems in Big Bang Falsifying Evolution	Week 4, 5, 9, 12 (4) Video + paper (5), (9) Video + Discussion (12) A challenging topic with selected papers discussion	Experimentation, types of testing: PowerPoint presentation Spontaneous human combustion: Video + papers Star formation: Video+ discussion Evolution in action: A famous paper Quantum biology
Interpreting information and problem solving	Moral and immoral sciences (4) A brief report (graded) Science vs. technology for a developed nations	Week 4, 14 Class debate (14) Video + Discussion	Ethics in science: Some papers and discussion debate Aliens and extra-terrestrial life: Video+Debate
Identifying the context of selected popular science	Activity: Elliot wave principle Activity Black holes: Short graded questions Activity Black hole information paradox Activity: Biological action principle	Week 5, 10, 15 (5), (10) Video + discussion (13) Brief report + Discussion	Risk management: an integrated approach selected papers Micro-universe: video+ discussion Mathematical biology: A scientific paper to be revealed

Burda, Z., Jurkiewicz, J., & Nowak, M. N. 2003, "Is Econophysics a Solid Science," arXiv:0302096

Carey, S. 2011, "A Beginner's Guide to Scientific Method," 4th ed., Wandsworth

Driessen, A. 2014, "Life and Quantum Biology, an Interdisciplinary Approach," Acta Philosophia

Grandpirre, A. 2008, "Biological Extension of Action Principle: Endpoint Determination Beyond the Quantum Level and the Ultimate Physical Roots of Consciousness," arXiv:0802-0601

Kahil, M. 2014, "Geometrization of Econophysics: an Alternative Approach for Measuring Elements of Risk Management of an Economic System," Hyperion International Journal of Econophysics and New Economy, 7, 61

Part V

Astronomical Heritage in Literature and Culture

Astronomical Heritage of the Middle East
ASP Conference Series, Vol. 520
Sona V. Farmanyan, Areg M. Mickaelian, J. McKim Malville, and Mohammad Bagheri, eds.

The Starmus Festival – Our Primary Goal: Science Communication

Garik Israelian
Instituto de Astrofísica de Canarias, 38200 La Laguna, Tenerife, Spain;
gil@iac.es

Abstract. *Starmus* is a global festival of science communication and art that brings together the most brilliant minds on the planet. The festival is a gathering of top artists and scientists celebrated every two years. The aim of the project is to inspire and educate the next generation of explorers and regenerate the spirit of discovery. Starmus combines art, music, and science to enhance science communication. During the four previous Starmus Festivals, a large number of world-famous musicians, scientists (among them 21 Nobel Laureates), and astronauts have performed and given lectures. Previous guests have included: Neil Armstrong, Stephen Hawking, Alexei Leonov, Brian May, Harold Kroto, Hans Zimmer, Neil deGrasse Tyson, Sarah Brightman, Brian Eno, Buzz Aldrin, Kip Thorne, Jean-Michel Jarre, and many other brilliant minds of our times.

The past fifty years have seen drastic changes in our lives in many ways: we travel more, there are more cars and machines in general, and the arrival of the World Wide Web made our world a very different place—it changed the way we see each other and communicate with each other. The knowledge accumulated by humanity doubles every two years or so. But there is one fundamental thing that has not changed: despite our being surrounded by gadgets, internet banking, and virtual identities, we are still profoundly human. And that means that we strive for basic necessities, such as food, shelter, and companionship. But we also strive for knowledge and inspiration, for understanding ourselves, and to follow the examples of the worthiest and smartest among us. But when we look for examples to follow, there seem to be really very few good options. Whom should we follow? Politicians? Pop stars? Famous actors? Sports stars? Indeed, many teenagers and young people tend to idolise public figures because they are successful, they are fabulous, they are in the headlines. We hold nothing at all against sportsmen, pop stars, or models, but how successful will their followers be? Will they become as successful and fabulous as their idols by following them? By no means. In order to become successful, you need to build values and accumulate knowledge. It is our firm belief that what our youngest generation needs are *values* and *knowledge*. They do not even need to be taught those, it would be sufficient just to show them the true values and fundamental knowledge that humanity has crystallised through millennia. And the sad truth is, values and knowledge are not as easily accessible as they seem to be in the 21st century. They get blurred by so many flashy events, new blockbusters, noise in the social media, and fake news.

Back in the 1960s and 1970s, almost every child or teenager dreamed of becoming an astronaut or a cosmonaut. Today, children rave about the top sportsmen and women, pop music stars, models, and Hollywood actors. Is it a coincidence that the generation of the 1960s and 1970s has been able to make such a breakthrough in science and technology? This generation has created the internet, virtual reality, satellite communications, electric and self driving cars, the smart home and the smart phone. What will the generation admiring a Playboy cover girl be able to create?

We might not be aware of it in our daily lives, but space exploration and the space race have largely shaped our reality and our civilisation. However, today there is no space race, and space exploration has slowed down a hundred times as compared to the mid-20th century. It was 66 years from the first flight of the Wright brothers to the first flight of the man to the Moon. In the following 60 years, nothing remotely as significant has been achieved in terms of space exploration.

However, there is perhaps not a single other task as pressing as that of exploring our Universe for several reasons. Firstly, exploring the Universe will give us an understanding of our origins, of how we evolved from star dust to human beings, creative and rational creatures capable of thinking, reasoning, exploring, and creating new realities. On the other hand, space exploration may provide us with a new home, new planets to colonise, to give our long-suffering and over-populated planet a break and save it from calamities. We might find a place to escape from those calamities if we find them unstoppable and imminent. In addition, understanding the laws of the Universe will help us stop those calamities, help protect our beautiful and unique planet, and increase our chances of survival on it. Finally, the history of the Universe is simply fascinating, and it is most rewarding and enriching to get to know about it both for young learners and adults.

But how can we do that? What with the lack of funding for science, the space race safely over, and our space legends deceased or heading for their 80th anniversaries, and very few new heroes to accomplish deeds similar to theirs, where will we find a new generation of heroes to inspire, motivate, and teach us how to overcome the hostilities of the Universe? It is not for the lack of potential heroes: humanity is now more prepared than ever to face new challenges and produce a generation of new supermen and superwomen who will guide us to the stars. It is for the lack of communication, the lack of knowledge, and the lack of understanding of truly important things. It is because of the perception of science as a time-consuming, unrewarding, and unfashionable activity.

At this point in time, science communication is of utmost importance. Communicating science will open new doors and expand our horizons for us as a civilisation. It will promote interest and true devotion in young people and push them towards exploring and achieving new space goals for the benefit of the whole human race. This is when communicating science can actually change the world in which we live, make it a world of opportunities, cooperation, and joined effort. This is what Starmus aspires to achieve, hence its lemma, “Discover the Cosmos and change the world!”

It all began with a peripheral study of the acoustic waves in the Universe by the founder of Starmus, Garik Israelian. That study was out of his usual scope of research, but he was keenly interested in capturing and measuring the sounds emitted by the stars, and he began to investigate. It turned out that our Universe was a very noisy place at its beginnings, but there was rhythm in that noise—the music of the stars!

It is surprising that we have for separated science from art for so long, perceived them as antagonistic in fact, when they are two forms of one and the same activity: exploration of the reality. Scientists study the reality through their logic and rationality, artists do the same thing employing their senses. Ultimately, we are after one and the same truth, approaching it from different angles, so why not try to combine them? After all, many famous scientists are also accomplished artists, and many artists are very good at science. Some of them even hold a PhD, like the Queen guitarist Brain May, who has a PhD in Astronomy.

That was the idea of Starmus, the Music of the Stars, combining art, science, space exploration, and, above all, a history of human effort and cooperation, an experience of heroic deeds and overcoming immense challenges, a history of trial and error, of human effort. A history of human civilisation. A history which is not painted in political colours, but with colours of rationality and creativity, and with the spirit of exploration. A history told to a curious and enthusiastic, but not necessarily very knowledgeable audience by Nobel prize laureates, top scientists, astronauts and cosmonauts—and also by artists. The presentations range from cosmic microwave background radiation, exoplanets, dark matter, dark energy, accretion disks, pulsars, black holes and gravitational waves, told in a comprehensible and engaging manner, to reflections of astronauts and artists. All this is spiced with discussion boards, round tables, debate panels, concerts, exhibitions, film screenings, etc.

The idea of combining science and art turned out to be very productive in a number of ways. The public responded with enthusiasm and keen interest. Furthermore, it was endorsed by such giants as Alexei Leonov, Brian May, and Stephen Hawking. Nobel prize winners, space explorers, and artists seem to be happy to share their knowledge and experience with a non-professional audience. The line-up of such powerful science communicators as Professor Hawking, Neil deGrasse Tyson, Richard Dawkins, Brian Cox, Nobel prize winners George Smoot, Jack Szostak, Kip Thorne, Edvard Moser, May-Britt Moser, Robert Wilson, John Mather, Harold Kroto, Adam Riess, Elizabeth Blackburn, Barry Barish, Eric Betzig, Brian Schmidt, David Gross, Joseph Stiglitz, Susumu Tonegawa, Stefan Hell, Chris Pissarides, Torsten Wiesel, Adam Smith, Tim Hunt, Arthur McDonald, and the performances of renowned artists such as Brian May, Brian Eno, Rick Wakeman, Tangerine Dream, No Sound, Katerina Mina, 51 Degrees, Alexandros Hahalis, Anthony McCarten, David Zambuka, MC Hawking, Anathema, Hans Zimmer, Sarah Brightman, Steve Vai, Devin Townsend, Grace Potter, Nuno Bettencourt, Ane Brun, The Pineapple Thief, Jennie Abrahamson, Marcus Reymann, Artone, BK Duke, DJ Favorite, and others, all of whom made Starums a highly relevant venue for science, astronomy, and art-lovers from all over the world.

It was also productive in the sense of cooperation between different speakers and performers who met in person for the first time and conceived joint projects thanks to Starmus.

Another outcome of Starmus to which we aspired from the beginning and which is slowly coming to life is the discovery of new talent, younger professionals who make impactful research or breath-taking pieces of art. Starmus becomes a scene where they can step in and present their achievements, share them with the public, and inspire the coming generation, just as the heroes of the past did.

As part of our commitment to promoting science among the younger generation, from the very beginning Starmus has offered a 50% discount to students in order to make the experience affordable for them. The first Starmus Festival was held in 2011 and featured two Nobel Prize winners, George Smoot and Jack Szostak (Kip Thorne, who was also speaker at Starmus I, won the Nobel prize six years later in 2017), several NASA and Apollo astronauts, Neil Armstrong and Buzz Aldrin among them, along with Russian cosmonauts Alexei Leonov, Sergei Zhukov, and Yuri Baturin. Only 300 people attended it.

Since the first Starmus, its growth has been exponential: the second edition of the festival was attended by 1500 participants, and Nobel laureates Harold Kroto, John Mather and Robert Wilson, as well as Professor Hawking and many other distinguished scientists who talked about recent advances in astronomy, cosmology, and related areas, while astronauts Walt Cunningham and Charlie Duke, along with their Russian colleague, the legendary cosmonaut Alexei Leonov, talked about new and old challenges of space exploration.

The third Starmus festival brought together seven Nobel prize laureates and a powerful constellation of astronauts, as well as distinguished artists and musicians. The most recent fourth edition of Starmus, held in 2017 in Trondheim, Norway, featured a stunning panel discussion of 11 Nobel prize winners, a powerful gathering of 8 space explorers, and a never-seen-before gathering of top artists and scientists, with an ever-increasing number of 3000 participants. Besides the main programme, Starmus IV also offered a series of side activities, workshops, lectures, and performances within the frame of the Trondheim City Programme, which was free and open to anyone. This event had an enormous social and cultural impact in Trondheim, where people of all ages and occupations were able to attend performances and lectures given by legendary figures.

Every edition of the Festival is organised around a broad theme, a general topic which serves as a centre of gravity for all the lectures, round tables, debates, panel discussions, concerts, exhibitions, performances, and off-festival activities. Starmus I was dedicated to the flight of Yuri Gagarin and was named "50 years of man in space," Starmus II set out to explore the most profound cosmological issues and was called "Origins of the Cosmos," Starmus III was a tribute to Stephen Hawking and carried the title "Beyond the Horizon," while Starmus IV was dedicated to the exploration of "Life and the Universe." Starmus V will be held in 2019 in Bern, Switzerland, to celebrate the 50th anniversary of the man's first step on the Moon. The theme of the festival is "A Giant Leap."

When Stephen Hawking kindly joined the supporters of Starmus, we decided to create a special award for the best science communicators of our times, the Stephen Hawking Medal for Science Communication. The medal was beautifully designed by Alexei Leonov, the first person to perform a spacewalk and also a talented artist, and its first receiver was Neil deGrasse Tyson, one of the most passionate and successful science communicators of our times.

Starmus brings together people of different fields and disciplines, different nationalities, different beliefs, and different ages. It was designed as an apolitical and non-discriminatory event to bring together the best minds and hearts of humanity to create something really important, beyond the news headlines and scandals which scatter our everyday life. It was designed as an effort of cooperation to unite all the outstanding achievements of the human race, and to promote knowledge and the spirit of exploration, to motivate and inspire the younger generation to be as extraordinary and daring as they can be.

While this may seem a very ambitious goal, we believe that it is achievable, and hope to be able to demonstrate that in the near future.

Astronomical Heritage of the Middle East
ASP Conference Series, Vol. 520
Sona V. Farmanyan, Areg M. Mickaelian, J. McKim Malville, and Mohammad Bagheri, eds.

Kushyar ibn Labban's Scientific Legacy

Mohammad Bagheri
History of Science Department, Encyclopaedia Islamica Foundation, Tehran, Iran

Abstract. Kushyar ibn Labban (Gilani) was an Iranian astronomer and mathematician who lived around 10 centuries ago. Little is known about his life. But his rich scientific legacy has been preserved. They are all composed in Arabic, the lingua franca of his age. He composed an influential treatise on Hindu reckoning, two *zijes* (collections of astronomical tables with explanations), a treatise on the astrolabe and another on astrology. A major part of his works are edited, translated, and analyzed. His revision of the Ptolemaic interpolation schemes for finding the ecliptical longitudes of the planets are regarded as an important achievement in mathematical astronomy. This paper gives an account of his scientific heritage and the secondary literature. I also displayed a collection of books (editions and translations of his works) and a collection of about 90 posters prepared from about 50 manuscripts of his treatises extant in different libraries and collections of different countries.

1. Pre-Islamic Astronomy in Iran

The first part of this paper discusses pre-Islamic astronomy in Iran. Then I will explain some major aspects of Iranian astronomy in the so-called Islamic civilization, which includes Kushyar's scientific heritage. Finally, I will provide an account of Kushyar's astronomical works.

The first topic is a reference to the sphericity of the earth, found in the sacred Iranian text, composed around 30 centuries ago. This book is called *Avesta* and it is the holy book of the Zoroastrianism, the dominant religion of Iranians before the introduction of Islam in Iran around 14 centuries ago. At present, there is still a Zoroastrian minority in Iran and India, beside those scattered all over the world. In later Zoroastrian texts, there are references to the names of constellations, stars, and planets.

In the first century AD, a group of Zoroastrian sages emigrated from the southeast of Iran to India. They took their beliefs and knowledge (including astronomy) into India and mixed them with the local culture. Apparently, two great Indian astronomers, Varahmihra (ca. 500 AD) and Brahmagupta (ca. 600 AD), were among their descendants. The present formal calendar accepted in India after independence is taken from the Iranian calendar through these Zoroastrian emigrants. It is said that their descendants are still living in India and have kept their identity.

The only astronomical work we know from pre-Islamic Iran is entitled *Zij-e Shah* which literally means "The Royal astronomical handbook with tables." It was composed around 450 AD for the Iranian King Bahram. One century later, another Iranian King, Anushirvan, wished to improve the astronomical work. So he asked his astronomers to compare Ptolemy's *Almagest* and the Indian astronomical treatise *Arkanda*.

They found the Indian work more desirable and composed a new version of *Zij-e Shah* based on *Arkanda*. No copy of *Zij-e Shah* has survived. However, fragments of it are preserved in the works of the Islamic-period Iranian astronomers, like Biruni who quotes a fragment of *Zij-e Shah* on the method of drawing the hour-lines on a sundial in his book on spherical trigonometry. *Zij-e Shah* existed until the ninth century AD, as we know from other texts composed at that time.

Pre-Islamic astronomy, which was influenced by Indian and Greek astronomy, in its turn influenced the Islamic period astronomy, and through it, influenced later development of astronomy in Europe. Not only methods and parameters, but also a few technical terms from ancient Iranian astronomy were thus transferred to Europe via Islamic period astronomy. For example, the ancient Persian word *zig* originally meant a thread or cord. By extension, the word came to stand for the set of parallel threads making up the warp of a fabric. Then from resemblance between the closely drawn vertical lines of a numerical table and the warp set up in a loom, the meaning was further extended to include numerical tables. And by a final extension, the word came to denote whole sets of astronomical tables with explanatory texts appended to them. The Persian word *zig* was Arabicized as *zij* and appeared in Medieval Latin as *zich* or *ezich*.

2. The "Golden Age"

After the conquest of Iran by Arabs in the seventh century AD, there was a two-century period of battle and armed resistance. In this period, scientific activities stopped, and if there was anything accomplished, it did not remain and was not reported. Anania Shirakatsi, an Armenian scholar of the seventh century AD who lived in this period, provides a valuable account of the Iranian calendar in the section "Tomar" of his encyclopedic treatise entitled *K'nnikon*. Grigor Brutian from Yerevan has published an English version of this account in his paper "Persian and Arabic calendars as presented by Anania Shirakatsi" (*Tarikh-e Elm*, Iranian Journal for the History of Science, Institute for the History of Science, University of Tehran, 2009).

After "two centuries of silence," the cultural and spiritual resistance against the Arab invasion started. A glorious manifestation of this reaction is the major role of the Iranian scientists and thinkers in the so-called Islamic period science. During the period from the ninth to the fifteenth century AD, in a vast territory of unified Islamic rule from Andalusia to western China, scientists shared and exchanged their scientific heritages using Arabic as their *lingua franca*. This period is sometimes referred to as the "Golden Age" of Islamic period science. Although it is sometimes called "Islamic science" or "Arabic science," the whole rich scientific heritage was achieved by scholars of different religion, nationality and language. Some of them were atheists or "unfavorable believers." However, a major part of them had—genuinely or apparently—accepted Islam. This "Golden Age" is sometimes regarded as a revival of the ancient Iranian civilization carried out by the Iranians who were in the service of the Muslim rulers.

In the period from the ninth to the fifteenth centuries AD, many astronomical works from the Greek and Indian traditions were translated into Arabic. However, the Islamic period astronomers also made important theoretical and observational contributions to the development of astronomy. Their main achievements were:

- analysis and modification of the Ptolemaic astronomical models;

- measurement and modification of the astronomical parameters of the Ptolemaic system through new observations;
- composition of around 200 astronomical handbooks with tables (*zijes*), about half of which have survived; and
- devising more complete versions of astronomical instruments, such as astrolabes and quadrants.

3. Arabic Scientific Works in Europe

During the Renaissance in Europe, many Arabic scientific works were translated into Latin, among which were some books whose original Greek texts were lost. For example, a part of Apollonius' *Conics* (from the third century BC) only remained in its Arabic version. This part was translated into Latin by Edmund Halley (d. 1742 AD), the discoverer of the famous comet. In the process of translating Arabic texts into Latin, many Arabic scientific terms were also introduced in Europe. For example, the star α-Piscis Australis is called Fomalhaut, which comes from its Arabic name *Fam al-hut*, which literally means "mouth of the fish." In the same way, the name of an Iranian astronomer and mathematician al-Khwarazmi from the ninth century AD was transformed to the term "algorithm" and the name "algebra," a mathematical branch which was first introduced in Europe through a Latin translation of al-Khwarazmi's book, is derived from the Arabic word *al-jabr*, which literally means "compensation." On the other hand, some Arabic texts were saved only in these Latin translations; an important example is al-Khwarazmi's book on Hindu reckoning. His astronomical handbook has also remained in the Latin translation of a summary prepared by a later astronomer in Madrid.

There are several possible reasons for the flourishing of science in the Islamic countries after the ninth century AD and its decline after the fifteenth century. Many social, economic and political factors are mentioned. However, there is no unanimously accepted theory for these phenomena, yet. After the fifteenth century, most scientific works in the Islamic civilization were mainly limited to rewriting and commenting on former works. As a formal definition of the "Islamic period astronomy," which started around the ninth century AD, its end is regarded as the introduction of modern European astronomy in the Islamic territories. A concrete instance of an early reflection of modern European astronomy in Iranian astronomy is found in an Iranian astrolabe kept in a private collection in Temse near Antwerp (Belgium). On the rim of the astrolabe, a Persian poem describing the parts of the astrolabe is engraved. The last line presents the date of construction of the astrolabe as the sum of the numerical values of the letters in line, based on the Arabic alphanumerical *Abjad* system. This gives the year 1070 AH, which coincides approximately with the year 1660 AD. On another plate of the astrolabe depicting the constellations, we read in Persian that "since there were some discrepancies in the preceding observations of the star positions, and the most accurate positions of the stars were [found] in the European observation records, we based the positions of the stars [here] on their observation which had been performed ten years before the construction of this [astrolabe]." So around 360 years ago, Iranians resumed the contacts with European astronomy.

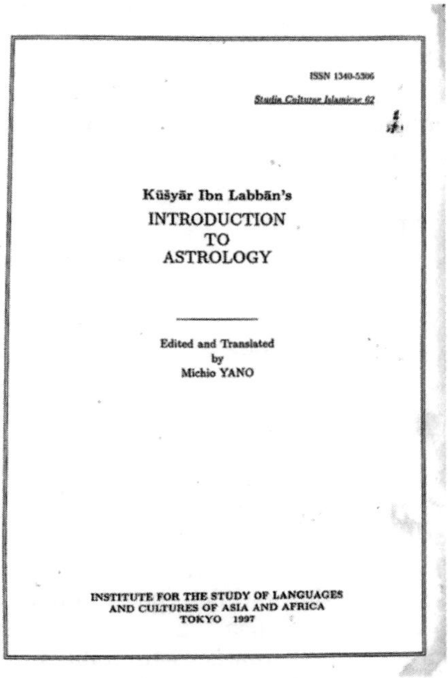
ISSN 1340-5306

Studia Culturae Islamicae 62

Kūšyār Ibn Labbān's
INTRODUCTION
TO
ASTROLOGY

Edited and Translated
by
Michio YANO

INSTITUTE FOR THE STUDY OF LANGUAGES
AND CULTURES OF ASIA AND AFRICA
TOKYO 1997

Figure 1. Title page to the English translation of Kushyar ibn Labban's *Introduction to Astrology.*

4. Kushyar ibn Labban

Kushyar ibn Labban was a typical eminent scientist from Iran who was active in the first half of the eleventh century AD. We know little about his life. As far as we know, he was born and grew up in Gilan province on the southern coast of the Caspian Sea. He later moved to the flourishing city Rayy near present Tehran. In Rayy, he was in contact with a circle of scholars like Biruni, the most famous Iranian scholar of the Islamic period. Kushyar later moved to Gorgan in the province east of Gilan, where Iranian local rulers supported scientific and cultural activities. He wrote his famous astronomical handbook entitled *al-Zij al-Jami'* in Gorgan. Kushyar's name is a symbol of a wise and knowledgeable person in Persian literature. His astronomical handbook is mentioned in a Persian poem as a sample of a very complicated text.

All of Kushyar's works were written in Arabic, the scientific *lingua franca* of his time in the Islamic territories. His book on Hindu reckoning describes decimal and sexagesimal operations based on the Hindu-Arabic numerals, which originated in India and found their way to Europe through the Middle East and North Africa. Editions of the Arabic text and translations of this work into English, French, Persian and Russian have been published. It has also been translated into Hebrew.

Kushyar's treatise on astrology was widely known in Iran and exists in several manuscripts. Its Arabic text with English and old Chinese translations were published in Japan 20 years ago. It has also been translated into Persian and Turkish. Manuscripts of the Arabic text written in Hebrew characters are kept in the Vatican and New York.

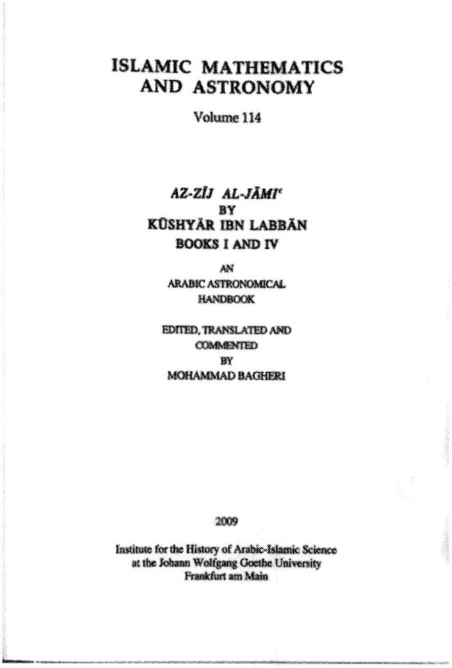
ISLAMIC MATHEMATICS
AND ASTRONOMY

Volume 114

AZ-ZĪJ AL-JĀMIʿ
BY
KŪSHYĀR IBN LABBĀN
BOOKS I AND IV

AN
ARABIC ASTRONOMICAL
HANDBOOK

EDITED, TRANSLATED AND
COMMENTED
BY
MOHAMMAD BAGHERI

2009

Institute for the History of Arabic-Islamic Science
at the Johann Wolfgang Goethe University
Frankfurt am Main

Figure 2. Title page for Kushyar's book *al-Zij al-Jami'*.

Kushyar also composed a treatise on astrolabes. An edition of its Arabic text has been prepared by Taro Mimura (Japan) as a Masters thesis. I have published an edition of an old Persian translation of this work which I found by chance in Tashkent (Uzbekistan) around 20 years ago. All these three items were published together in Tehran in 2014 on the occasion of the international gatherings in Tehran and Rasht (Gilan) for the 1050th anniversary of the birth of Kushyar.

Kushyar's major astronomical work is entitled *al-Zij al-Jami'* (the comprehensive astronomical handbook) in four volumes. I have published an edition of the Arabic text of books I and IV (the mathematical part) with English translation and commentary (Frankfurt, 2009). Books II and III include tables and an account of the configuration of the Universe (*Hay'at*), respectively. Two chapters of Book III relating to glossary of astronomical terms and the distances and sizes of the planets have also been published with English translation and commentary.

Kushyar's *zij* is mainly based on Ptolemy's *Almagest*. However, he innovated a new interpolation method and modified an astronomical parameter for the planet Mars. He has also added some very accurate calculations which are not of practical importance, but reflect his pure mathematical interests. Some traces of Indian astronomy are also found in this work. Kushyar's astronomical handbook is also important for its precious information about the old Iranian calendar and for its systematic presentation of calculation methods and their proofs.

There are three complete and about fifteen partial manuscripts of this work. Four partial manuscripts of the Arabic text written in Hebrew characters exist, which, when combined, cover the whole text. Two partial manuscripts in Egypt contain a very interesting new piece of information about the observation of a conjunction of Mars and Saturn in July 993 AD by Kushyar.

I am still working on the remaining parts of Kushyar's astronomical handbook. I have established a center to archive, research, and publish Kushyar's scientific heritage in the University of Gilan, Kushyar's birth-place. This center has offered honorary citizenship to Prof. Michio Yano from Kyoto Sangyo University (Japan), who published Kushyar's treatise on astrology, and to the late Prof. Khorshid F. Abdollazadeh from Khujand (Tajikstan) for his comprehensive book (in Russian) on Kushyar, after the publication of its Persian translation.

Figure 3. Title page to Khorshid Abdollazadeh's book on Kushyar.

I finish my paper by quoting a dictum from Kushyar: "If two persons are interested in the same thing, the one ignoring the defects of that thing is really unfair to himself."

Astronomical Heritage of the Middle East
ASP Conference Series, Vol. 520
Sona V. Farmanyan, Areg M. Mickaelian, J. McKim Malville, and Mohammad Bagheri, eds.

Some Astronomical-Cosmic Realities in the Armenian Epic "Sasna Tzrer"

Grigor Broutian
NAS RA V. Ambartsumian Byurakan Astrophysical Observatory (BAO), Byurakan 0213, Aragatzotn province, Armenia; gbroutian@gmail.com

Abstract. Some investigators have identified connections between the heroes of the epic *"Sasna Tzrer"* with the heavenly bodies. For example, for some of the heroes it is said that, unlike other children who grew up over years, these children grew by the day. The life of Mher (and some other heroes) is divided into main periods of seven years. At the age of seven he was grown and went to school. He also became a hunter and began to feed the city with his hunt for seven years. At the age of 14–15 he defeated the white devil, freed his bride, and married. At the end of his life he spendt seven years drunk in Egypt, then returned back to Sasoun, gave birth to his successor son Davit, and died. It is obvious that here we have the description of the lunar month with four main phases (1st quarter, full Moon, 3rd quarter and new Moon—the birth of a new Moon—and the next generation of heroes).

Similar relationships can also be seen concerning the other heroes of this epic. Thus, we can ascertain the fact of the existence of close attachments of heroes of the epic with the heavenly bodies. However, in addition to the connection of individual heroes with celestial bodies, there is also a *systemic* link between the heroes of the epic and constellations. In order to discover the *systematic correlation* between the epic heroes and the heavenly bodies, we have examined not only the lives and behavior of individual heroes, but also their family tree. It turns out that almost all the heroes of this epic have their parallel constellations in the Zodiac, and their family tree corresponds to the order of corresponding constellations in zodiacal circle.

1. Introduction

Today more than 160 variants of the Armenian epic *Sasna Tzrer* have been discovered and recorded, and about 70 have been published (Armenian Soviet Encyclopedia, v. 10, 1984, Sasna Tzrer, v. 1–4, 1936–1999). The (proposed) original text of the epic was published in 1939. Although the compiled original text basically corresponds to the content of different variants, nevertheless, in many details there are sometimes significant differences between the compiled text and different variants. Nevertheless, a certain family tree of the heroes of this epic is now accepted as the most trustworthy one. In the course of our research, we keep in mind the family tree presented in the compiled original text of the epic (Sasountzi Dawit 1936).

The epic *Sasna Tzrer* primarily includes the history of four generations of epic heroes, beginning with the twin brothers Sanasar and Baghdasar, and ending with Small Mher. The main branch of heroes is as follows:

- King Gagik,

- Tsovinar (and Caliph of Baghdad),
- Sanasar and Baghdasar,
- Tzran Vergo,
- Dzenov Ohan,
- Mets (Joj, Ariwts, Ariwtsadzew) Mher,
- Davit, and
- Phokr (Small) Mher.

There are also some other heroes not in this main family group who play significant roles in the epic. Among these other heroes are Q'eri Toros (Uncle Toros), the Old Woman, the owner of the field, Msra Melik, Holbashi (or Kozbadin), Ismil Khatoun, Paron Astghik, Charbakhar Q'ami, and others.

As we know, some investigators have noticed relationships between the heroes of the epic with the heavenly bodies (Orbeli 1936). We know, for example, that:

- some of the heroes grew by the day, unlike other children who grew up over a period of years;
- the life of Mher (and some other heroes) is divided into main periods of seven years;
- at the age of seven, Mher became big and went to school; he also became a hunter and began to feed the city with his hunt for seven years;
- at the age of 14 or 15, Mher defeated the white devil, freed his bride, and married; and
- at the end of his life he spent seven years drunk in Egypt, then returned to Sasoun, gave birth to his successor son Davit, and died.

It is obvious that the life of Mher represents the lunar month with four main phases (1st quarter, full Moon, 3rd quarter, and new Moon, the birth of a new Moon, or the next generation of heroes). It is obvious also that the heroes who were growing up in a few days in their childhood must be related with Moon, as their growth is similar to the development of lunar phases which also change by the day.

Similar connections can also be seen concerning the other heroes of this epic. Thus, we can ascertain the existence of close attachments of heroes of the epic with the heavenly bodies and look for deeper correlations between the whole epic and the starry sky.

2. The Family of Sanasar

To reveal a systematic correlation between the heroes of the epic *Sasna Tzrer* and the realities of the starry sky, we shall first examine the relationships of the main members of the family of Sanasar and try to find their star parallels. The members of this family are the main characters in the epic, and all main events are connected to them.

2.1. Sanasar and Baghdasar

These twin brothers are the founders of the heroic family. They were born from one and a half handfuls of the water of a wonderful spring which miraculously arose from a rock standing inside the sea. In order to find thier heavenly parallels, we have to take into account that:

- they are twins,
- they always act as classic twins,
- they are the founders of the city Sasoun,
- they defeated the dragon, freed the spring and the maiden, and saved the city from thirst,
- one of them is much stronger and wiser than the other, and
- only one of them has descendants.

All these characteristics of Sanasar and Baghdasar indicate that their star parallel (as heavenly symbol of twin brothers) must be the zodiacal constellation of Gemini, th Twins. So, we take as the heavenly parallel of these brothers the zodiacal constellation *Gemini*.

2.2. Vergo

The next representative of the family of Sasoun is Vergo, the elder son of Sanasar. There is a lack of information concerning this member of the family of Sasoun. We only know that he is called "Tzran Vergo." We know also that his son was named Paron Astghik (Master Astghik) or Astghik Tzolatsin (Astghik born of glare, or generating glare) where the name Astghik can be translated as Stella, or star. We know as well that Vergo's epithet "Tzran" is usually explained as coward, a cowardly man, and in the epic it is explained as if Vergo had a hernia. However the epithet of this hero helps us to find his heavenly parallel.

- Although the word "Tzran" usually is explained as *coward, a cowardly man*, nevertheless it can be explained also as having its origin from the Armenian word *tzir = Onager* (wild-ass).
- In old Armenian tradition, some stars in the constellation Cancer were known as *asses* (Ajarian 1973).
- In some Armenian fairy-tails, instead of Cancer we have an *Ass* (Broutian 2008).

Thus we can take as the heavenly parallel of Vergo the constellation *Cancer*.

2.3. Ohan

In the epic, Ohan is the middle son of Sanasar. He is also a giant (as were all the members of the family of Sasoun). He had three wonderful horses (white, red, and black) and his favorite weapon was the rooted poplar tree. He also observed the stars and checked the condition of his family members by means of their stars. Ohan was the keeper of the treasury of the state of Sasoun. His main distinguishing feature from other heroes was his unbelievably strong voice. His voice passed from one world to another. In order to find Ohan's heavenly parallel we take into account that:

- the epithet of Ohan is "Dzenov;"
- his main characteristic is his unusually strong voice—his voice could reach the most distant places;
- in order not to burst while crying, he wrapped himself with the leather of seven buffaloes and sometimes with the chains of plow as well; and
- he was very strong: he fought with a rooted poplar tree.

We see that while performing his main acts Ohan "entered" into the hides of buffalos as if being transformed into a buffalo himself. And what is more, sometimes he also wrapped himself with chains of plow. These chains also point to a correlation with the buffalo, as the most powerful animal used in Armenia to move a plow was a buffalo. In Armenian, the male buffalo and oxen in general are called "Tzoul," or Bull. The same name is used in Armenian for the constellation Taurus. All these characteristics and relations show that Dzenov Ohan is closely related to the buffalo, and this most powerful pack animal must be considered his zoomorphic parallel. Thus we can take as the heavenly parallel of Ohan the constellation *Taurus*.

2.4. Mets Mher

We know this hero as the younger son of Sanasar. As mentioned above, Mher is closely correlated with Moon as his life is divided in accordance with four main phases of Moon (1st quarter, full Moon, 3rd quarter and new Moon). Unlike other children who were growing up by years, Mher grew up in days. Such lunar (and solar) characteristics we can be seen in the lives and acts of almost all the main heroes of this epic. So, it is something "usual" for all of them. In order to find the exact heavenly parallel of Mher, we have to examine his deeds and characteristics as a whole:

- Mher killed the lion that had closed the road;
- he divided the lion into two halves and put them on either side of the road; thus his way passed through the Lion;
- due to this he was called *Ariwts Mher*, or Mher the Lion, or *Ariwtsadzew Mher*, Mher that shapes the Lion.

We see that defeating the lion and having the epithet Ariwts or Ariwtsadzew are the most significant distinguishing characteristics of Mets Mher. Thus we can take as the heavenly parallel of Mher the constellation *Leo* the Lion.

As a result of these parallels we see the following picture:

- Sanasar & Baghdasar: Gemini
- Vergo: Cancer
- Ohan: Taurus
- Mher: Leo

We can see that just as these heroes compose a family, the four corresponding constellations make together one integral 120-degree part of the zodiac. It is also important to note that according to Armenian folklore (and other sources), it was revealed that these four zodiacal constellations together compose a larger figure of a *heavenly dragon*. The head of this dragon corresponds to Taurus, while the tail corresponds to the Lion. So, it turns out, that Sanasar and his family completely coincide with the huge heavenly dragon.

It is important to notice that Sanasar, as the fighter and killer of the dragon, occurs as the direct heir of the dragon. And as a result of this heritage we see that all the members of his family together form the huge heavenly dragon.

3. The Ancestors of the Family of Sanasar

It is natural to examine also the ancestors of this family and try to find out other possible parallels between the heroes of this epic and the real stars. As ancestors of Sanasar's family,

- we know that the mother of the founders of this family is called *Tsovinar*;
- we know also that the father of Tsovinar was *King Gagik*;
- her uncle was *Q'eri Toros* (Uncle Toros); and
- the husband of Tsovinar was the *Caliph of Baghdad.*

3.1. Tsovinar

Let us examine first the mother of twin brothers Sanasar and Baghdasar. We know from the epic that:

- she was an extremely beautiful maid;
- she was the daughter of the King Gagik;
- her name is closely correlated with the Armenian word *tsov-see*;
- she became pregnant from one and a half handfuls of *water* that she drank before marriage from the spring inside the sea, and her twin sons were from this water; and
- she was married forcibly, but she was never with her husband, and in fact remaining a virgin.

Thus we can take as the heavenly parallel of Tsovinar the constellation *Virgo*.

3.2. Q'eri Toros

To identify the constellation associated with Q'eri Toros, we note that:

- he is the only hero of the epic that figures in all four generations;
- he was also a giant like the other heroes and, like Ohan, he fought with a rooted poplar tree (although he fought with a mace as well);

- he always rode his six-legged, fast-footed horse, *Vetzotnen Lezgin*;
- his most significant intervention in the events of the epic is connected with the battle of Davit and Melik; before Davit's last decisive blow, he warned that he will weigh and demanded that the two halves of the murdered Melik be exactly equal to each other.

This means that the main characteristic of Q'eri Toros is the *weighing function*. Thus we can take as the heavenly parallel of **Toros** the constellation *Libra*.

3.3. The Caliph of Baghdad

The husband of Tsovinar, the Caliph of Baghdad, figures as an *enemy* in the epic. What we have concerning this character is that:

- He first acted as the enemy of Tsovinar's father, the King Gagik. He wanted to marry King Gagik's daughter, Tsovinar, but King Gagik refused to give his daughter in marriage in deference to their faith: Armenians were Christians (*khachapasht*, or worshiping the cross), while the Caliph was an idolater. The Caliph fought against the city of King Gagik and tried to force him to give up his daughter.
- In different episodes, he tried to kill the children, the sons of Tsovinar, Sanasar and Baghdasar.
- He promised to sacrifice the twin brothers Sanasar and Baghdasar to his Great Idol and always tried to keep his promise.
- He also tried to kill Tsovinar.
- The brothers were forced to escape from him, and they finally killed him.

We see, thus, that the Caliph always acted as the most dangerous enemy, and the only way to escape his persecution was to kill him. The only heavenly character that can be interpreted as a dangerous enemy is the zodiacal constellation Scorpius, the scorpion. This constellation traditionally represents the figure of a huge scorpion that tries to bite the heel of the man standing above (Ophiuchus). Thus we can take as the heavenly parallel of the Caliph of Baghdad the constellation *Scorpio*.

4. The Descendants of the Family of Sanasar

Now, as we examined and found the heavenly parallels of the family of Sanasar and his ancestors, let us see what we can find by examining the descendants of this family. We have two main heirs of Sanasar's family, Davit, and his son Pokr (Small) Mher.

4.1. Sasountzi Davit

The only son of Mher is Davit. He can be regarded as the central hero of the epic, as the description of his life and deeds is the most detailed and complete. We know as well that some variants of this epic are called after him: *Sasountzi Davit*. He has no special epithet like his father. He is called mainly *Sasountzi Davit* or *Sasna Davit*. These epithets only show his origin from Sasoun (as they mean "originating from Sasoun").

As his life is described completely and in great detail, we can find numerous qualities that distinguish him from other heroes of the epic. We find that:

- Davit was a shepherd;
- Davit was also the master of wild animals;
- He was an unsurpassed ploughman;
- Davit was also an archer (he hunts with his bow and arrow); and
- He also had the job of measuring (he measured the gold of the treasury).

We see, thus, that Davit had several characteristics that distinguish him from other heroes of the epic. Two of these characteristics are correlated with constellations. The zodiacal constellation of Sagittarius (the Archer) can be set as parallel to his skill as an archer; alternatively, we know that the old Mesopotamian constellations *Iku*, the Field, and *Apin*, the Plough (Van der Warden 1946) can be correlated with Davit's quality as a ploughman. As we found heavenly parallels of the other heroes of the epic in the zodiac, it is natural to look for Davit's parallel also in the Zodiac. We have to conclude that the heavenly parallel of Davit is the constellation Sagittarius, the Archer, and ignoring the parallel with the Field and Plough, as these constellations are not part of the zodiacal circle (although the Field, the rectangle of Pegasus, is very close to Pisces). Thus we find the heavenly parallel of Sasountzi Davit to be the constellation *Sagittarius*.

4.2. Pokr (Small) Mher

The last representative of the family of Sasoun is the only son of Davit—Mher. He is also called also *Pokr Mher* ("Mher the Small" to distinguish from his grandfather Mets Mher, "Mher the Great"). In spite of his epithet, Mher is the most powerful and greatest among the giants of the family of Sasoun. He became so big and heavy that the earth could not hold his weight. His horse's feet are stuck in the ground. Let us look what we have concerning this last hero of the epic.

- In his various qualities, he was quite comparable to the rest of the heroes of the epic, with the only difference being that he was immortal and had no offspring.
- Mher was the only one among the heroes of the epic that fought against his father, Davit.
- Mher became immortal and infertile in accordance to his father's curse, which he received as a result of fighting with his father.
- Mher fought with the angel of God and did not win nor lose.
- The soil did not withstand his weight, and he began to stick in the soil. Since the earth could no longer hold him, Mher eventually closes the cliff known as *Agrawou Kar* (Raven's Stone).
- There was an episode where his uncle said to him that he was the last one of the *rams* of the family of Sasna Toun (Sasountzi Dawit, p. 350). In other words, he was a Ram.

We can see that only the last of the above characteristics of Mher correlates with the constellations of the zodiac. Thus we can take as the heavenly parallel of Small Mher the constellation *Ares* the Ram.

We have identified nine persons from the epic with the zodiacal constellations. It is noteworthy that these nine constellations constitute one integral part of the Zodiac and follow one after another mainly in the same sequence as the heroes of the epic. There are only two cases where there is a lack of correspondence between the zodiacal constellations and the epic heroes. The first one is in the case of Dzenov Ohan, and the second one refers to Davit. It is possible that these two discrepancies are due to some changes in the original text of epic. We conclude that, in general, the heroes and their actions in the epic are in good correspondence with the constellations of zodiac.

We note as well that the temporal sequence of the heroes goes exactly in the opposite direction of the zodiacal constellations. This can be explained by the difference between the two worlds—Earth and Heaven—analagous to how the motions of the heavenly bodies and their paths as seen on earth go in opposite directions.

References

Ajarian Hrachia 1973, Armenian etymological dictionary, v. 2, p. 117, Yerevan (in Armenian)
Armenian Soviet Encyclopedia, 1984, v. 10, p. 195, Yerevan (in Armenian)
Broutian Grigor, 2008, Ejmiatsin, No. 2, p. 49 (in Armenian)
Orbeli I. A. 1939, Preface, Sasountzi Davit, p. XXVI Yerevan (in Armenian)
Sasna Tzrer, 1936, v. 1, Yerevan (in Armenian)
Sasna Tzrer, 1944, v. 2, part 1, Yerevan (in Armenian)
Sasna Tzrer, 1951, v. 2, part 2, Yerevan (in Armenian)
Sasna Tzrer, 1979, v. 3, Yerevan (in Armenian)
Sasna Tzrer, 1999, v. 4, Yerevan (in Armenian)
Sasountzi Dawit, 1936, Yerevan (in Armenian)
Van der Warden B. L. 1949, Babylonian Astronomy. II, The Thirty-six stars, Journal of Near Eastern Studies, 8, 6

Astronomical Heritage of the Middle East
ASP Conference Series, Vol. 520
Sona V. Farmanyan, Areg M. Mickaelian, J. McKim Malville, and Mohammad Bagheri, eds.

Sunrise as Precondition for the Orientation of Grave-Tombs in Ancient Armenia

Hakob Y. Simonyan
Scientific Research Center for the Historical and Cultural Heritage of the Ministry of Culture, Republic of Armenia; haksimon@gmail.com

Abstract. In the beliefs and funeral rites of ancient peoples, great importance was placed on the journey of the deceased to the sacred side of the world. Each religion and culture chose a particular cardinal point that was considered sacred. The orientation of Bronze Age grave pits in Armenia gives us reason to conclude that the orientation of the grave pits was done by observing the sunrise. The shifting of the grave pits from the accepted geographical directions was conditioned according to the daily journey of the sun on the horizon. Astro-archaeological calculations give us the opportunity to determine, with a precision of three days, the month and the day when the pit was dug.

1. Introduction

According to different beliefs of ancient peoples concerning the afterlife, the soul of the deceased could reach the world of "eternal peace" only if all the funeral rites were performed properly; otherwise, their souls would either become wanderers finding no rest, or face an eventual permanent death.

In the mythology concerning the afterlife one, of the cardinal directions of the world was considered sacred. It was in that direction that the world of the dead was located, where the souls of the deceased would make their long and hard journey (Petrukhin 1980). Due to this perception, the funeral rites gave a special significance to the correct "pointing" of the way to the "netherworld." Every deviation from that orientation would have threatened the ability of the soul of the dead to reach the world of afterlife; thus, the lost soul would lose the ability to attain the blissful life in the netherworld. Even now, when many features of those rites are long forgotten, the rite of orientation is properly performed.

The grave was often viewed as the "home of the dead" as an intermediary between the real and the other worlds. Very often, the funerary structures were given shape of a carriage or a boat that would serve as a means of transportation to accomplish the final and the extremely dangerous journey (Mongayt 1974; Petrukhin 1980). Therefore, one of the essential elements of the funeral rites of the Bronze and Early Iron Ages was the orientation of the graves towards the respective sacred direction.

However, while studying the location of burial chambers in the Bronze and early Iron Age cemeteries of Armenia, we often come across grave pits that seemingly point in any particular orientation. Most of the burial chambers show deviations from more or less precise orientation towards the cardinal directions of the world. Based on such data, a number of archaeologists have suggested that the orientation of graves was chosen at will (Kleyn 1962; Khachatryan 1975).

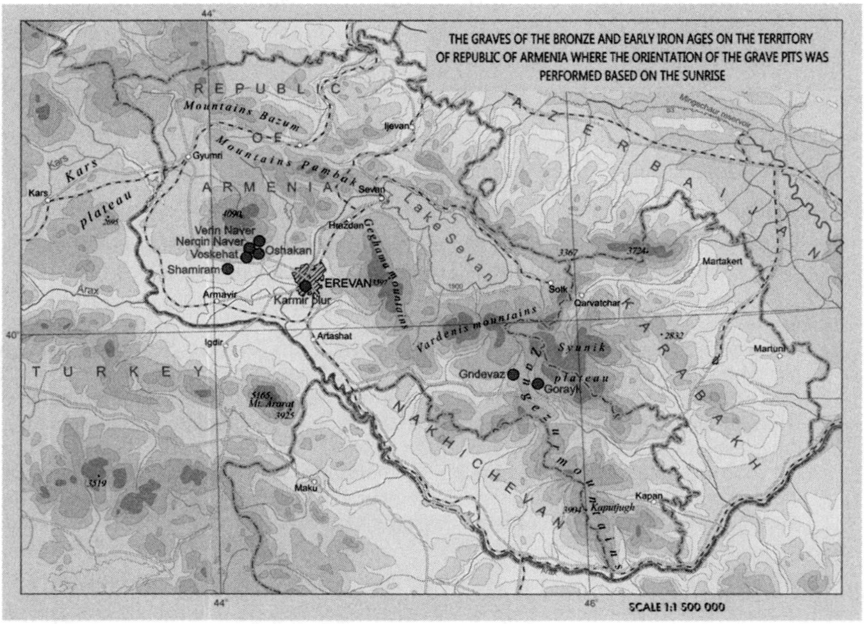

Fig. 1. Map of the Bronze Age cemeteries with burial orientation observations.

Figure 1. Map of Bronze Age cemeteries having burial orientation observations.

We have conducted excavations at a number of cemeteries in Armenia (Shamiram, Verin Naver, Nerkin Naver, Karmir Blur, Voskehat, Gndevaz, Gor Hayk and others) and measured the orientations of the grave pits (Fig. 1). Based on that research, we concluded that north was the distinct sacred direction for the inhabitants of the ancient Voskehat (early Bronze Age), the people of the Karmir Berd culture (middle Bronze Age), and the inhabitants of Shamiram town and Teishebaini (early Iron Age). As for the Treghk-Vanadzor culture (middle Bronze Age) of the same region, the burial chambers were oriented to the east. At Verin Naver, the burial pits narrowed on the northern side and ended in a curving wall, resembling a boat with its stern directed toward the north (Fig. 2). Meanwhile, at the Shamiram cemetery, small altars constructed of four vertical tuff slabs were attached to the northern side of the stone cists where the remains of the deceased were deposited. However, the longitudinal axes of the burial pits of these cemeteries, definitely pointed toward the north, and at the same time deviated towards both the northeast and the northwest. Thus, we need to reconcile the seeming contradiction between the obvious intent to orient the grave pits toward the north and the actual deviations recorded in the field.

2. Orientation of the Burial Pits According to the Astronomical Observations in Ancient Armenia

In some cases, burial pits were placed side by side, and even those located within one fenced area of the cromlech, deviated from the precise direction of the cardinal points

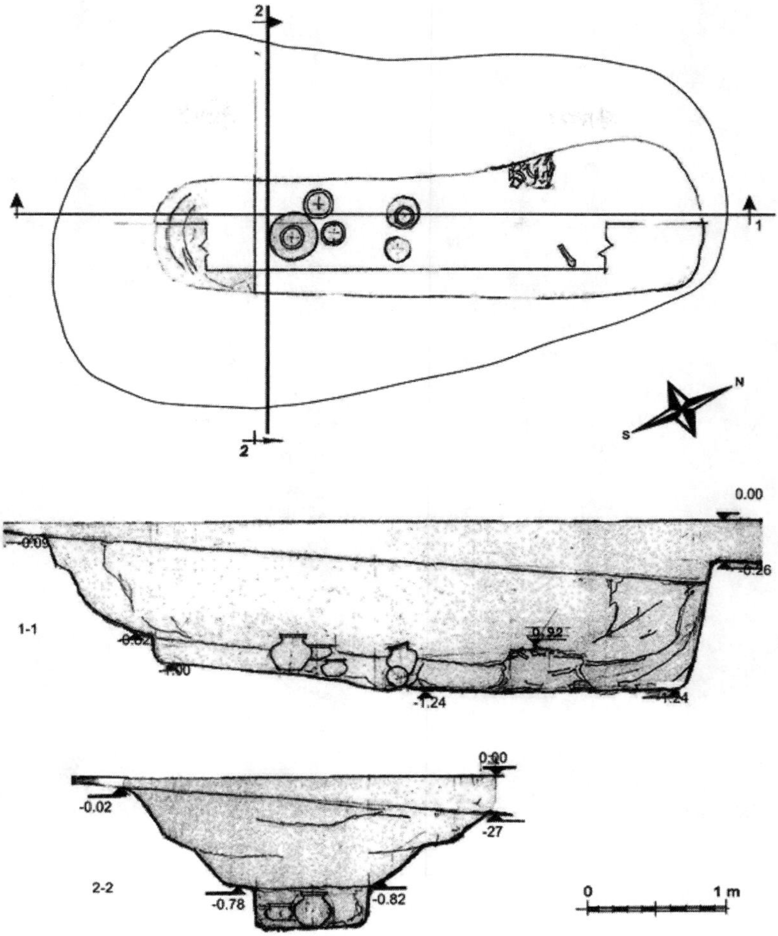

Figure 2. Burial No. 96 from Verin Naver cemetery with the rock-carved boat-shaped outline.

(Fig. 3). We may conclude that mountains or other major features of landscape that served as objects of worship could not have served as points for directional cultic orientation toward the north. There is sufficient textual, ethnographic, and archaeological evidence to conclude that ancient peoples worshiped celestial bodies and used those as points for orientation. Celestial bodies such as the Sun, the Moon, the planets and bright stars that almost unnoticeably changed their perceivable trajectories in the sky could have served as points for orientation. Nevertheless, those heavenly bodies should possess the following features:

- visibility during every season of the year;

Figure 3. Karmir Blur cemetery aerial view with marked orientations of the graves.

- perceived permanence of the point of rising over the horizon; and
- a regular cycle of oscillation of the trajectory of movement.

The planets are not visible in the sky all year round. In different years, they appear during various seasons. The visibility of stars also depends on seasons, except for a small group of them, which have insignificant oscillation and are not bright stars. Sirius appears in the spring and is visible only for six months. Polaris has no oscillation and, always pointing to the north, serves as an immovable point for orientation like the immovable objects on Earth (Bakulin 1981). Therefore, the planets and stars do not satisfy the requirements set for the reference point.

Among the possible moving celestial bodies, only the Sun possesses the required features. Only twice during the annual cycle, on the vernal and autumnal equinoxes, does the Sun rises exactly in the east. On all other days, because of the elliptical orbit and axial tilt of the Earth, the point of the sunrise over the horizon shifts cyclically during the year towards the north or the south depending on the season of the year. In the northern spring and summer, the Sun rises north of east, and towards the south in autumn and winter (Fig. 4).

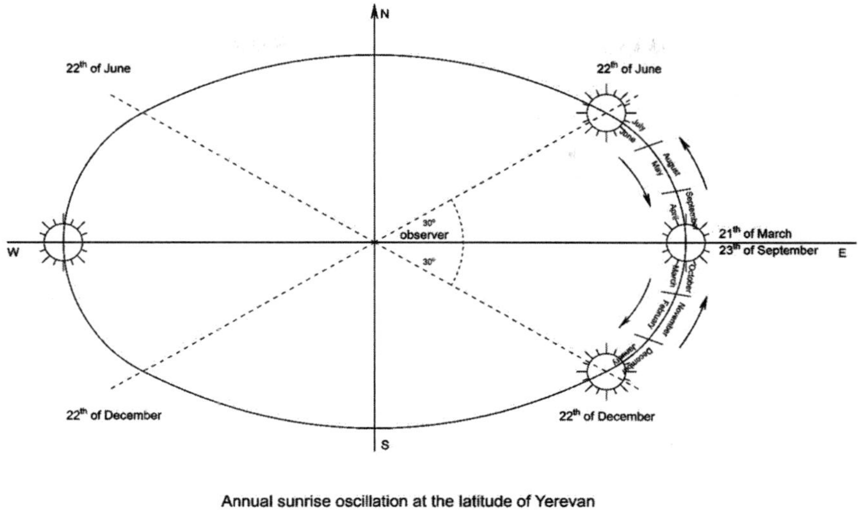

Figure 4. Annual sunrise oscillation at the latitude of Yerevan.

Hence, when using the Sun as a reference for cardinal directions, deviations in the orientation of grave pits were certain to take place. Naturally, the orientation of burials deviated in both directions from the cardinal points. However, the deviations of the axes of burial chambers from the cardinal points should correspond to the amplitudes of deviations in the points of sunrise according to the geographical latitude within the entire geographic area of a given culture.

In order to determine the largest amplitude of the rise of the Sun from the true east, we must detemine the azimuth of the sunrise—the angle between the geographic east and the actual points of sunrise during the winter and summer solstices. The annual cycle of divergence of the sunrise from the true east can be calculated according to well-known formulas of astronomy, by calculating the azimuth of the sunrise:

$$\cos\alpha = -\frac{\sin\beta}{\cos\gamma}$$

where α is azimuth, β is Sun deviation, and γ is the geographical latitude of the place. (Tumanyan & Vatyan 1960; Bakulin, Kanonovich, & Moroz 1983).

The maximal deviation of the sunrise from the true east happens at the solstices (usually June 21 and December 21). On these days the deviation of of the Sun is $23°26'$ north or south. The geographical latitude of Armenia is about 40° north. By inserting these into the equation above, we get 31° as the maximum angle of the deviation of the Sun on the horizon at the Armenian latitude. The deviations in the orientations of the Karmir Berd grave pits toward the north and the south from true east at the necropolis of Verin Naver are listed in Table 1. The data in the table indicate that not a single grave pit deviates from the E/W axis by more than 31°. Hence, we can conclude that the

Sun was the celestial object used to determine the orientation of the grave pits. Using the point of sunrise in orienting the graves of their deceased, people of the Bronze and early Iron Ages would have logically followed its seasonal deviations from true east.

Table 1. Measured deviations of the burial orientations of the Verin Naver graves from east, and the estimated time of their construction.

Burial number	Deviation NE	Deviation SE	Burial time
1	25°		10 May or 1–3 Aug
2		28°	10 Jan or 5 Dec
3	18°		18–23 Apr or 17–19 Aug
4		11°	25 Feb or 17–18 Oct
5	18°		18–23 Apr or 17–19 Aug
6		3°	
7	27°		16–17 May or 30 Jul
8		10°	
9	31°		1 Jun or 10 Jul
10	7°		1 Apr or 12–13 Sep
12	43° or 33°		
16	20°		
17	26°		
18	18°		
21		11°	25 Feb or 17–18 Oct
22		21°	2–3 Feb or 10 Nov
23		12°	22–23 Feb or 20 Oct
24		27°	15 Jan or 30 Nov
25		24°	25 Jan or 20 Nov
26	6°		30 Mar or 15 Sep
27	10°		
28		15°	15 Feb or 26–29 Oct
29	20°		
30	7°		
33	30°		
34	45°		
36		23°	
40		344°	
41		23°	
42		94°	
47		315°	
48		355°	
49	17°		
50		12°	
51	20°		
52		10°	
53		0°	
55		29°	
56		16°	

Based on this hypothesis, we may attempt to reconstruct the following procedure of orientation of graves at the cemetery of Verin Naver. The ancient observers (the priests) viewed the point of the sunrise as the eastern side of the world. At dawn the point of the sunrise was fixed. Drawing a straight line towards the perpendicular, they determined the needed northern reference point to dig the burial pits. The observable deviations from true north in the orientation of the grave pits was determined by the seasonal oscillation of the sunrise, which created the appearance of arbitrary orientations of burials, which took place at different times of the year.

A similar calculation can be made for other cultures, where the archaeological burials are oriented to the north or east. For cemeteries with burials oriented in different directions, the calculations can serve as one of the criteria for determining membership of individuals in a particular culture. This theory can be tested with DNA analysis.

The sunrise point moves on the horizon along a constant trajectory. Using the annual cycle of variations in the azimuth of the Sun and its deviation from the point of the true east, we can address the question concerning the specific time of the year when the burial ritual was performed. Measuring the magnitude of the deviation of the longitudinal axis of a burial pit from the point of the true east, we can determine the season, the month, and the date of construction of the burial pit with the accuracy of about three days (Table 1). Since the Sun appears at the same point on the horizon twice a year (while moving to the extremal point of its amplitude and on its way back), there is a specific number of days corresponding to each degree of deviation during different months of the year.

Since the period under discussion precedes the present time by about four thousand years, our calculations must take into account the shifts in the tilt of the ecliptic towards the equator. At present, the tilt of the ecliptic is approximately 23°26′. According to the calculations, the tilt angle was equal to 24° during the mentioned period, i.e., the differences in the azimuth of the sunrise at the latitude of Yerevan over the course of elapsed centuries were quite small (Table 2).

Table 2 presents the locations of the sunrise azimuths for the latitude of Yerevan. The eastern azimuths are usually accepted as negative values, while the deviations of the sunrise from east is arbitrary (we have adopted positive for north in Table 2).

At the same time, the impact of features of the landscape in a given place should be considered as a factor in the observation of the magnitude of the deviation of the point of the sunrise from the true east. The Sun observed in the mountains rises later than in the valleys at the same latitudes. In particular cases, major elevations of the landscape on the horizon can delay the observation of the sunrise. It is very important to take topography into account when determining the probable months and days when a particular burial was performed. When determining the orientation of a burial using a compass in the field, one must also take into account the difference between the location of the geographic poles of the Earth and the north magnetic pole. This difference is quite substantial in the northern regions of the Earth, but at the latitude of Yerevan, the correction barely reaches one degree.

Figure 5a shows the distribution of the death rate of the population according to the seasons of the year, based on the records from 23 burials excavated at the cemetery of Verin Naver. Since the variations in the orientation of burial pits corresponds to the deviation of the sunrise from the true east two times a year, we drew two possible versions of the figure based on the same data. At present, we draw no conclusions due to the lack of sufficient anthropological and statistical data. However, we find it quite appropriate as an illustration of the practical application of our method for determining the time of burial. The histogram (Fig. 5a) shows that the dead were buried during every season of the year at the cemetery of Verin Naver.

Table 2. Annual variation of the azimuth angle α (east of north is negative) and their deviations from true east β (with north of east being positive) at the latitude of Yerevan (N 40°5′, E 44°33′).

Days	January		February		March	
1	−58°.76	−29°.24	−68°.53	−21°.47	−80°.89	−8°.99
5	−61°.39	−28°.61	−70°.28	−19°.72	−83°.03	−6°.97
10	−62°.07	−27°.93	−72°.30	−17°.70	−86°.06	−3°.94
15	−63°.06	−26°.94	−74°.73	−15°.27	−88°.08	−1°.92
20	−64°.60	−25°.40	−76°.75	−13°.25	−91°.11	+1°.11
25	−66°.29	−23°.71	−79°.36	−10°.64	−93°.54	+3°.54
30	−67°.76	−22°.24	−81°.01	−09°.11	−96°.17	+6°.17
	April		May		June	
1	−96°.94	+06°.94	−111°.10	+21°.16	−121°.16	+31°.16
5	−99°.09	+09°.09	−112°.72	+22°.72	−121°.51	+31°.51
10	−101°.52	+11°.52	−114°.78	+24°.78	−122°.25	+32°.25
15	−104°.15	+14°.15	−116°.46	+26°.46	−123°.48	+33°.48
20	−106°.08	+16°.18	−117°.97	+27°.97	−123°.00	+33°.00
25	−108°.62	+18°.62	−119°.12	+29°.12	−123°.00	+33°.00
30	−110°.67	+20°.67	−120°.31	+30°.31	−123°.00	+33°.00
	July		August		September	
1	−122°.75	+32°.75	−115°.50	+25°.50	−102°.90	+12°.90
5	−122°.12	+32°.12	−113°.92	+23°.92	−100°.74	+10°.74
10	−121°.00	+31°.00	−111°.86	+21°.86	−98°.30	+08°.30
15	−120°.34	+30°.34	−109°.40	+19°.40	−95°.65	+05°.65
20	−119°.15	+29°.15	−107°.35	+17°.35	−93°.22	+03°.22
25	−118°.01	+28°.01	−105°.71	+15°.71	−90°.58	+00°.58
30	−116°.90	+26°.90	−103°.67	+13°.67	−88°.15	−01°.85
	October		November		December	
1	−87°.66	−02°.34	−72°.61	−17°.39	−62°.70	−27°.30
5	−85°.51	−04°.49	−71°.06	−18°.94	−62°.00	−28°.00
10	−83°.08	−06°.92	−69°.03	−20°.97	−61°.70	−28°.30
15	−80°.05	−09°.95	−66°.97	−23°.03	−60°.99	−29°.01
20	−78°.03	−11°.97	−65°.90	−24°.10	−60°.99	−29°.01
25	−76°.00	−14°.00	−64°.78	−25°.22	−60°.99	−29°.01
30	−73°.98	−16°.02	−63°.62	−26°.38	−60°.99	−29°.01

3. Conclusions

The proposed method of determination of burial pits' orientation and the season of their construction is based on data from the archaeological sites of Armenia. It can also be applied to other areas with developed religious systems where burials display analogous deviations from the sacred cardinal direction for the given culture. The geographical latitude and topography of the area of a particular archaeological culture should be considered as well.

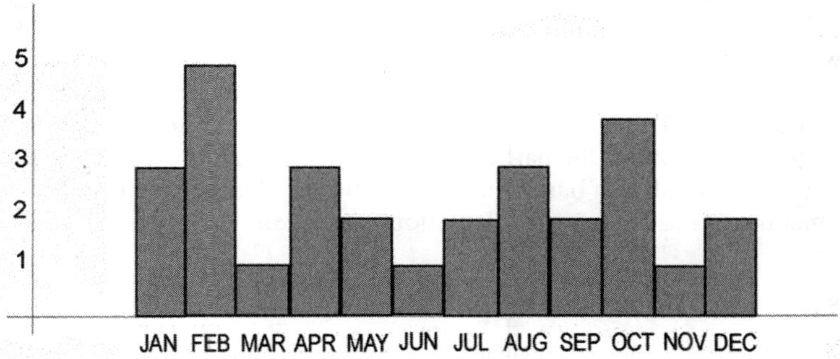

a. Distributions of the death rate of the population according to the data from 23 burials from Verin Naver cemetery

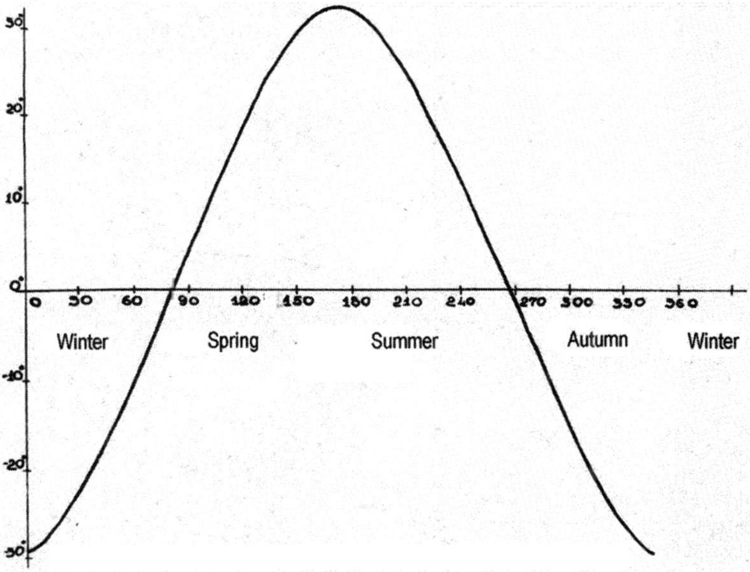

b. Annual cycle of deviation of the sunrise from the East

Figure 5. (a) Distributions of the death rate of the population according to data from 23 burials from Verin Naver cemetery; (b) Annual cycle of deviation of the sunrise from the east.

Our hypothesis was born around 40 years ago (Simonyan 1984). Back then, there were very little data based on methodically and properly excavated sites. In the course of time, I accumulated a great number of measurements from various Bronze and early Iron Age sites. The further we go, the more facts support our hypothesis.

Acknowledgments. Given the difficulty of this research, I had to counsel with outstanding scholars of the time period. I would like pay special tribute to the late scholar and astronomer Benik Tumanyan, the astrophysicist Yevgeni Andreev, as well as to express my gratitude to all of the participants of our long-lasting research expeditions, due to which a great number of burial pits were meticulously excavated, archeologically measured, making our astronomical calculations possible.

References

Bakulin P. I. 1981, Astronomicheskij kalendar'. Postoyannaya chast'. Seventh issue elaborated (Moskva: Nauka), 704 (in Russian)

Bakulin P. I., Kononovich E. W., Moroz V. I. 1983, Kurs obschej astronomii (Moskva: Nauka), 560 (in Russian)

Kleyn L. S. 1962, "Katakombnie pamyatniki epokhi bronzi I problema videleniya arkheologicheskikh kul'tur," Sovetskaya Arkheologiya, 2, 26 (in Russian)

Mongayt A. L. 1974, Arkheologiya Zapadnoj Evropi, Bronzovij I zheleznij veka (Moskva: Nauka), 408 (in Russian)

Petrukhin V. Ya. 1980, Zagrobnij mir, Mifi narodov mira, Entsiklopediya, vol. I (Moscow), 452

Wood J. E. 1978, "Sun, Moon and standing stones," (Oxford University: London, New-York)

Astronomical Heritage of the Middle East
ASP Conference Series, Vol. 520
Sona V. Farmanyan, Areg M. Mickaelian, J. McKim Malville, and Mohammad Bagheri, eds.

The First Cosmological Works (*Rasāʾil al-Hayʾa*) in the Islamic Period: a Comparison between al-Farghānī's *Jawāmiʿ ʿIlm al-Nujūm* and the 3rd Book of Kūshyār's *al-Zīj al-Jāmiʿ*

Hanif Ghalandari
Institute for the History of Science, University of Tehran, Iran;
hanif.ghalandari@ut.ac.ir

Abstract. *Hayʾa* is a general name for historical astronomical manuscripts that provide a geometrical description of the configuration of the world. Most of this kind of astronomical works were written by Islamic scholars most likely influenced by Ptolemy's *Planetary Hypotheses*. Most of Islamic hayʾa works have a similar structure and order of chapters, and were written in the 6th century AH (12th century AD). Nevertheless, some astronomical works from the 3rd–4th centuries AH (9th–10th centuries AD) have similarities with the hayʾa works, so we can call them the "predecessors" of the hayʾa. Al-Farghānī's *Jawāmiʿ ʿIlm al-Nujūm* and the third chapter of Kūshyār's *al-Zīj al-Jāmiʿ* are two important works among these predecessors. Considering the typical four chapter-organization of Hayʾa works, i.e., mathematical and physical introduction, configuration of the heavens, mathematical geography, and masses and distances (Abʿād wa Ajrām), we conclude in this article that these two treatises are impressive predecessors to the hayʾa, and that the third chapter of Kūshyār's *al-Zīj al-Jāmiʿ* is more similar to the subsequent Hayʾa works.

1. Introduction

Hayʾa works are a genre of astronomical literature which developed as early as 11th century in eastern Islam. The main goal of the authors of these works was to construct a picture of the world, the heavens, and the Earth. Naṣīr al-Dīn al-Ṭūsī in his *Tadhkira* written it in second half of 13th century, defines hayʾa as "the subject of astronomy of the simple bodies, both superior and inferior, with respect to their quantities, qualities, positions, and intrinsic motions" (F. J. Ragep, vol. 1, p. 91). In other words, hayʾa works focused on topics related to the configuration of the universe (*hayʾa al-ʿalam*) as a coherent whole: both the upper bodies ("cosmo-graphy") and the lower bodies of the terrestrial realm ("geo-graphy"). Hayʾa works have a definite structure. Although not all hayʾa books use the exact structure, most of these works, and certainly those was written in 13th century, contain four chapters or sections: (1) geometrical and physical principles, (2) on the heavens and the movements of the planets, (3) mathematical geography and (4) masses and distances of the Sun, Moon, and planets (for a brief history of the hayʾa texts, see Ragep 1993, pp. 29–40; Saliba 1994, pp. 15–39; Langermann 1997, pp. 185–213; Gamini 2014, pp. 243–290).

Although we can't know with certainty the date in which writing hayʾa books began, we can place the beginning of this genre somewhere around the year 1100 AD. The reason is that the probable date of the redaction of *Giyhān Shinākht* (Configuration of the universe) by Qaṭṭān Marwazī (in Persian) is 1104 AD and *Muntaha al-Idrāk fī*

Taqāsīm al-Aflāk (Ultimate comprehension of subdivision of celestial spheres) of ʿAbd al-Djabbār Kharaqī (in Arabic) was written in 1132–1133 AD. Nevertheless, from a historical point of view, we can include some other works written in previous centuries in this literary genre. We call this group of works "the predecessors of hayʾa works." The predecessors of hayʾa mean the works which were written before Ibn al-Hayatham's *Configuration of the World* (*fī Hayʾa al-ʿAlam*), probably written in the first half of the 10th century. This work influenced the hayʾa works that followed; Kharaqī, in the introduction to his *Muntaha al-Idrāk*, mentions that he adheres to *fī hayʾa al-ʿAlam* and says that the work is the first to mention solid orbs (Ghalandari 2012, pp. 19–22).

Some of the best known predecessor works are: Yaʿqūb ibn Ṭāriq's *Tarkīb al-Aflāk*, two works by Thābit ibn Qurra, *Tashīl al-Majisti* and *fī Dhikr al-Aflāk wa Khalqiha*, and Farghānī's *Jawāmiʿ ʿIlm al-Nujūm*, which is the most famous. There are no extant manuscripts of Yaʿqūb's treatise, and information about this book is only available through Bīrūnī's reports in his *Taḥqīq Mal al-Hind.* In addition, Tābit's works are very brief; the author tries to make some definitions in short sentences in order to prepare a summary on Ptolemy's works the *Almagest* and *Planetary Hypotheses* (Ragep 1993, vol. 1, pp. 29–30; Ragep 2016, pp. 45–55).

In this paper we add another important work to this list, the 3rd book of Kushyār's *al-Zīj al-Jāmiʿ*. Although this is not an independent treaise, the arrangement of contents of the chapters of this book is like the hayʾa works. Here, we compare these two treatises to show that one of these predecessors is more similar to the classical hayʾa works.

2. *Jawāmiʿ ʿIlm al-Nujūm*

Jawāmiʿ contains 30 chapters: a brief review of history (chapter one), six chapters on mathematical geography (chapters 6–11), two chapters about the masses and distances of planets, Sun, and Moon (chapters 21 and 22) and the rest about the movements of the heavenly bodies. We compare the contents with subsequent hayʾa works and observe the following:

1. Unlike classical hayʾa works, chapters of *Jawāmiʿ* are not ordered in a definite structure. Excluding the first chapter of *Jawāmiʿ*, the History, we can compare it to Ptolemy's *Almagest*. In other words, it is as if *Jawāmiʿ* is an abstract of *Almagest*.

2. Farghānī's chapter on history contains some of the famous chronicles, i.e., the Syriac calendar, the Islamic calendar, the Persian calendar, the date of their beginnings and their months. This information is not included in later hayʾa works.

3. Past scholars believed that the most of inhabited world in the northern hemisphere fell between 10 and 50 degrees in latitude (in some sources as far north as 67 degrees). In Islamic astronomical works, this area was devided into seven parallel parts called "climes." Each clime extended east to west in longitude while its extent in latitude was narrow, which resulted in a difference of half an hour in the length of the longest day. Chapters 8 and 9 of *Jawāmiʿ* talks about this division, as do the other classical hayʾa works.

4. Co-ascensions, or measuring the arc on the equinoctial that rises with a given arc on the ecliptic orb, is an important part in historical astronomical books. As the

variation of the co-ascensions depends the latitude, this subject was divided into two parts: co-ascensions in "sphaera recta" (as viewed from the equator, where latitude is zero) and co-ascensions "sphaera obliqua" (from any other latitude). Ptolemy included similar contents in the *Almagest*, but he first covered the geometrical demonstation of these calculations. The authors of the hayʾa works wrote these contents without any geometrical demonstrations or calculations. They discribed the co-ascensions for different latitudes on the Earth. Farghānī provides these descriptions in chapters 7 and 10 of *Jawāmiʿ*.

5. Description of the movements of planets is the main goal of the hayʾa works. Usually, in classical hayʾa books, these arguments were arranged in several chapters: (1) on the arrangement and order of the bodies, (2) on the motion of the Sun in ecliptic longitude, (3) on the motion of the Moon in ecliptic longitude, (4) on the motion of the upper planets (Mars, Jupiter, Saturn) and Venus, in ecliptic longitude, (5) on the motion of Mercury in ecliptic longitude, (6) on the motion of the planets, Sun, and Moon in ecliptic latitude. We do not see this arrangement in *Jawāmiʿ* and Farghānī considers these arguments in less detail. Also he did not describe the characeristics of each planet in one chapter, for example in chapter 14 he wrote about the movements of the planets in longitude but doesn't mention some important characteristic like the size of the epicycle radius of each planet, the motion of their epicycle orb, the eccentricity of their deferent orb, etc. Nevertheles ,he wrote some of this information in chapters 16 and 17. This is unlike the usual organization that we see in subsequent hayʾa works.

3. The Third Book of *Kūshyār's al-Zīj al-Jāmiʿ*

Kūshyār's *al-Zīj al-Jāmiʿ*, which was written in the middle of the 4th century AH (10th centure AD), is one of the famous astronomical tables (*zījes*) of the Islamic era. Kennedy reports that geometrical introduction and tables of *al-Zīj al-Jāmiʿ* is similar to Battānī's *al-Zīj al-Ṣābī* although *al-Zīj al-Jāmiʿ* includes some new precise observations. Moreover, Kūshyār adds a "Book" (*Maqāla*) to his treatise entitled "fī al-hay'a," which is about the configuration of the world and it is, like Farghānī's *Jawāmiʿ*, similar to the later hayʾa works. In the following we compare these two treatises according to what was mentioned in the case of *Jawāmiʿ* :

1. Unlike *Jawāmiʿ*, "fī al-hya'a" is well organized, and the four principle chapters of hayʾa works can be seen in it: (1) geometrical and physical principles (chapters 1–3), (2) on the heavens and the movements of the planets (chapters 14–21 and 23–30), (3) mathematical geography (chapters 4–13) and (4) masses and distances (chapter 22).

2. "fī al-hayʾa" does not have a distinct chapter on history.

3. Kūshyār does not write about the descriptive geography; in chapter 9 he only writes about the extention of the climes based on the length of longest day in each of them.

4. On the co-ascensions, there is no difference between the two treatises except that the order of chapters in "fī al-hayʾa" is more like the later hayʾa works.

5. Unlike *Jawāmiᶜ*, description of the movements of planets in "fī al-hay'a" is well ordered. Kūshyār writes about the movements of the Sun, Moon, and planets in longitude in chapters 16–19, and about the movements in latitude in chapter 20.

4. Conclusion

Among the predecessors to the hay'a works, these two books, *Jawāmiᶜ* and the 3rd book of *Zīj al-Jāmiᶜ*, are more similar to the standard works. By comparison, it is clear that the structure of Kūshyār's text and its chapters are more similar to hay'a books. This work is one of the important sources used by the writers of the hay'a books.

Acknowledgments. This paper is based on a comperehensive article by the author "Third Chapter of Kūshyār's *Zīj* and Farghānī's *Jawāmiᶜ ᶜIlm al-Nujūm*: a Comparison between Two Predecessors of Hay'a Works and Their Importance in Hay'a Books," previously published in Persian in *Tarikh-e Elm*, vol. 12 (1), pp. 39–72.

References

Bagheri, M. 2009, *Books I and IV of Kūshyār ibn Labbān's Jāmiᶜ Zīj* (Frankfurt)

DeYoung, G. 2007, "Farghanī," *The Biographical Encyclopedia of Astronomers*, Ed. T. Hockey et al. (New York: Springer)

al-Farghānī (Alfarganvs) 1669, *Elementa Astronomica*, Arabicè & Latinè, opera Jacob Golli, Amstelodami (Reprinted in Fuat Sezgin, 1997), *Islamic Mathematics and Astronomy*, Vols. 9–10 (Frankfurt)

al-Farghānī (Alfarganvs), *Jawāmiᶜ ᶜIlm al-Nujūm*, Mn. No. 135L (University of Princeton)

Gamini, A. M. 2014, "Diversity and Variety of Hay'a Books in Islamic Civilization," *Tarikh-e Elm*, 11(2), 243–290 (in Persian)

Ghalandari, H. 2012, "A Survey of the Works of *Hay'a* in the Islamic Period with a Critical Edition, Translation and Commentary on the Treatise *Muntaha al-Idrāk fī Taqāsīm al-Aflāk* witten by Baha' al-Dīn al-Kharaqī," Ph.D. dissertation, Institute for Humanities and Cultural Studies (IHCS), (in Persian)

Ghalandari, H. 2015, "Third Chapter of Kūshyār's *Zīj* and Farghānī's *Jawāmiᶜ ᶜIlm al-Nujūm*: a Comparison between Two Predecessors of *Hay'a* Works and Their Importance in *hay'a* Books," *Tarikh-e Elm*, 12 (1), 39–72 (in Persian)

Goldstein, B. R. 1967, "The Arabic Version of Ptolemy's Planetary Hypothesis," Transaction of the American Philosophical Society, vol.LVII (part4)

Kennedy, E. S. 1956, "A Survey of Islamic Astronomical Tables," Transactions of the American Philosophical Society, 46 (2)

Kūshyār Gīlānī, *al-Zīj al-Jāmiᶜ*, Mn. No. 748/3. Yeni Cami Library (Istanbul)

Langermann, Y. T. 1997, "Arabic Cosmology," Early Science and Medicine, 2(2), 185–213

Ptolemy 1998, *Almagest*, translated and annotated by G. J. Toomer (London)

Ragep, F. J. 1993, *Naṣīr al-Dīn al-Tūsī's Memoir on Astronomy* (*al-Tadhkira fī 'ilm al-hay'a*) (New York)

Ragep, S. 2016, *Jaghmīnī's Mulakhkhaṣ: An Islamaic Introduction to Ptolemaic Astronomy* (Springer)

Rosenfeld, B. A. & Ihsanoglu, E. 2003, "Mathematicians, Astronomers, and Other Scholars of Islamic Civilization and their Works (7th–19th cen.)," (Istanbul)

Sabra, A. I. 1976, "al-Farghānī." Dictionary of Scientific Biography. Edited by Charles Gillispie. vol. 4 (New York: Charles Scribner's Sons)

Saliba, G. 1994, A History of Arabic Astronomy (New York: New York University Press), 15–39

Suter, H. (J. Vernet) 1991, "al-Farghānī," Encyclopedia of Islam (2nd ed.) (Leiden: Brill)

Astronomical Heritage of the Middle East
ASP Conference Series, Vol. 520
Sona V. Farmanyan, Areg M. Mickaelian, J. McKim Malville, and Mohammad Bagheri, eds.

Transylvanian Astronomical Observatories in the 18th Century

Ferenc Szenkovits
Babeş–Bolyai University, Cluj-Napoca, Cluj, Romania;
fszenko@math.ubbcluj.ro

Abstract. The natural sciences in Transylvania came to the fore in the 18th century. At that time, the teaching of astronomical knowledge became more and more important. The first observatories suitable for astronomical observations were founded in this period in Cluj (Kolozsvár) and Alba Iulia (Gyulafehérvár). The famous astronomer Maximillian Hell was one of the Jesuits who contributed substantially to the foundations of these astronomical observatories. This paper offers a short overview of the beginnings of these two observatories, which are important parts of the European astronomical heritage.

1. Introduction

Interest in astronomy in Transylvania, part of Romania today, started centuries ago. We mention here only a few of the most important people and buildings related to the astronomical heritage of Transylvania; more details can be found in Dezső (1944) and Heinrich (1978).

- In this multi-secular heritage is important to mention the astronomical accomplishments of Ioan Vitéz de Zredna (1408–1472), Bishop of Oradea, who was in contact with the famous astronomers Peuerbach and Regiomontanus.
- The Transylvanian humanist and reformer, Johannes Honter (Honterus, 1498–1549) presented in his geographical book *Rudimenta cosmographica* (Kraków, 1530) the description of the Universe following Ptolemy. Honter also engraved and printed two stellar charts (Bazel, 1532).
- The most spectacular astronomical phenomena, the appearance of comets, were observed and recorded in books printed in Transylvania by authors Gaspar Heltai (Caspar Helter), Wilhelm Misocacus, and Stephanus Wolphard.
- Astronomical papers were published in the 17th century by Izrael Hübner and Samuel Kaposi.
- Astronomy was also present in the schools of the time. Famous professors and authors of textbooks include Johannes Henricus Bisterfeld (1605–1655), Johann Heinrich Alsted (1588–1638), and János Aptáczai Csere (1625–1659), who introduced astronomy in the schools.

- The presence of astronomical knowledge in Transylvania is also reflected in the medieval buildings and architectural heritage. There are dozens of medieval sundials in this region. One of the oldest sundials in Transylvania can be found on the south wall of the Cluj-Mănaştur Calvaria Church (see Figure 1). The original church of the Benedictine abbey, surrounded by defensive walls, was founded in the 11th century.

- As part of the astronomical heritage, we also mention the Wolphard-Kakas House, located on the main square of Cluj-Napoca, the city's most important Renaissance monument, built in the 16th century. This house built by Stephanus Wolphart, the city chief judge who was interesting in astronomy, was extended by later owner István Kakas, creating a zodiacal room decorated with twelve zodiacal symbols (Figure 2).

Figure 1. Sundial on the wall of the church in Manastur-Cluj (photos by the author).

Figure 2. Zodiacal ornaments in the Wolphard–Kakas house: Aries, Gemini, Cancer and Libra, from Heinrich (1978).

2. Designers of the Cluj Observatory

The special interest in astronomy in Transylvania led to the foundation of astronomical observatories. The most important ones were the Observatory of the Cluj University (1759–1766), and that of the Batthyaneum Library in Alba Iulia (1792–1794).

The first university college in Transylvania was founded in 1581 in Cluj by the Prince of Transylvania and King of Poland Stefan Bathory, under the direction of the Jesuits with the Italian Antonio Possevino as Rector. The establishment of the College was approved by Pope Gregory XIII in 1582, the year of the calendar reform. The College's old picture (Figure 3) can be found in the Vatican library (Heinrich 1978).

Figure 3. *Left:* The first building of the Bathory College near the church (around 1600); *Right:* The building of the Collegium Academicum (1740).

With the development of the Cluj University, the natural sciences, including astronomy, became increasingly important. As a consequence, the construction of an astronomical observatory became a necessity. Two Jesuits played a central role in astronomical education and the design of an astronomical observatory in Cluj: Nicolaus Ianosi (1701–1741) and Maximilian Hell (Maximiliano Höll) (1720–1792). Ianosi and Höll lived and worked in the new building of the Jesuit Collegium Academicum shown in Figure 3 (Heinrich 1978).

The Jesuit father Nicolas Ianosi, born in Cluj, was appointed professor at the University of Cluj (1734–1739) a few years after graduating from Vienna. In this period, he laid the foundations of a high-quality astronomical education and made the first plans for an astronomical observatory. He published a student book in trigonometry: *Trigonometria plana et spherica cum selectis ex Geometria et Astronomia problematis* (Janosi 1737). This book can be considered the first student textbook of mathematics and astronomy published in Transylvania. He completed and edited a version of the book originally written by the Jesuit Jacob Gooden (1670–1730), published initially in Lüttich in 1703. This Latin work was structured in two chapters dedicated to plane and spherical trigonometry, with problems from geometry and astronomy. Basic astronomical notions were also presented, e.g., the ecliptic is described as "circulus maximus, quam Sol unius anni spatio percurrit," i.e., the maximal circle described by the Sun in space through one year. The basics of classical fundamental astronomy were presented in this volume.

To help advance construction of the astronomical observatory, the Jesuits sent Maximillian Hell to Cluj in 1752. Hell started a new project to build an astronomical tower, directed the construction, and ordered astronomical instruments from Vienna for the future observatory.

In the subsequent three years Hell spent in Cluj, he taught mathematics and astronomy. He also started to write a series of mathematical monographs, under the title *Elementa mathematica naturali philosophiae ancillantia*. In 1755, he finalized and published the book dedicated to arithmetic and algebra, *Elementa Arithmeticae numericae, et literalis seu Algebrae* (Hell 1755a), complete with a book of problems (Hell

Figure 4. Maximilian Hell (Höll) and the inner front pages of his mathematical books written in Claudiopoli (Cluj).

1755b) (Fig. 2). He also printed a short booklet referring to the mathematical examination hold in 1755. During the period spent in Cluj, Hell had an intensely productive and valuable didactic and scientific period in his carreer (Heinrich 1978; Szenkovits 2001, 2005, see).

Unfortunately for the Cluj University, Hell was not able to finalize his dream to see with his eyes the completed astronomical observatory in Cluj, because he was moved to Vienna in 1755, before the new buildings of the university and the astronomical tower were finished. In the following years, as Imperial Astronomer and Director of the Astronomical Observatory in Vienna, Hell had a brilliant scientific career.

3. Astronomical Towers at the University of Cluj

The work started by Ianosi and Hell was finalized by Nandor Hartmann, professor of mathematics, who was taught in Cluj from 1755 to 1769. In this period, construction of the new building of the university, with the special tower dedicated to astronomical observations, was completed. The tower is shown in an engraving dated 1759 (Fig. 5, left panel). Hartmann continued to endow the new astronomical observatory with specific instruments. He ordered from Vienna a mobile quadrant, an astronomical clock, and telescopes.

The history of this first astronomical tower was not a long one. On 31 August 1798, a big fire ravaged the center of Cluj, devastating the University and Astronomical Observatory, together with the first instruments. Reconstruction of the University was finished in 1805. A new astronomical tower had been erected in the new building, and new instruments were procured (Fig. 5, right panel). The 9.3 cm telescope brought at this time from the Pössel Institute of Vienna still exists in the Báthory High School in Cluj.

The astronomical observatory of the University was dedicated mostly to didactical activities. The most important astronomical observations made in this tower were those

Figure 5. *Left:* Astronomical tower in Cluj in 1759, copper engraving by János Szakál. *Right:* tint-drawing with the second astronomical tower.

of the transit of the Venus, made in 1874 by Miklós Konkoly Thege (1842–1916), the famous Hungarian astronomer, and later the observations of variable stars made by Friedrich Schwab (1858–1931), in 1885 and 1886. The results of these observations of variable stars were published in the *Astronomische Nachrichten*.

The second tower dedicated to astronomical observations existed until 1893. In that year, the University started the reconstruction of all its central buildings. At athat time, conditions were no longer favorable for a new astronomical observatory in the center of the town. A new, modern astronomical observatory was built in Cluj later, starting in 1925, in a more favorable place at the margin of the town.

4. The Batthyaneum Observatory

Count Ignatius Batthány (1741–1789), Roman Catholic Bishop of the Transylvanian diocese (1780–1789), distinguished himself through his own scientific activities, which he conducted at the high level of contemporaneous scholarship. According to Kovács (2011), with the consist hard work and by promoting high academic standards in Transylvania, Batthány established an indispensable "laboratory," consisting of a library of natural sciences and antique collections, a print shop, a paper mill, and an astronomical observatory in Alba Iulia (Gyulafehérvár, see Figure 6).

Figure 6. *Left:* the Batthyaneum library. *Right:* the Astronomical Observatory in the Batthyaneum, Alba Iulia.

The most important primary source for the initial overview of the Observatory is the work of Antonius (Antal) Mártonffy (?–1799), entitled *Initia astronomica spec-*

ulae Batthyanianae Albensis in Transilvania (Mártonffy 1798, see). This is something like a promotinal postcard or brochure about the observatory put together by the first astronomer of the institute, a former student of Maximilian Hell in Vienna. *Initia astronomica* is composed of an extensive description of the Observatory and twelve prints showing the building with its decorations and the instruments of the observatory (Kovács 2011, for details, see). More details about the instruments of the Batthya-

Figure 7. The Batthyaneum and its astronomical instruments.

neum Observatory (Figure 7) are presented by Mártonffy (1798). The Observatory functioned for only a few decades, and was completely neglected in the last century.

References

Dezső, L. 1944, History of the Hungarian astronomy, Hungarian: A magyar csillagászat története. Múzeumi füzetek, Kolozsvár, 2. évf. 1

Heinrich, L. 1978, The first astronomical observatory in Kolozsvár, Hungarian: Az első kolozsvári csillagda, Kriterion, Bukarest

Hell (Höll), M. 1755 Elementa mathematica naturali philosophiae ancillantia. Tomus I: Elementa Arithmeticae numericae, et literalis seu Algebrae. Claudiopolitanae

Hell (Höll), M. 1755, Exercitationum Mathematicarum. Pars I. Exercitationes Arithmeticae, Claudiopolitanae

Hell (Höll), M. 1755, Materia Tentaminis Mathematici. Aula Academica S. J. Claudiopolitanae, die 14 Mensis Julii

Janosi, N. 1737, Trigonometria plana et spherica cum selectis ex Geometria et Astronomia Problematis, Sinuum Canonibus et Propositionibus ex Euclide magis necessariis, Claudiopoli

Kovács, A. 2011, The Batthyaneum Observatory of Alba Iulia a Less Known Iconographic Program, in Batthyaneum, Omagiu fondatorului Ignatius Sallestius de Batthyan (1741–1798), ed. Doina Biro Hende, Editura Bibliotecii Naţionale a României, Bucureşti, 195–212

Antonius (Antal) Mártonffy, A. 1798, Initia astronomica speculae Batthyanianae Albensis in Transilvania, Alba Iulia

Szenkovits, F. 2001, Maximilian Hell la Cluj, Lucrările Seminarului Didactica Matematicii, vol. 18, 159–162

Szenkovits, F. 2005, Hell Miksa (Maximilian Hell, 1720–1790), Műszaki szemle. Historia Scientiarum 2. 30, 2005, 13–26

Szenkovits, F. 2006, On the history of Cluj astronomical Observatory, Hungarian: A kolozsvári egyetemi csillagda történetéből. In A csillagászati tanszék negyed évezrede – évfordulós kötet. Publicationes of the Astronomy Department of the Eötvös University, ed. K. Petrovay, Budapest, Hungary, Padeu 16, 103–128

Szenkovits, F. 2008, The First Astronomical Observatory in Cluj-Napoca. In Exploring the Solar System and the Universe, Bucharest, Romania 8 – 12 April 2008, Editors: V. Mioc, C. Dumitrache, N. A. Popescu, American Institute of Physics Conference Proceedings 1043, 55–58

Wollmann, V. 2011, Zur Technischen Ausstattung der Karsburger Sternwarte (Specula) aus dem Jahre 1798, in Batthyaneum, Omagiu fondatorului Ignatius Sallestius de Batthyan (1741–1798), ed. Doina Biro Hende , Editura Bibliotecii Naţionale a României, Bucureşti, 213–223

Astronomical Heritage of the Middle East
ASP Conference Series, Vol. 520
Sona V. Farmanyan, Areg M. Mickaelian, J. McKim Malville, and Mohammad Bagheri, eds.

Vsevolod Viktorovich Stratonov (1869–1938): an Astronomer Driven by Fate from the Middle East to Prague

Martin Šolc and Petra Hyklová,
Astronomical Institute, Charles University, Prague, Czech Republic;
`martin.solc@mff.cuni.cz`

Abstract. Vsevolod Viktorovich Stratonov was born in Odessa, graduated from the local university, and began working there for two years at the university observatory. He spent the next two years at Pulkovo Observatory, where he acquired a large degree of skill in photographic techniques. In 1895 he moved to Tashkent and remained there as an astrophysicist at the observatory for ten years.

In 1905, Stratonov left his career as a practical astronomer due to an eye disease. He became the deputy manager of the tsarist directorium and worked in public service for the Caucasus mountain regions, reorganized the Polytechnical University in Tbilisi, and gave astronomy lectures at the high school for girls. In 1911 he was appointed the deputy director of the Imperial state-owned bank in the town of Tver. During the revolution of 1917 he lost this position. Afterwards, he was named Professor of Astronomy at the Moscow State University.

In Moscow, Stratonov successfully established the State Astrophysical Institute (GAFI). His hope was to set up a large astrophysical observatory and several branch observatories from Odessa to Vladivostok, and to include the observatory in Tashkent. After his forced emigration in 1922, the project continued under the leadership of Fesenkov, and led to the creation of the Sternberg Astrophysical Institute in 1931.

In 1923, the Stratonov family moved to Czechoslovakia, via Berlin. Stratonov became a teacher at the Russian People's University and held numerous popular lectures in and outside of Prague, and also abroad. After acquiring Czech citizenship in 1938 he was appointed as a professor of Astronomy at the Czech Technical University in Prague.

Decreasing mental abilities, or perhaps other causes led Stratonov to the final chapter in his life—his death by suicide. He found his final rest at the Russian cemetery in Prague. During the Soviet period, Stratonov's name was erased from the official history of astronomy. Only the past decades have come to show the significant role of this astronomer played. He represented a significant link between the Middle East and Central Europe.

1. Prologue

In 1995, the PhD student Konstantin Ivanov from Tula State Pedagogic University visited us in Prague. He was searching for evidence of scientists who had been refugees from the countries of the former Soviet Union. His particular interest was Vsevolod Viktorovich Stratonov, an astronomer. I remembered the name of Stratonov because I possessed his textbook on astronomy. We found this book and obituaries in the Czech journal Říše hvězd, The Realm of Stars (Svoboda 1938) and in the German Astronomische Nachrichten (Svoboda 1939). Since it was not much, we consulted the Prague phone book, found the name of Oleg Stratonov, and dialed the number.

The person who answered the phone was Dr. Stratonova, an old woman speaking Czech with a heavy Russian accent. She was the first wife of Oleg Stratonov, the son of Vsevolod Viktorovich. She was a physician, now almost a hundred years old, but sanguine and of a nimble mind.

We accepted with pleasure her kind invitation and enjoyed her dramatic reminiscences about her life in Russia and 1918 escape from Odessa, which was war-torn by the civil war between Bolshevik troops and the White Army. She married Oleg about two years later, and after 1923 they joined the family of Vsevolod Viktorovich Stratonov in Prague. Dr. L. Běloševská, a research worker from the Institute of Slavonic Studies of the Academy of Sciences of the Czech Republic, joined us for our next visit. We acquired some photographs, and during the following three visits we recorded Dr. Stratonova's narrative (in Russian) on tape (Belosevska 1995).

Figure 1. V. V. Stratonov in ca. 1923 (left), in 1917 (passport photo), and in 1936 (right).

Konstantin Ivanov returned home, wrote his Ph.D. thesis, and later D.Sc. thesis, became a profesor of the history of science, and published a monograph about the origins of Soviet astronomical research institutions (Ivanov 2008). Our doctoral student Petra Hyklová and I continued the search for relevant documents in archives, and translated into Czech the roughly 120-page memoir of Dr. Stratonova.

2. Stratonov and Tashkent Observatory

Vsevolod V. Stratonov (b. 4 Apr 1869, d. 6 Jul 1938) was born in Odessa. His father, the director of a secondary school in Odessa, soon recognized his son's talent for the exact sciences and provided him with the best education available. Vsevolod graduated summa cum laude at Odessa University; his thesis *The Transit Instrument and the Determination of the Geographic Coordinates* was awarded a gold medal by the university.

He then began working as an observational astronomer at the Observatory of Odessa University (1891–1892), later at Pulkovo Observatory (1893–1895), and finally for ten years at Tashkent Observatory. At Tashkent, he led the astrophysical laboratory and devoted most of his work to astrophotography using the Carte du Ciel standard refractor; he exposed more than 400 plates. Photography was also what brought him to stellar astronomy. He inspected the dark belt that seemingly divides the Milky Way and found that the distribution of stellar images on the plates failed to represent the real distribution of stars in space because the dark belt, caused by the clouds of some absorbent cosmic substance, attenuates the light of distant stars. He described the nebulae of the Pleiades, Lyra, Perseus, and Scutum, performed observations of variable stars, and studied the Leonid meteor shower. With his 13-inch photographic refractor (installed by Repsold-Henry in 1894), he discovered the emission nebula IC 1990 near the Pleiades in May of 1896. The results of his work at Tashkent filled five volumes.

His roughly 50 books, essays, and scientific papers were written in Russian, French, German, and later in Czech. His book *Sur le mouvement des facules solaires* was honored with the Grand Imperial Award and the Golden Medal of the Tsar. Another of his books, *The Stars* (in Russian, Moscow 1918), won a renowned international prize for popular science literature. His richly illustrated 248-page monograph *The Sun*, published in Tiflis (Tbilisi) in 1910, was awarded a 500-ruble prize by the Tsar; its 2nd edition—the author's edition—was published in St. Petersburg in 1910. Stratonov contributed to the reorganisation of the first Middle East State University in Tashkent, was elected as the Dean of Faculty of Physics and Mathematics, and was named a lifelong Member of Honor of the Faculty of Medicine.

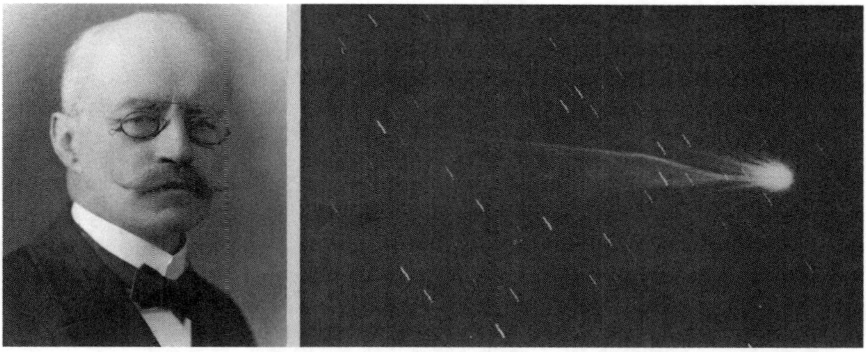

Figure 2. Left: Professor Josef Sýkora in 1930. Right: Comet Morehouse on November 14, 1908, Tashkent Observatory. Photo: J. Sýkora.

3. Josef Sýkora: Stratonov's Successor at the Tashkent Observatory

Stratonov worked at the Tashkent observatory from 1895–1905. He was succeeded by the Czech astronomer Josef Sýkora (1874–1944), who stayed there until 1911.

Josef Sýkora was born on January 16, 1870, in Kharkiv as the son of a Czech teacher who had been invited to teach Latin and classical Greek at a secondary school. In 1892, Josef graduated in Physics and Mathematics in Kharkiv and received a job at the university observatory there; he later moved on to the observatories first in Tartu, and then in Moscow. His main fields of interest were sunspots and facules; his work was recognized by the Russian Astronomical Society in 1898. Another field of expertise was geodesy, and he participated in the Spitsbergen/Svalbard expedition in 1899–1900.

Sýkora studied auroras as well—he successfully acquired the first auroral spectra in 1899 and correctly identified the emission lines of nitrogen. He also detected the 5577 Å line, which was identified as the "forbidden line" of oxygen 25 years later. In 1902–1903, Sýkora founded several astrophysical stations on the Kola peninsula, where aurora observations were made until 1911. He processed those observations in Tashkent, and therefore papers from 1910–1913 concerning the auroral spectroscopy appeared in the publications of Tashkent observatory, where auroras were never seen.

In Tashkent, Sýkora, who had come there as a renowned and experienced astronomer, addressed a broad range of scientific problems. On his journey to Tashkent, Sýkora was ordered to repair seismic instruments at petroleum locations in Baku. That sparked his interest in seismology. Since the only sesmic station in the region was in Tashkent, Sýkora decided to build a system of multiple seismic stations.

Sýkora observed and photographed many bright comets in Tashkent: the comet 1907 IV Daniel, the comet 1908 III Morehouse, the Great Southern Comet 1910 I, and Halley's comet 1910 II. The observation of the transit of the cometary nucleus over the solar disc was attempted at many observatories. However, Sýkora was the only astronomer in the world to reach the incorrect conclusion that he had seen this in fact non-observable effect (the nucleus is too small to be seen against the disc of the Sun).

Figure 3. The expedition of the Tashkent observatory to observe the total solar eclipse on January 14, 1907, at Chernayevo. Sýkora is standing by the left telescope. Photo: Sergei Mikhailovich Prokudin-Gorskij (1863–1944), `http://www.veinik.by/foto/1801-1900/18411.jpg`

Meteor photography was Sýkora's most successful area of research. He followed in the long-lasting tradition of meteor observations at Tashkent and improved the observational method by using short focal-length cameras instead of a large astrograph. In 1906, he organized the establishment of observation stations in Iskandar and Chimgan. On August 11, 1909, he succeeded in photographing one bright bolide from all three sites simultaneously. In 1911, Sýkora left the observatory and became the director of the Tashkent Girls' School. In the same year, he was invited to Lithuania to teach

Physics at Shiauliai at the secondary school. In the subsequent years, he held positions at various schools in Poland, Perm, and Yekaterinburg and as a lecturer of astronomy at Kharkov University (1920–1921).

Figure 4. The founders of the State Astrophysical Institute (GAFI) in August 1922. From left to right: B. M. Stschigolev, P. J. Davidovich, V. G. Fesenkov, V. V. Stratonov, V. N. Mikhailovich, and S. V. Orlov.

After the Czechoslovak legions moved against Soviet Russia, the Soviet government began to expel Czechoslovak citizens from the country. Sýkora was deported to Czechoslovakia in 1921 with no regard to his merits in Russian science and education.

Back in Czechoslovakia, Sýkora took a position at Ondřejov Observatory (40 km south of Prague) in 1923 and returned to the scientific work that he had been doing in Tashkent. He joined the Czech Astronomical Society and initiated the establishment of Meteor Section. At the General Assembly of the International Astronomical Union in Leyden 1928, he was elected to Commision No. 22, *Meteors, meteorites and interplanetary dust*, and was reported to be one of the best meteor observers.

Sýkora died on February 23, 1944, and is buried at the cemetery in Ondřejov. One glacier at Spitsbergen and Sykora Island in the Kara Sea have been named in his honor.

4. Stratonov in Tbilisi, the Caucasus Region, Tver, and Moscow

In 1905, an incurable eye disease forced Stratonov to withdraw from observational astronomy. Thanks to his social connections, he found an appointment as Vice-Director of Administration for the Caucasus region. In addition to this, he organized the military geodetic service, reformed the Technical College in Tbilisi, and gave astronomy lectures at the girls' school. Beginning in 1911, he served as the deputy head of the Russian State Bank in the city of Tver.

The 1917, the Bolshevik revolution removed him from this position, but Moscow State University immediately elected him as a full professor of astronomy. In Moscow, he established a small institution, which he named The Main Russian Astrophysical Observatory (GRAFO, in 1920) and intended to build, in addition to a large mountain astrophysical observatory, numerous branch astrophysical observatories from Odessa to the Vladivostok meridian. By 1922 he had managed to transform GRAFO into the State Astrophysical Institute (GAFI) with minor branch observatories in Kutshino (near Moscow) and Novotsherkatsk (on the River Don). Throughout this period, Stratonov was trying to change the status of Tashkent Observatory, which had been a military institution during the times of the Russian Empire. He succeeded, and the observatory became a part of the university, but Stratonov had actually intended from the very beginning to attach it to GRAFO and later to GAFI.

Until the expatriation of intellectuals in 1922, he succeeded in establishing the Turkestan University in Tashkent, the fledgling nation's largest scientific publishing house in the state in Moscow, and the meteorological service for soil improvement in 14 governorates of the Upper Volha region. Furthermore, he served as a consultant for the Commissariat for Public Enlightenment and as the head of the Department of Physics and Mathematics of the Main State Library in Moscow. He was elected as the Dean of the Faculty of Physics and Mathematics of Moscow State University (MGU).

5. Exile and Life in Prague

In 1922 Lenin personally selected a list of about 160 undesirable intellectuals and entrusted Stalin with the job of ensuring their expatriation. The list included Stratonov, who had initiated a strike of MGU professors against Soviet Commissariats's meddling in the internal affairs of the university. Stratonov was arested on August 17, 1922, like many others in the following days. After his release, he resigned from all of his positions and duties and arranged with Vasily Grigorievich Fesenkov (1889–1972) for the continuation of establishing GAFI. After many negotiations and mishaps, the GAISh (the Sternberg Astronomical Institute) was ultimately created in 1931, essentially from GAFI and the observatory of Moscow University (Ivanov 2008).

Stratonov's family left Russia for Germany out of St. Petersburg on board a German ship that was later nicknamed the "Philosophers' Steamboat" (Stratonov 1938). In Berlin, along with a few colleagues, Stratonov established a Russian scientific institute to enable Russian emigrants to continue their studies and research. He led the scientific department of this institute as its Dean.

In 1923, Stratonov and his family moved to Prague. He lectured at the newly established Russian National University of Prague, and at many other places in Czechoslovakia, Latvia, Lithuania, and Estonia. Hovever, his primary income came from writing and publishing books. He was also active in the Prague and Vienna stock markets, where he both served as a consultant and directly participated in transactions. His office, situated in the city centre, was a visible proof of his success as a trader. Stratonov received Czechoslovak citizenship in 1938. His close friend Prof. Jindřich Svoboda, renowned for his studies of meteor showers and Rector Magnificus of the Czech Technical University, helped him get an appointment as a professor of Descriptive and Practical Astronomy, beginning on March 1, 1938.

Figure 5. Popular Prague publications by V. V. Stratonov: *To the Faraway Worlds, Astronomy* (Czech and German editions), and *Venus as the Future Colony of Earth, Atlantis, The Invisible Inhabitants of Heaven, On the Fairy Ocean of the Sun.*

Professor Stratonov was a person of a great charisma, a "lounge lizard," and an excellent storyteller and tutor. He was constantly surrounded by many people. Yet perhaps in anticipation of the coming war, or perhaps in anticipation of his own impending mental decline, he took his own life on July 6, 1938. He found his final rest at the Olšany Russian cemetery, near its Dormition of the Virgin Mary Church.

During his life in Prague, Stratonov published a popular textbook *Astronomy*. It was the largest textbook on astronomy in the Czechoslovak Republic in 1918–1938, suitable both for teachers and serious amateurs. It was 640 pages long and richly illustrated. Its Czech translation by Josef Sýkora came out in several editions; meanwhile the German edition was translated by Adalbert Prey, the director of the Astronomical Institute at the German University of Prague. Fig. 5 shows a few of Stratonov's most popular works.

Stratonov and Sýkora both made it their habit to provide numerous popular-science lectures on a variety of topics, including Persian astronomy, Al Suphi's *Book of Fixed Stars*, or the Ulugh Beg observatory. It should be appreciated that they both, primarily in the Czech country, raised awareness of the astronomical traditions of the Middle East and of Central Asia, where they had spent substantial parts of their lives.

Acknowledgments. Our research was supported by Charles University, project GA UK No. 101-10/250343, and Faculty of Mathematics and Physics.

References

Bělolševská L., 1995, personal communication

Ehgamberdiev, S. 2017, *A Czech Astronomer in the Service of the Tsar* (in Czech), Astropis No. 4, 28

Ivanov, K. 2008, *Heaven in the Earth's Reflection* (in Russian), Territoria buduschtsevo, Moscow, 479

GAISh 1986, *History of the Astronomical Observatory of the Moscow University and of the Sternberg Astronomical Institute (GAISh) (in Russian)*, Publ. House of the Moscow University, 190

Stratonov V. V. 1927, *Nouvelles nebuleuses dans les Pleiades*, Astronomische Nachrichten, 3366, 103

Stratonov V. V. 1927, *Astronomy* (in Czech), Publ. House B. Kočí, Prague, 640

Stratonov, V. V. 1938, *On the waves of life. Memoirs* (in Russian), State Archive of Russian Federation GARF f. p-5881, op. 2, storage place No. 667, 668, 669

Svoboda, J. 1938, *Profesor Vsevolod V. Stratonov*, obituary, Říše hvězd (The Realm of Stars) Vol. 19, pp. 172-174

Svoboda, J. 1938, *Vsevolod Viktorovič Stratonov*, obituary, Astronomische Nachrichten 6420, 191

Astronomical Heritage of the Middle East
ASP Conference Series, Vol. 520
Sona V. Farmanyan, Areg M. Mickaelian, J. McKim Malville, and Mohammad Bagheri, eds.

Heavenly Bodies and Phenomena as a Basis for Poetical Images in the Works of Missak Metsarents

Karine Arakelyan
NAS RA H. Acharian Institute of Language; karine.arakelyan.2016@mail.ru

Abstract. From ancient times, heavenly bodies, stars, and astronomical phenomena were an inexhaustible source for poetical inspiration. Western Armenian poetry at the beginning of the 20th century is not an exception in this sense. The metaphorical applications of heavenly bodies and phenomena in the poetry of the famous Western Armenian poet Missak Metsarents are very diverse and varied. They often help to reveal one's inner life, emotions, expectations, and feelings. Many times astronomy is the basis for building new, poetical, and uniquely beautiful words.

The stars and heavenly phenomena have inexhaustible sources for poetical inspirationfrom ancient times. Artists praised the creations of God, admired the beauty of the stars, and perceived man as a part of nature who is closely connected with all other parts of nature. Heavenly bodies and phenomena often serves as a means to describe one's inner world, including one's path, future hopes, and expectations. Western Armenian poetry at the beginning of the 20th century also drew upon heavenly phenomena for inspiration. Remarkable examples of this theme are found in the poetry of Siamanto, Daniel Varoozhan, Rouben Sevak and, especially, Missak Metsarents.

It was often said that Missak Metsarents loved and had deep feelings for Dame Nature; in fact, most of his poems are devoted to descriptions of beauties of nature. Heavenly bodies, stars, and celestial phenomena are of particular interest in these descriptions, and it is no mere chance that we find in his poetry a clear-cut singleness of purpose. In a letter sent to his sister's husband, priest Vardan Arslanian, Metsarents answered his question and brought to light one of the very important features of his works: "I am touched once more while reading your letter apropos of 'Rainbow.' 'Why don't you speak with the Spirit of universe?',— you write. But evidently almost all my works seek to this big conversation, even if it is not a face to face conversation with God. Isn't it true, that God is everywhere? In 'Pure Winter Night' whispering prayer and my entreaty applied to Sun were for Him. What a happiness it is to see the secret of universal power in the rustle of leaf and from this little detail go to infinity" (Metsarents 1981).

We shall review the facts, which are very talkative. The word "Sun" in the poetry of Missak Metsarnts is found very often, 62 times at all, not including its different synonyms, and there are many words built from this root. In the titles of his poems, Metsarents often referred to the heavenly bodies, and to the Sun, as in the poems "Stars," "In the Morning Sun," "To My Dear Sister Moon," "Sun is Going to His Easter," etc.; there is even a poem devoted to the noon luminary entitled "To the Sun," in which the Sun is represented as a charitable sister who heals a man and gives him recovering warmth. Many critics of his time spoke about this poem but couldn't understand it. Metsarents wrote about it and answered them in the article "Criticism." He explained the lines

"Let the sluggish children go in turn through my soul." The author said: "Here child is a generalized metaphor of young feelings, hopes, wishes, which were being sluggish for a long time, and they need sun's vivifying rays to come to life" (Metsarents, p. 259).

In another poem named "Evening," Metsarants, while creating the image of evening, utilizes either metaphor (animating the sun), or simile (a farm-hand returning home at sunset): "And the sun returns home turning round above the mountains, As a hired man comes down to village" (Metsarents, p. 100).

The synonyms of the word Sun, and words built from this root, make the speech even more lyrical and rich, and endow it with various shades of originality. Here we want to recognize the accidental synonymy. Many linguists think, that "in the poetical texts one may meet more peculiar and subtle phenomenon — appearance of accidental synonymy. The rapprochement of meanings bringing to this type of synonymy happens in two ways: with the help of semantic context or with the help of words grown nearer phonetically" (Pelevina 1980). In the poem "Sun is going to His Easter," the word *sun* has two accidental synonyms, as "*Big fire-haired virgin* came from the sea and became the bride of the blue" (Metsarents, p. 124) or "The dear and glorious brilliant is gifted to the *first luminary*" (Metsarents, p. 124).

Siamanto, the other famous Western Armenian poet at the beginning of the 20th century, often invoked a theme in his poems that was very timely in his day, and unfortunately is a problem of today too: wandering. As a result of emigration of the masculine population, the Mother land loses defenders and working people, families lose their support, and young girls find themselves in isolation. In the poem "The Dream of the Bride," the young bride remembers the day when she saw her husband for the last time. It is interesting that while describing this hard psychological situation, Siamanto utilizes many similes, and one component of them all is the word *sun*:

Do you remember how bright the sun was

On the day you departed?

Its glare was as plentiful as my tears and as hot as my kisses.

It was as kind as your promises and as swift as your return.

Do you remember the sun and my prayers

On the day of your departure? (Markarian 2015)

Daniel Varoozhan also wrote about the wanderers. In his poems, we meet heavenly bodies too. The poem "A Letter of Longing" describes the sufferings of a mother, whose son travels around the world. She writes a letter to him waiting for his home-coming. The poem starts:

My mother writes, "my bantookht son,

How long would you roam under strange moons?

How long shall wait to have your tired head

Rest on my warm chest?" (Markarian 2015)

In the poem "To My Dear Sister Moon," Metsarents sketches an image of the moon. Who is, after all, the night luminary? Is it really unattainable beauty, whose distant brightness is enough to give a man happiness, or is it a desired sister, whom one can reach, even if she is hidden behind the clouds? The essence of the moon is in the close connection with the theme of the night, which was also the center of the author's attention.

In the article "An Attempt of self-criticism as regards 'Rainbow'," Metsarents wrote, "Isn't it in the night, that dreams have brighter and faster flights? In the night, when the magnificent view opens up, and expectation is so delightfully poignant" (Metsarents, p. 242).

The beauty of night is closely connected with the moon, as, "I looked at the flower-garden in front of me, at the blue lilacs, and the moon was stretching out her beautiful, coquettish, careless wrist" (Metsarents, p. 128). The luminary is not cold and lifeless, but having spirit and soul, taking part in one's emotional state. With the help of the moon's animation (not only in this example, but also in some other poems), the author creates unique images.

The theme of the moon has attracted the attention of poets of all times and nations. One of them is the famous English poet George Gordon Noel Byron. In one of his poems, Byron wrote about lost happiness. The text is constructed by comparing emotions and the heavens at night. Here the light of the moon reminds him of joyful feelings lost. Light is an object of simile and grief is the image. While Metsarents' moon is a lovely sister, Byron's moon is clear, but cold, who shines, but does not warm. Here one can see not only the presence of moon, but even more the absence of the sun:

> Sun of the sleepless, melancholy star!
>
> Whose tearful beam glows tremulously far.
>
> That show'st the darkness thou canst not dispel,
>
> How like art thou to joy remember'd well!
>
> So gleams the past, the light of other days,
>
> Which shines, but warms not with its powerless rays;
>
> A night-beam Sorrow watched to behold,
>
> Distinct, but distant-clear-but oh how cold! (Byron)

From ancient times, mankind's thinking about stars was accompanied with love or expectation of love. This theme has long been expressed in works of art, especially in poetry, including Western Armenian poetry. Siamanto, Varoozhan, and other poets wrote about the stars. Metsarents often invoked this theme, making stars the basis of poetical imagery. In the poem "Stars," the simile of stars and lonely hearts leads to involved psychological generalizations: "The stars are softly melting In the darkness till morning, Like bleeding hearts, That are left alone" (Metsarents, p. 17).

Besides the examples above, the poems of Missak Metsarents also refer to other to heavenly bodies and phenomena. One of them is the meteorite. It is utilized twice as a component of simile, being bound up with metaphorical meanings, such as in the poem "Irrigation," at the description of irrigation: "And lasts the light-silver ribbon of this speech, as milky pedicles of escaping meteorite" (Metsarents, p. 94).

As we see in this short paper, the metaphorical applications of heavenly bodies and phenomena in the poetry of the famous Western Armenian poet Missak Metsarents are very diverse. They often help to reveal the poet's feelings about his inner life, emotions, expectations, and feelings. Many times these references are the basis for building new and poetical words. They create unique beauty.

While, for astronomers, heavenly bodies and phenomena are the objects of scientific investigation, for writers, and especially for poets, they are an inexhaustible source of inspiration, and they give readers the opportunity to enjoy the greatest aesthetic pleasure, that of looking at the luminous sphere.

References

Bryon, 1933, The Complete Poetical Works of Byron (Boston) 220
Markarian, H. 2015; The Martyred Armenian Writers 1915–1922 (an anthology), NY, 162
Metsarents, M.; Complete Works, Yerevan, 1981, p. 288 (Arm)
Pelevina, N.; Stylistic Analysis of Belles-lettres, L., 1980, p. 49 (Rus)

Part VI

Astronomical Heritage in "Language"

Astronomical Heritage of the Middle East
ASP Conference Series, Vol. 520
Sona V. Farmanyan, Areg M. Mickaelian, J. McKim Malville, and Mohammad Bagheri, eds.

Armenian Names in the Universe

Areg M. Mickaelian and Sona V. Farmanyan
NAS RA V. Ambartsumian Byurakan Astrophysical Observatory (BAO);
`aregmick@yahoo.com, sona.farmanyan@mail.ru`

Abstract. This paper discusses the naming conventions for celestial objects, and how Aremian language and culture have influenced those names.

1. Introduction

Astronomical names are assigned in different ways. Planets, stars and constellations mostly carry historical and mythological names, which are very often arbitrary. Currently, the International Astronomical Union (IAU, founded in 1919) is considered to be the only body responsible for naming celestial objects, and it has developed a special policy for the naming of space bodies. In particular, the IAU does not recognize the naming of a particular star by individual private organizations (such as when people pay to have a star named after them, which is actually only a piece of paper in the form of an unrecognized certificate). Besides stars, there are many other cosmic objects, and approaches to naming them differ.

2. Constellations

Constellations are usually the first way people get acquainted with the starry sky. Of the 88 constellations in the sky, some of the more famous are those located near the northern pole (Ursa Major and Ursa Minor, Cassiopeia, and others) and, of course, the 12 zodiacal constellations of the ecliptic zone, with which astrology is connected. The Ancient Greek philosopher Hipparchos, in the 2nd Century BC, decided to catalog the stars, dividing them into 48 constellations; later, thanks to geographical exploration, many constellations of the southern sky were added during the 16th–19th centuries, often with the names associated with geographical discoveries, astronomical instruments, or other technical devices.

The Armenian names of the constellations have been used since at least the 5th century, but some of them were forgotten during the Soviet times (Mickaelian 1989). Knowledge of them has now been restored, including the constellations of Hayk and Vahagn (Orion and Hercules, respectively; see Figure 1), which were named after the Armenian gods. However, there are no original Armenian names among the official constellation names.

Figure 1. The constellation Orion, or Hayk in Armenian.

3. Solar System Objects

Planets and moons have been named from ancient times on the basis of Greek mythology, from which the Roman mythology also originated. In fact, the Latin names have been preserved: Mercury (Greek Hermes), Venus (Aphrodite), Mars (Ares), Jupiter (Zeus), Saturn (Cronos), Uranus (Uran), Neptune (Poseidon), and Pluto (Hades, although Pluto is currently considered a dwarf planet). The Sun, the Earth, and the Moon were accordingly called Helios, Gea (or Gaya) and Selena by the ancient Greeks. It is interesting to note that before the mythological names were used, the Greeks also used names equivalent to the ancient Armenian ones (Harutyunian 2014). The Armenian names of the five planets that are visible by naked eye are still applicable (Paylatsu, Arusyak or Lusaber, Hrat, Lusntag, and Yerevak).

In the case of planetary moon names, the Greek mythology mostly continues its tradition, although the moons of the planets (apart from the Moon) were discovered only in 1610 or later. Currently 175 moons are known (around Jupiter 69, Saturn 62, Uranus 27, Neptun 14, Mars 2 and Earth 1).

3.1. Lunar Craters and Other Formations on the Moon's Surface

There are many features on the surface of the Moon, including mountains, seas, and craters. There is also the Ocean of Storms; in fact, there is no liquid water on the surface of the Moon, that is to say "seas" and the "ocean" are actually solid rock (maria). Shirakatsi Crater (Figure 2; Harutyunyan 2012) was named for Anania Shirakatsi, our greatest medieval scientist; he was an Armenian philosopher, mathematician, astronomer, geographer, and alchemist, and his name has been properly immortalized in the heavens. The diameter of the crater is 51 km. It is located on the opposite side of the Moon, and on the right side it overlaps with the Dobrolovski Crater. There is also the Sisakian Crater on the Moon, again on the side that is not visible from the Earth, which has a diameter of 34 km. It is named after the great Armenian biochemist, and one of the founders of space biology, Norayr Sisakyan (1907–1966).

Figure 2. Anania Shirakatsi (left) and the crater on the Moon named in his honor (the left one of the two merged craters).

3.2. Mars, Venus, and Mercury Surface Features

In the Solar System, Mars, Venus, and Mercury are the best studied planets. Their surface formations were also named, but, until recently, none had Armenian names. On 24 April 2015, one of the smaller craters on Mars was given an Armenian name, Garni, in recognition of the Astronomical Institute of Garni created by Grigor Gurzadian. The diameter of this crater is only 2.57 km, but it became very popular after the discovery of the possible existence of water in it.

3.3. Asteroids

Asteroids were also known as small planets until 2006. The naming of asteroids is the responsibility of the IAU. The first asteroids (found in 1801), like planets and moons, were given mythological names, such as Ceres, Pallas, Juno and so on. However, the number of asteroids found in the 20th century grew radiply, so that mythological names were not enough. At present, there are more than 700,000 known asteroids, 50,000 of which have names.

There are a number of asteroids having Armenian names: *Armenia* (#780, in honor of Armenia), *Ani* (#791, named after the Armenian medieval capital), *Ambartsumian* (#1905, in honor of Viktor Ambartsumian), *Shavarsh* (#3027, recognizing the swimmer Shavarsh Karapetyan), *Stepanian* (#3444, in honor of the astronomer Jivan Stepanian), *Aghekian* (#3862, named for the astronomer Tadevos Aghekian), *Parajanov* (#3963, in recognition of the film-maker Sergey Parajanov), and *Khachatryan* (#4802, honoring the composer Aram Khachatryan). In 2016, #21057 asteroid was named *Garikisrayelyan* in honor of the Armenian astronomer Garik Israelian.

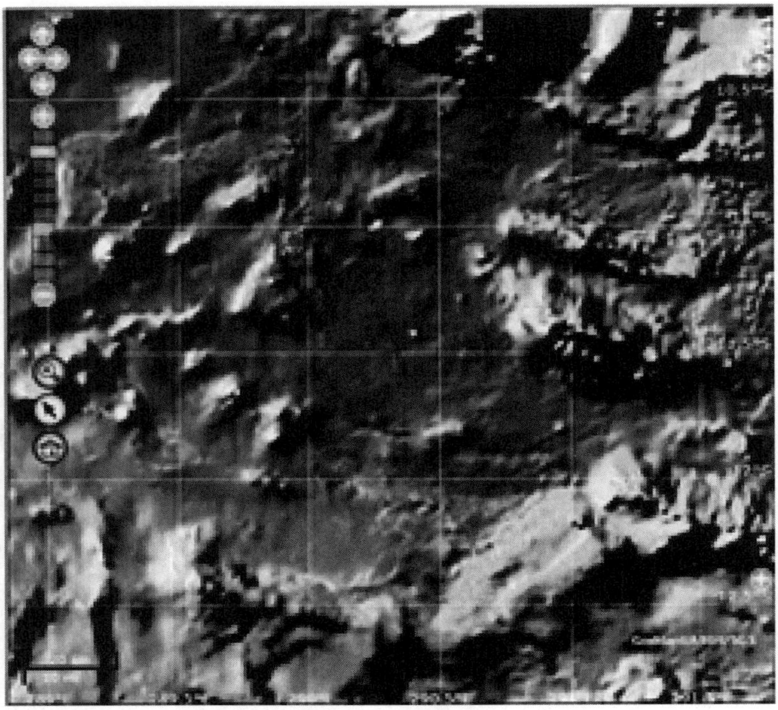

Figure 3. Garni Crater on Mars.

3.4. Comets

According to tradition, comets are named after the discoverer or discoverers. Moreover, it is noteworthy that many of them are not famous astronomers, and very often comets are discovered by amateur astronomers. Among the most famous comets are: Halley, Hale-Bopp (named for the amateur astronomers Thomas Bopp and physicist Alan Hale), Shoemaker-Levy 9 (David Levy has discovered 22 comets, the most discovered by any one person), Churyumov-Gerasimenko (famous thanks to the Rosetta Mission) comets. Periodic comets are more famous and important for research, which have approached the Sun and Earth more than once. As Byurakan Astrophysical Observatory does not deal with the study of the Solar System, accordingly there are no Armenian names among the comets.

4. Stars, Nebulae, and Galaxies

There are 375 stars with historical names, but only 313 are included in the current IAU list. Greek and Arabic names of stars are common. The Byurakan Astrophysical Observatory has always focused on finding and exploring stars, nebulae, and galaxies.

4.1. Viktor Ambartsumian

The most famous astronomical concept associated with the name of Viktor Ambartsumian (1908–1996) is the *Ambartsumian hypothesis* on star formation and and stellar evolution, on the basis of which *Ambartsumian* or *Byurakan Cosmogonical School* was created. It assumes the emergence of stars from superdense protostellar matter and is diametrically opposite to the classical theory, according to which the stars emerge from gas and dust material. The names of scientists often appear in astronomy in terms of concepts, formulae, functions, effects, phenomena, methods, models, theories, and so on. There are a number of versions, theories, formulae, functions and definitions connected with the name of Ambartsumian. These are *Ambartsumian's invariance theory* or *principle*, *Ambartsumian's method* of the determination of the temperature of the central star of the planetary nebulae, and *Ambartsumian's ϕ function* (1942).

Another object named after Ambartsumian is in the galaxy NGC 3561, which is also known as Arp 105. It is a pair of colliding galaxies that are part of the Abell 1185 galaxy cluster. They are also called the "Guitar." The system contains a small tidal dwarf galaxy known as *Ambartsumian's knot* that is believed to be the remnant of the extensive tidal tail pulled out of one of the galaxies.

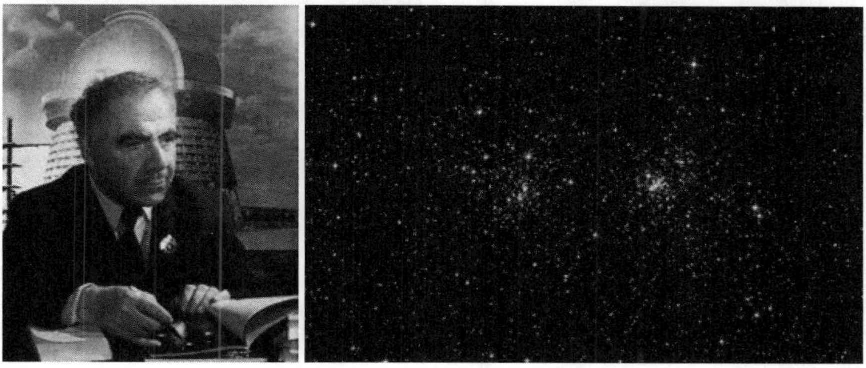

Figure 4. Viktor Ambartsumian and the famous star cluster h & χ Persei supporting his star formation hypothesis.

The discovery and naming of *stellar associations* are also related to the name of Ambartsumian. This term was first used in Armenian (literally "spreading of the stars"), then translated into other languages and currently the English version "stellar association" is most used (Ambartsumian 1948). Ambartsumian also introduced several other astronomical terms, such as Herbig-Haro objects, Fuors, etc.

4.2. Beniamin Markarian

Beniamin Markarian (1913–1985) is Armenia's second famous astronomer. He discovered ultraviolet (UV) emission in 1515 galaxies, now known as *Markarian Galaxies*, which are famous among astronomers.

A chain of galaxies is also named after Markarian, and the most famous galaxies among them (NGC 4438 & NGC 4435) are called *Markarian's "eyes"*.

The Markarian Survey is often called the First Byurakan Survey (FBS), as there is also the Second Byurakan Survey (SBS). The objects discovered by these surveys

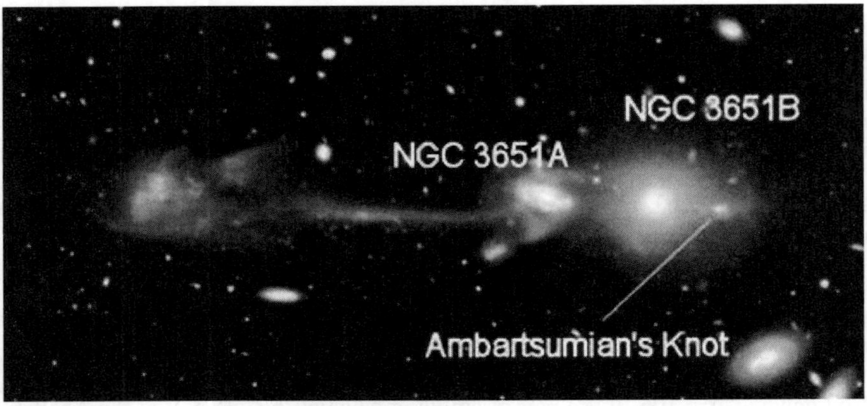

Figure 5. Ambartsumian's knot in NGC 3651B.

Figure 6. Beniamin Markarian at Byurakan Astrophysical Observatory with the 1 m Schmidt telescope. The most famous galaxies in Markarian's chain of galaxies are "Markarian's eyes," NGC 4438 (left) and NGC 4435 (right).

are called FBS and SBS objects: FBS galaxies are Markarian galaxies, but there are also FBS Second Part objects, about 1100 blue stellar objects, and about 1050 late-type (red) stars, which were discovered by H. V. Abrahamyan, A. M. Mickaelian, and K. S. Gigoyan. The Second Part of the FBS is the continuation of Markarian Survey for stellar objects, where Markarian's criteria were used for point sources.

Similarly there are Arakelian high surface brightness and Kazarian UV emission galaxies.

4.3. Marat Arakelian

Marat Arakelian (1929–1983), by analyzing the surface brightness of emission line galaxies (including Markarian and Seyfert galaxies), developed a new method for finding high-surface brightness galaxies. As a result, in 1975 he published the *Catalogue of High Surface Brightness Galaxies*, which includes 621 northern sky objects. These galaxies are called *Arakelian Galaxies* and are designated "Akn." The catalog is a source for discovery of many new AGN. They were observed at different observatories around the world, as well as with space telescopes, at optical, radio, X-ray and other wavelengths. The most famous Arakelian galaxies are Akn 120 and Akn 564; both are strong optical and X-ray variable AGN (Figure 7). Akn 564 is a narrow line Seyfert 1 (NLS1), which over the time changes greatly in emission line intensity. Akn 79, Akn 81, Akn 347, Akn 479, Akn 524, Akn 539, and Akn 557 are also famous.

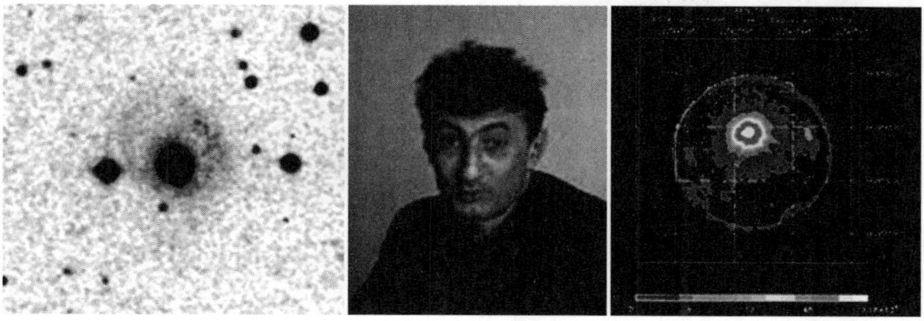

Figure 7. A direct image of the galaxy Akn 120 (lwft), M. A. Arakelian (center), and an X-ray image of Akn 564 (right).

4.4. Misha Kazarian

In the 1970s and 1980s, Misha Kazarian (b. 1932) discovered 706 new UV excess galaxies, which are now called *Kazarian galaxies* with the prefix "Kaz.". The entire catalog was published in 2010. 120 of these objects were studied in Byurakan and at the Special Astrophysical Observatory in Russia; as a result, new active galaxies were discovered. The most famous are Kaz 102, a variable quasar, and Kaz 163, a double galaxy containing a Seyfert component.

4.5. Romela Shahbazian

Romela Shahbazian (1925–2012) discovered the *Shahbazian compact groups of compact galaxies*, which can play a major role in understanding the evolution of galaxies. They were called *Shahbazian groups* and are labelled "Shk." Based on published papers from 1973–1979 (with the more recent papers published by German scientists F. W. Baier and H. Tiersch), in 1996 the entire catalogue of 377 Shk objects was published.

4.6. Cometary Nebulae

Elma Parsamian (born in 1929) published the first catalogue of cometary nebulae in 1965. These objects are called *Parsamian nebulae* (Parsamian 21 is especially famous). They are important from the point of view of star formation and stellar evolution, as they are related to young stars. In 1979, Parsamian, together with V. M. Petrosian, published a new catalogue of 106 cometary nebulae, which are classified into 4 morphological classes: conical, double conical, coma, and bow-like.

Using the Palomar Observatory Sky Survey images, 60 new cometary nebulae were discovered since 1977 by Armen Gyulbudaghian and Tigran Magakian, which are called *Gyulbudaghian-Magakian objects* and are designated as GM1, GM2, and GM3 (from the three lists that were published). In 1978, 37 other nebular condensations and Herbig-Haro objects were discovered. The discovery was made by Gyulbudaghian

Figure 8. Misha Kazarian, Romela Shahbazian, and Elma Parsamian.

and Russian astronomers Glushkov and Denisyuk, and the objects were designated with "GGD" (Gyulbudaghian-Glushkov-Denisyuk). It is noteworthy to mention that in 2003 Magakian compiled and published a catalog of 913 reflecting nebulae, which is currently the most complete list of such objects.

5. Space Telescopes

In 1987, the space telescope *Glazar* (in Russian "Glaz Armenii" and in English "Armenian Eye") operated on the Soviet "Mir" space station. The telescope was designed by Hrant Tovmassian (born in 1929) and also carries the Armenian name. As a result, a number of O-type stellar associations were studied in the UV, and new ones were discovered. It is worth mentioning that Grigor Gurzadian (1922–2014) is the founder of the Soviet Space Astronomy, who made valuable studies with his Orion-1 and Orion-2 UV space telescopes.

6. Byurakan Objects

The optical identification of infrared objects by Areg Mickaelian resulted in the discovery of 1279 galaxies and 276 stars. To promote the name of Byurakan, he called them *Byurakan Infrared Galaxies* (BIG) and *Byurakan Infrared Stars* (BIS). They are also included in the most important astronomical databases (SIMBAD, NED, HyperLEDA). The catalog of optical identifications of X-ray sources published by Mickaelian is called the *Byurakan-Hamburg-ROSAT Catalog* (BHRC, 2696 sources), as the work was done in collaboration with the Hamburg Astrophysical Observatory. Thus, thousands of cosmic objects are named after Byurakan.

7. Agop Terzan

There are Armenian astronomers all around the world. The French-Armenian astronomer Agop Terzan (born in 1927) is one of the outstanding scientists of the Armenian diaspora. He has worked in the Lyon Observatory and is the author of numerous astronomical studies. He discovered 710 variable stars in the immediate vicinity of 14 globular clusters, 11 new globular clusters (*Terzan stellar clusters*), 158 diffuse nebulae, 124 galaxies (of which 25% appear to be active galaxies of Sy2 type), 4430 red variable stars in the direction of the Galactic center, 1428 high proper motion stars, 26 planetary nebulae, and 122 diffuse galaxies in the direction of the center of our Galaxy.

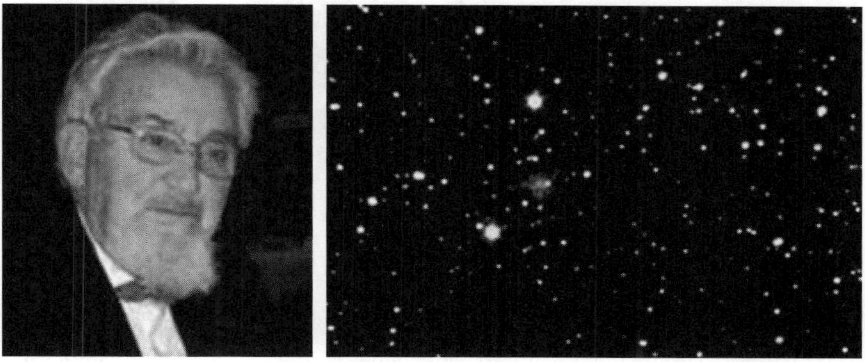

Figure 9. The French-Armenian astronomer Agop Terzan (left) and the Terzan 1 globular cluster (right).

8. Vahe Petrosian

The Armenian-American astronomer Vahe Petrosian (born in 1938) is known in the fields of cosmology and particle acceleration. Methods and concepts named after him include the *Petrosian radius* and the corresponding *Petrosian magnitude*, which are used to calculate the magnitudes of extended objects. They were especially important for calculating the object magnitudes in the large Sloan Digital Sky Survey (SDSS).

9. Zadig Mouradian

Zadig Mouradian (born in 1930) is widely known among theoretical astronomers and Solar physicists. He is the author of a number of discoveries, one of which is named after him (*Mouradian's Effect*"), according to which the Solar atmosphere velocity field creates a Doppler shift that reduces the optical width in the line cores of the optically thick spectral lines.

10. Tabetha Boyajian

In 2015, the American astronomer Tabetha Boyajian (born in 1980) discovered a strange pattern of light from the star KIC 8462852, nicknamed "Tabby's Star" after Boyajian. Boyajian observed strange, extended dips in the light from the star observed by the Kepler Space telescope. Although people have speculated about the origin of the dips, Boyajian does not think that extraterrestrials or a Dyson sphere are likely explanations.

Figure 10. Vahe Petrosian, Zadig Mouradian, and Tabetha Boyajian.

11. Exoplanets and Stars

Recently, the IAU initiated a new project "NameExoPlanets" aimed at naming newly discovered exoplanets and their host stars. As a result, dozens of stars and planets were given names. ArAS applied for Armenian names for 55 Cnc and its five planets, but another proposal was accepted instead. In 2017, the IAU also accepted new offers for names of bright stars, and, as a result, 86 stars received new names. Interestingly, among these new names there are names from the Australian, Indian, Chinese, Coptic, Hindu, Mayan, Polynesian and South African cultures. Armenia plans to send its proposals to the relevant IAU commission to immortalize the ancient Armenian culture in the Universe.

12. Catalogs

The modern astronomy is based on a large number of Big Data systems, the development and analysis of which lead to new discoveries. In particular, the list of important astronomical catalogs includes VizieR, the largest astronomical database situated in Strasbourg, containing more than 18,600 catalogs (as of 2019). Here too, Armenian astronomers have made a significant contribution. Besides VizieR, several other catalogs have been prepared and published by H. V. Abrahamian, H. V. Abrahamyan, K. S. Gigoyan, M. V. Gyulzadian, A. A. Hakobian, S. A. Hakopian, T. Y. Magakian, A. P. Mahtessian, N. D. Melikian, A. M. Mickaelian, A. R. Petrosian, and others. Each of the catalogs contains hundreds or thousands of objects. The Armenian-American scientist Kevork Abazajian is one of the authors of the large Sloan Digital Sky Survey (SDSS), a catalog containing some 1 billion objects.

In closing, for the readers' convenience, we present a summary table of the Armenian names of cosmic bodies with short descriptions.

Armenian astronomers have made a significant contribution to the study of the Universe, which has left its trace in the names of the cosmic bodies. The table shows that there are 13,407 heavenly bodies bearing Armenian names, including craters on the Moon and Mars, asteroids, stars, star clusters, nebulae, galaxies, groups of galaxies, and quasars. At present, Armenian astronomers continue their discoveries; Markar-

Table 1. Concise list of the Armenian names of cosmic objects.

Cosmic Objects	**Abbr.**	**Description**	**Number**
Solar System bodies		Moon and Mars craters, asteroids	12
Markarian Galaxies	Mrk	UV-excess galaxies	1515
Arakelian Galaxies	Akn	High surface brightness galaxies	621
Kazarian Galaxies	Kaz	UV-excess galaxies	706
Shahbazian Groups	Shk	Compact Groups of Compact Galaxies	377
Parsamian Nebulae	Pars	Cometary nebulae	106
Gyulbudaghian-Magakian objects	GM	Nebulae	60
Gyulbudaghian-Glushkov-Denisyuk objects	GGD	Herbig-Haro objects	37
Second Byurakan Survey Galaxies	SBS	Galaxies in 7 Deep Fields	1863
Second Byurakan Survey Stellar Objects	SBS	Quasars and stars in 7 Deep Fields	1700
First Byurakan Survey Blue Stellar Objects	FBS	Quasars, Seyfert galaxies, white dwarfs, hot subdwarfs, cataclysmic variables	1103
First Byurakan Survey Late Type Stars	FBS	M class and carbon stars	1471
Terzan clusters	Terzan	IR discovered optically hidden star clusters	11
Byurakan Infrared Galaxies	BIG	Optical identifications of IRAS point sources; galaxies and quasars	1279
Byurakan Infrared Stars	BIS	Optical identifications of IRAS point sources; stars	276
Byurakan-Hamburg-ROSAT Catalogue X-ray objects	BHRC	X-ray sources identified in the optical range	2696

ian, Arakelian, and Kazarian galaxies, Shahbazian Groups, Parsamian Nebulae, Terzan star clusters, and many other Armenian names will remain forever in the Universe, reminding us of the accomplishments of Armenian scientists. While the contributions of Armenian scientists in other spheres is also great, the Armenian names of astronomical bodies have been widely accepted and recognized by the astronomical community and will remain a lasting legacy.

Part VII

Astronomical Heritage for Outreach and Tourism

Astronomical Heritage of the Middle East
ASP Conference Series, Vol. 520
Sona V. Farmanyan, Areg M. Mickaelian, J. McKim Malville, and Mohammad Bagheri, eds.

Astronomical Heritage as a Separate Direction in Scientific Tourism

Davit Tashchyan
Econom Tour; davit.tashchyan@gmail.com

Abstract. The historical heritage of Armenian astronomy is rich, but it is not yet fully explored. Scientific tourism gives visitors an opportunity to travel, discover, and contribute to the historical knowledge about Armenian astronomy.

1. Introduction

Human beings have always been interested in exploring the heavens. It's not accidental that human beings from the earliest times of their existence paid special attention to the stars and tried to understand the heavenly lights. The feelings of mystery and curiosity made people explore it deeply. Through time and development, mankind's investigation of the stars led to deeper understanding. People came to understand the patterns of day and night, the phases of the Moon, and create caldendars of days, months, and years. These investigations led to the establishment of the first observatories.

The heritage of Armenian astrology in the context of Middle Eastern science is rich, but it has not yet been fully explored. In its role, astronomical tourism gives astronomers an opportunity to travel, to make new discoveries, to find volunteers, and to make their activities more impactful.

Armenian astrology, the most important locations are the Metsamor observatory, the Complex of Karahunj (Zorats Karer), Tatev twisted pillar, and the Byurakan observatory. The most important modern-day astronomers of Armenia are Anania Shirakatsi, Azaria Sanetsi, Viktor Ambartsumian, Tadevos Aghekyan, and Grigor Gurzadyan, and others.

As a commercial tour operator, our company offers tour packages that give vistors a chance to investigate the heritage of astronomy based on research that gives them the opportunity to explore, with the help of professional astronomers and scientific experiments. Besides visiting monuments, we offer astronomical packages where the participants have the opportunity to explore the stars and planets with telescopes, to learn about them, and to meet and learn from astronomers.

2. Scientific Tourism

Scientific tourism is a special type of tourism where the scientist and the citizen ineract. This meeting gives the scientist an opportunity to share freely and informally. The visitor has the chance to get closely acquainted with the scientist and to delve into the science to enrich his knowledge of astronomy. In this way "scientific tourism" is targeted to students and those who have some knowledge about astronomy and want to explore it more. The exchange of information is done in a setting where it is pleasant and effective. Scientific tourism should be regarded as a teaching program held under unusual circumstances and locations.

The heritage of Armenian astrology has not been fully explored. In its role, astronomical tourism gives astronomers the opportunity to travel, to make new discoveries, to find volunteers, and to make their activities better known to the public.

Astronomy is a science that explores the movement of heavenly bodies and their features. It is regarded as one of the oldest natural sciences. In ancient times, astronomy was a rich part of the culture of Armenians who lived in the highlands. Metsamor observatory, the Complex of Karahunj (Zorats Karer), Tatev twisted pillar, and Byurakan observatory are examples of Armenian observatories. Famous Armenian astronomers include Anania Shirakatsi, Azaria Sanetsi, Tadevos Aghekyan, Grigor Gurzadyan, and Viktor Hambardzumyan.

3. Observatory of Metsamor

The Observatory of Metsamor is a part of Metsamor's ancient historical-archeological reserve. Metsamor is an unique ancient castle residence and world history monument. It is located 35 km to the southeast of Yerevan, not far from the village Taronik on the shore of the river Metsamor.

Many excavations have been done in Metsamor staring in 1965 until the present. The excavations date the buildings to the iron and bronze ages. The excavations prove that in the bronze age, Metsamor was a great cultural center. Metsamor was a great residence of 10.5 hectares in size, and it had an acropolis surrounded with gates.[1]

In the early era (11th–9th centuries BC), Metsamor was a city. The acropolis, the observatory, and the residence occupied about 100 hectares of space. Among the things that were found in this place, the most prominent is the carved onyx weight in the shape of a frog (16th century BC) with the name of Babylonian King Ulam Burarich (or Ulamburiash) inscribed, and the Egyptian hieroglyphic sardonyx stamp of another Babylon king (15th century BC).[2]

4. Karahunj (Zorats Karer)

Karahunj occupies about 7 hectares. It has monumental structures and many buildings. To the west there are many tombs, and to the east near the gates there is circle with many roads and stones, with interesting holes in the stones. While exploring the territory, Paris Herouni came to the conclusion that it was an observatory (Herouni 2006). Excavations continue to confirm this hypothesis.

5. The Twisted Pillar of Tatev Monastery

The unique Tatev twisted pillar is the subject of serious discussions.The age of the pillar is considered to be about 1100 years old (S. Orbelian 1910). The meaning of the pillar is

[1]`http://hushardzan.am/671/`

[2]`http://hushardzan.am/en/archives/category/metsamor`

interpreted in different ways. From the spiritual point of view it is meant to be a divine transmitter. One of the most famous interpretations is that the pillar began twisting while Hovannes was praying; the other interpretation is that the pillar began twisting when the army of the enemy was approaching the monastery. Of course these versions are valuable but they do not provide a scientific explanation for the twisted pillar. There is also the third meaning of pillar, which is that the pillar has an astronomical purpose as a stellar indicator. The architecture of the pillar is very important. The anchor consists of two huge basalt blocks. The base of the basalt has a square form, and the upper part of the basalt turns to an octagonal structure. The anchor is located on it, and the middle of the octagonal section is divided into two parts, one 2.66 m high and the second part 2.13 m high (Ayvazyan & Badishyan 2016). The pillar is decorated with a sculpted octagonal cap on which stands the cross with sculpted pedestal. Some parts of the pillar are sculptured with crosses and on the southern part of the pillar there is a sun clock.

6. Modern Observatory of Byurakan

The Astrophysical Observatory of Byurakan (BAO) was founded in 1946 on the southern slopes of mount Aragats, near the village Byurakan. During construction of the towers, Victor Hambardzumyan began preparing the specialists at Yerevan State University. At that time, in the open air, the first observations were done with simple telescopes. The formal opening of the Observatory was in 1956. Through important discoveries and surveys, the Observatory has become known around the world.

Victor Hambardzumyan was the Director of BAO from its opening until 1988. During this period, the Observatory was equipped with the most famous telescope of the Europe at that time, the 2.6 m diameter Schmidt telescope, with many other scientific instruments. They gave the astronomers the opportunity to do important research on stars and galaxies.

7. Scientific Tours

As a tour operator, we offer tour packages that give visitors a chance to investigate the heritage of astronomy in Armenia. Tours are based on research that give visitors a chance to explore with the help of scientists ad scientific experiences.

We make scientific trips throughout Armenia. All trips have three components: excursions, lectures, and scientific work. The durations and locations of trips are different. The types of tours include short day trips, overnight trips, 2–10 day tours, summer schools, camping tours, and evening trips to the Byurakan Observatory. The aim of our activities is to study the heritage of astronomy, increase understanding, and advocate for protection of ancient sites. Our tour packages are based on scientific research to satisfy all those who are interested, and they also provide opportunities to make discoveries using scientific methods. Besides the studying of astronomical monuments, we also offer astronomical tour packages that give visitors a chance to make observations with telescopes in the Observatory, study the planets and stars, and increase their knowledge of astronomy.

References

Ayvazyan, Samvuel 2016, Amiran Badishyan, Charter N12
Byurakan Observatory, Armenian Encyclopedia, http://www.encyclopedia.am/pages.php?bId=1&hId=118
Herouni, Paris 2006, Armenians and Ancient Armenia, Yerevan
"Metsamor" Historical and Archeological Reserve Museum, 09/02/2012. http://hushardzan.am/
Orbelian, Stepanos 1910, History of the Province of Sisian, Tiflis
Zamarovsky, Voitekh 2002, Astronomy of ancient societies

Astronomical Heritage of the Middle East
ASP Conference Series, Vol. 520
Sona V. Farmanyan, Areg M. Mickaelian, J. McKim Malville, and Mohammad Bagheri, eds.

Astronomical Tourism Sites in Armenia

Sona V. Farmanyan and Areg M. Mickaelian
NAS RA V. Ambartsumian Byurakan Astrophysical Observatory (BAO), Byurakan 0213, Aragatzotn province, Armenia;
`sona.farmanyan@mail.ru, aregmick@yahoo.com`

Abstract. Armenia is rich in scientific and astronomical tourist sites, including astronomical archaeological sites, medieval universities, modern scientific institutions, and science related museums. Examples of archaeological sites, ancient observatories, and astronomical petroglyphs (rock art) are found in Armenia, as well as intangible items of astronomical heritage, such as Armenian calendars derived from ancient astronomical knowledge. Modern observatories and astronomical institutions can also be considered astronomical tourist sites as well. Armenia also has an astronomy and space science museum. Despite the fact that astronomical tourism is a new field, it has great popularity, and Armenia has great potential in this field. It is very important to introduce Armenia to the world from this perspective. In this paper we introduce the major astronomical tourism centers of Armenia.

1. Introduction: Scientific Tourism

Scientific Tourism is a new area not only in Armenia but also in the world. Scientific tourism involves visiting science-related centers. In order to be well organized, it is necessary that the scientific center has proper infrastructure to be able to present the center to public in an interesting way. Despite the fact that scientific tourism is a new field, it has great potential, and Armenia has great potential in this field. It is very important to present Armenia to the world from this perspective, including both scientific archaeological sites and modern institutions and museums.

Scientific tourism centers can be divided into 5 groups:

- scientific archaeological sites,
- medieval educational institutions (universities, etc.),
- nature sites,
- research institutions, and
- scientific or science-related museums.

Armenia has astronomy-related sites in four of these categories: archaeological sites, mediaval universities, modern research institutions, and museums. In this paper we give brief descriptions of the Armenian astronomical tourism centers, including those that are not yet active but may serve as sites for future scientific tourism.

There are many petroglyphs (rock art) spread throughout the territory of Armenia. Armenia is unique from this point of view. Ancient "observatories" like Metzamor and Zorats Karer are sites where astronomical observations were made are also among the astronomical heritage of Armenia. Armenia is also famous for its modern professional observatory, the Byurakan Astrophysical Observatory (BAO). Armenia is unique among dozens of countries to have its own space museum.

2. Astronomical Archaeological Sites

The most important astronomical archaeological sites in Armenia are the Metzamor historical-archaeological reserve museum, the Zorats Karer (Karahunge) ancient astronomical observatory, Ukhtasar, and the Voskehat rock art.

Zorats Karer (Karahunge)

The most fascinating historical astronomical building in Armenia is *Karahunge* (the "Armenian Stonehenge") derived from *kar* for "stone," and possibly meaning "singing stones" (it is also known as *Zorats Kar*). It is a megalithic assemblage 200 km from Yerevan and 3 km from town Sisian at an altitude of 1,770 m. The northern latitude is 39°34′, and the eastern longitude is 46°01′. It is an assemblage of many stones arranged in a circle with a few arms radiating from it. As with many other such buildings, Karahunge was thought to be a religious site. Only in the middle of 1980s was Karahunge first interpreted as an archaeoastronomical monument and studied by Prof. E. S. Parsamian (1999) and Prof. P. M. Herouni (1998). The age of Karahunge is estimated to be between 4000 and 7700 years.

Figure 1. The ancient observatory of Zorats Karer (Karahunge).

At the site there are 222 stones with a total extent exceeding 250 m, including 84 with holes of diameter 4–5 cm. Dozens of astronomical stone instruments with accuracy of 30 arcsec have been found. 40 stones form the central ellipse (45×36 m), with a ruined stone-cluster in the center. There is a 8 m wide 8-stone road to the northeast. Some stones were used to find the directions to definite stars. Observations of definite stars suggest that the observatory was used between 7700–2200 BC, a period of about 5500 years. According to many researchers (e.g., Bochkarev & Bochkarev 2005), a comparison of the present state of the monument with its situation a hundred years ago reveals considerable degradation. Thus, the monument urgently needs protection. The monument is unique in the Trans-Caucasian region and could be the oldest known observatory in the world. If the estimated age of Karahunge is confirmed by archaeological methods, it clearly should be included in the UNESCO World Heritage list of the most important cultural memorials of our planet.

Metzamor Historical-Archaeological Reserve Museum

The ancient settlement of Metzamor is one of the unique monuments of world culture in which ancient copper and bronze complexes, sanctuaries, tombs, a possible observatory, etc. were excavated. It is located 35 km southwest of Yerevan in Armavir province, not far from Taronik village on the shores of the Metzamor River.

Metzamor is the site of one of the possible ancient observatories in Armenia. Metzamor was an ancient town. A settlement has existed there since the 5th millennium BC. It was first interpreted as an archaeoastronomical monument in the middle of the 1960s by Prof. E. S. Parsamian (1985). He found an observatory outside of the fortress with a probable age of 4600 years. As with Zorats Karer (Karahunge), Metzamor also needs more study and proper support and protection both from the Armenian government and the world archaeoastronomical community.

Rock art

Studies of the Armenian rock art present in the territory of modern Armenia (historic Armenia was ten times larger, having 300,000 square km area) show that the Armenians were interested in heavenly bodies and phenomena. The Earth, the Sun, the Moon, planets, comets, Milky Way, stars, and constellations are reflected in these pictures drawn on rocks in mountains around Lake Sevan and elsewhere in Armenia. These pictures and drawings are being studies by a number of historians, archaeologists, and astronomers. However, there is not enough governmental support to organize large-scale studies or at least try to catalog and preserve these ancient treasures.

Figure 2. Examples of the Armenian rock art.

2.1. Research Institutions

Among the modern institutions, those that are considered scientific tourist sites where equipment or laboratories present attractions include the NAS RA V. Ambartsumian Byurakan Astrophysical Observatory (BAO) and the Aragatz Cosmic Particles Station (operated by the Alikhanyan National Laboratory, Yerevan Physics Institute, YerPhI).

NAS RA V. Ambartsumian Byurakan Astrophysical Observatory

The BAO is one of the most important places in Armenia and attracts tens of thousands of visitors every year, both from Armenia and from abroad. It is no coincidence that for many years a visit to the Observatory is included as a mandatory program for Armenian schoolchildren, and an offer to visit the Observatory is almost always made by travel agencies.

The BAO was founded in 1946 and it is the only professional astronomical observatory in Armenia. It has a unique architectural collection of buildings and rich botanical garden. It is commonly used to represent Armenia to visitors. The BAO is situated in the Aragatzotn Province, on the southern slope of Mt. Aragatz, near the village of Byurakan.

Viktor Ambartsumian's house, where he lived for 46 years from 1950–1996, is now a museum located on the site of the Observatory. It was established in 1998 and introduces visitors to the life and activities of the great scientist.

Figure 3. The 2.6 m telescope of the Byurakan Astrophysical Observatory (BAO).

Visitors can see the largest (2.6 m) telescope and get an introduction to its operation, view the 1 m Schmidt telescope, and participate in observations with a dedicated small 32 cm telescope during the evening or hours. One of the most important accomplishments of the BAO is the famous and unique Markarian spectroscopic survey and its digitized version, which has been included in the UNESCO's *"Memory of the World"* International Register of documentary history. It is worth mentioning that it is one of 11 UNESCO heritage items from Armenia, very few of which are scientific in nature.

The Observatory has a conference hall where many international and local symposia, conferences, workshops, seminars, summer schools for young astronomers, and other events have been held. Viktor Ambartsumian's office is being used for smaller events, which has been preserved in its original state and is being used for Scientific Council meetings, solemn meetings, seminars, lectures, etc.

There is a hotel (Guest House) just inside the boundaries of the Observatory that can accommodate 25–30 guests, designed in the Armenian national architectural tradition by the famous architect Samvel Safaryan. The hotel has a kitchen and laundry. The Observatory also has a canteen, where the hotel guests can take meals; receptions, dinners, and other events may be organized there for up to 70 people.

Several types of visits to BAO are available, including astronomical lectures given by professional astronomers. Full information may be found on the BAO Scientific Tourism webpage.[1]

2.2. Science Museums

Among museums having displays related to astronomy and space sciences are the Space Museum, the Matenadaran museum of ancient manuscripts (where many manuscripts of astronomical nature are preserved), and the Viktor Ambartsumian house museum in Byurakan (discussed above).

The Space Museum was founded by Grigor Gurzadyan in Garni in 2001, and in 2015 it was moved to Yerevan. The museum exhibits early samples of space research, including landing equipment returned from the Cosmos, other spacecraft, and a recreation of the ORION-2 Observatory, which was exhibited in Hannover EXPO 2000. The museum collection includes documents and manuscripts authored by cosmonauts, Vostok and Soyuz space engineers, scientists, and from NASA. One of the most valuable examples is a manuscript by Nobel Prize laureate Hans Bethe, written jointly with Grigor Gurzadyan.

---

[1] `https://www.aras.am//SciTourism/eng`

Astronomical Heritage of the Middle East
ASP Conference Series, Vol. 520
Sona V. Farmanyan, Areg M. Mickaelian, J. McKim Malville, and Mohammad Bagheri, eds.

Byurakan Astrophysical Observatory as a Tourism Center

S. V. Farmanyan, V. L. Harutyunyan, G. A. Mikayelyan, A. M. Mickaelian
NAS RA V. Ambartsumian Byurakan Astrophysical Observatory (BAO), Byurakan 0213, Aragatzotn province, Armenia; sona.farmanyan@mail.ru

Abstract. The Byurakan Astrophysical Observatory (BAO) is one of the most important organizations in Armenia. It is a National Value officially recognized by the Republic of Armenia (RA) Government in 2012. The BAO plays not only a great role in Armenian science and its scientific achievements, but also in Armenian culture, education, and science outreach. BAO has a unique architectural ensemble of buidlings and a rich botanical garden recently recognized as dendropark by an agreement with the RA Ministry of Ecology. BAO is one of the most frequently visited touristic centers in Armenia. The famous Markarian Survey, a unique spectroscopic sky survey that led to many discoveries, was included in the UNESCO "Memory of the World" International Register in 2011. BAO is a symbol that represents the country to the rest of the world.

1. Introduction: Byurakan Astrophysical Observatory

The Byurakan Astrophysical Observatory (BAO) is one of the most important scientific institutions in Armenia and one of the most famous observatories in the former Soviet Union and Euro-Asian continent. It was founded in 1946 by the great Armenian scientist Viktor Ambartsumian (1908–1996) and at present it is one of the 34 research institutes of the Armenian National Academy of Sciences (NAS). Many scientific achievements have been accomplished here, including the discovery of stellar associations (Ambartsumian, 1947), the hypothesis on the activity of galactic nuclei (Ambartsumian, 1954–1956), the famous Markarian Survey (also known as the First Byurakan Survey, or FBS; Markarian et al. 1965–1980), discovery and study of hundreds of flare stars (Ambartsumian, Mirzoyan et al. 1968–1980s), the Second Byurakan Survey (or SBS, Markarian, Stepanian et al. 1978–1991), and others. New ideas on the birth and evolution of stars and nebulae in stellar associations, activity of galactic nuclei, and new theoretical approaches have been developed by Ambartsumian and other Armenian astrophysicists. Numerous active galaxies and quasars, flare and T Tauri type stars, novae and supernovae, and many other objects have been discovered in Byurakan.

BAO is actively involved in international collaborations with some hundred observatories from dozens of countries (especially productive are the collaborations with France, Russia, Germany, Italy, and USA). A number of scientific meetings have been held in Armenia, including the International Astronomical Union (IAU) 6 Symposia and Colloquia, and the all-European annual astronomical meeting in 2007. High-level politicians, Nobel prize winners, other outstanding scientists, and many other honorable guests have visited the BAO.

At present, the Observatory has 106 associates, including 40 researchers. Due to new approaches and seminal scientific results, Byurakan astronomers have achieved international recognition and respect, and the BAO is recognized by the scientific community as one of the main centers for astrophysical research in the world. The ideas proposed in Byurakan have motivated further research at many observatories, and the objects discovered in Byurakan have been observed worldwide by famous astrophysicists.

BAO is a national treasure officially recognized by the Republic of Armenia (RA) Government in 2012. This includes not only the great role it plays in Armenian science and its scientific achievements, but also its role in Armenian culture, education, and science outreach. BAO has unique architecture and a rich botanical garden recently (2017) recognized as a "dendropark" through an agreement with the RA Ministry of Ecology. BAO is one of the most frequently visited tourist centers in Armenia. The famous Markarian Survey, a unique spectroscopic sky survey that led to many discoveries, was included in the UNESCO *"Memory of the World"* International Register in 2011.

In 2015, the BAO was recognized by the IAU as a regional center to coordinate astronomy development in southwest and central Asia. BAO is a distinctive place that represents Armenia with people from all around the world.

2. Byurakan Astrophysical Observatory's Architectural Ensemble and Garden

BAO is distinguished by its unique *architectural ensemble* of buildings and structures built with the finest traditions of Armenian architecture. The architect of the complex, which was started in 1947 and basically completed in 1956, is the outstanding architect Samvel Safaryan (1902–1969). Unlike the majority of observatories in the world, beauty was not sacrificed for practical use here. Safaryan was the founder of functionalism in Armenian architecture, and the BAO complex was adapted and optimized to meet the demands of scientific work, while remaining a masterpiece of Armenian architecture. Being one of our greatest architects, Safaryan also designed and constructed the NAS RA complex in Yerevan, which proves insight into his attitudes about science. The oldest Safaryan buildings of the complex are the main administrative building, V. Ambartsumian's house, the hotel, the towers of small and medium-size telescopes, the residential building for BAO staff, and several other smaller buildings.

From their student years, the future famous architects Martin Mikayelyan and Sargis Gurzadyan participated in the BAO design work. In 1960–1980, Gurzadyan became the principal architect of the second phase of BAO's development. Among the newer buildings are the Conference Hall, which has attached to it a canteen and shop, the new laboratory building, the 2.6 m telescope building, the living quarters for staff members constructed next to BAO's campus, and several other smaller buildings.

The Observatory's campus is a unique *botanical garden* (dendropark) surrounding the scientific laboratories and telescope towers with more than 120 species of trees and other plants from around the world. In order to enrich the park, Viktor Ambartsumian brought various seeds from different countries, which were planted and processed on the BAO property. BAO collaborates with the NAS RA Institute of Botany, whose specialists visit from time to time to work on and catalog the trees. There are more than 10 types of fir trees, including the rare silver fir tree. There are also many types of fruit trees.

Armenia is also famous for its diversity of birds; it is known that 349 out of 535 species of Euoropean birds come to Armenia and to the BAO gardens. Here you can

Figure 1. BAO's New Laboratory Building and a part of its surroundings.

find woodpeckers, hoopoes, thrushes, starlings, jays (wild magpies), crows, magpies, owls, turtle-doves, pine-finches, tits, chaffinches, goldfinches, wagtails, nightingales, sparrows, swallows, sometimes orioles, bullfinches, and larks. Other animals can be found as well, including rabbits, hedgehogs, beetrootes, moles, etc.

Interestingly, following BAO's initiative, a number of other observatories around the world have created gardens to make an attractive place of work for scientists and for the entertainment of visitors.

3. Byurakan Astrophysical Observatory as Tourism Center

BAO is a historic landmark of high cultural value and one of the most important places in Armenia. It attracts tens of thousands of visitors every year, both from Armenia and from abroad. It is no coincidence that for many years a visit to the Observatory is considered mandatory for Armenian schoolchildren. Travel agencies almost always offer visits to the Observatory.

Viktor Ambartsumian's house museum is located on the Observatory grounds, where he lived for 46 years from 1950–1996. The museum was established in 1998 and introduces visitors to the life and activities of the great scientist.

Visitors can see the large 2.6 m telescope and get an understanding on its operation. The 1 m Schmidt telescope is also open for visitors. A dedicated 32 cm telescope is used for visitor observations during the evening or night hours.

One of the most important contributions of the BAO is the famous and unique *Markarian spectroscopic survey* and its digitized version, which has been included in the UNESCO's *"Memory of the World"* International Register of heritage documents. It is one of the 11 UNESCO heritage items from Armenia. It is one of the few entries on the Register of scientific nature.

The Observatory has hosted many international and local symposia, conferences, workshops, seminars, summer schools for young astronomers and other events in the Conference Hall. Viktor Ambartsumian's office is used for smaller events, which has been preserved in its original state, and is used for Scientific Council meetings, seminars, lectures, etc.

The subjects of popular lectures at BAO include:

- Our understanding of the Universe from ancient times to the present.

Figure 2. The largest (2.6 m) telescope of the Byurakan Astrophysical Observatory.

Figure 3. The Conference Hall of the Byurakan Astrophysical Observatory.

- Byurakan Astrophysical Observatory and its achievements.
- Viktor Ambartsumian's life and achievements.
- Beniamin Markarian and the Byurakan surveys.
- Grigor Gurzadyan and space astronomy.
- Armenian astronomers in the world.
- Armenian archaeoastronomy.
- The starry sky, astronomy, and mythology.
- The Solar System: from the Sun to space debris.
- Extrasolar planets and astrobiology.
- Stars and nebulae.
- The world of galaxies.
- Large-scale structure of the Universe and cosmology.

- Cosmic catastrophes and "End of the World."
- Extraterrestrial civilizations, UFOs, and paleovisitology.
- Astronomy and astrology.
- An astronomer's tools: telescopes and observatories.
- Space astronomy: the invisible Universe.
- Big Data in astronomy and Virtual Observatories.
- Astronomy and culture.
- Practical importance of astronomy; astronomy and other sciences.

Astronomical films are also shown with commentary by professional astronomers.

The hotel (or Guest House) just inside the Observatory property houses about 25–30 guests. It was built in the Armenian national architectural tradition by the famous architect Samvel Safaryan. The hotel has a kitchen and laundry. The Observatory also has a canteen where the hotel guests can eat, and where receptions, dinners, and other events are held.

Several types of visits have been organized:

- "Sky and Telescope." Short daytime visits (1.5 hours), including a visit to the 2.6 m telescope and Viktor Ambartsumian museum.
- "From Byurakan to Space." Longer daytime visits (4 hours), including visits to the 2.6 m telescope, the V. A. Ambartsumian museum and office, lunch, and a popular lecture or astronomical film.
- "Night Sky Walk." Short evening visit (1.5 hours), including observations with small telescopes.
- "Byurakan and Sky Stars." Long evening visit (4hours), including a popular lecture, dinner, and observations with small telescopes.
- "Between the Rest and Knowledge." 2-day visit (from 11:00 to 17:00 the next day), including an introduction to BAO, V. A. Ambartsumian museum and office, 3 or 4 popular lectures or astronomical films, a visit to the 2.6 m telescope, observations with small telescopes at night, a visit to the Byurakan church, and accommodation in the BAO guest house, including 4 meals.
- "From Armenian Letters to Cosmic Elements." 3 to 7-day visit aimed at regional tourist attractions, including the BAO (similar to "Between the rest and knowledge" package), but also including Oshakan, Aghtsk, Tegher Monastery, Amberd, the Orgov 54 m radio telescope, Aragatzlake, and the Cosmic Ray Station.
- "Astronomy in your Room." This tour includes on-site lectures (1.5 hours) in hotels, meeting rooms, and elsewhere to allow the tourists to combine learning and rest.

- "Constellations and observations." An evening visit (4 hours), including outdoor observations at various locations outside big cities to avoid light pollution. Participants are acquainted with the night sky (constellations and mythology, planets, Milky Way and bright stars) through a popular lecture by a professional astronomer, and include observations with a Celestron Astro Master 130 mm amateur telescope.

Programs include a professional tour manager, specialist guides, night operators for observations, professors and other lecturers, popular materials, and souvenirs. A number of souvenirs are offered, including amateur telescopes, books, sky maps, BAO 2.6 m telescope models, cups, calendars, postcards, bookmarks, pens, etc. For two-day visits, visitors should contact BAO at least 7 days in advance, and 3 days in advance for short visits. Groups may have up to 15 people.

4. Byurakan and Surroundings

It is worth mentioning that Byurakan and its neighboring region are rich with nearby sightseeing attractions. Several other sightseeing attractions are located on the southern slope of Aragatz and adjacent areas (e.g., Amberd, Tegher Monastery, Kari Lake, Armenian kings' grave in Aghtsk, Mashtots grave in Oshakan, Hovhannavank, Saghmosavank, etc.), which makes it even more attractive to rest in the Observatory and visit the surrounding picturesque sites. The most convenient road to climb the Aragatz peaks (the highest is the Northern one, 4090 m above sea level) also passes through Byurakan; it is possible to reach Kari Lake (3207 m altitude) from Byurakan, and then the southern peak (3879 m) becomes more accessible. There are two medieval churches in Byurakan village: St. Hovhannes and Artavazik. For more than 1000 years, the summer residence of the Armenian Catolicoses (Head of the Armenian Church has been located between Byurakan and the nearby neighboring village Antarut. Several other scientific centers are situated in this area, including: the Institute of Radiometrical Measurements near Orgov village, with its unique 54 m radio-optical telecsope (ROT-54, Paris Herouni); the Alikhanyan National Laboratory (Yerevan Physics Institute) Amberd science station and high-altitude cosmic ray station (3250 m altitude) near Aragatz lake (founded by Alikhanyan brothers); and the meteorological station located on the same site. Ashtarak town has two more scientific-research institutes: the NAS RA Institute of Physical Research and NAS RA Institute of Radiophysics and Electronics. The monument to the Armenian Alphabet is located on the road from Byurakan to Aparan. Recently, the Byurakan Art Academy and the Museum of Armenian Traditional Festivals have been founded. The Agarak ancient archaeological site is also worth mentioning.

A comprehensive website has been constructed for the BAO as a Tourism Center,[1] where all necessary information is given.

[1] `https://www.aras.am//SciTourism/eng`

Part VIII

Astronomical Heritage for Development

Astronomical Heritage of the Middle East
ASP Conference Series, Vol. 520
Sona V. Farmanyan, Areg M. Mickaelian, J. McKim Malville, and Mohammad Bagheri, eds.

IAU Regional Office of Astronomy for Development: Coordination in Southwest and Central Asia

A. M. Mickaelian, S. V. Farmanyan and G. A. Mikayelyan
NAS RA V. Ambartsumian Byurakan Astrophysical Observatory (BAO), Byurakan 0213, Aragatzotn province, Armenia; aregmick@yahoo.com

Abstract. The International Astronomical Union (IAU) announced its Strategic Plan on *Astronomy for Development* in 2009, during the International Year of Astronomy (IYA). One of its main components was the creation of the Office of Astronomy for Development (OAD) and corresponding Regional Offices (ROADs) for implementation and coordination of its aims. The OAD was created in Cape Town, South Africa and later on ROADs were created in 8 regions. Since 2015, Armenia has hosted one of them, IAU South West Asian (SWA), later renamed to South West and Central Asian (SWCA) ROAD. At present, six countries have officially joined (Armenia, Georgia, Iran, Kazakhstan, Tajikistan, and Turkey), but the Office serves a rather broad region, from Eastern Europe to Central Asia. Armenia's geographical location and its historical role in astronomy (both for well-known archaeoastronomical heritage and the presence of the famous Byurakan Astrophysical Observatory, BAO, founded by Viktor Ambartsumian in 1946) serve as a link between Europe and Eastern Partnership countries, and the Middle East and Asia in general. We run activities in 3 directions, with three Task Forces: *Universities and Research*, *Children and Schools*, and *Public Outreach*. We present our projects and all other accomplishments, and discuss the role of our ROAD in maintaining contacts and development of astronomy in the region, as well as contacts between Europe and the Eastern Partnership countries. The most up-to-date information about the IAU SWCA ROAD is available on its webpage: http://iau-swa-road.aras.am/eng/index.php.

1. Introduction

The International Astronomical Union (IAU) developed and adopted in 2009 a decadal Strategic Plan (SP), now entitled *"Astronomy for Development"*. The SP significantly increases the role of astronomy for development of other sciences, technology, culture, and society, as it is tightly linked to all these (Figure 1). This plan resulted from an extensive process of consultation beginning with a meeting of stakeholders in Paris in January 2008 and including feedback from key stakeholders during the various drafts of the SP. It was endorsed by the General Assembly of the IAU in Rio de Janeiro in August 2009, and builds on the momentum of the very successful International Year of Astronomy 2009 (IYA). The objective of this SP is to use astronomy to stimulate development in all regions of the world. Crucial to the implementation of the SP was the creation of a global "Office of Astronomy for Development" (OAD). The OAD is tasked with establishing and strategically coordinating the efforts of Regional Offices (ROADs) and Language Expertise Centers (LOADs) around the world, as well as establishing three Task Forces (TF), namely (i) *Astronomy for Universities and Research*, (ii) *Astronomy for Children and Schools*, and (iii) *Astronomy for the Public*.

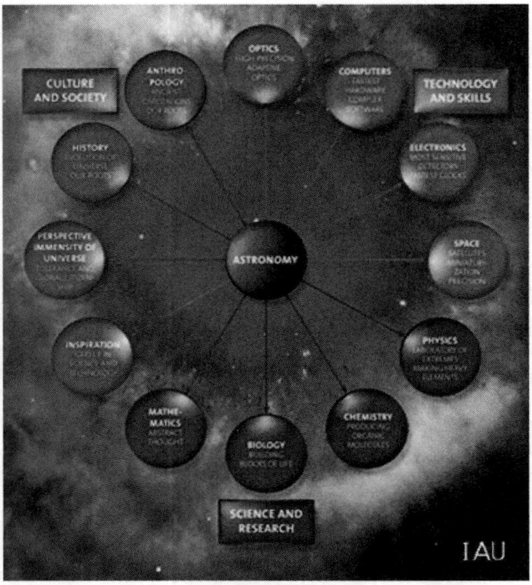

Astrophysics
Astrochemistry
Astrobiology
Astrogeology
Space research
Instrumentation
Astroinformatics
Astrostatistics
VO / e-Science
Astronomy in Culture
Archaeoastronomy
Astro tourism
Astro journalism
Astronomical Education
Amateur Astronomy
Popular Astronomy

Figure 1. The relation of Astronomy to other sciences, technology, culture, and society. A number of inter- and multi-disciplinary sciences and other fields are listed which have tight relations to astronomy.

Through a strong partnership with the South African government, the OAD[1] is hosted at the South African Astronomical Observatory, a facility of the National Research Foundation. Furthermore, eight regional offices were created, one of them in Armenia.

2. The Establishment of the IAU SWA ROAD

Armenia showed interest in establishing of one of the regional IAU offices since the beginning of the initiative (2009–2011). Armenia's proposal to host a ROAD was strong because of its available facilities and ongoing activities. The proposal was finally approved on 10 June, 2015 and an agreement was signed between the IAU and the Byurakan Astrophysical Observatory (BAO) on 6 August 2015. The office was formally established on 1 September 2015 and was called IAU South West Asian (SWA) ROAD. Dr. Areg Mickaelian became its first Director. The SWA Regional Steering Committee was created in September 2015, as well as the SWA webpage[2] (Figure 2). Staff members joined on 1 October 2015: Naira Azatyan (TF1 coordinator), Sona Farmanyan (TF3 coordinator), and Gor Mikayelyan (webmaster). Finally, the Inauguration

---

[1] `www.astro4dev.org`

[2] `http://iau-swa-road.aras.am`

Ceremony was held on 13 October 2015. The IAU OAD/SWA ROAD Workshop was attached to this event. Representatives from the IAU (General Secretary Piero Benvenuti, former Vice-President George Miley, and OAD Director Kevin Govender) and regional representatives from other countries (Georgia, Iran, Turkey, Russia, Israel, and Jordan) were present.

Like our long tradition of organizing joint Armenian-Georgian (or Byurakan- Abastumani) workshops (colloquia) since 1974, we conducted a new series of Armenian-Iranian Astronomical Workshops (AIAW). The first one was organized on 13–16 October, 2015 in Byurakan, so that Iranian guests were able to participate in the IAU SWA ROAD Inauguration Ceremony.

Georgia and Iran were the first countries to officially join the SWA ROAD with Armenia. During many centuries, these countries have had numerous relations in all areas, including science and astronomy.

Figure 2. IAU South West and Central Asian ROAD webpage.

Astronomy in Georgia is generally represented by the Abastumani Astrophysical Observatory (AbAO), founded in 1932. It is one of the leading scientific institutes in the country. The main fields of research at AbAO are Solar System bodies (including near-Earth asteroids), various aspects of Solar physics, stellar astronomy (including binary stars and open clusters), extragalactic objects (AGN), theoretical astrophysics, cosmology, atmospheric, and Solar-terrestrial physics. Research in these fields is carried out in other institutions in Georgia as well, including the Institute of Theoretical Physics at Ilia State University, Javakhishvili Tbilisi State University, and the School of Physics at Free University of Tbilisi. At AbAO, several telescopes are operational today, as well as instruments for atmospheric studies. At present, in AbAO the 70-cm Maksutov meniscus telescope, 53-cm azimuthal reflector, 22-cm reflector ORI, 40-cm double astrograph, 53-cm large and 11.5-cm small solar coronagraphs are operational. They are equipped with CCD cameras, and spectroscopic and photometric observations are carried out. The instruments for atmospheric studies include: the aerosol lidar M-10 system, GRIPS-5 to measure the mesopause temperature, an all-sky imager for measurements of some parameters of the nightglow, and the ozonometer. A new magnetometer to monitor the Earth's magnetic field has been installed recently. Astroclimatic conditions at Abastumani makes the location favorable for installation of advanced astronomical instruments, which is one of the main goals of the Observatory for the near future. In 2007, the Observatory was integrated with Ilia State University, merging scientific research and education, which facilitated the growth of a new generation of researchers. There are groups of astronomers and astrophysicists in other Georgian universities and institutions as well. Georgian scientists collaborate with research centers and universities worldwide. Research groups participate in various international scientific projects. In the recent years the interest in astronomy in Georgia has grown, which increases the future potential of its development in the country.

Figure 3. Abastumani Astrophysical Observatory in Georgia.

Astronomy in Iran dates back many thousands of years. When Cyrus the Great, the founder of the Persian Empire, captured Babylon in 539 BC, Magi who migrated there transformed Babylonian astronomy. They were the first to record planetary motion through constellations. In turn the Magi learned from the Babylonian astronomers and translated Babylonian books into the early Persian language. The Andromeda galaxy was first identified by a Persian astronomer Abd al-Rahman al-Sufi, calling it a little cloud in his book of fixed stars around 964. In the 13th century, Maragheh Observatory was established. It holds a unique place in the history of medieval astronomy, representing a new wave of scientific activities in the Islamic world. It played a key role in the development of sophisticated pre-Copernican non-Ptolemaic systems for explaining the planetary motions, and it was the model for several observatories that were built in Persia, Transoxiana, and Asia Minor up to the 17th century.

In the present day, research in astronomy and astrophysics is conducted in a number of universities in Iran. There are two dedicated research institutes in Iran, the IPM School of Astronomy in Tehran and the Research Institute for Astronomy and Astrophysics in Maragheh (RIAAM). Among the most active universities in astronomy are: IASBS in Zanjan; Ferdowsi University of Mashhad (FUM); the University of Tabriz; Shiraz, Birjand, Kerman and Zanjan universities; Amirkabir University of Technology; and Sharif University of Technology, both in Tehran. With nearly 400 members, the Astronomical Society of Iran (ASI) is a non-governmental organization that represents the Iranian astronomy community internationally and with IAU. It coordinates various activities for professional and amateur astronomers and organizes training workshops throughout the year. Iranian astronomers gather twice a year to share their ongoing research activities. The ASI also publishes an International Journal of Astronomy and Astrophysics (IJAA), which is a peer-reviewed open-access journal dedicated to original research and invited/review articles by distinguished astronomers around the world. Astronomy is also popular in Iranian schools. The International Olympiad of Astronomy and Astrophysics (IOAA), in which Iranian teams do very well, see participation of a few thousand students in a national exam.

The Iranian National Observatory (INO, Figure 4) is under construction at an altitude of 3600 m at Gargash summit, 300 km south of Tehran. The site selection was concluded in 2007 and the site monitoring activities have begun since then, which indicates that it is a high quality site with a median seeing of 0.7 arcsec through the year. One of the major observing facilities of the observatory will be a 3.4 m Alt-Azimuthal Ritchey-Chretien optical telescope which is currently under design. This f/11 telescope will be equipped with high resolution medium-wide field imaging cameras and medium and high-resolution spectrographs.

Figure 4. Iranian National Observatory (INO) at Gargash summit.

Since the beginning, the IAU SWA ROAD has developed activities in all three Task Forces. A number of meetings and schools were organized and a number of other projects were accomplished. Especially successful were the activities related to Scientific Tourism in Armenia and in the region. We were awarded two grants in 2016: an OAD grant for the development of Astro Tourism in Southwest Asia, and a Swiss SDC grant for the development of Scientific Tourism in Armenia.

SWA ROAD representatives, the Director Areg Mickaelian, and BAO Director Haik Harutyunian took part in the IAU Arab World ROAD opening in December 2015. In February–March 2016, Areg Mickaelian and Sona Farmanyan took part in the OAD/ROADs meeting in Cape Town, South Africa. The whole SWA team (Areg Mickaelian, Susanna Hakopian, the new TF1 coordinator, Sona Farmanyan, and Gor Mikayelyan) visited Georgia in March–April 2016 to tighten the collaboration and exchange scientific knowledge. In August 2016, Areg Mickaelian and Sona Farmanyan were invited to RIAAM, Maragheh, Iran to participate in a workshop and give talks.

We hold regular telecons both with OAD/ROADs and with the SWA ROAD Steering Committee to give short reports on accomplishments and to discuss all current and future matters.

3. Expansion of the IAU SWCA ROAD

In May 2016, Areg Mickaelian and the Scientific Secretary of BAO Elena Nikoghosyan visited three Central Asian countries, Uzbekistan, Kazakhstan, and Tajikistan to renew our former contacts and collaborations. As a result, during June 2016, Kazakhstan and Tajikistan officially joined our ROAD. This significantly strengthens our center and expands its sphere of activities to Central Asia.

Kazakhstan. The Fesenkov Astrophysical Institute (APhI, founded in 1941), in Almaty, Kazakhstan is the main astronomical center. At present, Dr. Rashit Valiullin is the Director. One of the outstanding scientists, Prof. Eduard Denissyuk, was one of the first astronomers to spectroscopically observe Markarian galaxies and has collaborated with BAO astronomersfor many years. There are three observatories attached to APHI. The Kamenskoe Plateau Observatory (altitude 1450 m), has an AZT-8 70 cm telescope, a Zeiss-600 60 cm telescope, a Hertz 50 cm reflector telescope, and a wide aperture 50 cm Maksutov meniscus telescope. The Tian-Shan Observatory (TShAO, altitude 2735 m) has two 1 m telescopes. Finally, the Assy-Turgen Observatory (altitude 2750 m) has a 1 m telescope). The research subjects addressed at these observatories are: the physics of stars and nebulae; physics of the Moon and planets; cosmology, stellar dynamics and computational astrophysics; nuclear astrophysics; artificial Earth satellites; and advanced astrophysical research. The Kazakh National University (KazNU) with 16,000 students is the main university preparing professional astronomers and astrophysicists.

Figure 5. Main building of Fesenkov Astrophysical Institute in Almaty, Kazakhstan.

Tajikistan. The Institute of Astrophysics (founded in 1932) in the Academy of Science of the Republic of Tajikistan in Dushanbe is the main astronomical center in Tajikistan. At present, Dr. Gulchehra Kokhirova is the Director. Profs. Pulat Babadjanov and Khursand Ibadinov, former directors and eminent Tajik astronomers, are still active and strongly support the collaboration with IAU ROAD. There are three observatories associated with the Institute: Hissar Observatory (HisAO, altitude 730 m), hosting the AZT-8 40 cm astrograph; the observatory "Sanglokh" (altitude 2300 m) in the Dangara area, which has a 1 m telescope and 60 cm Carl Zeiss); and its branch, the observatory "Pamir" (altitude 4350 m) in the Murghab district of Badakhshan, with a 70 cm telescope). Sanglokh Observatory was recently reopened and the President of Tajikistan was present at the opening ceremony. The Institute's primary research subjects include comets and asteroids, experimental astrophysics, meteor astronomy, ionospheric physics, astrometry, variable stars, and the structure and dynamics of stellar systems.

Figure 6. The 1 m telescope of Sanglokh Observatory, Tajikistan.

Due to the involvement of Kazakhstan and Tajikistan, our regional centre was renamed the **IAU South West and Central Asian (SWCA) ROAD**. Corresponding changes and additions were made to our webpage.

Turkey. On November 17, 2017, Turkey officially joined the South West and Central Asian ROAD office and expressed its desire to carry out its activities for the development of Astronomy by adopting Armenia's coordinating role. It is important to state that this was a big step politically for Turkey to accept and join the ROAD and to have Armenia recognized as an astronomical center by Turkey, even though it was organized by the IAU. Now our cooperation will be closer, with more visits and exchange of scientific experience. We hope to work on various astronomical topics to develop astronomy, to organize conferences, and to carry out all the activities that this cooperation assumes.

4. Other Collaborations

Beside activities for the regional countries, IAU SWCA ROAD also encourages and strengthens collaboration with other regions and countries. Especially promising are contacts with two neighbouring ROADs: East Asian (based in Beijing, China) and Arab World (based in Amman, Jordan). The first ROAD is coordinating efforts over a huge region (China, Mongolia, North Korea) and is especially interested in the development of Silk Road projects that also relate Armenia. The Arab World ROAD covers some of the same large region (Middle East) and is our closest neighbor. We have dis-

Figure 7. TÜBITAK National Observatory with its 1.5 m telescope at 2450 m altitude. Bakırtepe, Antalya Province, Turkey.

cussed possibilities to tighten collaboration in all areas (and TFs) and to establish new programs.

In addition, recently we have discussed the possibility of collaborating with Indian astronomers, as India is not currently a member of any ROAD; the closest one is SWCA.

5. Projects and Other Activities

It is our rule that both senior and young scientists from regional countries (especially Georgia and Iran) participate in our meetings, conferences, schools, and other events. We also have many outreach intiatives with the regional countries, again most often Georgia and Iran.

Armenia is active in organizing astronomical meetings, schools, conferences, workshops, and other events. Among the most important meetings and schools held in Armenia are the IAU Symposia and Colloquia: IAU S029: *Non-Stable Phenomena in Galaxies* (1966), IAU S129: *Observational Evidence of Activity in Galaxies* (1986), IAU S137: *Flare stars in Star Clusters, Associations and Solar Vicinity* (1989), IAU S194: *Activity in Galaxies and Related Phenomena* (1998), IAU C184: *AGN Surveys* (2001), and IAU S304: *Multiwavelength AGN Surveys and Studies* (2013). Another large event was the all-European annual astronomical meeting in 2007, JENAM-2007 (Joint European and National Astronomical Meeting), held in Yerevan, Armenia. It was the biggest scientific event ever held in Armenia. We have had many joint meetings with individual countries, such as the Byurakan-Abastumani (Armenian-Georgian) Colloquia in 1974–2013, the Armenian-French Workshops in 1995 and 2009, and the Armenian-Iranian Astronomical Workshop in 2015, as well as many meetings dedicated to the anniversaries of BAO, Viktor Ambartsumian, Anania Shirakatsi, Beniamin Markarian, Ludwik Mirzoyan, and others. Many guests from the region were present at BAO's 70th anniversary meeting in September 2016. Our office has been awarded an OAD grant for organization of a Regional Summer School and Workshop in September 2018, at which many regional guests are expected.

Among the summer schools, the Byurakan International Summer Schools (BISS) are very famous. We started this initiative in 2006 and so far, have organized five such events: 1BISS in 2006, 2BISS in 2008, 3BISS combined with the IAU International School for Young Scientists (ISYA) in 2010, 4BISS in 2012, and 5BISS in 2016. 6BISS will be organized in September 2018.

Byurakan Summer Schools for YSU students (BSS) are our local schools. They have been organized in 1995, 2005, 2009, and 2013. For the students, on the initiative of Sona Farmanyan, we organized BAO Science Camps (BSC); four such events were held in 2014–2017.

The ArAS School Astronomical Lectures program was started in 2012 on the initiative of Yervant Terzian and Areg Mickaelian. During 2012–2017 we organized four such programs. In addition, in 2016 and 2017, we had the "My Universe" contest and the winners visited BAO.

Our ROAD representatives are active in participating in many international and regional meetings and have given numerous talks at events related to the scientific, educational, public outreach, and in general about our SWCA ROAD activities.

In relation to our collaboration between Armenia and Iran, on 15–18 August 2016, Areg Mickaelian (IAU SWCA ROAD Director) and Sona Farmanyan (IAU SWCA ROAD TF3 Programme Coordinator) were invited to the Research Institute for Astronomy and Astrophysics of Maragha (RIAAM). During those days, the 8th Advanced Astrophysics Workshop of Maragheh was organized. Areg Mickaelian delivered two lectures on *"Astronomical Surveys, Catalogues, Databases, Archives, and Virtual Observatories"* and *"Multi-Wavelength Studies of Active Galaxies."* Sona Farmanyan delivered two talks on *"Archaeoastronomy and Cultural Astronomy in South West and Central Asia"* and *"Ancient Mythology and Cosmology"*. Sona also presented the IAU SWCA ROAD activities during 2015–2016. A number of discussions were held with Iranian astronomers, including Prof. Hossein Ebadie (RIAAM Director) and Prof. Pantea Davoudifar, with whom there is a collaboration in co-supervising of two Iranian students.

The fourth Middle East and African IAU Regional meeting (MEARIM IV) was jointly hosted by the Entoto Observatory & Research Center (EORC) and the East African Regional Office of Astronomy for Development (ROAD-IAU) from 22–25 May 2017 in Addis Ababa, Ethiopia with the theme of *"Exploring our Universe for the benefit of Humankind."* The purpose of the meeting was to strengthen and build capacity of Middle East and African Regional development in Astronomy, to provide a forum and network for scientists and students in the region to exchange ideas and experiences, to share research output and knowledge, to provide updates on the current status of astronomy development in the region, to build joint working groups and research networks, to provide training opportunities for young scientists, to encourage education and cross border collaboration and sharing of resources, and to enhance cooperation between Middle Eastern and African countries in furthering Astronomy and Space Science in the region. During the symposium there were activities such as scientific paper presentations, plenary and invited speaker sessions, meetings, and discussion on the general progress and assessments of astronomy and astrophysics development in the Middle East and Africa. From Armenia, Areg Mickaelian (Director of BAO and

SWCA ROAD) and Sona Farmanyan (Public Outreach Coordinator of SWCA ROAD) participated in the meeting and presented five talks on various subjects, including one about the SWCA ROAD.

The international conference *"Astronomical Heritage of the Middle East"* approved by the UNESCO Director General within the UNESCO Participation Program for 2016–2017 was devoted to the role of Astronomy in Culture and other fields of human activity. It was held on 13–17 November, 2017 in Armenia and brought together over 70 participants from 20 countries. A discussion of the problems of astronomy-related interdisciplinary sciences and further possible collaborations was organized. The proceedings of the meeting will be published by the Astronomical Society of the Pacific Conference Series (ASPCS), which is a world-wide publication, and the papers will appear in all libraries and electronic databases (including ADS).

We have given many presentations on our ROAD and its activities at a number of international meetings, including NKAS (Belgrade, Serbia, 2015), IAU GA XXIX (Honolulu, Hawaii, USA, 2015), EWASS-2016 (Athens, Greece), EWASS-2017 (Prague, Czech Rep.), Armenia-Brandenburg Workshop (Nor Amberd, Armenia, 2017), and MEARIM-IV (Addis Ababa, Ethiopia, 2017).

We maintain a tight relationship between the IAU SWCA ROAD and Eastern Europe. This includes collaboration with European countries, ArAS affiliation to the European Astronomical Society (EAS), ArAS collaboration with the Euro-Asian Astronomical Society (EAAS), Armenia's participation in South-East Regional European Astronomical Committee (SREAC), H2020 program (Armenia has been an associate member since 2016), collaboration between the Armenian Virtual Observatory (ArVO) and EuroVO, VO-France (also VO-Paris), GAVO, and VObs.it. We also have contacts with other European organizations, including Euroscience, SEAC, etc.

6. Summary

The ROAD regional astronomical center supports the regional and international contacts and collaborations, both for Armenia and the neighboring countries; in this way Armenia, by hosting such a center, strongly contributes to collaborations at all levels: professional, educational and popular. Astronomy plays an important political role in establishing and strengthening friendship and cooperation among the regional nations, which is especially important in our complicated area.

Among our future plans, we envisage enlarging the number of participating countries (involving more Central Asian and maybe some others), working harder for fundraising for implementation of various projects (in all three task force areas), and organizing regional meetings, workshops, schools, and camps.

Acknowledgments. The IAU SWCA ROAD team would like to thank the IAU OAD, BAO Administration, and the NAS RA Presidium for supporting the office.

References

Mickaelian, A., Azatyan, N., Farmanyan, S. & Mikayelyan, G. 2016, IAU South West Asian ROAD, Astronomy in Focus, presented at the IAU XXIX General Assembly, 2015. Proc. IAU, 29A, Cambridge Univ. Press. (CUP), 422

Mickaelian, A. M., Khosroshahi, H. G. & Harutyunian, H. A. (Eds.) 2016, Proceedings of the Armenian-Iranian Astronomical Workshop, held on 13–16 October 2015 in Byurakan, Armenia, Yerevan, NAS RA "Gitutyun" Publishing House, 1

Mickaelian, A. M. 2017, Regional activities related to IAU strategic plan and integration of Armenia in European astronomy, Publ. Astron. Observ. of Belgrade (PAOB), 96, 343

Mickaelian, A. M., Hakopian, S. A., Farmanyan, S. V. & Mikayelyan, G. A. 2019, IAU South West and Central Asian ROAD, Proc. Intern. Conf. "Astronomical Heritage of the Middle East," held on 13–17 November 2017, Astron. Society of the Pacific Conf. Series (ASPCS)

Astronomical Heritage of the Middle East
ASP Conference Series, Vol. 520
Sona V. Farmanyan, Areg M. Mickaelian, J. McKim Malville, and Mohammad Bagheri, eds.

Astronomy Development in Perú Through Two Generations: the Good and the Bad

José K. Ishitsuka[1] and Mutsumi Ishitsuka[2]
[1]*Universidad Nacional del Centro del Perú, Facultad de Ingeniería Eléctrica y Electrónica, Av. Mariscal Castilla No. 3909, Huancayo, Junin, Perú;* `i.jose617@gmail.com`

[2]*Instituto Geofísico del Perú, Calle Badajoz No. 169, Urb. Mayorazgo IV Etapa, Ate, Lima, Perú;* `mutsumi.ayako@gamail.com`

Abstract. Ancient Peruvian cultures probably had advanced knowledge of astronomy, but after Spain colonized Perú, unfortunately almost nothing to develop astronomy was done. In 1957, Mutsumi Ishitsuka, a young solar physicist from Kyoto University in Japan, traveled to Perú to establish a Solar Coronagraphic Observatory. In October of 1987, the Cosmos Coronagraphic Observatory was inaugurated, the highest in the world. Unfortunately, shortly after it began operations, the observatory was occupied by terrorists and finally destroyed by dynamite in October of 1988. Mutsumi Ishitsuka, who was in charge of the observatory, had to move to Lima to save his life. After some years, when terrorist activities were better controlled by the Peruvian Government, the Japanese Government donated a new planetarium to the Geophysical Institute of Perú, in part to compensate for the loss of the coronagraph constructed by Mutsumi Ishitsuka. A fund was organized by Dr. Takehiko Kuroda of Nishi-Harima Astronomical Observatory of Japan to buy and send a 60 cm telescope to Perú, to establish an Educational Astronomical Observatory near Lima, the capital of Perú; the project was led by Mutsumi Ishitsuka. The 60 cm reflector telescope arrived in Perú in 2010, the same year a Solar Flare Monitoring Telescope arrived from the Hida Observatory of the University of Kyoto to observe solar activity from Ica at a Solar Observatory also established by Mutsumi Ishitsuka at the Ica National University. In 2006, Mutsumi Ishitsuka's son José Ishitsuka returned from Japan to Perú to establish the first radio astronomical observatory in Perú in Huancayo, a place situated in the mountains 300 km east of Lima. A 32 m parabolic antenna previously used for telecommunications was transformed into a radio telescope with the strong support of National Astronomical Observatory of Japan. The 32 m radio telescope is equipped with a 6.7 GHz receiver to monitor and search for new methanol maser emission from young stellar objects. Unfortunately, since the end of 2016 when the directives of the Geophysical Institute of Perú changed, the new directives did not recognize Mutsumi and Jose Ishitsuka's efforts and work, and the Peruvian government decided to close the Radio Astronomical Observatory.

1. Introduction

Ancient Peruvian cultures date back 5000 years to the Caral Culture (Shady 1997); before Columbus discovered the Americas in 1492, the Inca Empire had thrived for hundreds of years, with the pre-Incan cultures (including Chavin, Paracas, Nasca, etc.) dating back much further. From the earliest times of the Caral Culture, astronomical knowledge was an important element of people's lives, for agriculture and fishing, for instance. Some buildings were oriented for celebrating important dates based on solstices and equinoxes.

The oldest solar observatory in the Americas (Ghezzi & Ruggles 2007) is located at Chankillo, a hill with 13 stone towers (Figure 1) that date back 2300 years. Malville (2014) argued that it was for ceremonial purposes. However it was used, the people needed to measure time, and ancient Peruvians did so in some way, using the solstices and equinoxes.

Unfortunately, this basic and perhaps quite advanced astronomy knowledge was not documented or preserved by the Spaniard conquerers, whose mission was to establish Catholicism in the new continent. As Cusco was the capital of the Incan Empire, conquerers built Catholic churches on Incan sanctuaries and temples (as shown in Figure 2).

No relevant astronomy development was done, either during the Spanish occupation or after Independence in 1821. Nevertheless, at the end of 19th century, Harvard University established an astronomical observatory in Arequipa, 1000 km south of Lima, and significant results were achieved, such as Henrietta Leavitt's measurements of Cepheid variable stars in the Magellanic Clouds (Leavitt 1908). The telescope was moved to South Africa in 1927.

Figure 1. Thirteen towers in Chankillo, a pre-Inkan building 2300 years old.

1.1. Solar Observations

To observe the Total Solar Eclipse of June 8, 1937, a group from Kyoto University traveled to Trujillo, Perú, 700 km north of Lima, and made successful observations from

Figure 2. Coricancha, an Incan temple over which the Santo Domingo church was built.

Trujillo (Yamamoto 1937). After that success, they returned to Japan and considered establishing a major solar observatory in Perú due to the excellent clarity of the sky. In 1957, Mutsumi Ishitsuka, a graduate student from Kyoto University moved to Perú and worked for the Geophysical Institute of Perú (IGP) to etablish Huancayo Observatory as the responsible of Solar Physics Department. After searching for many years for an adequate place to construct a Solar Coronagraphic Observatory, construction began in 1972 and it was inaugurated in 1978. A successful observation of the Corona of the Sun was done in 1988. Unfortunately, a few weeks later, the observatory was occupied by terrorists and then destroyed (Ishitsuka 2015). While Mutsumi Ishitsuka was in charge of Solar Physics Department in Huancayo, he also installed a new Heliograph that replaced the Hale-type Heliograph, and then installed a 6.9 Radio Polarimeter (Ishitsuka 1975).

1.2. After the Nightmare

Since Mutsumi Ishitsuka's life was threatened by the terrorists, he had to hide in Lima. Some years later, in 2008, when he was the Director of Ancon Observatory of the IGP, the Japanese Government donated a new planetarium to Geophysical Institute of Perú, in part to make up for the destruction of the coronagraph by terrorists, (Kitamura 1999). As scientific culture was not well established, few students came to visit the Planetarium, but installing a 3D projection room using Mitaka 3D Visualization Software (Kokubo et al. 2005) helped bring more visitors (Nemoto et al. 2013). In 2015 the Planetarium had almost 20,000 visitors.

Dr. Takehiko Kuroda, the Director of Nishi-Harima Astronomical Observatory of Hyogo Prefecture of Japan, began a fund to send a 60 cm reflector telescope to Mutsumi Ishitsuka. It became the main telescope for the Educational Astronomical Observatory that Mutsumi Ishitsuka was planning to construct near Lima. In 1995, the Minister of Education of Perú proposed to construct an Astronomy Education Observatory near Lima, but sky conditions thereare not good, so an adequate place was finally found 300 km south of Lima near the city of Ica. In 2010, the telescope was sent to Perú and installed temporarily in the Ica Solar Observatory until the Educational Astronomical Observatory would be ready. Students from Ica National University use the telescope for observations and outreach.

1.3. Solar Observatory: Capacity Building

While Mutsumi Ishitsuka searched for a place for the Educational Astronomical Observatory near Ica, he went to Ica National University and installed a telescope for sunspot observations. As student interest in solar observations increased, Ica University allocated a large space to construct a Solar Observatory within the university campus. Meanwhile, when the Solar Observatory was ready, Prof. Kazunari Shibata)Hida and Kwasan Observatory of Kyoto University in Japan) moved the Flare Monitoring Telescope (FMT) to Ica (Ishitsuka et al. 2007). It arrived in 2010. FMT is part of the Continuous H-alpha Imaging Network (CHAIN; Ueno et al. 2007). This important collaboration with Hida Observatory produced significant results with the first publication appearing in the *Astrophysical Journal* using FMT data (Cabezas et al. 2016). It is remarkable that the first author of the publication is a Ph.D. student at Kyoto University.

2. Radio Astronomy

Mutsumi Ishitsuka visited Huancayo Observatory in 2002, where he was informed that the satellite communications station of the telephone company was no longer being used and the telephone company was looking for an institution that could use it. At that time, José Ishitsuka, Mutsumi Ishitsuka's son, was in Japan doing radio astronomical research. Mutsumi Ishitsuka told his son about the communications antenna, and José asked to collaborate with the National Astronomical Observatory of Japan (NAOJ). That year he traveled to Perú to visit the telecom station. Dr. Shinji Horiuchi from the Jet Propulsion Laboratory (JPL) in Pasadena, CA, joined José for the first visit to Huancayo. Some months later, a team from NAOJ visited the telecom station to see if the 32 m antenna would be useful for radio astronomy. The evaluation was positive, and conversations with the telephone company began to have the whole telecom station donated. In 2004, José moved to Perú as a Visiting Researcher to the IGP to lead the transformation of the 32 m antenna into a radio telescope. In 2006 he became a Scientific Researcher at the IGP, and took charge of the Astronomy Division. In 2008, the telephone company donated the telecom station to the IGP.

Between 2000 and 2006, when the telecom station was practically abandoned, electrical cables containing copper were stolen. Getting the 32 m parabolic antenna working again was one of the hardest issues for José Ishitsuka. In 2009, the first radio signals from a methanol maser were detected at 6.7 GHz (Ishitsuka et al. 2017), but without tracking the star because at that time the electric cables were not fully restored.

In December of 2016, the IGB directives changed. It is usual for changes when the Central Government changes, but new leadership did not understand the relevance of making the 32 m Radio Telescope, the only one in Perú, fully operational. Even though the last step of the transformation (the tracking system) was almost complete, the new leaders decided to close the Sicaya Radio Astronomical Observatory as of June 2017, and decided not to hire José Ishitsuka to work for IGP.

Before the closing of Sicaya Radio astronomical Observatory, the interface for the tracking system was ready and successfully tested (Figure 3). The Antenna Control System (ACU) Interface was designed and constructed by a graduate student at a local technical school. From the beginning, NAOJ contributed significantly to the radio telescope transformation and operation.

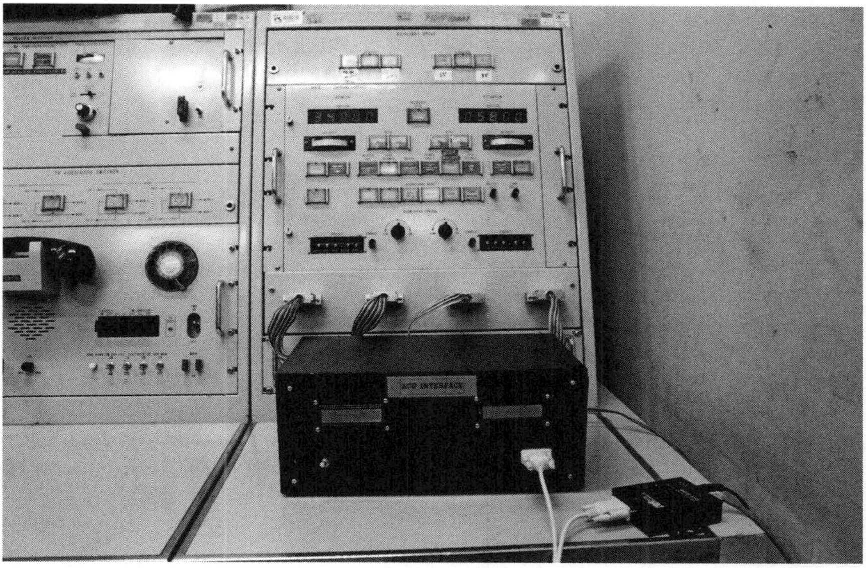

Figure 3. Antenna Control Unit and the ACU Interface, connected to a PC through the Internet.

2.1. Conclusions

After retiring from the IGB in 2006, Mutsumi Ishitsuka became an Emeritus Scientific Researcher at the IGP, but in spite of his contributions to the development of astronomy and Solar Physics for the institution and for Perú, his contributions have not been recognized by the new authorities. After the destruction of the Cosmos Observatory destruction, Mutsumi Ishitsuka continued working. His most relevant achievements are:

- Construction of a new planetarium in Lima.
- 15 cm refrator telecope for sunspot observations in Ica.
- Construction of the Spectrum Helioscope in Ica.
- The Flare Monitoring Telescope.

- Installation of the new 60 cm reflector telescope for astronomy education.
- Transformation of the 32 m radio telescope equipped with 6.7 GHz receiver.

These accomplishments represent a significant heritage for astronomy development in Perú, as it never had before, but sadly it has not been adequately recognized. Perú is not an underdeveloped country, but development in the sciences has been delayed compared to other areas of economic development. Much work has to be done at schools and universities to help young people understand the importance of science for the development of the country.

Acknowledgments. Special thanks to the organizers of the event, the LOC and volunteers that made the meeting productive. My gratitude to the National Astronomical Observatory of Japan; without their support, the transformation of the telecom antenna into a radio telescope would have been impossible.

References

Bailey, Solon I. 1922, The Harvard Astronomical Observatory in Peru, HUG 1191, Box 1, Folder 10. Harvard University Archives, Cambridge, Mass.

Cabezas, Denis P., Martínez, Lurdes M., Buleje, Yovanny J., Ishitsuka, Mutsumi, Ishitsuka, José K., Morita, Satoshi, Asai, Ayumi, UeNo, Satoru, Ishii, Takako T., Kitai, Reizaburo, Takasao, Shinsuke, Yoshinaga, Yusuke, Otsuji, Kenichi, Shibata, Kazunari 2017, " 'Dandelion' Filament Eruption and Coronal Waves Associated with a Solar Flare on 2011 February 16," ApJ, 836, 33, 11

Ghezzi & Ruggles 2007, "Chankillo: a 2300-year-old solar observatory in coastal Peru," Science 2007 Mar 2: 315, 5816, 1239

Ishitsuka, J. K., Kobayashi, H., and Miyoshi, M. 2017, "Capabilities and Present Status of The Sicaya Radio Telescope in Peru," Revista Mexicana de Astronomía y Astrofísica (Serie de Conferencias), 49, 59

Ishitsuka, Mutsumi 1975, "Solar patrol at 9400 MHz," Final Report, 1 Oct. 1970 - 30 Jun. 1975 Instituto Geofisico del Peru, Lima.

Kitamura, Masatoshi, "Provision of astronomical equipment for developing countries through ODA of the Government of Japan," Seminars of the UN Programme on Space Applications: Selected Papers from Activities Held in 1999.

Kokubo, E., Hayashi, M., Kato, T., Takeda, T., Kaifu, N., Miyama, S., Miura, H., Takahei, T. 2005, "4-Dimensional Digital Universe Project," Journal of the Korean Astronomical Society, 2005, 38, 153

Leavitt, Henrietta Swan 1908, "1777 Variables in the Magellanic Clouds," Annals of Harvard College Observatory 60 (1908): 87?108.

J. McKim Malville 2014, "The Casma Vally of Peru: A Cradle of Pre-Inca Astronomy," Mediterranean Archaeology and Archaeometry, 14, 3, 169

Nemoto Shiomi, Itoh Shoichi, Agata Hidehiko, Zegarra Mario, Ishitsuka José, Choque Edwin, Quispe Adita, and Kato Tsunehiko, 2013, "Stereoscopic 3D Projections with MITAKA An Important Tool to Get People Interested in Astronomy and Space Science in Peru," Sun and Geosphere, 2014, 9(1-2), 115

Shady, R. 1997, "La ciudad sagrada de Caral-Supe en los albores de la civilizacioń en el Perú," Lima: UNMSM, Fondo Editorial

Ueno, Satoru, Shibata, Kazunari, Kimura, Goichi, Nakatani, Yoshikazu, Kitai, Reizaburo, Nagata, Shin'ichi, "CHAIN-project and installation of the flare monitoring telescopes in developing countries," Bulletin of the Astronomical Society of India, 35, 697

Yamamoto, I. 1937, "Observations of the Solar Eclipse of June 8, 1937," Nature, 140, 501

Astronomical Heritage of the Middle East
ASP Conference Series, Vol 520
Sona V. Farmanyan, Areg M. Mickaelian, J. McKim Malville, and Mohammad Bagheri, eds.

Under the Kyrgyz Sky: from Astronomy to Development

Ewelina Gradzka
Ignatianum Academy, Cracow, Poland; ewelina.gradzka@yahoo.com

Abstract. Astronomy outreach used for the purpose of development operates in a complex environment of foreign aid actors, purposes, and forms of application. Working in the field of education, especially informal education, may require one to face critics and reconsider the goals and arguments in favor of and against it. Therefore, it is worthwhile to conduct a basic analysis to gain a general view of the discussion, which should strengthen the success of the projects. Additionally, a different attitude towards projects' aims is offered. We recommend using the Ancient Greek philosophers' idea of *theoria*, which invites one to look at the world and explore it just for the sake of knowing without focusing solely on the practicality of the activity. This should work well for astronomy as it has a natural potential to make people wonder. This philosophy is the foundation of the "Under the Kyrgyz Sky," a project prepared to train teachers and establish astronomy clubs in their schools so that pupils could gain new skills, knowledge, and contemplate the beauty of the Universe.

1. Introduction

The "Under the Kyrgyz Sky" project was created three years ago with the purpose of popularizing cutting-edge knowledge about the Universe as well as teaching skills using modern technology like the *Stellarium* program and telescopes. It is also inspired by the ancient Greek ideas of *arête, paideia,* and *theoria* which give a more humanistic component to the project. It is believed that, given contemporary focus on STEM subjects, it is important to remember humans' unique capacity to contemplate the reality that surrounds them. This enriches them as reasoning human beings who desire to understand just for the sake of knowing. Surprisingly, such an attitude gives positive results and participants of the project appreciate this. However, in the course of running the project, critics questioned the effectiveness of social work and education, issues related to cultural relativism, and in what circumstances it is sensible to teach astronomy (i.e., does not an underdeveloped country have bigger problems than astronomy?). Additionally, the whole foreign aid environment is heterogeneous, and we experienced how unjust arguments are used against various types of aid, which motivated us to examine foreign aid to help clarify the discussion.

2. Foreign Aid Complexity

Foreign aid aims to improve the economic, political, and social climate in the receiving country. However, it is a controversial topic and faces many challenges. According to the *Encyclopedia Britannica*, foreign aid is "the international transfer of capital, goods,

Figure 1. Logo for the "Under the Kyrgyz Sky" project.

or services from a country or international organization for the benefit of the recipient country or its population." Critics focus on the problems of effectiveness, proper allocation, affordability, creation of a dependence mentality, lowering competitiveness through the Dutch Disease effect, its use as a tool of control to open markets and influence policy, and the difficulty in proving the a link between foreign aid and economic development. Doubt about whether development aid works and if it should be sustained or not are reasonable issues to discuss, but it is important to emphasize that the subject is multilayered and complex, and we must take into account its history and methods used to evaluate it.

2.1. History

The history of foreign aid can be traced far back in time as it refers to any military help to an ally; a more recent example is 18th century Prussia, which subsidized its allies. In the 19th and 20th centuries we find examples of European powers improving the infrastructure of their colonies to increase their economic output. The modern day idea of foreign aid is related to the Marshall Plan implemented after the Second World War to help the ruined European countries, and the establishment of international organizations such as the United Nations. During the Cold War period, foreign aid flourished with aid given to "third-world countries" in order to stabilize them and avoid their switch in political ideology. The end of the 20th century and the beginning of the 21st century witnessed changes in world politics and attitudes towards application of foreign aid which will be presented below.

2.2. Division

For the purpose of this paper, foreign aid will be divided into two basic categories: humanitarian aid and development aid. Aid is channeled through International Organizations (IOs), Regional Organizations (ROs), Intergovernmental Organizations (IGOs) like the United Nations, the World Bank, and the International Monetary Fund (IMF), unilateral assistance by the foreign-aid and development agencies or militaries of individual countries, and nongovernmental organizations (NGOs).

Humanitarian aid is usually a response to man-made or natural disasters and is composed of material or logistical assistance. It is short-term and focuses on saving lives. Whereas, development aid is long-term and responds to systematic problems related to economic, social, and political development. It can be dived into the subcategories of budget support and project aid. Budget support can be conditional or unconditional. The idea of unconditionality is based on the assumption that the recipient's government knows better the needs of its own country; there is no crowding-out effect and transaction costs are lower. However, it faces transparency issues, and the aid may go to inefficient, bureaucratic, or corrupt systems. Conditional aid, on the other hand, can be directed to different social or political needs like improvement of working conditions or regulation of elections. But the effects of this kind of aid on the economy are difficult to measure. Project aid involves transfer of knowledge and know-how, or donation of materials and personnel to the location where they are needed. The donors decide what, how, and where the work should be done. Potential drawbacks to this kind of aid are lack of coordination amongst donors, projects may lack sustainability, donors' preferences may be different from the receiving country's ideas, and there can be fungibility problems (fungibility defines how easily resources can be transferred from one sector to another, e.g., financial funds are the most fungible resources). Project aid funds can crowd out the recipient's own funds. Project aid can be made transparent and conditional if desired.

The discussion about foreign aid types must also consider on which stage each type of aid works best. Both forms of aid have their own positive effects, but we propose that project aid might be more effective during the initial stages of foreign aid whereas budget support could be more appropriate in the later stages of development. There is also an ongoing discussion about whether developmental aid has direct influence on economic growth or not, and how to measure it.

2.3. The Funding Gap: LRRD

A funding gap between humanitarian aid and development aid has been observed between the moment when the first is withdrawn and the latter has not yet arrived. As a response, a new model was proposed linking relief, rehabilitation and development (LRRD; Ramet 2012). It is understood that the transition between the two types of aid is a non-linear process and often a continuous cycle where populations are constantly moving from relief to development and back to relief again. The change is chaotic and unpredictable. A continuous approach was adopted where the necessity of simultaneous use of different aid instruments is incorporated. Although there has been little progress on the implementation of LRRD so far (even though it was accepted by the European Union and the OECD/DAC showed interest), it is a new model and an important change in approach, rejecting simplistic views, and acknowledging the complexity of foreign aid programs.

2.4. Economic Growth

Each type of foreign aid has its own advantages and disadvantages; little has been settled in the debate about the effectiveness of foreign aid and its influence on economic growth (Burnside & Dollar 2004; Easterly 2003; Minoiu & Reddy 2010). It is claimed that aid of the right kind is good for growth. Findings show that whereas aid may have unmeasureable effects on macroeconomic scales, it can have positive effects in microeconomic studies (Minoiu & Reddy 2010). Moreover, it is believed that various components of aid (developmental aid such as investment in physical infrastructure, human capital development, and organizational support, or non-developmental aid) may have different effects on growth (positive, zero, or negative). Therefore, an adequate analysis is required that does not treat foreign aid as a homogenous issue. The policies of the donors matter too (e.g., Nordic countries tend to be more development-friendly whereas France and the UK tend to be more geostrategic-oriented donors, and are less effective). Important to the purpose of this article is the finding that "aid which is aimed at supporting democracy, the environment, health, and education is likely to have a long-term impact on growth, but do not statistically identify its effect" (Minoiu & Reddy 2010). To summarize, if aid is oriented toward development and not motivated by geostrategic interests, it is likely to be effective and have positive influence on economic growth, but often this is only visible over a long period of time.

2.5. Quality of Life and Education

Economic growth is an important aspect of human well-being, however its negative effects are often overlooked: "creative destruction" (there are winners and losers to the changes; Schumpeter 1942), natural social tension, health challenges, increases in income inequality, increased pollution, depletion of natural resources, family distortions, changing cultures, depression, and stress are few of them. In some instances, quality of life factors such as education, healthcare, and social and political liberties, do not improve even though we observe economic growth (e.g., India enjoyed strong GDP growth between 1990 and 2012, but inequality deepend). For the purpose of this article, it is important to mention that many theoretical and empirical analyses see the major role human capital plays in ensuring economic growth. Besides, it was not until 2000 when social progress (and not just economic growth) started to be emphasized (UNESCO 2015, p. 277). This led to the involvement of NGOs in providing education and not leaving it strictly to governments. Although it was already stated in the Universal Declaration of Human Rights,[1] the change in the approach was confirmed again in the Millennium Developmental Goals. The second goal aims at universal primary education. Establishment of the "Astronomy for Development" as an office of the International Astronomical Union in 2011 is a sign of that change. However, a great deal of work remains to be done. Education remains under-financed. Governments fall short in prioritizing education in national budgets, and donors since 2010 have reduced their contributions to this issue. Therefore another re-affirmation of the idea was the "Dakar Framework for Action, Education for All: Meeting Our Collective Commitments." It acknowledged that the problem of education does not end with children but refers to young people and adults (goal 3), and that we should also care about the quality of education (goal 6).[2]

---

[1] Article 26: (1) Everyone has the right to education. Education shall be free, at least in the elementary and fundamental stages. Elementary education shall be compulsory. Technical and professional education shall be made generally available and higher education shall be equally accessible to all on the basis of merit. (2) Education shall be directed to the full development of the human personality and to the strengthening of respect for human rights and fundamental freedoms. It shall promote understanding, tolerance and friendship among all nations, racial or religious groups, and shall further the activities of the United Nations for the maintenance of peace. (3) Parents have a prior right to choose the kind of education that shall be given to their children.

[2] *The Dakar Framework for Action,* Goal 1: Expand early childhood care and education; Goal 2: Provide free and compulsory primary education for all; Goal 3: Promote learning and life skills for young people and adults; Goal 4: Increase adult literacy by 50 percent; Goal 5: Achieve gender parity by 2005, gender equality by 2015; Goal 6: Improve the quality of education (UNESCO 2015).

Interestingly, investment in education is considered "life-saving," and for decades the educational sector has been trying to convince humanitarian aid donors. In many cases humanitarian help is not short-term as would be expected, but continues for years. The Central Emergency Response Fund (CERF) supplies school tents, education and recreational materials, emergency repair of education facilities, and teacher training in emergencies and provision of lifesaving skills. However, this vision of what educational needs are is narrow. These few examples show that quality education, and not only for children but also for adults, is crucial to social and economic growth, and should form an important part of development and humanitarian aid. Moreover, it affects our subjective well-being. For many years it was believed that it is nurture that influences our perception of "happiness" and that the idea of universal aspects of human nature is dated. Presently, however, it is claimed that learning new things, among other aspects like respect of others and supportive social relationships,[3] is one of the universal needs (it is beyond the scope of this article to analyze the long debate about universal human nature and cultural relativism). In consequence, this means that in all cultures the same circumstances are necessary for a high quality of life. The problem, it seems, should be focused on how to provide quality education for children and adults with respect to local cultures.

3. Ancient Greek Education and Philosophy

Various documents treat education as part of Human Rights, and access to education at a basic level is one of the Millennium goals. Education is an important contributor to social and economic growth of a country and therefore a recommended element of developmental aid projects. Eduation influences our level of subjective well-being. Educational quality and adult education are increasingly important. Investment in education should be treated as "life-saving," even in humanitarian aid situations. These contributions are fundamental and relevant. However, there is something more that we can do to inspire shaping of developmental aid in the area of education which comes from the broad ancient Greek concepts of *arete, paideia,* and *theoria*. Due to the limited scope of this paper, readers may wish to refer to Jeager 1967 or Reale 1990 for a good introduction to the topic.

In the beginning of ancient Greece, the main influences in education came from its great epic poets, Homer (the legendary author of the *Iliad* and the *Odyssey*) and Hesiod (*Works and Days, Theogony*, and *Shield of Heracles*). From Homer's works the idea of *areté* crystallized, which refers to moral virtue ("excellence"), the act of living up to one's full potential. It is related to bravery, effectiveness, and prudence. Heroes and nobles (males and females) were noted for possessing *areté*. On the other hand, Hesiod praised the dignity of farmers' hard work.

This was poetry with its crucial significance for Greek culture. However, philosophy gradually appeared with its rational investigation of the *kósmos* (Universe). Pre-

[3] A study run in 123 countries by the University of Illinois at Urbana and the Gallup Organization assessed the relative importance of the following needs: food, shelter/safety, security/social support, love/feeling respected, pride in activities/self-direction, autonomy, mastery—doing what one does best—and learning new things (met when a person had the experience of learning something and did what he/she does best at work). There are also likely individual and cultural differences in what people desire and find rewarding, but these can coexist with the universal needs (Tay & Diener 2011, pp. 354–365).

Socratic philosophers started asking questions about the nature of the world (Thales of Miletus is considered to be the first), about the *arche*, the "beginning" or "origin" of the Universe (Anaximander was the first to employ this term). What is it made of? They were searching for the element that constitutes all things, the first principle. Their interests were motivated purely theoretically and their method of explanation (though not yet well developed) was different from mythological. The word *philosophia* was used for the first time (by Pythagoras) meaning the "love of wisdom" to distinguish from *sophia* (wisdom). The philosopher is one who loves wisdom, seeks it (asks questions), but does not possess it (in contrast with the wise man).

Another interesting moment in Greek history was the apparition of sophists with their vision of common education which, came to play a key role in the functioning of the *polis* (city-states). Their philosophical interests lay in understanding the human being as a citizen, and not part of the *kosmos*. They educated adult citizens of Athens (this was a novelty) to be able to participate in the governance of their *polis*. It was believed that political *arete* can be taught and does not belong to nobility only. The idea of *paideia* (cultural education) offered by sophists included theoretical and practical elements. In general, they taught rhetoric, dialectic, eristic, grammar, semantics, and epistemology. Students were taught how to use their knowledge for the good of the polis and their own good. In fact, they put human intelligence, the formative power of knowledge and education, into the center of attention.

It was also commonly thought that to live a life of *arete*, the mind, body, and soul all had to be developed, and therefore athletics played an important role in *paideia*. The presence of a *gymnasion* (a kind of sports hall or athletic stadium) in Plato's Academia and Aristotles' Lykeion is a good example.

Finally, Plato conceptualized philosophical "theorizing." He adjusted the traditional model of practice of *theoria*, where a person (*theoreos*) sets out on a journey or pilgrimage abroad (to detach from home) to witness some important event (like religious festivals) or other spectacles, and then return home (reentry; see Nightingale 2004). This enabled him to identify philosophy with the specific kind of cultural practice, as part of the society and its politics. Theoretical activity, which requires disconnection from the daily issues, is crucial for the flourishing of a *polis*. The "Allegory of the Cave" is an example of such a perspective. A new, theoretical kind of rational gazing to see the truth arose. Moreover, such a contemplative journey transforms the individual, who should then apply the things he/she learned to the benefit of the *polis*.

It is worth mentioning that Philip of Opus, who belonged to Plato's Academy, believed that people should contemplate the stars, not forms. Astronomical *theoria* makes people virtuous and supremely good here on Earth (Nightingale 2004).

Platos's and Philip's idea of *theoria* seems very practical. However, Aristotle saw this differently. For him, philosophic theorizing is a pure and disinterested, even useless activity. It is done for its own sake, unlike other technical or practical activities that give particular results. Consequently, he claims *theoria* to be the most desirable human endeavor as it contemplates the first principles. "Wisdom is not useful or advantageous, for we call it not advantageous but good, and it should be chosen for the sake of any other thing, but for itself. For just as we go to the Olympian festival for the sake of the spectacle, even if nothing more should come of it—for the *theoria* itself is more precious than money" (Aristotle, *Protrepticus* B44). He even considered it to be the true source of happiness.

These few ideas relate to education, and the value of knowledge gives hints about how educational projects could be shaped. Moreover, in Book 1 of *Methapysics*, Aristotle states that "All men by nature desire to know" (Aristotle, *Metaphysics*, Book 1). Myths that each culture possesses as their heritage are expressions of this desire. Plato, on the other hand, claimed that philosophy begins in wonder.[4] It might be worth including "wonder" in educational projects more often.

4. Under the Kyrgyz Sky – from Astronomy to Development

A project operates among the complexities of foreign aid with its positive and negative effects. It is important to be aware of the discussion and take it into consideration during preparations. However, when working with educational projects, one faces criticism and disappointments. The significance of education for the effectiveness of development aid exposed in the documents mentioned above is uplifting. The "Under the Kyrgyz Sky" project was developed considering these complexities. However, it was also inspired by the Greek ideas which claim that "All men by nature desire to know" and that, even given the practical considerations, wondering about the *kosmos* is a valuable theoretical activity for its own sake. It is believed that these two frames of reference—practical and theoretical—when used together, give a multilayered platform to work with people as rational human beings for their benefit. Rejection of one or the other of them is a harmful simplification.

4.1. Under the Common Sky

"Under the Common Sky" is an association established in Gdynia, Poland, to apply the same ideas into projects which use astronomy as a base. It employs the practical part of astronomy to interest children in studying math, physics, geography, and engineering, and to give teachers modern methodology and equipment. However, it also tries to provide a broader view. First of all, it emphasizes the unity of humanity. We live under one and the same sky, and we admire the same constellations, not to mention the Sun, the Moon and the planets. Secondly, it reminds us that, from a cosmic perspective, we live on a very small "pale blue dot" (as Carl Sagan put it). State borders and nationalities do not matter. We should rather consider ourselves "Earthenians." Moreover, we contemplate the sky just as the Greek philosophers did—to wonder, ask questions, and look for answers just for the sake of it.

4.2. Training for Teachers and Astroclubs

The project is designed so that it can be applied in any region of the world. It possesses the qualities described above, which makes it a successful way for using astronomy for development. The first part of the project, which was envisioned initially for Kyrgyzstan, consisted of training science teachers and establishing astronomy clubs in their schools. Usually, math, physics, informatics, and geography teachers were invited to participate as it should be easiest for them to acquire relevant knowledge and skills. However, it was not limited to these subjects, and it happened that teachers of history

---

[4]Soctrates: "I see, my dear Theaetetus, that Theodorus had a true insight into your nature when he said that you were a philosopher, for wonder is the feeling of a philosopher, and philosophy begins in wonder," Plato, *Theaetetus* 155c-d.

or PE were often more engaged. The training requires up to 30 hours of workshops, including hands-on activities, usage of the *Stellarium* program, lectures, and 15 hours of night sky observation with sky maps and telescopes. The emphasis is put on engaging the teachers in then activities and inspiring them with the new knowledge using the philosophical perspective explained above. Schools are chosen after they send an application form and successfully pass the recruitment process. After being accepted, an agreement is signed that requires the school to provide teachers for the full training (without absences) and to run the astronomy clubs for a minimum of two years. If they fulfill the requirements, they receive a telescope. If there is sufficient funding (which is often not the case), a computer is given to the club for the independent use of *Stellarium*. If the school fails to meet the conditions, the telescope is not granted or is withdrawn. In this way the project is transparent and focuses the participants on the required tasks. There are currently fifteen schools with astroclubs in the south of Kyrgyzstan in the Jalal-abad region; there will be five more schools in the north of the country in the near future.

During the training, a work plan for astroclubs is developed. The idea is to organize astroclub meetings twice a month throughout the school year. The work plan is flexible and the teachers decide on the order of the activities and age group.

It is interesting to observe how the teachers evolve during the training and how their attitude toward astronomy and the project changes. Many admit being forced to come by their director, but in the end they enjoy the program, the philosophical approach, and the methodology. They seem refreshed, inspired, and empowered as teachers and as humans. One of them once said that before the training he used to look at his feet while walking, but now he cannot resist looking up and figuring out how the sky looks in the evening.

After applying the methodology during club meetings, teachers claim children are engaged, motivated, and they start to see the sense in learning more about science. This leads to growth in the teachers' self-esteem, as well as recognition from the director, parents, students, and colleagues. Another participant recounted that children point at her during breaks and say she is the one who knows the stars. Additionally, computer skills are developed thanks to the *Stellarium* program, which is a valuable tool to convince pupils that there is more to computers than games and internet chatting. *Stellarium* and usage with the telescope motivates students who misbehaved or paid little attention in science classes because they thought they were boring and useless. Surprisingly, girls are often highly inclined to operate the telescope, and there have been cases where the whole club is female or where they deny boys access to the telescope. Interestingly, parents get involved as well, as they accompany their children during nighttime stargazing activities. It happened frequently that parents did not let students observe, as they themselves were too curious. Surely, it was much more *theoria* than *praxis*.

4.3. Astrocamps

The second part of the project was designed to strengthen the existing astroclubs with new skills. In this part, pupils are at the very center of efforts involving the ideas of *arete* and *paideia*. For students to participate in Astrocamps, they have to fulfill assigned tasks. The approach is to develop social skills such as: working for the local community, caring for those in need and the less privileged (for example, they could organize an event for an orphanage or old people's home), sharing gained knowledge and skills (so they could learn to become more dependable, as discussed above, and to give as well as to receive), practicing public performance and team work, being responsibile for the equipment; and considering ecological issues. Students declare that it is a memorable, new, and mind-opening experience. Presenting knowledge about astronomy to others inspires them and gives them a different perspective on what they can do and how rewarding sharing is. The club members bond with each other, and teachers establish out-of-classroom relationships that strengthen their positive influence on students. At the end of a half-year of work, there is a test of basic astronomy. Teachers grade involvement and specialists appraise the tests. To avoid bias and assess different capabilities, two different types of evaluation are used on two different tasks. Students with the highest scores are invited to a five-day astronomy camp in summer. At present, 150 students have participated in the camps over the last two years (and there will be one hundred more this summer).

The camp itself can be considered another part of the program, as it has additional aims. It is organized on the banks of the Issyk-kul lake in the northeast part of Kyrgyzstan. Students gain knowledge and skills, which they use for the benefit of the camp. It is an intensive time where mind, body, and collaboration are at the center of the program. The first day is fully devoted to teamwork activities that require strong cooperation and creativity. It is interesting to observe the struggle in the beginning when all want to dominate, and later, after their failure, they finally learn how to become a team and start to listen to and work with each other. Another attraction is a half day-trip to the local Barskoon valley, which has a waterfall and a Soviet monument of Jurij Gagarin. This is to show them what hiking means (especially since this is a country that is 80% covered by the mountains of Tien-Shan). For many of them, this is the first time they have made a trip like that, and they learn about a new active form of spending leisure time. Hiking and morning warm-ups refer to the Greek idea of cultivating both body and mind. On the last day there is also an entire day trip to the local town Karakol, where a "city-game" is run. Each group gets a map and they have to reach set points where volunteers wait for them with tasks related to astronomy—in the zoo they look for southern hemisphere sky map animals, in the park they try to teach passers-by how the sky-map works, in the historical Dungan mosque they analyze the solar system painted on the exterior wall, and in the town museum they try to find connections between a yurt (a traditional nomad housing) and astronomy. This is the first time they have used maps to see a town and its touristic attractions. Additionally, each evening there are deep-sky observations thanks to the marvelous darkness and remoteness of the place. During the daytime, hands-on activities related to astronomy are provided. On the last evening, clubs present a small performance of a Greek myth related to a constellation assigned to them on the first day of the camp. This is a creativity task that includes some humanistic and historical perspectives. It is astonishing how innovative and entertaining their ideas can be.

5. Conclusions

Astronomy is an efficient tool for education of adults as well as children and teenagers. The project "Under the Kyrgyz Sky" is a good example of how astronomy can be used not only for providing modern knowledge and technology to encourage studying STEM subjects, but also as a means for developing social skills like working for the local community and team work/cooperation. Additionally, it creates a great opportunity to invite people to start wondering about and contemplating the marvels of the Universe, like the Ancient Greek philosophers, and asking general questions just for the sake of it. This might be an enriching task in developed countries where the importance of studying science is already highly appreciated, but the contemplative part of wonder seems to be lost, which is an important part of being human—human beings that can easily be called "Earthenians" who live "Under the Common Sky" and who have much more that unites them than divides them (like one sky and home planet). When faced with critics, it is worth keeping in mind how complex foreign aid issues are and that not every type of help gives the same positive or negative results. Moreover, high-quality, life-long education is coming to be appreciated more and it should be emphasized in shaping future projects. Even in humanitarian crisis situations, the need for learning, which is a significant factor of human well-being, should not be forgotten.

References

Aristotle, *Metaphysics*

Aristotle, *Protrepticus*

Burnside, C. & Dollar, D. 2004, "Aid, Policies, and Growth: Revisiting the evidence," World Bank, World Bank Policy Research Working Paper 3251

Easterly, W. 2003, "Can foreign aid buy growth?," Journal of Economic Perspectives, 17, 3, 23

Jeager, W. 1967, "Paideia: The ideals of Greek Culture," Volume I-III, Oxford

Minoiu, C. & Reddy, S. 2010, "Development Aid and Economic Growth: A positive long-run relation," Quaterly Review of Economics and Finance, 50, 2

Nightingale, A. 2004, "Spectacles of Truth in Classical Greek Philosophy. Theoria in its Cultural Context," Cambridge

Plato, *Theaetetus*

Ramet, V. 2012, "Linking relief, rehabilitation and development. Towards a more effective aid," Policy briefing for Directorate-general for external policies, European Parliament

Reale, G. 1990, A history of ancient philosophy, Volume I-IV, New York

Schumpeter, J. 1942, Capitalism, socialism, and democracy, New York

Tay, L. & Diener, E. 2011, "Needs and subjective well-being around the world," Journal of Personality and Social Psychology, 101, 2, 354

UNESCO 2015, "Education for all 2000-2015: Achievements and Challenges," EFA Global Monitoring Report, UNESCO Publishing

Astronomical Heritage of the Middle East
ASP Conference Series, Vol. 520
Sona V. Farmanyan, Areg M. Mickaelian, J. McKim Malville, and Mohammad Bagheri, eds.

Armenian Astronomical Society (ArAS) Current Activities

A. M. Mickaelian, S. V. Farmanyan, and G. A. Mikayelyan
NAS RA V. Ambartsumian Byurakan Astrophysical Observatory (BAO), Byurakan 0213, Aragatzotn province, Armenia; aregmick@yahoo.com

Abstract. This paper provies a review of the activities and achievements of the Armenian Astronomical Society (ArAS), and of Armenian astronomy in general during the last few years. ArAS membership, ArAS electronic newsletters (ArASNews), the ArAS webpage, Annual Meetings, the Annual Prize for Young Astronomers (Yervant Terzian Prize) and other awards, international relations, presence in international organizations, Byurakan International Summer Schools (BISS) and local schools, Byurakan science camps, astronomical Olympiads and other events, and other matters related to astronomical education, astronomical heritage, amateur astronomy, astronomy outreach and ArAS future projects are described and discussed.

1. Introduction

The Armenian Astronomical Society (ArAS) was created on 22 June 1999; on that day the founding meeting was held, the first database of Armenian astronomers was collected, and By-Laws were developed. The Society was officially registered by the Armenian Ministry of Justice two years later on 29 August 2001. Before that, actions were taken to involve astronomers in the organization, and to identify and organize the future activities of the ArAS. After the official registration, the first steps were taken to establish affiliation with the European Astronomical Society (EAS) in September 2001, create the ArAS website in 2002, prepare and distribute electronic newsletters ("ArASNews") starting in 2002, and begin holding the ArAS annual meetings in 2002. Since then, almost all events related to astronomy in Armenia have been connected with ArAS, and in the 2000s Armenian astronomy has reached and in some ways exceeded the previous successes achieved during the Soviet Union years.

2. Membership

ArAS is an organization of exclusively professional astronomers (though a few exceptions are allowed) that aims to establish close ties between Armenian and foreign astronomers, inviting them to join and cooperate with astronomers in Armenia. There were 16 founding members of ArAS from the Byurakan Astrophysical Observatory (BAO). ArAS currently has 100 members (including 14 founding members) representing 49 institutions from 21 countries[1], including: Armenia – 50 (including 29 from BAO and 6 from YSU), USA – 15, Germany – 6, Russia – 5, France – 4, Mexico – 3, Italy and Spain – 2 each, Bulgaria, Canada, Chile, Greece, Hungary, India, Ireland, Netherlands, Portugal, Romania, Switzerland, Thailand, and UK – 1 (each). Among the most famous of teh ArAS members are Michel Mayor (Switzerland), Brent Tully (USA), Vahe Petrosian (USA), James Houck (USA), Daniel Weedman (USA), Yervant Terzian (USA), Igor Karachentsev (Russia), Zadig Mouradian (France), Daniel Kunth (France), Massimo Turatto (Italy), and many others. We have created for each member a personal webpage containing basic personal and professional data. ArAS has 3 Co-Presidents: Haik Harutyunian, Areg Mickaelian (Acting President), and Yervant Terzian; the ArAS Vice-President is Tigran Magakian; the ArAS Scientific Secretary is Elena Nikoghosyan; the Treasurer is Marietta Gyulzadyan; the ArASNews Editor is Sona Farmanyan; and the ArAS Webpage Administrator is Gor Mikayelyan.

3. ArAS Electronic Newsletter

Until recently, the ArAS periodically released electronic newsletters (ArAS Newsletters, ArASNews[2]) typically 8 times per year, every 6 weeks. The Editors are: Tigran Magakian (2002–2004, issues #1–12), Lusine Sargsyan (2005, issues #13–16), Lilit Hovhannisyan (2006–2007, issues #17–24), Areg Mickaelian (2008–2014, issues #25–76), and Sona Farmanyan (2015–present, issues #77–112). Starting in 2015, the Newsletter has been published monthly, 12 issues per year. As of the end of 2017, 112 issues had been prepared and released. In total, 929 articles have been published, totaling 1495 pages. ArAS Newsletters give news, announcements, articles on Armenian astronomy and BAO, a list of new members, international and local meeting information, summer/winter schools and participation of Armenian astronomers in them, astronomical education in Armenia, articles about Armenian archaeoastronomy and astronomy in culture, anniversaries, professional and popular scientific articles, other informational materials, etc. A Reference List[3] of ArASNews articles has been created allowing easy access to all articles.

---

[1] `http://www.aras.am/Members/members.html`

[2] `http://www.aras.am/ArasNews/arasnews.html`

[3] `http://www.aras.am/ArasNews/arasnewsreference.html`

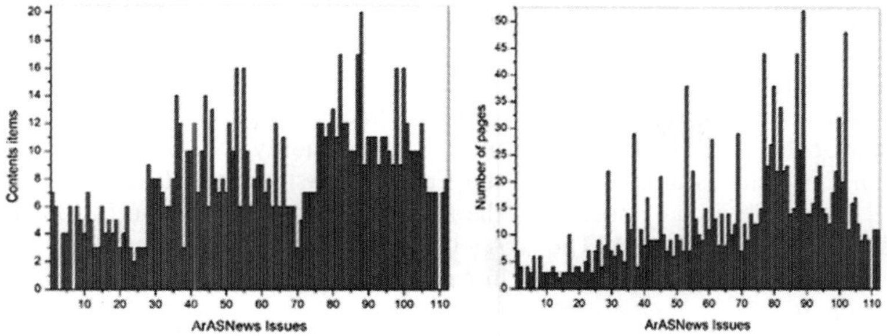

Figure 1. ArASNews statistics. Number of articles by issue (left) and number of pages per issue (right).

4. ArAS Webpage

The ArAS webpage[4] was created in February 2002 to give necessary information about ArAS, including ArAS objectives and forms of activities, lists of ArAS members, information about annual meetings and other events, etc. In early 2009, in connection with

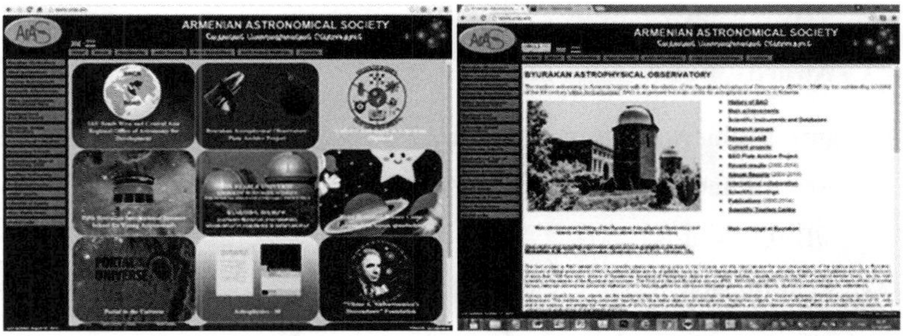

Figure 2. Examples of some of the ArAS webpages.

the International Year of Astronomy (IYA-2009), the ArAS webpage was redesigned and renewed to become a repository containing full information on Armenian astronomy. There is information about BAO history and achievements, current projects and international collaborations, grants received, and all publications since 2000 have been listed. All ArAS meetings have individual pages, and new sections have been opened for other scientific institutions in Armenia related to astronomy. Web pages have been created for 21 famous Armenian astronomers, and also a database of 257 Armenian astronomers throughout the world. Other pages are dedicated to astronomical education, amateur astronomy, archaeoastronomy, etc. Without exaggeration, in terms of the amount of information, the ArAS webpage is the richest one among all Armenian scientific organizations and one of the best in the world. We also published Calendars of Astronomical Events of 2014–2018 on the ArAS website.

---

[4]`http://www.aras.am/`

5. Participation in International Organizations

Armenia is one of 73 member-states of the International Astronomical Union (IAU). The ArAS is recognized by the IAU. Areg Mickaelian is the Secretary and the Acting President of the IAU National Committee for Astronomy (NCA) in Armenia. ArAS is one of 25 societies affiliated with the European Astronomical Society (EAS), and its level of activity is only second to the most powerful countries in Europe (the UK, Germany, France, Italy, etc.). The ArAS is also the official representative of the Euro-Asian Astronomical Society (EAAS) in Armenia. Areg Mickaelian is one of the EAAS Vice-Chairs and is a member of the EAAS International Bureau, and Tigran Magakian is a member of the EAAS Scientific-Technical Committee. Together with the Ministry of Education and Science, ArAS is also one of the official Armenian representatives to the International Astronomical Olympiads (IAO). Armenia joined the international Galileo Teachers Training Program (GTTP) with its official representatives being Areg Mickaelian (Ambassador in Armenia) and Marietta Gyulzadyan (Armenian Coordinator). Armenia is an associate member of the Sub-Regional European Astronomical Committee (SREAC) acting in Southeast Europe and an associate member of the International Planetary Data Alliance (IPDA). The EAS Annual Meeting (European Week of Astronomy and Space Science, EWASS) was held in April 2018 in Liverpool, UK; ArAS Acting President Areg Mickaelian and other Armenian astronomers took part from Armenia.

6. ArAS Awards

ArAS recognizes achievements in astronomy with five types of awards and certificates:[5]

- The ArAS Annual Prize for Young Astronomers (Yervant Terzian Prize).
- ArAS/BAO Awards (2009).
- ArAS Certificates.
- ArAS/OxArm (Oxford Armenian Society) Awards for scientific journalists.
- Galileo Teachers Training Program (GTTP) Certificates.

Since 2004, the ArAS Annual Prize for Young Astronomers (renamed the Yervant Terzian Prize in 2009) has been awarded to the most active young astronomers (younger than 35) considering all annual activities. The USD 500 award is sponsored by ArAS Co-President Prof. Yervant Terzian and the BAO, and it is awarded with an ArAS Certificate. The prize winner in 2017 was Naira Azatyan. In 2017, the GTTP international certificate was awarded to Armine Patatanyan.

7. ArAS Meetings and Other Events

The ArAS has organized regular annual meetings since 2002, some of which were associated or combined with other conferences and events where ArAS acted as co-organizer.

---

[5] `http://www.aras.am/Prize/awards.htm`

Table 1. List of annual ArAS meetings, and other related conferences.

Year	No.	ArAS annual meeting description and other related conferences
2002	I	Scientific Sessions and Annual Report
2003	II	Scientific Sessions combined with Armenian-Georgian Colloquium
2004	III	Scientific Sessions and Annual Report
2005	IV	Scientific Sessions and Annual Report
2006	V	Scientific Sessions combined with BAO 60th Anniversary Meeting
2007	VI	Scientific Sessions combined with 15th EAS Annual Meeting (JENAM-2007)
2008	VII	Scientific Sessions combined with Ambartsumian's 100th Anniversary Meeting
2009	VIII	Scientific Sessions related to IYA 2009 (incl. Astrobiology and Archaeoastronomy)
2010	IX	Scientific Sessions and Annual Report
2011	X	Scientific Sessions and Annual Report
2012	XI	Scientific Sessions combined with Anania Shirakatsi 1400 meeting
2013	XII	Annual Report and Summary of Astronomical Year 2013
2014	XIII	Scientific Sessions combined with RASCS meeting
2015	XIV	Scientific Sessions combined with Armenian-Iranian Astronomical Workshop
2016	XV	Annual Report and Summary of Astronomical Year 2016
2017	XVI	Scientific Sessions combined with the International Conference "Astronomical Heritage of the Middle East"
2018	XVII	Scientific Sessions combined with the IAU SWCA Regional Workshop

The ArAS XIII annual meeting was unprecedented in its theme. It took place on 7–10 October 2014 in Byurakan and was called *"Relation of Astronomy to Other Sciences, Culture and Society"* (RASCS[6]). In addition to the regular astrophysical topics, it also included sessions on Philosophical Problems of Astronomy, Astrobiology, Astroinformatics, Astronautics, Archaeoastronomy, Astronomical Education, Amateur Astronomy, Scientific Tourism, and other topics related to astronomy. It brought together astronomers, philosophers, historians, archeologists, philologists, artists, and representatives of other fields.

In addition to the meetings in Table 1, the ArAS has dedicated a few small-scale events (one-day conferences or seminars) to the anniversaries of famous astronomers.

8. Summer Schools and Science Camps

The ArAS and BAO have organized regular Byurakan International Summer Schools (BISS[7]) once every two years since 2006. Local summer schools for YSU Physics Department students have been organized since 2005 to promote interest to astronomy,

[6] `http://www.aras.am/Meetings/RASCS/index.html`

[7] `http://www.aras.am/Meetings/meetingsSummerSchools.html`

continuing the initiative started in 1995. ArAS also organized a Conference for Young Astronomers in 2011.

Table 2. List of summer schools sponsored by ArAS.

Year	Summer school name	Short
2005	1st Byurakan Summer School for YSU PhysDep students	1BSS
2006	1st Byurakan International Summer School	1BISS
2008	2nd Byurakan International Summer School	2BISS
2009	2nd Byurakan Summer School for YSU PhysDep students	2BSS
2010	3rd Byurakan International Summer School IAU International School for Young Astronomers	3BISS/ISYA
2011	FSU countries Young Scientists Conference	FSU-YSC
2012	4th Byurakan International Summer School	4BISS
2013	3rd Byurakan Summer School for YSU PhysDep students / IAU S304 training	3BSS
2016	5th Byurakan International Summer School	5BISS
2018	6th Byurakan International Summer School	6BISS

Young Armenian astronomers regularly attend several international schools and conferences, including the IAU International Schools for Young Astronomers (ISYA), European NEON/OPTICON schools, Vatican Observatory Summer Schools (VOSS), Virtual Observatory (VO) training schools, and a number of others. The ArAS website holds a directory of the most important regular international schools[8] which helps young astronomers and students to follow and find their respective schools based on the level of the students, location, subject, etc. (see also Table 2).

Figure 3. Advertisement banner for the Byurakan Science Camp.

Starting in 2014, on the initiative of Sona Farmanyan, the ArAS has also organized Byurakan Science Camps (BSC) for school pupils aged 12 to 15. The Fund for Armenian Relief (FAR) partially sponsors these camps. In 2018, the Fifth Byurakan Science Camp was organized.

9. Astronomical Olympiads

1996 The International Astronomical Olympiads (IAO) and International Olympiads on Astronomy and Astrophysics (IAOO) have been organized since 1996 and 2007, respectively. Armenian pupils have regularly taken part in both of them and have achieved

[8]http://www.aras.am/SS2010/ss_other_schools.htm

excellent results. In total, they have won 9 Gold, 5 Silver and 20 Bronze medals. Marietta Gyulzadyan has made an especially large contribution to the achievements of our team; she has been the team leader since 2006. Among the most successful participants are: Zhirayr Avetisyan and Mkrtich Soghomonyan (each have won one first, one second, and one third rank prize), Tigran Shahverdyan (one second and one third rank prize), and Tigran Nazaryan, Hayk Saribekyan, Hayk Tepanyan, Hayk Hakobyan and Edgar Vardanyan have one first rank prize each. Igor Chilingarian from Russia has won first rank prize as well.

10. ArAS School Lectures

ArAS Co-President Yervant Terzian and some other Armenian-American sponsors co-finance a program of ArAS School Lectures, which runs in collaboration with the Armenian Ministry of Education and Science (MES). The lecturers are given by professional astronomers and their visits leave a vivid impression on the school children. Schools are also given astronomical materials (books, pamphlets, sky maps, calendars, etc.). One purpose of the program is to establish connections with gifted children. In 2012, 2013, 2014, 2016, and 2018 programs were held in Armenian and Artsakh schools.

Figure 4. Professional astronomers present exciting astronomy lectures to Armenian and Artsakh school children.

11. Other Matters of Astronomical Education

The ArAS contributes to astronomical education matters in Armenia. Summer Schools, Science Camps, and astronomy Olympiads are examples of ArAS activites in this field. Other activities include providing various educational information on the ArAS webpage and preparation and release of educational CD/DVDs, such as the recently-released *"Astronomy for schools"* and *"Astronomy for students"* DVDs, which are collections of astronomical textbooks, books, dictionaries, encyclopedias, articles, reports, photos, videos, software, etc. In November 2014, through the initiative of Sona Farmanyan, the ArAS founded the Facebook group *"Junior Astronomer's Club"* (JAC). The "Viktor Ambartsumian's Descendants" Educational Charitable Foundation was founded in August 2014. Yervant Terzian is the Chair of its Board of Trustees, Areg Mickaelian is its Coordinator, and Robert Sargsyan is the Director of the Foundation.

Figure 5. The webpage of the "Viktor Ambartsumian's Descendants" Educational Charitable Foundation.

12. Archaeoastronomy and Astronomy in Culture

To coordinate issues related to Archaeoastronomy, an appropriate section was opened on the ArAS webpage.[9] Questions on archaeoastronomy are currently receive great attention in the world, and are the particular focus of international organizations such as UNESCO (the *Astronomy and World Heritage* project), the IAU (the *Astronomy and World Heritage* woking group), the International Council of Monuments and Sites (ICOMOS), the European Society for Astronomy in Culture (Société Européenne pour l'Astronomie dans la culture, or SEAC), the *Starlight* initiative, and others. In 2014, the IAU created a new working group *"Archaeoastronomy and Astronomy in Culture"* (AAC). Several related meetings have already been organized in Armenia (2011, 2012, 2014, 2015, 2016, and 2017). The ArAS Newsletter now has a permanent section *"Archaeoastronomy and Astronomy in Culture"*.

13. Astronomy Outreach

Armenia actively participates in outreach programs and the IAU Astronomy for Development (we have established a regional office in Armenia; see the corresponding article in this volume for more details). The Armenian Coordinator for astronomy outreach is Sona Farmanyan.

The ArAS gives importance to the development of amateur astronomy in Armenia. The *Goodricke John* amateur astronomers organization was created by Ruben Buniatyan, which succeed in achieving recognition of 18 September (Viktor Ambartsumian's birthday) as Astronomy Day in Armenia. ArAS created on its webpage a section for amateur astronomy (establishing the Armenian Amateur Astronomical Society, ArAAS), as well as a Facebook group where one can register as an amateur astronomer.

In December 2010, to promote scientific and astronomical journalism, the group *Scientific Journalists in Armenia* was formed, which includes more than 100 journalists. ArAS periodically prepares and disseminates press releases and organizes press conferences, interviews, and scientific journalism seminars. Press releases contain Ar-

---

[9] `http://www.aras.am/Archaeoastronomy/astronomyancientarmenia.html`

menian and international astronomical news, events organized by ArAS and BAO, past and future celestial phenomena, as well as recognition of the anniversaries of noted scientists. Astronomers hold public lectures for different organizations. On the ArAS webpage, the online *Astghagitak* ("Astronomy Expert") popular astronomical journal was created, where materials in Armenian are being published for those interested in astronomy.

ArAS publishes a number of books (Proceedings of its Annual Meetings), popular booklets, calendars, sky maps and other promotional material.

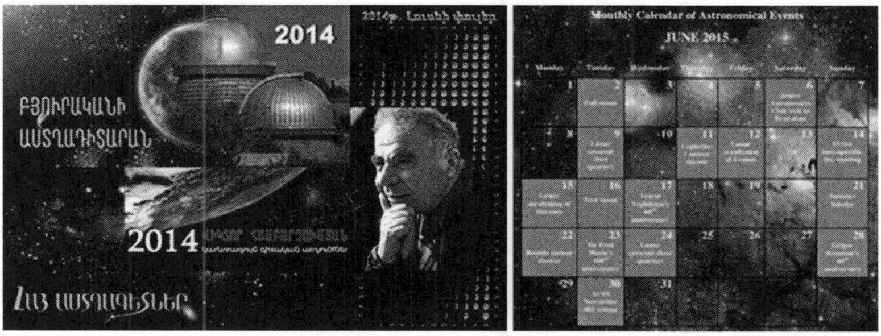

14. Further Plans

The ArAS continues to expand and grow. We plan to give serious attention to the *Astronomy and World Heritage* program in particular. We plan to implement a global program on Archaeoastronomy and History of Astronomy, Astronomy in Culture, as well as global programs on Amateur Astronomy and Astronomy Education. Areas that need more resources and attention at present include amateur astronomy, amateur telescopes, observational programs, amateur astronomers meetings, educational programs, and wider publicizing of astronomy. Our global programs are aimed at coordinating astronomy activities for the whole area of Armenia. We intend to create new sections on the ArAS webpage for Armenian versions of the *"Portal To The Universe"* (PTTU) and other webpages.

Author Index

Arakelyan, K., 229

Bagheri, M., 185
Broutian, G., 191

Capaccioli, M., 151
Cohen, A., 45

Erastova, L. K., 105, 117

Farmanyan, S. V., 3, 235, 253, 259, 267, 295
Fujiwara, T., 163

Ghalandari, H., 209
Gigoyan, K. S., 105, 117
Gradzka, E., 285
Gyulzadyan, M. V., 117

Hady, A. A., 33, 171
Harutyunyan, V. L., 259
Hirai, M., 163
Hovhannisyan, L. R., 105
Hyklová, P., 89, 221

Ishitsuka, J. K., 279
Ishitsuka, M., 279
Israelian, G., 181

Kahil, M. E., 171
Kapoor, R., 75
Konsul, K., 83

Malville, J. M., 17
Mickaelian, A. M., ix, 3, 105, 117, 235, 253, 259, 267, 295
Mikayelyan, G. A., 105, 117, 259, 267, 295

Nasiri Gheydari, S., 27

Ohanian, G. B., 117

Paronyan, G. M., 117
Polcaro, A., 135
Polcaro, V. F., 135

Sargsyan, L. A., 105
Simonia, I., 59, 127
Simonia, Ts., 127
Simonyan, H. Y., 199
Sinamyan, P. K., 105, 117
Šolc, M., 221
Szenkovits, F., 213

Yeghiazaryan, A. A., 71

ASTRONOMICAL SOCIETY OF THE PACIFIC

The Astronomical Society of the Pacific is an international, nonprofit, scientific, and educational organization. Some 130 years ago, on a chilly February evening in San Francisco, astronomers from Lick Observatory and members of the Pacific Coast Amateur Photographic Association—fresh from viewing the New Year's Day total solar eclipse of 1889 a little to the north of the city—met to share pictures and experiences. Edward Holden, Lick's first director, complimented the amateurs on their service to science and proposed to continue the good fellowship through the founding of a Society "to advance the Science of Astronomy, and to diffuse information concerning it." The Astronomical Society of the Pacific (ASP) was born.

The ASP's purpose is to increase the understanding and appreciation of astronomy by engaging scientists, educators, enthusiasts, and the public to advance science and science literacy. The ASP has become the largest general astronomy society in the world, with members from over 70 nations.

The ASP's professional astronomer members are a key component of the Society. Their desire to share with the public the rich rewards of their work permits the ASP to act as a bridge, explaining the mysteries of the universe. For these members, the ASP publishes the Publications of the Astronomical Society of the Pacific (PASP), a well-respected monthly scientific journal. In 1988, Dr. Harold McNamara, the PASP editor at the time, founded the ASP Conference Series at Brigham Young University. The ASP Conference Series shares recent developments in astronomy and astrophysics with the professional astronomy community.

To learn how to join the ASP or to make a donation, please visit `http://www.astrosociety.org`.

ASTRONOMICAL SOCIETY OF THE PACIFIC

MONOGRAPH SERIES

Published by the Astronomical Society of the Pacific

The ASP Monograph series was established in 1995 to publish select reference titles. For electronic versions of ASP Monographs, please see `http://www.aspmonographs.org`.

INFRARED ATLAS OF THE ARCTURUS SPECTRUM, 0.9-5.3μm
eds. Kenneth Hinkle, Lloyd Wallace, and William Livingston (1995)

VISIBLE AND NEAR INFRARED ATLAS OF THE ARCTURUS SPECTRUM 3727-9300Å
eds. Kenneth Hinkle, Lloyd Wallace, Jeff Valenti, and Dianne Harmer (2000)

ULTRAVIOLET ATLAS OF THE ARCTURUS SPECTRUM 1150-3800Å
eds. Kenneth Hinkle, Lloyd Wallace, Jeff Valenti, and Thomas Ayres (2005)

HANDBOOK OF STAR FORMING REGIONS: VOLUMES I and II THE NORTHERN and SOUTHERN SKIES
ed. Bo Reipurth (2008)

TWENTY YEARS OF ADASS
ed. Ian Evans (2013)

SCIENCE WITH A NEXT-GENERATION VERY LARGE ARRAY
ed. Eric J. Murphy and the ngVLA Science Council (2018)

A complete list and electronic versions of ASPCS volumes may be found at `http://www.aspbooks.org`.

All book orders or inquiries concerning the ASP Conference Series, ASP Monographs, or International Astronomical Union Volumes published by the ASP should be directed to:

Astronomical Society of the Pacific
390 Ashton Avenue
San Francisco, CA 94112-1722 USA
Phone: 800-335-2624 (within the USA)
Phone: 415-337-2126
Fax: 415-337-5205
Email: `service@astrosociety.org`

For a complete list of ASP publications, please visit `http://www.astrosociety.org`.

essentials of early childhood education

Second Canadian Edition

essentials of early childhood education

Carol Gestwicki
Central Piedmont Community College

Jane Bertrand
George Brown College

THOMSON
NELSON

Australia Canada Mexico Singapore Spain United Kingdom United States

THOMSON
NELSON

Essentials of Early Childhood Education, Second Canadian Edition

Carol Gestwicki and Jane Bertrand

Editorial Director and Publisher:
Evelyn Veitch

Executive Editor:
Joanna Cotton

Marketing Manager:
Chantal Lanning

Developmental Editor:
Joanne Sutherland

Production Editor:
Natalia Denesiuk

Senior Production Coordinator:
Hedy Sellers

Copy Editor:
Karen Rolfe

Creative Director:
Angela Cluer

Interior Design:
Opus House

Cover Design:
Angela Cluer

Cover Image:
Anne Bradley

Compositor:
Carol Magee

Printer:
Transcontinental

Printed and bound in Canada
2 3 4 05 04 03 02

For more information contact Nelson, 1120 Birchmount Road, Scarborough, Ontario, M1K 5G4. Or you can visit our Internet site at http://www.nelson.com

National Library of Canada Cataloguing in Publication Data

Gestwicki, Carol, 1940–
Essentials of early childhood education / Carol Gestwicki, Jane Bertrand. — 2nd Canadian ed.

Previous ed.: The essentials of early education.
Includes bibliographical references and index.

ISBN 0-17-622318-5

1. Early childhood teachers—Canada. 2. Teaching—Vocational guidance—Canada. 3. Early childhood education—Canada. I. Bertrand, Jane, 1951– II. Title. III. Title: The essentials of early education.

LB1775.6.G47 2002 372.21'0971
C2002-902235-5

CONTENTS

Preface ... xi
Introduction ... xvii

Section One
Early Child Development Programs Today ... 1

CHAPTER ONE
Early Child Development Programs ... 3

Early Child Development Defined ... 4
Early Childhood Education and Care ... 7
- Child Care Centres ... 8
- Nursery Schools ... 12
- Family Child Care ... 13
- In-Home Child Care ... 16
- Early Childhood Intervention ... 17
- Recreation and Leisure Programs ... 22
- Kindergartens ... 22
- Family Support Programs ... 23
- Early Years Communities ... 27

CHAPTER TWO
Early Child Development Curriculum ... 33

Curriculum Models ... 34
- Developmentally Appropriate Practice ... 34
- Montessori Schools ... 35
- High/Scope ... 35
- Reggio Emilia ... 39
- Problem-Based Play ... 41

Curriculum Issues ... 41
- Mixed-Age Groupings ... 42
- Adult and Child Focus ... 43
- Readiness for School Learning ... 44
- Monitoring Readiness for School Learning ... 48
- Inclusion in Early Childhood Education and Care Programs ... 49
- Antibias Curriculum ... 51
- Technology in Early Education ... 54

CHAPTER THREE
Quality in Early Child Development Programs ... 62

Thinking about Quality ... 62
A Bird's-Eye View of Quality ... 66
Defining Quality for Early Childhood Practitioners ... 67
Dynamics of Quality ... 69
Respect ... 69
Responsiveness ... 72
Reciprocity of Learning ... 73
Framework of Quality ... 75
Underlying Developmental Understanding ... 76
Number of Children ... 78
Integration of Learning Experiences ... 79
Physical Environment for Active and Cooperative Learning ... 80
Context of Quality ... 82
Work Environment ... 83
Funding ... 84
Regulation ... 86
Voluntary Standards ... 89
National Standards for Quality Child Care ... 90
Evaluating Quality ... 92
Evaluation Tools ... 92
Not Found in Quality ... 92
Institutionalization ... 93
Failure ... 94
Indifference ... 94
Partners in Quality ... 94

Section Two
The Early Childhood Workforce ... 101

CHAPTER FOUR
Early Childhood Practitioners ... 103

Early Childhood Practitioners ... 103
The Early Childhood Workforce Defined ... 104
A Diverse Workforce ... 105
Men in the Early Childhood Workforce ... 105
Early Childhood Practitioner Roles ... 107
Providing Protection ... 107
Establishing Relationships ... 108
Designing Learning Opportunities ... 109

Connecting with Communities ... 110
What Early Childhood Practitioners Do ... 111
What Is This Early Childhood Practitioner Doing? ... 111
A Puzzle, a Picnic, and a Vision: Family Day Care at Its Best ... 112

CHAPTER FIVE
Becoming an Early Childhood Practitioner ... 117

Why Become an Early Childhood Practitioner? ... 117
Enjoyment of Children ... 118
Making a Difference to Children and Families ... 118
Variety and Challenge ... 120
Demand for Early Childhood Practitioners ... 121
Growing as an Early Childhood Practitioner ... 121
Self-Knowledge ... 121
Disposition, Knowledge, and Experiences ... 122
Stages of Early Childhood Practitioner Development ... 124
Stages of Student Early Childhood Practitioner Development ... 125
ECE Postsecondary Education ... 127
Diversity in Early Childhood Practitioner Preparation ... 127
Regulation and Certification ... 129
Equivalency ... 129
College and University ECE Programs ... 130
ECE Certificate and Diploma Programs ... 130
Guidelines for ECE Programs ... 133
Other Professional ECEC Preparation Programs ... 135
Mentoring ... 136

CHAPTER SIX
The Work Environment ... 141

Career Directions ... 141
Careers That Serve Children Directly ... 142
Careers That Serve Families Directly ... 143
Careers That Organize Services ... 144
Careers That Provide Information ... 145
Careers That Provide Goods and Services ... 147
Career Pathways ... 147
Working Conditions ... 148
Compensation ... 149
Employment Contracts ... 150
Health and Safety ... 151
Stress and Isolation ... 151
Respect and Recognition ... 156

Facing Challenges ... 157
- Knowledge ... 159
- Support ... 159
- Alliances ... 159
- Supervisor Support ... 160

Section Three
The Early Childhood Workforce Comes of Age ... 163

CHAPTER SEVEN
The Roots of Early Child Development Programs in Canada ... 165

Reasons for Understanding the History of Early Education ... 166
- Philosophical Views of Early Childhood ... 167
- The Study and Science of Psychology ... 167
- Developmental Health ... 168
- Social Context ... 169

Early Beginnings: Education and Social Welfare ... 169
- Infant Schools ... 169
- Crèches, Day Nurseries, and Salles d'Asile ... 170
- Public Schools ... 172
- Kindergarten ... 174

Moving into the Twentieth Century ... 177
- Children's Playgrounds ... 177
- Public Health ... 178
- First Nursery Schools ... 179
- Progressive Education and the Child Study Movement ... 180
- The Child Study Movement in Canada ... 181

Working Women: Day Nurseries to Early Childhood Education and Care ... 183
- War-Time Day Nurseries ... 184
- Baby Boom Time ... 184
- Times A' Changing ... 186
- Compensatory Preschool Programs ... 187
- The Child Care Decade ... 188

CHAPTER EIGHT
The Modern Profession ... 193

Is the Early Childhood Workforce a Profession? ... 194
- Criteria of a Profession ... 196

Professionalization of the Early Childhood Workforce in Canada ... 200
Core Knowledge ... 200
Codes of Ethics ... 202
Standards of Practice ... 204
Credentialing ... 206
Voluntary Certification ... 207
Self-Governance ... 208
Professional Development ... 208
Publications ... 209
Professionalization Dilemmas ... 209
Early Childhood Education and Care Organizations ... 213
Canadian Child Care Federation (CCCF) ... 213
Provincial/Territorial Early Childhood Education and Care Organizations ... 213

CHAPTER NINE
Advocacy ... 220

Personal Advocacy ... 221
Public Policy Advocacy ... 221
Understanding the Links: Quality, Compensation, and Accessibility ... 222
Public Policy Options ... 230
Taking Action ... 234
Informational Advocacy ... 237
Essentials for Early Childhood Practitioners ... 238
Attitudes and Dispositions ... 238
Collegiality and Connection ... 239
Reflective Teaching ... 241
Fine-Tuning Skills and Deepening Knowledge Base ... 241
Self-Evaluation ... 242
Personal Philosophy of Education ... 242
Rhythms of Working with Young Children ... 242

Appendix ... 247
References ... 267
Index ... 283

PREFACE

> In this century, children are cherished and recognized as our nation's future, at least rhetorically. Research shows us that the first years of life and early brain development are critical for later learning, health, and well-being. During those early years, children need adults—parents, other family members, teachers, and others—who support their healthy growth and development in many important ways. A competent early child development workforce that is able to work with children, parents and communities is critical to integrate and build on the early years programs now scattered across communities (McCain & Mustard, 1999).

The Early Years Study (McCain & Mustard, 1999) recognized that competent early child development program staff are crucial to implementing programs that are sensitive to the needs of young children and their families. Such a workforce will be a powerful agent in breaking down the barriers (professional, regulatory, and funding) that exist between staff now working in child care centres, family child care settings, nursery schools, parenting centres, family resource programs, and kindergarten.

"*The Essentials of Early Childhood Education* points to the issues that must be addressed in order to prepare individuals to work with young children and families in early child development settings" (J. Fraser Mustard, January 2002).

Why Was This Book Written?

This book is written from the authors' strong convictions that the care and education of children in their earliest years must go beyond rhetoric, to provide optimum experiences and environments for real children in our very real world. Those who enter the early childhood workforce find both enormous challenges and immense gratification as they nurture the healthy development of children and families. As child development research continues to discover the importance of the experiences, interactions, and environments that support children in the first years of their lives, it becomes increasingly clear that the adults responsible for children's care must have knowledge, attitudes, and skill to provide the best opportunities for children. As the need and demand for early child development programs for children from birth through the elementary school years continues to expand in unprecedented dimensions, so too does the need for educated early childhood professionals.

This demand is related to a second strong conviction from which this book grows: not just anyone can, or should, work with young children, and moreover, young children do not need "just anyone." They need particular people, with specific characteristics, knowledge, skills, and attitudes, who have thoughtfully and deliberately prepared to enter into caring relationships with young children and their families. They need people who have committed themselves to following career paths in early childhood education, and who understand and accept the realities of the profession as it has evolved to this point. They need people who have decided to touch young lives, and are willing to stay the course.

Two Main Themes

These are the two main themes of this book: (1) that early childhood practitioners do important, meaningful, valuable work, supporting children and families during the most critical period of development; and (2) that only persons who are willing to accept the need for thoughtful and careful professional preparation will be able to help children reach their full potential. This book, then, proposes to examine the world of early child development and to assist the process of professional growth for those who are considering it as their future, a future that impacts generations to come. It is assumed that students will use this text in a course that introduces them to concepts of early child development, near the beginning of their professional education, whether in a two-year or four-year program. Because the text is designed for students who will continue with other courses in an ECE college program, the specifics of theoretical perspectives and of curriculum ideas and activities are left for those later courses.

Since the intention of the text is to help students begin active construction of themselves as early childhood practitioners, the style is both informative, so that students may truly understand the current field, and introspective, so that students may actively juxtapose their personal knowledge, goals, and experiences to consider the professional roles and possibilities of working with young children. So, however you as a student came to study the essentials of early childhood education, you are invited to reflect on the profession.

Coverage

The *Introduction* outlines the social and political context for early child development programs in Canada.

Section One, *Early Child Development Programs Today,* introduces you to the field in Canada.

Chapter 1, *Early Child Development Programs,* defines the parameters of the field and explores the diversity of program structures, age groups that early childhood practitioners work with, special population groups, and program sponsors.

Chapter 2, *Early Child Development Curriculum,* discusses different approaches to organizing early childhood settings for young children.

Chapter 3, *Quality in Early Child Development Programs,* describes quality in early childhood programs.

The second section of the book, *The Early Childhood Workforce,* introduces the roles and responsibilities of early educators.

Chapter 4, *Early Childhood Practitioners,* explores the various roles of early childhood practitioners who work with children and families in a wide spectrum of settings.

Chapter 5, *Becoming an Early Childhood Practitioner,* demonstrates that good early childhood practitioners are not born but grow actively, with much personal effort and thought. It considers the characteristics, skills, knowledge, and experiences that are important in creating oneself as an early childhood practitioner.

Chapter 6, *The Work Environment,* discusses the career opportunities now open to early childhood practitioners. It also examines some of the current challenges you will face as you enter the early childhood workforce.

Section Three is entitled *The Early Childhood Workforce Comes of Age.* As we enter the twenty-first century, early child development is at an exciting time. The combination of a seemingly unending need for early childhood practitioners, clear knowledge about what contributes to quality programs, and demands for professional standards and acceptance offers us positive directions for our immediate future.

Chapter 7, *The Roots of Early Child Development Programs in Canada,* describes the history of early child development. In this chapter, you will see the multiple traditions and various philosophies and historic/social influences that have shaped the modern world of early child development. Many of the names and events that you will read on the historical Timeline will appear in this chapter to describe our historical roots.

Chapter 8, *The Modern Profession,* considers the current emphasis on professionalism, which will shape your introduction to joining the early childhood workforce.

Chapter 9, *Advocacy,* considers the role of advocacy for early childhood practitioners, both within early childhood settings and in the world at large. It leaves you with the challenge to become an early childhood practitioner and advocate for young children and their families, early child development programs, and the early childhood workforce.

Changes from the First Canadian Edition

This is the second Canadian edition of *The Essentials of Early Childhood Education.* The changes in the second edition reflect increased public interest in early child development and the expanding opportunities open to ECE graduates.

Throughout the text there are updated statistics and references to new Canadian studies. Because of recent research and policy reports, it is possible to provide a more complete picture of the Canadian situation. This edition relies much less on American research and, throughout the chapters, updated Canadian content replaces many of the American references found in the first edition.

The Introduction is a new addition that describes the social and economic environments that are propelling attention to, and investment in, early child development programs. This section points to the growing convergence of evidence from the biological and social sciences that supports increased investment in early child development. Chapter 2 offers more detailed coverage of various program models. Chapter 3 includes references to recent major Canadian child care and child development studies. Section Two has been reorganized to present a fuller picture of the opportunities now open to ECE graduates. Chapter 7 provides more in-depth information about how programs developed in Canada. Chapter 9 presents an expanded vision of advocacy for early childhood practitioners.

Features

Each chapter features objectives, review questions, study activities, and suggested readings. The key terms are defined at the end of each chapter.

Enjoy early childhood facts that spark discussion and thought, as they appear in a *Timeline* in the margin of the pages throughout the book.

Watch for the *Theory into Practice* feature, which highlights examples of current Canadian early childhood care and education policies and practices. The *Making It Happen* boxes feature early child development programs from across the country. *Meet the Early Childhood Workforce* features early childhood practitioners who began with studies in early childhood education.

Most of the photographs were taken on location at child care centres, family child care homes, schools, and hospitals, and feature children who responded spontaneously, in order to provide accurate "snapshots" of real world early childhood education.

We welcome you to consider studies in early childhood education. We hope that you and your fellow students will discuss the topics and issues in depth, and work through some of the additional readings and assignments at the end of each chapter. There are specific activities that will encourage you to explore the Nelson ECE Resource Centre Web site that accompanies this book: www.ece.nelson.com. The Web site is a wonderful resource that contains free downloadable activities and a wide variety of ECE-related Web links.

We need and want you to stay. Early childhood practitioners, families, and communities can join together to nurture, stimulate and educate young children. At no time in the history of Canada have children and families needed more the support, care, and expertise that early childhood education has to offer. Together we can prepare today's youngest Canadians for a rapidly changing and diverse world. The future may be uncertain and world tensions overwhelming, but we can be certain that a strong start is the best foundation for life.

Acknowledgments

Over the years we have heard the stories of many students, early childhood practitioners, and colleagues who have made their own discoveries of the essential truths and pleasures of working with young children. This book is dedicated to them with thanks for their friendship along the way.

This Canadian edition is indebted to the generous permissions I received to reprint materials. Thanks to Canadian Child Care Federation, Early Childhood Educators of British Columbia, Childcare Resource and Research Unit, *IDEAS Journal*, SpeciaLink, and the Early Childhood Education program at Red River College. A special thanks to Noel Young for his patient and knowledgeable assistance in selecting the historical photographs in Chapter 7.

Since I completed the first Canadian edition of this book, I have had the privilege of working with J. Fraser Mustard and the Hon. Margaret Norrie McCain on the Ontario Early Years Study. This wonderful experience has influenced my perspective and expanded my understanding.

In addition, I appreciate the efforts and responses of the editorial and production staff at Nelson Thomson Learning, and the helpful comments and suggestions of reviewers: Diane Bergeron, George Brown College; Judith Boyd, New Brunswick Community College; Cindy Brandon, Centennial College; Lee Hartlen, Seneca College; Zeenat Janmohamed, George Brown College; Kathryn Markham-Petro, St. Clair College; Gretchen Reynolds, Algonquin College; Bonnie Rikkonen, Confederation College; Judy Wainwright, Mount Royal College; and Lynn Wilson, George Brown College.

INTRODUCTION

Welcome to the Early Years

> What we envision will be a first "tier" program for early child development, as important as the elementary and secondary school system and the post secondary education system (McCain & Mustard, 1999).

Society and the Early Years

Public interest in the early years of life is exploding. Newspaper headlines, magazine articles, Web sites, and television newscasts blast out messages such as "the first years last forever" or "Canada's future prosperity lies in today's cribs." Pamphlets and guidebooks about why the early years are important and what parents and others can do to support healthy child development proliferate. Corporate business leaders join early childhood educators (Bay Street meets *Sesame Street*) to raise public awareness about what our youngest children need. Local, provincial/territorial, and federal governments are considering and implementing public policies to increase the resources available for the early years.

An interesting constellation of events has focused the spotlight on early child development in Canada. American and Canadian media reports of revolutionary new brain research (that confirmed what nurturing parents and early childhood practitioners have always known) pointed to hard evidence that early experiences influenced early brain development. Educators and employers are alarmed about poor youth literacy rates and Canada's ability to compete in a competitive world economy. More and more Canadian parents with young children are working outside the home and need to find regular nonparental care arrangements. Provincial/territorial and federal governments began to see budget surpluses after years of fighting deficits and cutting back on spending.

In 1999, Dr. Fraser Mustard and the Hon. Margaret McCain delivered a landmark report, *The Early Years Study: Reversing the Real Brain Drain*, to the Ontario government. The report marshals the knowledge from science, communities, practitioners, and families. The recommendations call for significant investment in the early years and the establishment of a new system for early child development and parenting. The timely report was recognized by governments now able to consider increased social spending, business leaders ready to consider long-term strategies to improve productivity, and families and communities struggling to support parents' need to work. The Early Years Study continues to influence public interest and investment in early child development programs and policies across Canada.

Canada's exuberance about the early years echoes what early childhood practitioners have always known. Early environments shape the course of early development. Early child development programs and strategies that support young children and their families are wise investments and simply the right thing to do.

It's good to see others join us and value our work with young children and their families.

As you look at the Timeline, you will see that there have been various forms of early child development programs in Canada over the past two centuries. The earliest public expenditures on children included funds for schools to teach primary academic skills. Public schools then expanded to include kindergartens, which began as products of a particular philosophy of early childhood. With their inclusion in the public schools, kindergartens have undergone changes in philosophy and format, and the debate about these changes is ongoing. The first child care centres sprung up at the end of the nineteenth century to provide care and protection for poor mothers seeking employment. The nursery schools that middle-class Canadian children began attending in the 1920s reflected both the societal value of early education and the cultural acceptance of the importance of childhood.

Canada does not yet have a cohesive policy of support for the early years, but most Canadians believe that the healthy development of young children is key to our future. There is consensus that the early years agenda is an important one. The agenda is supported by the science of early child development, concern about economic prosperity, and changing demographics and families. Momentum is growing to create a first "tier" system for the early years. It is an exciting time to enter the early childhood workforce.

The Long Reach of Early Childhood on Learning, Behaviour, and Health

The science of early child development draws together research from biology, psychology, sociology, medicine, anthropology, and neuroscience. The findings from all disciplines point to the impact of early experiences on the development of children's brains and on their lifelong health, well-being, competence, and abilities to cope (Doherty, 1997; Keating & Hertzman, 1999; Shonkoff & Phillips, 2000, Bertrand, 2001).

Human beings have known that the brain is an important organ for a long, long time. However, now technology allows us to have a greater understanding of how the brain develops. What children experience in early life becomes embedded in their biological and developmental pathways. New experiences, opportunities, and challenges will certainly affect later development but they will always interact with the pathways laid down by past experience.

We know from biological and observational studies of animals and humans that brain development is rapid, interactive, and dramatic.

- The neuron or nerve cell is the basic building block of the brain and nervous system. Neurons appear in many forms, but all possess the same basic structure—a cell body, an axon, and dendrites. Neurons communicate to each other by sending messages across the gaps between the axons and neurons. They must connect together to make neural pathways to function.
- Brain development begins during the prenatal period and is dramatic during the first years. (Of course, while early brain development sets down the foundation,

brain activity and development continues throughout the life cycle.) Early brain development is based on two processes—wiring and sculpting. The neurons build connections or are "wired" to communicate with each other. Excess connections are discarded or "sculpted" to construct efficient neural pathways. Wiring and sculpting activity is most active during the early period of development, but continues into adolescence.

- The human brain is shaped by experience. Experience leads to neural changes in the brain. The remoulded brain facilitates the embrace of new experiences—and so the process goes on. The daily experiences of early life become embedded in our biology and influence later development.

The evidence from animal and human studies in biological studies and the social sciences is consistent: the effects of the social and physical environment in the early years of human development affect our risk for physical and mental health problems in adult life. This is coupled with a growing body of evidence that the early period of child development affects cognition, learning, and behaviour in the later stages of life. The converging evidence creates a broad consensus about the fundamental importance of the early years of development.

"Early brain and child development sets a foundation for health, well-being, behaviour and learning but later development plays a significant role in building on the base" (Bertrand, 2001, p. 17).

Concern about Economic Prosperity and Social Well-Being

Canada's economy and society are undergoing significant change as we shift to a "knowledge-based" economy, which depends on ideas and innovation rather than natural resources or manufacturing. To be successful in the global market, Canada must have a competent, well-educated workforce. Success in school is much more likely when children enter school ready and motivated to learn—physically, emotionally, socially, and cognitively. Quality early childhood care and education experiences contribute to readiness for school learning and school success (Doherty, 1997).

Canada is in the midst of a technological revolution driven by the advent of computer systems. Over the past twenty-five years, electronic technology replaced human beings across all sectors of work. Automatic Teller Machines (ATM) replaced bank tellers. Artificial intelligence systems, rather than pilots, now guide commercial jets through takeoff and landing. University and college students can now take courses on-line, participate in seminar discussions, submit assignments, and take tests without leaving home or having direct contact with a teacher. Of course, all of the electronic technology requires human intelligence for creation and operation. The demands of the technology have created new job areas that were unknown twenty-five years ago—Web master, e-commerce, and dot-com companies. The new jobs in the knowledge economy require people with more skills and abilities than did many of the jobs that were replaced in this transformation.

Recognizing the need for a competent workforce, public attention turns to reports of school achievement and academic testing echoes. Now government and business leaders are starting to understand the connection between how well children are doing when they enter school and later school success. Children who are able to take turns, wait to have their needs met, listen to storybooks, and play for extended periods of time find it easier to adapt to school environments and are more likely to do well at school. Early childhood settings help to develop these critical skills so children can get along with other children and adults.

Changing Demographics

Demography is the study of human population trends. One of the greatest influences on trends in early child development programs is the demographic data that portray changing patterns of family life and employment. In the past thirty years, two interrelated social patterns have appeared.

- The number of single-parent families has increased owing to rising divorce rates and to the increase in the number of children born to single parents.
- Census data reveal that the majority of all parents, whether in one- or two-parent families, are in the workforce. The need of these families for some form of supplemental child care for children from birth on is changing the face of early childhood care and education.

One of the most obvious effects of the changing demographics of families and the workplace has been the enormous increase in the amount of nonparental care needed. Various family supports are recognized as necessary for the well-being of

FIGURE 1.1

Working Mothers with Children under Six Years in Canada

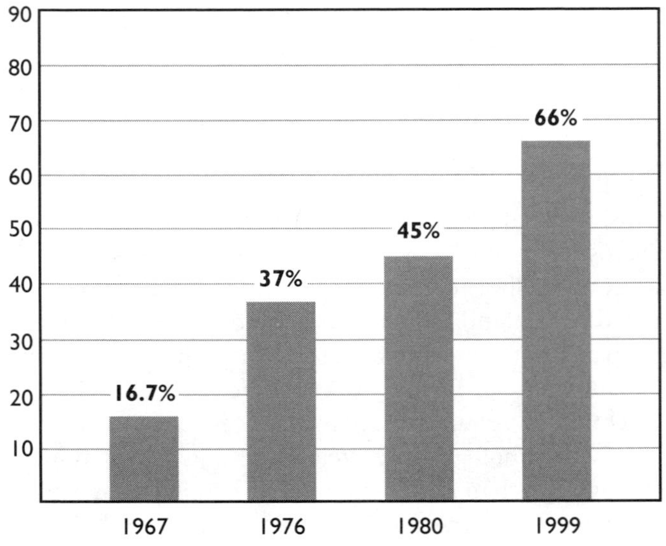

Sources: Beach, Bertrand, & Cleveland, 1998; Johnson, Lero, & Rooney, 2001.

children, families, and communities. Schools and child care programs are increasingly being asked to assume helping roles for parents by providing social support as well as education. Early years programs are now seen as a necessity both for the support of family needs and for the care and early education of young children. Early childhood practitioners are in a vital position to provide this community support in a time of changing social patterns.

Looking forward, the demand for child care to support parents who are working will probably continue to increase, although the number of children from birth to twelve years will not change much over the next decade. The participation rates of mothers in the labour force continue to increase, although the increases are smaller than those over the past three decades.

Families' needs for nonparental care arrangements and their awareness of the value of early child development programs are likely to increase. Public awareness of the importance of development in the early years is creating recognition of, and demand for, a variety of early child development programs that combine early education, care, and support to families.

Today's families come in all sizes and shapes.

How Are Canadian Children Doing?

Are children who enter Grade One in September, 2002, doing better—healthier, more ready for school, more outgoing and co-operative—than children were fifty years earlier in 1952?

We know that children are born healthier and are less likely to die from disease or injury today than they were fifty years ago. But we don't know if the numbers of children with difficult behaviour or emotional problems have increased or decreased. There are more reported child abuse incidents, but this may reflect greater public awareness and less tolerance.

Parents, teachers, child welfare workers, and doctors often report that children have more behaviour and emotional difficulties today than in the past. But we cannot verify the anecdotes with evidence, because this kind of information was not collected fifty years ago.

Today researchers and governments are collecting more information and we can draw a detailed picture about Canada's children—how they are doing and what their lives are like (Ross, Scott, & Kelly, 1996; Ross, Roberts, & Scott, 2000).

- Most children (85 percent) live in two-parent families.
- Most children seem to enter school with the competencies and skills need to be successful, but as many as 25 percent do not.
- About 20 percent of all children seem to have a behaviour or emotional problem.
- More than 15 percent of children live in relative poverty.

Rights of Childhood

"Action now will put our children and our future on a firmer foundation for the future. This action is necessary, not only to keep a reasonable standard of living, but also because it is the right thing to do for our young children" (McCain & Mustard, 1999, p. 2).

The Convention of the Rights of the Child was adopted by the United Nations in 1989. It includes the idea that children are citizens with rights and are entitled to resources. The child's individual rights to resources should be respected, quite apart from the child's value as an asset to society.

Canada supported both drafting the Convention and mobilizing the world's nations to sign it. The fifty-four clauses of the Convention include children's rights that have direct implications for early child development programs (Friendly, 2000). The Convention states that all children have the right to child care services while their parents work and to education opportunities that will prepare them to participate in society. Early child development programs are critical if we value children and respect their rights to their share of our resources.

Implications for Early Childhood Practitioners

In Canada, a continuum of early child development programs evolved from earlier early childhood education programs, child care arrangements, kindergarten, and family support programs. It includes a broad range of interconnected programs and services to support children's optimal development and families' child-rearing skills. Educators who make up the early childhood workforce are equipped to work with young children in all of these settings.

On September 11, 2000, the first ministers (provincial and territorial premiers and the prime minister) signed the federal/provincial/territorial Early Child Development Agreement. The agreement stated

> First Ministers agree to work together so that young children can fulfill their potential to be healthy, safe and secure, ready to learn, and socially engaged and responsible. In support of this common goal, governments will improve and expand early childhood development programs and services over time. Governments will work with families and communities to help meet the needs of young children and their families. Governments will report regularly on their progress and will continue to build knowledge and disseminate information to parents, communities and service providers to help them to give children the best possible start in life (Canadian Intergovernmental Conference Secretariat, 2000).

The federal government also promised $2.2 billion to be used for expanded early child development programs. Since then, provincial and territorial governments have announced new and expanded early child development programs.

Early child development programs are entering a new period of development in Canada. There is much work to do and not all new initiatives will best serve the interests of young children and their families. But the overall direction is unmistakable—early child development programs and related opportunities for early childhood practitioners are expanding. Canada's young children and families come in all sizes and shapes, so early child development programs must be sensitive to ethnic, linguistic, and religious diversity. The early child development programs that grow up in the twenty-first century will differ tremendously and will be dependent on an equally diverse early childhood workforce.

SUGGESTED READINGS

Bertrand, J. (2001). *Summary of research findings on children's developmental health.* Ottawa: Canadian Institute of Child Health & Canadian Child Care Federation.

Johnson, K., Lero, D., & Rooney, J. (2001). *Work-Life compendium 2001: 150 Canadian statistics on work, family and well-being.* Guelph: Centre for Families, Work and Well-Being, University of Guelph.

McCain, M. & Mustard, J.F. (1999). *Early Years Study.* Toronto: Ontario Children's Secretariat.

United Nations. (1989). *Convention on the rights of the child.* Ottawa: Human Rights Program Department of Canadian Heritage.

Section One

Early Child Development Programs Today

In this section, you are introduced to the field of early child development in Canada. Chapter 1 will define early child development programs in Canada, and Chapter 2 considers how early childhood settings are organized to provide both care and education to young children. Chapter 3 will explore the issue of quality from the perspective of children, parents, early childhood educators, and society, and will discuss specific components of quality in early child development programs.

CHAPTER ONE Early Child Development Programs

CHAPTER TWO Early Child Development Curriculum

CHAPTER THREE Quality in Early Child Development Programs

A Professional and Life Choice

GANDINI ➤ It seems that you made a choice to dedicate your life to the education and care of young children. When did you make this life choice?

MALAGUZZI ➤ I could just avoid answering, as others have done before, by saying that when you don't ask me I know, but when you ask me, I do not know the answer anymore. There are some choices that you know are coming upon you only when they are just about to explode. But there are other choices that insinuate themselves into you and become apparent with a kind of obstinate lightness, that seem to have slowly grown within you during the happenings of your life because of a mixing of molecules and thoughts. It must have happened this latter way. But also World War II, or any war, in its tragic absurdity might have been the kind of experience that pushes a person toward the job of educating, as a way to start anew and live and work for the future. This desire strikes a person, as the war finally ends and the symbols of life reappear with a violence equal to that of the time of destruction.

I do not know for sure. But I think that is where to look for a beginning. Right after the war I felt a pact, an alliance, with children, adults, veterans from prison camps, partisans of the Resistance, and the sufferers of a devastated world. Yet all that suffering was pushed away by a day in spring, when ideas and feelings turned toward the future, seemed so much stronger than those that called one to halt and focus upon the present. It seemed that difficulties did not exist, and that obstacles were no longer insurmountable.

It was a powerful experience emerging out of a thick web of emotions and from a complex matrix of knowledge and values, promising new creativity of which I was only becoming aware. Since those days I have often reassessed my position, and yet I have always remained in my niche. I have never regretted my choices or what I gave up for them (Edwards, Gandini, and Forman, 1994, p. 49).

CHAPTER ONE
Early Child Development Programs

OBJECTIVES

After studying this chapter, students will be able to

- describe developmental stages of early childhood,
- define early child development programs,
- identify different types of early childhood settings that provide early child development programs, and
- discuss and compare the availability of early child development programs found in different regions across Canada.

Each day almost 1.5 million children, from **infants** to **school-age children,** are cared for and educated by people other than their families, outside the regular school system. As they climb onto buses, streetcars, and trains, are strapped into car seats and strollers, or wave goodbye as their parents leave home, these children will be cared for and educated by adults who are designated as "teachers" or "caregivers." Some children will go to **child care centres** or after-school programs, which might be located in school buildings, work sites, community centres, or churches. Other children will be cared for and educated in home settings, either the child's own home or the caregiver's home. Some children will spend only a few hours a week in early childhood settings, whereas others will be there for most of their waking hours.

Other programs help parents and other family members to participate fully in their young children's early learning and development. Young children and their mothers, fathers, grandparents, uncles, aunts, or older brothers and sisters may attend play groups or drop-in centres together. Child care centres offer parents opportunities to learn about their children's new interests and achievements and other community resources, as well as access information about child development. Prenatal and postnatal programs offer support and resources to expectant and new parents.

All of these children and their families participate in situations that fall into our definition of **early child development programs. Early childhood practitioners,** the teachers and caregivers of young children, can choose to work in a variety of settings and with children in several different developmental phases. Early childhood practitioners also work with parents and other family caregivers to better understand each family's values and strengths and to offer support. In this chapter we shall examine the many facets and faces of early child development programs.

Early Child Development Defined

In the introduction of this book, you learned about the Early Child Development Agreement in Canada. Under this agreement, the Government of Canada and the provincial and territorial governments agreed to improve and expand programs for young children and their families in four key areas:

- healthy pregnancy, birth, and infancy;
- parenting and family supports;
- early childhood development, learning, and care; and
- community supports.

Early child development programs include programs that care for and educate young children, and support parenting, families, and communities. The provinces and territories have allocated the first installment of the federal early child development funding to a variety of programs and initiatives for young children and their parents. (The Government of Quebec does not participate in the agreement, but is receiving its share of funding and expanding early child development programs in that province.)

THEORY INTO PRACTICE

Early Child Development Programs

- Ayesha lives in downtown Toronto. She goes to kindergarten in the local public school every morning and goes home at lunch with a neighbour, a regulated family child care provider, who cares for five children in the afternoon. Her mother or father picks her up around 5:30 p.m.
- Carmen is four months old. She lives in a small village outside Halifax. Her mother receives maternity/parental leave benefits and is staying at home to care for Carmen for six months. Carmen and her mother meet other parents and infants in a local church basement every Friday morning. The group grew out of a prenatal support group offered by the local public health department; each Friday, the parents set up a floor play space for the infants, with equipment and play materials provided by public health.
- Simon lives in Calgary. He is three years old and goes to a nursery school three mornings a week. His father works a night shift and his mother works at the shopping mall on Friday nights and all day Saturdays. One parent is at home with Simon all of the time except for Friday night when he stays with his Grandma.
- Carla is twelve months old and goes to a child care centre at the university where her dad works. She is often there by 8:00 a.m. and goes home with her dad around 5:00 p.m.
- Elisheva is a four-year-old who is severely brain damaged. She can sit only with support, cannot speak, and needs to be fed. For a year, she has been in a municipally-operated child care centre where a specially trained Resource Teacher provides the extra help that allows her to be part of her group. Her mother and father are in the labour force.
- Ian's parents are both professionals working full-time. Almost every day, Ian goes to a community-based child care centre run by a parent board. His special group of friends includes Tyson, whose mom is struggling to move from welfare to work, Katie, whose parents are both autoworkers, and Liam, a psychiatrist's son.

Early Child Development Programs (cont'd)

- Jessica, who is two and a half, goes with her mother and younger brother to a family resource program two or three mornings a week. Sometimes her mother stays and sometimes she will leave for a couple of hours and then come back for Jessica and her brother.

Ayesha, Simon, Carla, Ian, Elisheva, Carmen and Jessica are all in early child development settings. Although the funding, management and administration of the programs differ, their daily activities are similar. If the settings are of high quality, the children are exploring rich social and physical environments that support their healthy development.

In each of these settings, the physical environment is set up for children. Carla's infant room has a large open play area with lots of soft cushions and pillows. The wall is lined with a cruising rail for beginning walkers and there are low shelves with several bright toys for shaking, poking, pushing and pulling. Jessica's favourite time at the family resource program is the circle time, when one of the staff or a parent reads stories from picture books, and all the children and parents join in songs and child-centred activities. Ayesha's kindergarten, Simon's nursery school, Ian's and Elisheva's child care, and Carla's infant room have a circle time too. Simon often plays in the dramatic play centre at his nursery school; there are dress-up clothes and lots of dishes and pots and pans. Simon likes to pretend he is a busy chef at a big restaurant. Ayesha's kindergarten looks a lot like Simon's nursery school. Her favourite place is the art easel with its fresh pots of paint each day.

In each of these settings, there are adults who both educate and care for each child as they build relationships that are both responsive and respectful to children's growing competence and abilities to cope. They make sure their needs for food, physical safety, sleep and toileting are met. These adults know that children are not isolated individuals but are part of families, and support to those families is critically important. There are also other children to play with in each of these settings and adults who play a role in setting the stage and encouraging the play among the children.

Children do not come to an early childhood setting merely to do their "learning". Rather, they "live" in these settings for several hours each week. Just as they are learning at home when they discover how the flusher on the toilet works or how to use simple tools to fix things or how soothing it is to hear dad's voice sing them to sleep, each hour in an early childhood setting is filled with new information and experiences.

Source: Reprinted with permission from Beach & Bertrand (2000), *More than the sum of the parts,* Childcare Resource & Research Unit, pp. 7–8.

All of the children described in the box "Early Child Development Programs" participate in early childhood settings that are guided by early childhood practitioners—adults who have studied **Early Childhood Education (ECE)** or have a combination of equivalent education and experience. Even the parent-run Friday drop-in program that grew out of a public health prenatal support program has access to an early childhood practitioner who offers suggestions and resources. If the settings are of high quality, the children are exploring rich social and physical environments that support their healthy development. Although the funding, management, and administration of the programs differ, their daily activities are similar.

Children have opportunities to play together at early child development programs (courtesy of Yes I Can! Nursery School).

The term "early child development program" includes a variety of programs available to young children and their families today in Canada. Early child development programs support children's early development, learning, care, and families' capacity to participate fully in their children's early development. Early child development programs include all forms of nonparental care and early education programs (with the exception of the formal education system that starts at Grade One in Canada) for children from infancy through middle childhood and many programs that also support pregnancy and early infancy, parenting, families, and communities. High-quality early child development programs ensure that the programs are geared toward young children, families, and communities and involve the participation of early childhood practitioners. Quality programs must be responsive to each child's individual development, each family's values and child-rearing practices, and each community's cultural context.

This definition gives us a sense of the variety of settings that early child development programs encompass. The definition conveys the concept that early child development programs include various structures that meet individual needs for different families. It is obviously important to include both care and education concepts in the definition. Early child development programs expand the definition to include programs that specifically support and nurture families' active participation in their children's early development and promote communities' capacity for early child development.

In this chapter, we will take explore different types of early child development programs. In later chapters, we will uncover the historical roots of early childhood programs that were derived from education, social welfare, recreation, and health concerns and that defined a need for young children to receive care and education outside their families. All good settings for young children provide educational activities to encourage children's competence, as well as care to support children's development of coping skills and their overall well-being. Good programs for chil-

1628
John Comenius, a Czech educator, writes *The School of Infancy*, referring to the "School of the mother's lap," in which a child from birth through age six would achieve the rudiments of all learning.

dren provide opportunities for learning, assistance with personal routines in safe, healthy, and nurturing environments, and, most important, positive relationships with adults and other children. Therefore, the joint functions of care and education are included in the term "early child development." Good programs also recognize that families have the most powerful influence on early learning and development. By increasing the capacity of family caregivers to nurture and stimulate children, we amplify our efforts. But we can support families only when we begin with full respect for individual childrearing values and practices and the strengths that each family has.

Early Childhood Education and Care

There are almost 5 million children from newborn to eleven years of age in Canada. Nearly 2 million children under twelve years of age are involved in some form of nonparental care (outside kindergarten and elementary school programs) while their parents are at work or school (Canadian Council on Social Development, 2000). Over 500 000 preschool children attend school kindergarten programs, and 300 000 children, with at least one parent or guardian not in the paid labour force, attend early child development programs outside the home or school environment. There are also more than 14 000 children who participate in provincial or territorial early intervention programs.

The need for nonparental care arrangements while parents are working or attending school often determines the type of **early childhood education and care (ECEC)** programs families choose. Some programs, such as child care centres, are designed to accommodate the needs of working parents and promote children's early development and learning. Other types of nonparental care arrangements often incorporate activities that support early child development. In many instances, children will participate in more than one setting. For instance, a child who is cared for in family child care may attend a local playgroup, family resource centre, or kindergarten. Because there are so many different types of programs and no organized system for early child development, it is difficult to determine exactly how many children and families are participating. However, Table 1.1 offers a broad overview of how many children participate in various early child development programs and/or nonparental care arrangements. Keep in mind that one-third of all children whose parents are working are in more than one early childhood education and care arrangement (Johnson, Lero, & Rooney, 2001).

The rest of this chapter looks at various types of early childhood care and education programs and arrangements. Table 1.2 gives an overview of seven different categories. Some programs could be included in more than one category or overlap with other types of early child development programs. We need to understand the terms and concepts used to describe each type of program, as well as each program's potential in meeting the needs of young children and their families. As early childhood practitioners entering the field, you can gain a better understanding of potential career opportunities by becoming familiar with early child development programs available across Canada.

1693
John Locke, in England, writes *Some Thoughts Concerning Education.*

TABLE 1.1

Early Childhood Education and Care Arrangements for Newborns to Eleven Year Olds in Canada

Arrangements Related to Parents' Employment	*Children Attending*
Centre-based program*	390 000
Regulated family child care	70 000
Unregulated family child care**	750 000
In-home care**	400 000
Self/sibling care	345 000
Arrangements Not Necessarily Related to Employment	
Centre-based program* (one parent not in paid labour force)	300 000
Kindergarten	512 900
Aboriginal Head Start	12 000
Provincial/territorial Early Intervention Programs	15 000

*Includes regular full-time or part-time participation in nursery schools, child care centres, preschool centres, and after-school programs.

** Includes both relative and nonrelative home-based nonparental care arrangements.

*** Some of the children in the arrangements not necessarily related to employment may also be in another arrangement related to parents' employment.

Unregulated family child care is the most common form of nonparental child care used by families in Canada.

Sources: Adapted from Johnson, Lero, & Rooney (2001); Beach, Bertrand, & Cleveland (1998); Childcare Resource and Research Unit (2000); Health Canada; Human Resources Development Canada; and Indian and Northern Affairs Canada (2001).

Parents who work often require full-day care for their children (courtesy of Eglinton Public School).

Child Care Centres

Child care centres include group or centre-based programs outside regular schooling for children from as young as three months to those up to twelve years of age. Most child care programs offer **full-day programs** five days a week (although some children may attend **part-day programs**), as well as programs before and after school and during school holidays for school-age children. The majority of child care centres in Canada offer service for preschool children two to five years of age.

Child care centres serve the needs of parents who are working outside the home or who are pursuing further education or training and, thus, need supplemental care for their children for eight or more hours a day. Many full-day programs operate from early morning (for example from 7:00 or 7:30 a.m.) until 6:00 p.m. or later, to allow parents time to drop off their children, put in a full work or study day, and return to the centre to pick up the children. Child care cen-

TABLE 1.2

Early Childhood Education and Care

Centre-based child care	Full-day early child development programs for groups of children under school age or for school-age children during out-of-school hours. Organized to provide stimulating, nurturing programs for children that accommodate the parents' work schedule.
Preschool/nursery school	Part-day, often part-week, early child development programs for children between the ages of approximately 2 1/2 and five years. Usually less than four hours per day.
Family child care	Nonparental care for small groups of children from newborn to twelve years in a caregiver's home. May incorporate early child development activities.
In-home care	Care provided by a caregiver or nanny in the child's own home. May incorporate early child development activities.
Early childhood intervention	Early child development programs and activities available to young children and families who have developmental challenges or who are at risk of developmental delays.
Recreation and leisure	Early child development programs that are organized around specific physical or creative activities for young children.
Kindergarten	Early child development program available to all five-year-old children through the education system. In some parts of Canada, kindergarten is also available to four-year-old children.

tres usually have staff scheduled to work at staggered times, to provide adequate coverage over the full period.

Some child care centres that serve particular corporations, health care institutions, or businesses offer several shifts of child care to accommodate the needs of parents who work during the evening or night. However, most programs offer regular daytime care only. Child care centres may accept children on a part-time basis, but seasonal or emergency care or services for parents who work shifts and irregular hours are scarce across the country.

Most child care centres in Canada are **nonprofit;** they are operated to support child and family needs and well-being. A nonprofit centre might be a stand-alone organization with a voluntary board of directors, which can include parents, community members, and individuals with child care expertise. Or it might be operated by a larger agency or institution. The YMCA in Canada, for instance, operates over seventy-five centres. Approximately eighty community colleges directly operate child care centres, which also provide model sites for ECE students (Beach, Bertrand, & Cleveland, 1998). **Commercial** child care centres, also called proprietary or for-profit centres, are privately owned businesses. About 23 percent of Canada's child care centres operate as commercial businesses, ranging from small, owner-operated programs to large chains (Childcare Research and Resource Unit, 2000). In Newfoundland and Alberta, most child care centres are proprietary, whereas other provinces or territories, such as Manitoba, have no commercial child care centres.

THEORY INTO PRACTICE

Provincially and Territorially Regulated Child Care Spaces

Many early child development programs are regulated by provincial and territorial governments' child care legislation. Each province has established its own approach to organizing and licensing child care and preschool or nursery school programs, and it is difficult to make direct comparisons among the different jurisdictions. However, Table 1.3 provides an overview of what types of services are regulated, how many children attend, and what the different services are called. The regulated services are the anchor of Canada's early child development programs, although less than 15 percent of all Canadian children are enrolled in these programs.

TABLE 1.3

Regulated Early Childhood Education and Care Programs

Province/Territory	*Regulated Programs*	*Children Attending (1998)*
Newfoundland and Labrador	Child care centres (full-time & part-time)	3 740
	School-age child care	535
	New infant/toddler child care & regulated family day care spaces not included	
Prince Edward Island	Early childhood centres (full-time & part-time, including kindergarten, nursery school, & child care centre)	3 045
	School-age child care centres	482
	Family day care homes	39
Nova Scotia	Child care centres (full- & part-time, including nursery schools, preschools, and child development centres)	10 994
	Family day care homes	169
New Brunswick	Day care centres & school-age programs (full- & part-time)	9 048
	Community day homes	156
Quebec	Centres de la petite enfance (full- & part-time centre & family child care)	60 540
	Milieu scolaire (school-age child care)	92 700
	Milieu familial (family child care)	21 761
Ontario	Child care centres (full- & part-time, including nursery schools & school-age child care)	149 090
	Supervised private home day care	18 000
Manitoba	Day care centres (full- & part-time), nursery schools & school-age child care centres	13 104
	Family day care homes	3 489

Provincially and Territorially Regulated Child Care Spaces (cont'd)

Province/Territory	*Regulated Programs*	*Children Attending (1998)*
Saskatchewan	Child day care agencies (full-time)	4 884
	School-age child care centres	919
	Family child care homes	2 235
Alberta	Day care centres (full-time) & nursery schools, parent co-ops, kindergartens & drop-in (part-time)	40 528
	Family day homes	6 505
British Columbia	Group day care centres (full time)	16 505
	Preschools (part-time)	15 500
	Out-of-school care	16 404
	Family child care	17 357
	Child-minding & occasional care	2 470
Northwest Territories	Day care centres (full-time), nursery schools (part-time), after-school care, & family day homes	1 145
Nunavut (April 1, 1999)	Day care centres & family day homes	206
Yukon	Child care centres (full- & part-time)	665
	School-age child care	226
	Family day homes	416

Note: Regulated group programs and family child care may have different names and age categories in various parts of the country, but they share a number of commonalities in the care and education they provide.

Source: Adapted from Childcare Research and Resource Unit (2000).

Nonprofit, government-operated child care centres, which are directly operated by a government, are far less common in Canada. However, Ontario municipalities operate close to one hundred child care centres, and some Alberta municipalities operate school-age child care programs. In Quebec, the provincial department of education administers school-age child care centres, which are located in elementary schools.

In Canada, regulated early childhood programs may be operated by nonprofit organizations, commercial or independent operators, and public organizations (such as local government). The majority of regulated child care in Canada is operated by nonprofit organizations.

Child care centres are licensed and regulated by provincial or territorial governments. Specific requirements may concern staff qualifications, **adult-child ratios** (the maximum number of children allowed for every staff member), maximum number of children, physical space regulations, including the minimum

1815
The first Maternal Association is established by a group of ministers' wives in Portland, Maine, to encourage the moral and religious training of children.

FIGURE 1.1
Sponsorship of Regulated Child Care Centre Spaces in Canada

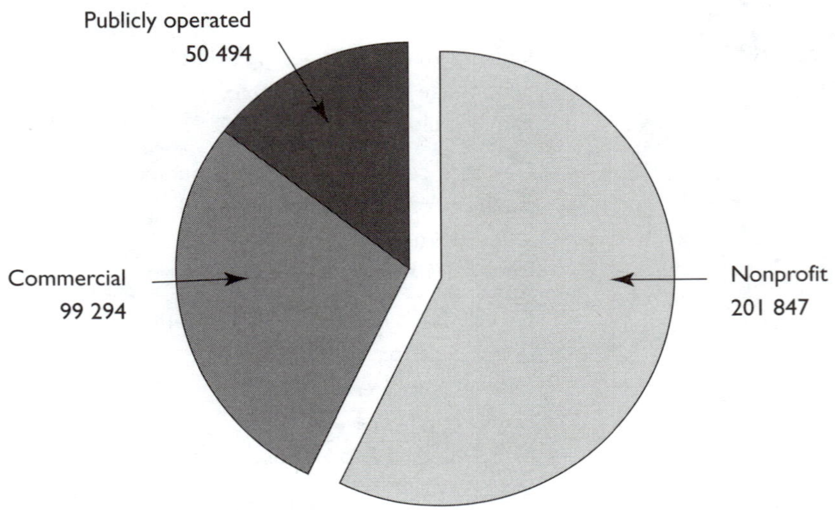

Source: Adapted from Childcare Resource & Research Unit (2000) and Johnson, Lero, & Rooney (2001).

amount of space necessary for each child, daily care routines, and program activities. We will examine these requirements in more detail in Chapter 3.

Parents' fees, which may be eligible for government subsidies, pay for the majority of the costs of child care centres. In some jurisdictions, provincial or territorial governments provide grants directly to child care centres. These funds may include operating, start-up, and capital funding, funding for children with **special needs,** and funds to enhance wages. Chapter 3 examines the funding of regulated child care centres in more detail.

Nursery Schools

Nursery schools offer two- to three-hour care for preschool children (two to six years old) during the school year (September to June). These programs may not function during a full week (for example, a group of three year olds might attend classes from nine until noon, on Monday, Wednesday, and Friday mornings). The usual purpose of these programs is to offer stimulation, activities, and group experiences to help the children develop optimally. Many of these programs have their roots in early nursery school or kindergarten education (see Chapter 7). Today, in some parts of Canada, nursery schools are called "preschools" or "preschool programs."

In most provincial/territorial jurisdictions, nursery schools are regulated through the same legislation as child care centres and must meet similar structural, funding, and organizational requirements. However, nursery schools in Saskatchewan and Quebec are not covered by child care legislation. Nursery schools and preschool programs may be nonprofit or commercial operations, or they may be publicly run.

There are no statistics available on how many children attend nursery school programs or how many programs exist. Some provinces and territories include nursery schools in their report of regulated child care centre spaces, whereas others exclude them. Sometimes nursery schools are considered part-time child care centre programs and are combined with school-age child care programs.

We do know that 300 000 children whose mothers are not in the paid labour force are in centre-based early child development programs (Statistics Canada, 1992). It is quite likely that at least half of these children are attending nursery school or preschool programs. Some of the children whose parents participate in the paid labour force or in training and education programs also attend nursery school programs. Parents might use nursery school services in combination with other nonparental child care arrangements, or they might juggle their work and study responsibilities around the nursery school hours.

Family Child Care

Family child care refers to the small groups of children that a child care provider cares for in a private residence. The residence is usually the home of the provider. Family child care may be regulated through provincial/territorial legislation or may operate outside licensing requirements.

MAKING IT HAPPEN

Rural Roots Children's Centre

Rural Roots Children's Centre, located in the Slate River Valley near Thunder Bay, is the first rural child care centre in Northwestern Ontario. We opened our doors in September 1992 in response to parental concerns for children's safety before and after school. A provincial initiative to locate child care centres in new schools made it possible to place Rural Roots in Valley Central School.

The current program evolved as Rural Roots responded to the needs of the community. In the early days it was necessary to combine groups when enrollment was low. Today mixed-age grouping is a part of a variety of services offered, including full-time care, part-time care, drop-in, and emergency care. Farm families, seasonal workers, and families diverse in structure can all find a program to suit their needs at Rural Roots Children's Centre. Families with children eighteen months to twelve years are enrolled at the Centre.

At the Centre, quality care and education are provided not only by the early childhood educators on staff but also, in a very natural way, by the children who spend time with us.

Children learn from children, and children care about each other.

The centre feels like a large family where the older children play an important part in looking after the younger children. Teachers are nurturers. In mixed-age groups, children are free to move between rooms, and there is flexible grouping at lunch. Siblings are often in the same playroom regardless of age. Children are as likely to seek out another child as they are an adult.

The Centre has created a home-like atmosphere to meet the children's needs. We are proud of our Centre's sensitive response to the families in our community.

Source: Adapted from Broere (1995). Used with permission.

Government regulations, policies, and funding shape the organization of regulated family child care services in each province or territory. Each jurisdiction sets out the maximum number of children who may be cared for by an individual in a regulated or unregulated home setting. If there are additional children, the program must meet the province or territory's requirements for regulated child care centres. The provincial and territorial governments either license and regulate caregivers or license a family child care agency that supervises individual family child care programs.

Unregulated family child care is the most common type of remunerated child care arrangement in Canada; it is also the least visible and most variable model of child care. It can be an early childhood education and care program, in which an early childhood practitioner who has access to resources and support in her community provides nurturing care and stimulating education to children in a home setting. It also includes custodial child care arrangements, in which children are left in the care of an individual who "keeps an eye on the kids" while carrying out other housekeeping tasks.

THEORY INTO PRACTICE

Family Child Care in Canada

The structure and organization of family child care varies among the provinces and territories.

Newfoundland: In 1999, the new *Child Care Services Act* regulated family home child care. The province has extended financial support to family resource programs that support family home child care. Up to four children (including the caregiver's own children under seven years) are permitted.

Prince Edward Island: Regulated family child care homes are licensed and monitored by the provincial government. Caregivers are allowed up to seven children (including the caregiver's own, under ten years of age), with a maximum of three children under two years of age in regulated homes. In unregulated family child care settings, a total of five children, including the caregiver's own preschool children, are allowed, with a maximum of two children under two years or a maximum of three children if all three children are under two years of age.

Nova Scotia: Regulated family child care homes are supervised by licensed agencies. Six children, including the caregiver's own preschool children, are permitted in either regulated or unregulated family child care. If all the children, including the caregiver's own, are school-age, eight children are allowed.

New Brunswick: Regulated family child care homes are licensed and regulated by the provincial government. Regulated caregivers are permitted to care for one of the following groups of children: a maximum of three infants, five children between two and five years of age, nine school-age children, or a mixed-age group of six children, including the caregiver's own children under twelve years. An unregulated family child care home is allowed a maximum of four children, including the caregiver's own children under twelve years, and a maximum of two children under two years.

Quebec: Regulated family child care settings are supervised by early childhood agencies (centres de la petite enfance) that also offer centre-based programs and are licensed by the provincial government.

Family Child Care in Canada (cont'd)

Caregivers in regulated and unregulated family child care are permitted to care for up to six children, including the caregiver's own children under twelve years. If the caregiver is assisted by another adult, he or she may care for a maximum of nine children.

Ontario: The province licenses private home day care agencies, which supervise family child care homes. Care for a maximum of five children up to twelve years of age is permitted in a home that is supervised by a licensed agency. No more than two of the children may be under two years, and no more than three of the children may be under three years, including the caregiver's own children under six years. An unregulated caregiver is allowed to care for up to five children, not including her own.

Manitoba: Regulated family child care homes are directly licensed by the province. One licensed caregiver may care for up to eight children under twelve years, including her own children under twelve years. No more than five children may be under six years, and no more than three children may be under two years. Unregulated caregivers are permitted to care for up to four children, including the caregiver's own children, and no more than two children may be under two years.

Saskatchewan: Regulated programs are licensed and monitored by the provincial government. A regulated caregiver is permitted to care for up to eight children between six weeks and twelve years, including her own children under thirteen years. Only five of the children may be younger than six years, and only two may be younger than thirty months. Unregulated caregivers are allowed to care for up to eight children, including the caregiver's own children under thirteen years.

Alberta: The provincial government does not license family child care homes or agencies but does enter into contracts with agencies to approve and monitor caregivers according to provincial standards. Regulated caregivers are permitted to care for up to six children under eleven years, including the caregiver's own children under eleven years. A maximum of three children may be under three years, and no more than two children may be under two years. In unregulated homes, a caregiver is allowed to care for up to six children, including her own children under twelve years. A maximum of three children may be under three years.

British Columbia: Family child care homes are individually licensed by the provincial government and are permitted to care for up to seven children under two years, including the children living in the home. Of the seven children, there may be no more than five **preschoolers** and two school-age children, no more than three children under three years of age, and no more than one child under twelve months. Unregulated caregivers are permitted to care for two children, not including the caregiver's own children.

Northwest Territories: Regulated family child care homes are individually licensed by the territorial government. A maximum of eight children under twelve years, including the caregiver's own children, is permitted. No more than six of the eight children may be five years or under, no more than three children may be under three years, and no more than two children may be under two years. An unregulated caregiver may care for up to four children, including her own children under twelve years.

Yukon Territory: Regulated homes are individually licensed by the territorial government. Regulated caregivers may care for up to eight children, including the caregiver's own children under six years. No more than three infants are permitted if there are also three preschool or school-age children. If there are two caregivers, an additional four school-age children are permitted. Unregulated caregivers may care for up to three children, not including the caregiver's own children.

Sources: Childcare Resource & Research Unit (2000); Beach et al. (1998); Beach & Bertrand (2000).

1816
Robert Owen establishes an infant school providing care for children of mill workers in Lancaster, England.

The **Canadian Child Care Federation (CCCF)** (a national organization, discussed in Chapter 8) describes a variety of models of family child care in Canada and distinguishes not only between regulated and unregulated family child care but also between supervised and unsupervised and supported and nonsupported settings. The models include

- licensed family day care agencies,
- licensed family day care providers,
- registration of family day care providers,
- resource-supported, regulated family day care providers,
- resource-supported, unregulated family day care providers, and
- unsupported, unregulated family day care providers (Canadian Child Care Federation, 1991, p. 17).

Regulated or unregulated, family child care programs can often be adapted to meet the needs of working parents and diverse families in isolated rural and urban settings. At the same time, these programs can accommodate an individual child or a small mixed-age grouping from infancy to early adolescence.

In-Home Child Care

In-home child care arrangements include nannies and caregivers who provide care and education in the child's own home. They may live in the child's home, or they may live elsewhere and come for designated work periods. Live-in caregivers are typically called "nannies" in Canada (Kaiser & Rasminsky, 1991). Many fami-

FIGURE 1.2
Ages of Children Attending Regulated Family Child Care in Canada

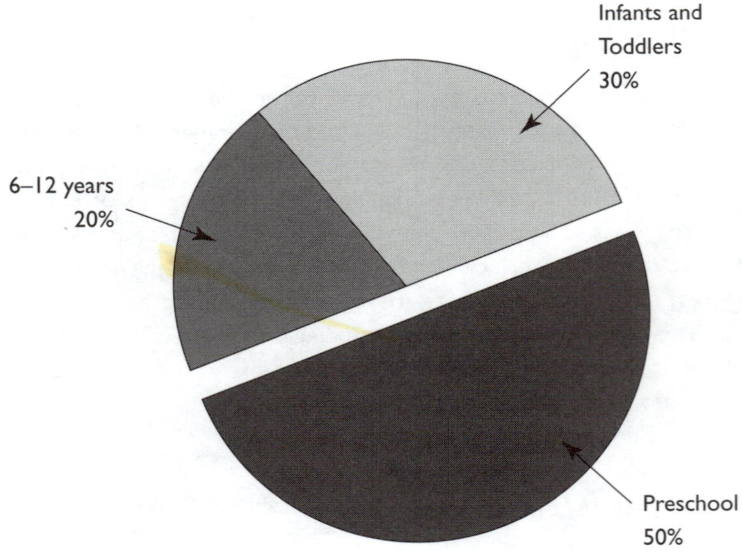

Sources: Adapted from Goss Gilroy (1998) and Beach & Bertrand (2000).

lies specifically seek out early childhood practitioners to provide in-home child care, hoping to ensure nurturing and safe care and stimulating experiences for their children.

In-home child care can provide greater flexibility for parents who commute long distances or work long hours. Children remain in their own home and have the constant care of one individual in their parents' absence. In many instances, children will also participate in other early child development programs, including nursery schools, family resource programs, and kindergarten.

Parents often find someone to provide in-home child care in their local communities through family resources programs, community information services, or postsecondary services. Alternatively, parents might purchase the services of an agency that selects candidates to provide in-home child care. In Canada there are no governmental regulations or requirements for such agencies, although in British Columbia, Ontario, and Quebec, the agencies must obtain a license to operate.

Many live-in nannies are from other countries and are able to move to Canada through the Live-in Caregiver Program. This is a federal immigration program that allows future employers to sponsor immigrants who wish to enter Canada to provide in-home child care. The nanny must live in the family's home for a two-year period; at the end of the period, she may apply for landed immigrant status. These arrangements are usually conducted through an agency.

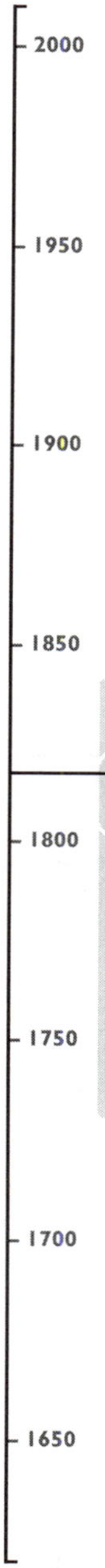

1819
Johann Pesalozzi, a Swiss educator, becomes the first recognized early childhood teacher in Yverdon, Switzerland.

Early Childhood Intervention

Specific public policies, programs, and activities have been designed to meet the needs of young children and their families with specialized needs in order to promote healthy development during early childhood (Shonkoff & Meisels, 2000). These initiatives are included under the broad term "early intervention." Early childhood intervention draws on the knowledge and expertise of early childhood education, child and family development, education, health, and **social services.** Early childhood intervention services include a wide range of programs and approaches offered to families with children who have developmental challenges or who are at risk of developmental delays owing to disability or negative environmental conditions.

An early childhood intervention service refers to "any non-medical, non-protection service, primarily developmental, that is used to assist a young child and his family. Thus attendance in an appropriate nursery school or daycare program can be seen as an early intervention" (Irwin, 1995, p. vii).

In Canada, early childhood intervention services include the following.

- **Inclusive early childhood programs** are "programs that include children with disabilities in the same programs they would attend if they did not have disabilities" (Irwin, 1995, p. vii). Inclusive programs encourage and facilitate the full participation of children with disabilities or special needs, not only their physical presence (Irwin, 1995). We will discuss the curriculum elements and challenges of inclusive early childhood education and programs in Chapter 2.
- **Compensatory programs** are designed for children up to six years who have developmental delays or who are at risk, or likely to be at risk, of developmental delays because of environmental conditions.

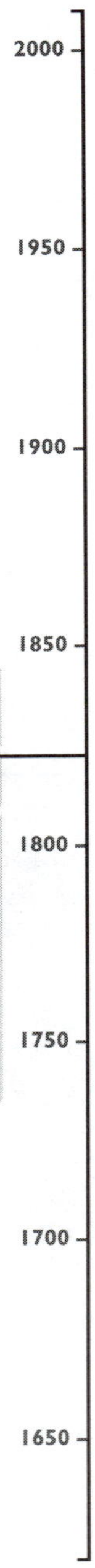

- **Early Intervention (EI)** programs support families whose children up to three or six years have special needs or are at risk of developmental delays owing to disability or psychological or social factors (Irwin, 1995).

Early Intervention programs may be discrete services or integrated into another health, social service, or education program. These other early childhood education and care programs include centre-based programs, family child care services, nursery schools, and family support programs.

As you can see, the delivery of early intervention programs varies. But the provision of early intervention programs does follow two basic principles. They are interdisciplinary, meaning that they include services and strategies from more than one field. They are also family-centred, which means that the family is the primary focus of the services, rather than an individual child out of the context of her or his family (Meisels & Shonkoff, 1990).

Compensatory programs. These programs are designed to support families whose children's development is delayed or at risk because of environmental conditions, particularly conditions related to economic disadvantage and poverty. **At risk** means there are concerns about developmental delays, learning abilities, or challenging behaviours. Compensatory programs provide specific interventions to families to reduce the child's vulnerabilities and to increase the likelihood of success in formal schooling. While this intervention may include a high-quality early childhood education and care program that stimulates and supports preschool children's development, it also offers specific supports and services to other family members.

Canada has a number of early childhood intervention services that are compensatory programs. In some instances, they provide an early child development program for preschool children that also involves their parents and a number of other health and social services. These programs may be identified as **Head Start** programs. (Head Start programs originated in the United States during the 1960s; we will learn more about them in Chapter 7.) The following compensatory programs focus on improving parents' and other family members' abilities to nurture and stimulate their children:

- *Community Action Program for Children (CAP-C).* This federal initiative involves community coalitions, which deliver health and community services, including early education, to children up to six years of age who are living in conditions of risk. Project initiatives include home visits, Head Start programs, parent training, nutrition education, counselling, collective kitchens, and traditional Aboriginal healing programs. There are approximately 500 CAP-C initiatives in over 300 communities across Canada. Over 100 000 children, parents, and other caregivers visited CAP-C projects in 2001 (Johnson et al, 2001) and as many as 28 000 of them visit each week (Guy, 1997).
- *The Aboriginal Head Start program.* This program is another federal government early intervention initiative that was developed to "provide comprehensive experiences for Indian, Metis and Inuit children and their families … based on caring, creativity and pride flowing from the knowledge of their traditional beliefs within a holistic and safe environment" (Childcare Resource & Research Unit, 2000). Aboriginal Head Start began in 1994 with a focus on Aboriginal

1822
Robert Owen moves to New Harmony, Indiana, and establishes an infant school and day care centre as part of the community.

families living off-reserve in cities and large northern communities (Beach, Bertrand, & Cleveland, 1998). In 2001, there were 114 Aboriginal Head Start projects across Canada. In 1998, Aboriginal Head Start was extended to First Nations communities (Childcare Resource & Research Unit, 2000). First Nations Head Start is an early childhood intervention program intended to prepare children for their school years and to establish early child development programs that are created and controlled by First Nations (on-reserve) communities. In 2001 there were 168 First Nations Head Start programs serving 305 communities (Johnson, Lero, & Rooney, 2001).

- *Better Beginnings, Better Futures.* This project, located in Ontario, is the first early childhood intervention project of its kind in Canada. The project focus is on children up to eight years of age living in eight low-income communities in Ontario. The communities themselves have defined what services they need to promote their children's development and to alleviate the impact of economic disadvantage. Each project is unique, but most now offer early childhood intervention services, such as home visits, family support programs, and school and child care centre enrichment, which are coordinated with other early child development services in each community. Better Beginnings has an exciting research component, whereby the impact on children, families, and communities will be monitored and a group of children and families will be followed for twenty years to measure the long-term effects of the program. Short-term outcomes reported that "where programs for children 0–4 were sustained from infancy (home visiting) through the preschool period, with parent-child playgroups and quality child care, children started school with less anxiety, fewer behavioural problems, and more ready to learn" (Peters, 2001, p.3).

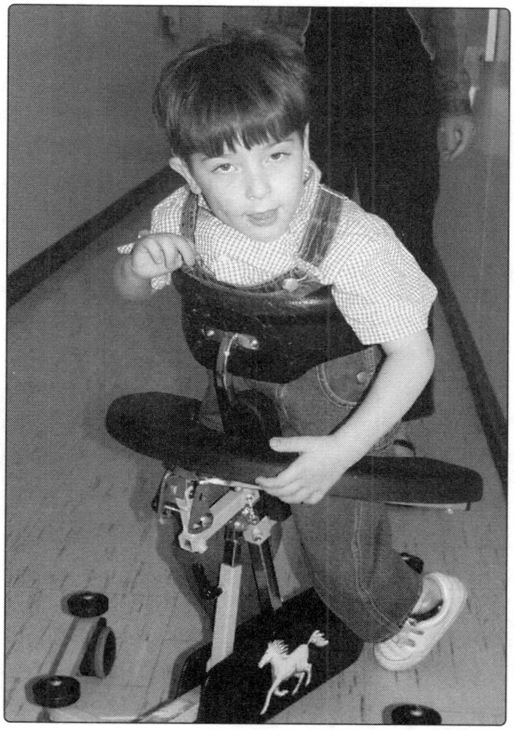

Early Intervention programs are directed to children with developmental challenges (courtesy of Play and Learn).

Early Intervention (EI) programs. These programs deliver services to children (from birth to age three or six, depending on the province or territory) who have developmental challenges or who are at risk because of a disability or psychological/social factors. EI programs usually include a home-based component, early identification assessment, program planning (family service plans and individual program plans), and specialized equipment. The programs may also offer family support and early childhood services such as family resource programs, parent/caregiver support groups, nursery schools, and toy-lending libraries.

EI programs, like regulated child care programs, lie within provincial/territorial jurisdiction. Each province and territory organizes the structure of the EI services and determines how they are funded. Some provinces refer to these services as Infant Development Programs or Direct Home Services (Irwin, 1995). There are over 170 specific EI programs across Canada, although, in some instances, a single EI program is province-wide with several separate delivery sites (Human Resources Development Canada, 1998).

THEORY INTO PRACTICE

Moncton Headstart Inc.

In 1974, Moncton Headstart began in a small apartment on Fleet Street in Moncton. The program offered free day care for a few children with parents who could not afford outside care and who were unskilled in their parenting role. Its initial goal was to help prepare children of low-income families for success in Grade One of the public school system. In 1992, Moncton Headstart Inc. adopted the following mission statement:

> to provide socially, emotionally, and educationally disadvantaged parents and preschool children with a professional, nonthreatening and accessible learning environment that will enable the family to become whole, self-sufficient, and contributing members of the community.

The Children's Program

The children's program seeks to foster age-appropriate self-help, motor, cognitive, and communication skills; to expose the child to the outside world; to stimulate thinking and problem-solving skills; to develop the child's sense of security, personal control, and personal worth, and to foster feelings of being loved, of belonging, and of being capable.

Activity centres engage children in both cognitive and motor activities. A loft provides a quiet area, a family centre encourages sociodramatic play, a circle area is used for group activities, and a playground provides for healthy outdoor play. Good nutrition and positive experiences in the larger community, through planned outings and activities, are emphasized.

Family Activities

Family outings and special events give parents an opportunity to enjoy recreation and learning experiences with their children. A growing sense of community with other families is a potential benefit.

Periodic family meetings are held at the centre, with van transportation provided. During these meetings, children spend time in the classroom with the assistant teacher while the parents meet with the head teacher and group leader to set and make plans to meet goals.

Participation of the parents in both the child development program and in their own development program is mandatory. It involves a weekend retreat, two "Days In" per month (participating in the children's program), parent sessions, and workshops.

Moncton Headstart Adjunct Programs

Moncton Headstart is founded on the belief that basic physical needs (food, shelter, and safety) must be met before one can deal with concepts of self-development. This program offers the following services:

- Adult Literacy Initiative;
- Mapleton Teaching Kitchen. This special education program for families looks at basic nutrition, economic food provision, and home management skills;
- Lorne Preston Educational Fund. The fund awards scholarships to assist Headstart parents and former Headstart children to attend college or university;
- Future Horizons. This service offers subsidized housing to integrate economically disadvantaged Headstart families in stable neighbourhoods; and

Moncton Headstart Inc. (cont'd)

- Moncton Headstart Recycles Inc. This small-business initiative offers education and support to people who depend primarily on government social programs. Through working in the recycling business, participants acquire employment and other skills. Used clothing is channelled in two directions: wearable clothing is exported to developing countries' markets for sale, and discards are converted to cotton rags.

Too often, children and families who are caught up in the cycle of generational poverty experience exclusion and segregation within their community. The programs of Moncton Headstart are designed to teach skills that promote inclusion and participation. Caring for others is the main message.

Source: Adapted from Bradshaw (1997).

TABLE 1.4

Provincial and Territorial Early Intervention Programs

Province/Territory	*Number and Type of Programs*	*Children Served Per Year*
Newfoundland	1 program, with District Offices of Social Services throughout the province	400
Prince Edward Island	No stand-alone program; families supported through five regional Health Authorities	n/a
Nova Scotia	11 provided by community agencies	200
New Brunswick	16 provided by community agencies	1 500
Quebec	2–3 rehabilitation centres in each region of the province, coordinated through community health centres	n/a
Ontario	60 provided by community agencies	5 500
Manitoba	1 province-wide program	900
Saskatchewan	16 provided by community agencies funded by Indian and Northern Affairs	465, plus 125
Alberta	19 provided by health units, school boards, and nonprofit societies	750
British Columbia	48 provided by voluntary agencies	4 000
Yukon	1 outreach program—community-based Child Development Centre	220
Northwest Territories	5 Regional Health Boards, Regional Social Services offices, and Divisional Boards of Education	n/a

Source: Adapted from SpeciaLink (1996). Used with permission.

Recreation and Leisure Programs

You are probably familiar with recreation and leisure programs in your community—from swimming classes at the community centres, to preschool storytime at the local library, to T-Ball baseball leagues at the local park. Most Canadian neighbourhoods offer physical and creative activities to young children and their families. Public libraries, children's museums, and out-of-school recreation programs are three examples of recreation and leisure programs.

Public libraries encourage children to enjoy books and learning to read from an early age. Collections include a designated children's section that features children's picture books, easy-to-read books, first novels, video and audio tapes, and computer activities. Many offer regular parent/caregiver and child storytimes.

Children's museums are interactive displays designed to engage children in displays about a particular topic or concept. They may be part of larger museums or may be established as separate museums for children.

Out-of-school recreation programs offer a variety of regular physical activities to school-age children. They may be organized to accommodate parents' work and provide regular nonparental care arrangements.

Recreation and leisure programs include many activities that support early development. What kinds of programs are available in your community and how do they complement other early child development programs?

Kindergartens

In all provinces and territories most **kindergarten** programs are offered through the public school system and are operated under provincial/territorial education legislation. In Prince Edward Island, publicly funded kindergarten programs are provided by community early childhood education and care programs. Some kindergarten programs operate outside the public school system on a fee-for-service basis, either as private schools or as part of licensed child care programs.

Adult and child music programs and storytimes help parents and other caregivers participate in their children's early learning (courtesy of Play and Learn).

Provincial/territorial education legislation sets out requirements for children's minimum age, the number of instructional days and hours, and teacher qualifications. There may also be specific adult-child ratio and maximum group size requirements.

Most kindergarten is offered on a part-time basis, either half days (morning or afternoon) or two to three full days (school days) per week, although there are some full-time, full-day kindergarten programs that operate within public schools. Half-day programs are usually two to three hours in length, and full-day kindergarten programs are four to five hours in length (excluding lunch programs). In New Brunswick, Quebec, and Nova Scotia, kindergarten for five year olds comprises full days. In all jurisdictions except New Brunswick, attendance is voluntary. Families are not required to pay fees for their children's participation in publicly funded kindergarten.

Age limits for attending kindergarten programs vary slightly across provincial/territorial jurisdictions but most require children to be at least four years and eight months old in September (five years by December 31) to attend kindergarten or senior kindergarten programs, and at least three years and eight months old in September (four years by December 31) to attend junior kindergarten programs.

Ontario has **junior kindergarten** programs for four-year-old children and **senior kindergarten** programs for five-year-old children. Kindergarten programs in other jurisdictions are generally available only to five-year-old children, although there are a few kindergarten programs for four-year-old children in Quebec, Winnipeg, and British Columbia. In Saskatchewan, schools operate publicly funded preschool programs (for three and four year olds) in some communities. There are approximately 383 000 children attending public education kindergarten or senior kindergarten programs and 121 000 children attending junior kindergarten programs in Canada.

Kindergarten programs often intersect with child care services. Many four- and five-year-old children who attend public kindergarten programs also attend licensed child care programs or informal family child care, or they are cared for by an in-home caregiver. Licensed child care programs are often located in public school buildings. Kindergarten programs are intended to provide educational and social experiences for young children in preparation for formal schooling. At the same time, many Canadian families use kindergarten programs as part of a child care package. Kindergarten programs and other early childhood services for young children often share many of the values and assumptions that underlie curriculum approaches. There are a number of initiatives between child care and kindergarten programs that have attempted to coordinate these programs, including school board policies and practices to encourage increased collaboration, shared physical space, and common curriculum planning and professional development opportunities.

Kindergarten programs also intersect with early intervention services and family support programs, particularly as provincial and territorial governments explore options for integrated children's services.

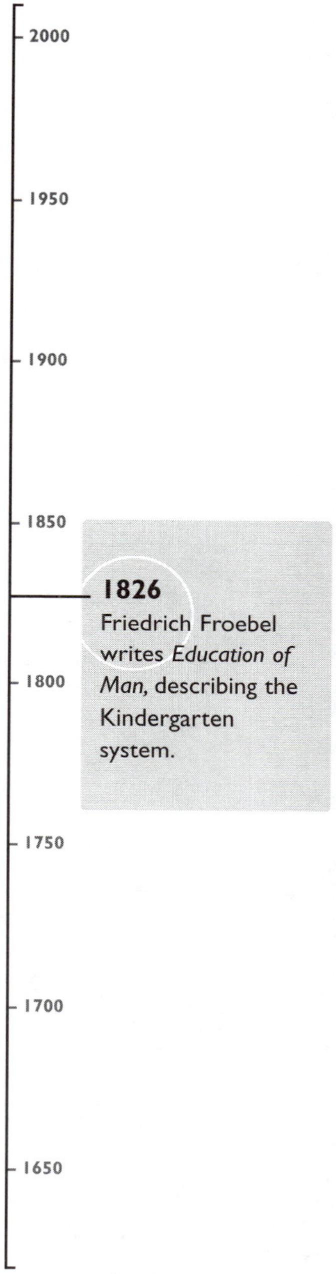

Family Support Programs

The category of family support programs is a broad one that includes a variety of services, strategies, and activities. Family support programs complement and sometimes enrich a family's existing strengths and resources. These programs may address existing problems or aim to prevent potential problems and also be considered as early childhood intervention. Three specific types of family support programs are found in most parts of Canada today: family resource programs, home-visiting, and family literacy. They may be offered as separate programs or in combination with each other or with other early child development programs.

Family resource programs are family-focused; that is, they offer activities that are directed to young children and to parents or other caregivers. There are approximately 2000 such programs across Canada, with representation in each province and territory (Beach et al., 1998). They offer a range of services in diverse physical settings and with different types of funding and sponsorship. Nevertheless, family resource programs can be identified by common principles, functions, and types of activities.

Family resource programs share the following set of principles:

THEORY INTO PRACTICE

Family Literacy in British Columbia

The family literacy movement in B.C is growing; Literacy BC identified more than 80 family literacy programs/services in that province in 2000. There has been a significant increase in the number of schools, libraries, family resource programs, social service agencies, health boards, and government organizations that have contacted Literacy BC for information, consultation, and support regarding family literacy program development, practitioner training, and resources.

These groups are rapidly becoming more aware of the value and benefits of family literacy programs and services, and are looking for leadership, partnership, and direction.

Literacy BC's annual **Family Literacy Day/Week Campaign** raises awareness and builds partnerships and capacity for family literacy. In fact, staff report that they can barely keep up with the requests for information, resources, funding, and offers of partnerships for the next year's events!

Other family literacy activities supported by Literacy BC include

- support to new and existing family literacy initiatives through consultation, referral, and provision of resources;
- ongoing development of a provincial family literacy resource collection and provision of on-loan access;
- maintenance of the family literacy directory of programs and services in B.C., including the B.C. component of the national family literacy online directory;
- facilitation of the family literacy electronic conference in order to provide on-line support and training to family literacy practitioners and groups in B.C.;
- encouragement of good practice in the family literacy field through the distribution and promotion of the B.C. Framework of Principles and Standards of Good Practice in Family Literacy; and
- designing and delivering a one-day training workshop for family literacy practitioners.

Source: J. Rasmussen (2000). Personal communication.

- an ecological approach to services for children, families, and communities;
- an emphasis on prevention and wellness of families;
- a recognition of the need for social networks to support families;
- an emphasis on interdependence in families' needs and abilities to give and receive support;
- a view of parenthood as an important stage of adulthood;
- an acceptance of cultural diversity in approaches to childrearing;
- a recognition that play is essential to optimum child development (Kellerman, 1995).

Based on these principles, family resource programs provide services that have voluntary attendance and involve the entire family, including nonfamily caregivers (Kyle, 1994, reported in Kellerman, 1995). Parents and program staff work together to determine which services will be offered. Family resource programs encourage peer support groups among mothers, fathers, other family members, and caregivers. Family resource programs are multidisciplinary; programs cut across health, social services, education, recreation, and child care service categories. Program staff bring experience and credentials from different disciplines, and programs establish working links with other community services. Family resource programs may direct specific services toward problems, but the overall approach is to provide support with the realization that all families experience difficulties from time to time and that their strengths can be used to work through the difficulties.

Most family resource programs focus on children up to six years of age, their families, and their caregivers. The programs may serve a number of functions, including

- parent support and education;
- support to in-home and family child care providers;
- promotion of optimal child development;
- material support (such as food or clothing) to families;,
- support to high-need families (such as parents of children with special needs, recent immigrant or refugee families, or parents with substance abuse problems);
- resources for early childhood practitioners and social service professionals;
- support and resources for informal social networks; and
- community development, planning, and advocacy (Kyle, 1993, reported in Kellerman, 1995).

As you can see, this is an overwhelming list of functions that could be carried out in many types of specific services. In practice, most family resource programs focus on a few primary functions and then develop activities that will best carry out the function in the community (Kellerman, 1995). For instance, parent support may take the form of a parenting course, or parents may observe and interact with other parents and program staff as part of a drop-in program. Some activities serve several functions. Toy-lending libraries promote optimal child development, provide material support, draw parents together to encourage informal social networks, and provide adaptive play materials and activities for children with disabilities and their families.

Family and in-home child care providers are the main users of over half of the family resource programs in Canada. Support to this group of users may include

1830s
Infant schools are introduced in Halifax by factory owners.

1836
Massachusetts passes the first child labour law (seldom enforced) in the United States, prohibiting children under fifteen from working in factory mills unless they had three months of schooling in the previous year.

in-service training such as workshops, drop-in programs, child care information and referral services, and toy-lending libraries.

A 1994 survey of family resource programs in Canada found that the most common family resource program activities are

- playgroups and drop-in programs;
- parent support groups;
- parenting, caregiver, and early educator courses and workshops;
- toy-lending libraries;
- special events for children and families;
- child care information and referral services;
- respite care for parents and caregivers and child care while parents/caregivers participate in other family resource program activities;
- "warm-lines" (noncrisis telephone support or electronic discussion group);
- crisis intervention and counselling (including informal and peer counselling);
- early childhood intervention services for families and their children with developmental challenges or at risk of developmental delays;
- services and support programs to meet specific needs, such as prenatal and postnatal support for teen mothers and fathers, support groups for survivors of violence, ESL classes, life-skills courses, and literacy programs; and
- community development initiatives, including planning and advocacy for children's and family services (Kellerman, 1995).

1837
First public normal school is established at Lexington, Massachusetts.

Friedrich Froebel establishes the first kindergarten (children's garden) for children ages three to six years in Blankenburg, Germany.

Family resource programs offer program activities that best serve their own mandate, primary functions, and community. Think about how a family resource program could best support families with young children in your community. What would be the program's primary function? What types of activities should be included? Can you identify the role of early childhood practitioner as part of the program staff?

Not only do the program activities vary from one family resource program to another but also are found in a variety of settings and under different sponsorship and funding across Canada. A 1994 survey noted that almost two-thirds of the programs are not-for-profit organizations, governed by boards of directors (Kellerman, 1995). Almost a quarter of the programs are part of other community, recreation, health, or social service organizations such as community centres, YMCAs or YWCAs, public libraries, and community health centres. About 8 percent are offered in conjunction with other types of ECEC programs, including child care centres and/or regulated family child care, and another 7 percent are operated by social service agencies, such as a child welfare or children's aid association.

There are also variations in the way family resource programs are defined and organized in each province or territory (Kellerman, 1995). Programs that focus on support to caregivers in unregulated child care are more likely to be recognized as family resource programs in Ontario than they are in British Columbia. In Alberta, programs offering support to parents of school-age children are considered family resource programs. Nevertheless, the broad definition of family resource programs we have used in this book is typical of the direction of the family resource program network across Canada.

In Ontario, some family resource programs receive funding from the provincial government to provide support for unregulated family child care providers.

Provincial funding, which is a part of the child care budget, is made available to about half of the family resource programs. In Newfoundland, family resource programs are funded to support both regulated and unregulated family child care.

In British Columbia, a separate system of family resource programs, called child care support programs, is funded by the provincial government to support caregivers by offering services such as child care registries and family child care training courses. The government of British Columbia also funds another group of family resource programs, known as Family Places, which focus on services to families with young children. A Family Place typically offers drop-in and play-groups services, which are not offered by the child care support programs. Although the programs are intended for parents, other caregivers often participate (Kellerman, 1995).

The federal Community Action Program for Children (CAP-C), discussed earlier in this chapter, provides funding support to high-risk families with young children. Many family resource programs across Canada are part of CAP-C projects.

Family literacy initiatives "recognize the influence of the family on the literacy development of family members and try to support families in literacy activity and in accessing literacy resources" (Thomas, 1998, p. 6). Specific family literacy programs and activities include storytelling and music circles for young children and parents or caregivers, adult education using family experiences, book bags for infants and young children and their parents, and the creation and use of play materials that encourage the acquisition of skills necessary to support literacy.

Home-visiting is a family support program that bring expertise and resources to the homes of families with young children. Many home-visiting programs are offered as early childhood intervention strategies and are part of early intervention programs for children with developmental difficulties or compensatory programs.

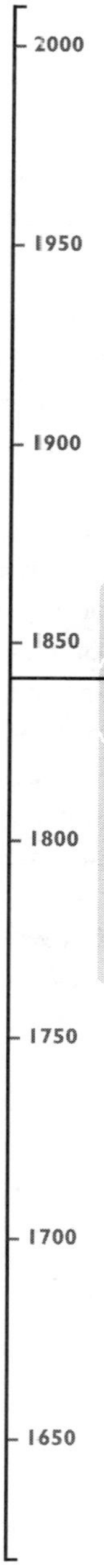

1846
Common School Act is passed in Canada West (later Ontario) and brings about curriculum standardization.

Early Years Communities

Many communities are trying to draw together and integrate different early child development programs to support young children and their families. Both public awareness and the overall number of early child development programs are growing. The Early Child Development Agreement has made resources available for new and expanded early child development programs.

Across Canada, many communities are taking advantage of the climate of opportunity and experimenting with community-based early child development models. Some of these initiatives grow out of CAP-C collaborations (mentioned earlier in the section on compensatory programs).

While each community is unique and their efforts reflect local context and provincial/territorial policies, two main strategies stand out:

- community-wide mobilization to improve environments for all young children; and
- weaving together of different types of early child development programs to create a seamless early child development program to meet the needs of young children and their families.

MAKING IT HAPPEN

Understanding the Early Years—Building Community Capacity for Early Child Development

Understanding the Early Years (UEY) is a national research initiative. It is based on the belief that communities will use community-specific research to make the case to allocate resources to provide opportunities for young children. Developed by the Applied Research Branch (ARB) of Human Resources Development Canada (HRDC), *Understanding the Early Years* project emerged in response to a growing recognition that (1) increasing our understanding of the factors that help or hinder child development and (2) increasing community tracking of how well children are developing are crucial to ensuring the best possible start for Canada's children.

One of the main purposes of UEY is to help determine the extent and nature of community influences on child development and how this might vary from child to child. Data is collected in the community about what resources are available, children's readiness to learn at school entry (using the Early Development Instrument) and child, family, and community context using the National Longitudinal Survey of Children and Youth. The three independent but complementary data collection components allow for more detailed monitoring and reporting at the community level. Together, this information helps fill gaps in our understanding of the community factors that affect early child development and the ways a community can best support the growing needs of young children and their parents.

Putting together the information collected from these three components provides a framework for analysis that will not only tell us more about *what* is working well or less well but also will give some indication as to *why* services and neighbourhood resources work the way they do. This analysis will also provide the basis for community-wide discussions on how to both develop community strategies and allocate resources with the goal of optimizing child development outcomes.

The thirteen communities in Canada involved in this initiative are Abbotsford, British Columbia; Saskatoon, Saskatchewan; South Eastman, Manitoba; Niagara Falls, Ontario; Montreal, Quebec; Hampton, New Brunswick; Mississauga, Ontario; Coquitlam, British Columbia; Prince Albert, Saskatchewan; Winnipeg, Manitoba; Prince Edward Island (province-wide); southwestern Newfoundland; and north Toronto, Ontario. In each of the communities, there is a coalition of individuals and organizations who want to make changes that will improve outcomes for children. They are hoping to use the information as a tool to help them both reallocate and increase investment in their communities that will benefit young children.

Sources: Janus & Offord (2000), Human Resources Development Canada (2001), and Garrison (2002).

South East Grey Community Outreach

The South East Grey community is a small rural community in rural southwestern Ontario comprising five villages and 25 000 residents. Farming and tourism are the major industries, and a high proportion of the community is self-employed. Parents commute long distances to work and at-home parents are isolated. (It takes about an hour and a half to drive from one end of South East Grey to the other.)

For fifteen years, South East Grey has been the site of an innovative, community-based program for children and families. Kids 'N Us South East Grey Community Outreach (SEGCO) started in 1985 when a group

South East Grey Community Outreach (cont'd)

of residents organized to meet the need for rural child services. SEGCO evolved into a comprehensive system of support for families.

Carol Gott, SEGCO's founding executive director, reflects the philosophy: "We've learned a lot of important lessons. We've learned that if you want to develop community services in a way that truly meets your community's needs, then you have to

- start from your own base and define where you want to go as a community;
- challenge conventional views on how to deliver services;
- learn how to use all your community resources, including government policies and funding;
- figure out how to fit the whole picture together in a way that treats communities and families as a whole instead of as targeted, isolated groups" (South East Grey Community Outreach, p. 3).

In 1992, SEGCO established an "integrated hub model" of service delivery. Seven "hub" sites operated throughout South East Grey and provide licensed child care centres, drop-in and playgroup programs for parents and family caregivers and their children, resource and toy-lending libraries, outreach home visits, and information and referral services are blended together. Prior to school board amalgamations, SEGCO piloted a junior kindergarten program with the board of education that had early childhood practitioners and school teachers working together. The staff teams at each hub deliver all of the service options. Rather than separating program options and staff according to funding streams—and asking families to fit themselves into different funding categories—SEGCO distributes its funding on a geographic basis to the hubs. By blending funding, the organization has been able to be more cost-effective and more creative.

Children come to play with other children and take part in an exciting environment that is guided by early childhood practitioners. Their hours of participation may change as their parents move in and out of the workforce, change jobs, or have babies. However, the children's environment remains consistent.

Source: Adapted from *The Early Years Study* (McCain & Mustard, 1999).

SUMMARY

The definition of early child development programs includes a wide variety of programs serving children from birth through age six and from six through twelve in out-of-school programs. These programs also serve the children's families. Early child development programs include full- and part-day programs; nonprofit, commercial, and public enterprises; and centres located in schools, communities, and workplaces. Home-based child care arrangements may be early child development programs that encourage children's optimal development and early learning and seek out opportunities for children to play with other children.

The recent federal/provincial/territorial Early Child Development Agreement is creating opportunities to expand and connect existing early child development programs.

REVIEW QUESTIONS

1. Define early childhood development programs and early childhood education and care.
2. Describe what is meant by each of the following terms: full-day care; part-day care; commercial centres; nonprofit centres; family child care; special needs; early childhood intervention, Head Start.
3. Compare similarities and differences in early childhood education and care programs in the provinces and territories.

STUDY ACTIVITIES

1. Read one of the following books for insight into working with a particular age group or in a particular setting. See the references at the end of the book for bibliographic information.

   Ashton-Warner, Sylvia. *Teacher*. (preschoolers in New Zealand)

   Ayers, William. *The Good Preschool Teacher.* (includes profiles of infant and toddler caregivers, family child care providers, kindergarten teachers, and a teacher of homeless children)

   Hillman, Carol. *Teaching Four-Year-Olds: A Personal Journey.*

   Kidder, Tracy. *Among Schoolchildren.* (elementary school)

   Roemer, Joan. *Two to Four from Nine to Five.* (family child care)

   Wollman, Patti G. *Behind the Playdough Curtain: A Year in My Life as a Preschool Teacher.*
2. Talk with a variety of early childhood practitioners from particular settings. Your instructor may invite some to class. Summarize their responses to the following questions: What contributes to their job satisfaction? Job dissatisfaction? What is their specialized training? What are three or four typical events during their day with the children?
3. Spend some time in the early childhood setting of at least one of the following: infant care, toddler care, a preschool setting, a kindergarten classroom, a classroom of one of the early grades, a family child care home, or a Head Start program. Listen to and discuss the reports of others about their visits.
4. Check your phone book to see what child care programs and preschools are located in your community. What types of program do you suppose they are, judging from their advertisements?

KEY TERMS

adult-child ratios: The number of children of a particular age who may legally (by regulation) or optimally (by accreditation standards) be cared for by one adult.

at risk: Used when concerns exist about developmental delay owing to negative environmental or physical conditions.

Canadian Child Care Federation (CCCF): Largest Canadian professional ECEC organization; established in 1987. Current membership is over 10 000 with affiliated provincial/territorial member organizations. Mission is to work within communities to improve the quality of child care for all Canadian children.

child care centre: Programs that provide nonparental care and early education to groups of children in a setting outside the children's home.

commercial: Also called proprietary and for-profit. Early childhood programs established to earn profit for their owners.
compensatory programs: Preschool programs designed to ameliorate the impact of social or economic disadvantage through an enriched environment.
early child development programs: All programs designed for young children, their families and communities to promote children's healthy development and early learning and to provide nonparental care.
early childhood education and care (ECEC): Early child development programs that are organized to provide care and learning opportunities for children from infancy through middle childhood.
Early Childhood Education (ECE): Postsecondary education programs to prepare early childhood practitioners.
early childhood practitioner: Individual with a combination of ECE credentials and expertise who works with young children and their families in early child development programs.
early intervention (EI): Programs that work with children whose development is delayed or at risk, often offering comprehensive services.
family child care: Arrangements for child care within small groups (often less than six children), usually in the care provider's home.
family literacy: Literacy programs for young children and their parents and caregivers.
family resource program: Family-focused programs that offer activities and services to young children and their families and caregivers.
full-day programs: Programs that provide care for children throughout the day, including meal and sleep arrangements.
Head Start: Established in 1965, an educational program for preschoolers in families below the poverty level. Comprehensive services include education; family support through social services and parent education; and medical, dental, and nutritional services for children.
home-visiting: Brings services and expertise to children's homes. May be directed to children, adults, or both.
inclusive early childhood programs: Early child development programs that facilitiate children's full participation, regardless of abilities/disabilities.
in-home child care: Nonparental child care provided in the child's own home.
infants: Children from birth through the first year to eighteen months of life.
junior kindergarten: Half-day programs for four-year-old children operated in Ontario's school system and in some Quebec at-risk communities.
kindergartens: Programs for five year olds, now generally operated in every school system, and originated by Friedrich Froebel.
nonprofit: Early childhood programs subsidized by government or agency funds, in which any surplus funds are used for program improvement.
part-day programs: Programs for children that operate for only a half-day, or for only several days a week.
preschoolers: Generally children aged three to five.
school-age children: Children in grades one through six.
senior kindergarten: Half- or full-day programs for five-year-old children operated in most Canadian school systems (except Prince Edward Island).

social services: The branch of professional services that helps provide for basic needs for families and individuals.

special needs: When development and learning do not follow typical patterns, individuals are said to have special needs requiring intervention through modifications in the environment or teaching techniques to help them develop optimally.

toddlers: Children from the time they become independently mobile (about age one) through to their third year (age three).

SUGGESTED READINGS

Assembly of First Nations. (1995). *National overview of First Nations child care in Canada.* Ottawa: Assembly of First Nations.

Beach, J. & Bertrand, J. (2000). *More than the sum of the parts: An early child development system for Canada.* Occasional Paper No. 12. Toronto: Childcare Resource and Research Unit, University of Toronto.

Canadian Council on Social Development. (1997). *The progress of Canada's children 1997.* Ottawa: CCSD.

Cleveland, G., & Krashinsky, M. (1998). *The benefits and costs of good child care: The economic rationale for public investment in young children.* Toronto: Childcare Resource & Research Unit.

Health Canada, Human Resources Development Canada, and Indian & Northern Affairs Canada (2001). *Federal/Provincial/Territorial Early childhood development agreement: Report on government of Canada activities and expenditures 2000–2001.* Ottawa: Minister of Public Works and Government Services Canada.

Kyle, I., & Kellerman, M. (1998). *Case studies of Canadian family resource programs: Supporting families, children and communities.* Ottawa: Canadian Association of Family Resource Programs.

Ontario Coalition for Better Child Care. (2000). *A guide to child care in Ontario.* Toronto: OCBCC.

CHAPTER TWO
Early Child Development Curriculum

OBJECTIVES

After studying this chapter, students will be able to

- identify different curriculum approaches used in early child development programs;
- discuss common concerns regarding early academics, readiness, and achievement testing;
- identify what is meant by inclusion and discuss the benefits for all children and adults involved;
- discuss the concepts that lie behind antibias curriculum;
- describe some of the benefits of mixed-age groupings in early education settings; and
- discuss some ideas regarding the use of technology in early child development programs.

Looking back at our definition of early child development, we can see that it also takes into account how programs are organized for children. A program's organization, or curriculum, in early child development includes all that the child experiences. This chapter considers different curriculum approaches and how you can begin to identify what approaches best suit your understanding about how children develop and learn.

This chapter also introduces some of the current issues facing early childhood practitioners in their day-to-day implementation of curriculum in family child care settings, child care centres, nursery schools, family support programs, early childhood intervention initiatives, and kindergartens.

As we discussed in Chapter 1, early child development describes programs and services for children within the early and middle period of childhood (birth to twelve years). The term can also refer to a **philosophy** of curriculum that is developmental rather than custodial (Doxey, 1990). A developmental curriculum is "based on knowledge about child development and is geared to the optimal development of the whole child, in contrast to a curriculum that concentrates on intellectual development for academic achievement" (Doxey, 1990, p. x).

Curriculum, in early child development, includes both the care and the learning that occur when the child participates in an early child development setting. It is everything that is part of a child's day or hours spent in a home- or centre-based program. Curriculum includes daily schedules and routines, the physical environment, play materials, learning experiences, and, most importantly, the people who are part of the early child development setting.

This chapter will examine a few specific early child development **curriculum** models to give you an idea of the many different ways that early child development programs can provide care and education to young children and their families.

Curriculum Models

Just as there are many forms or models of early child development services, there are a number of curriculum approaches that can be adapted to different types of settings for children of varying ages.

All begin with a philosophy that recognizes optimal child and family development as the most important goal. You will discover that curriculum philosophies will make different assumptions about how children develop and learn and about what is worth knowing. Curriculum philosophies are usually based on educational or psychological theories, a number of which you will consider in Chapter 7 and, in more depth, in your other early child development studies. Each one has its own ideas or images of the child and childhood.

Cultural values, beliefs, and practices shape our understanding of what is worth knowing. Early child development programs are social settings that guide children in learning about the world around them. Early childhood curriculum models reflect differences in culture as much as they are influenced by educational and psychological theories.

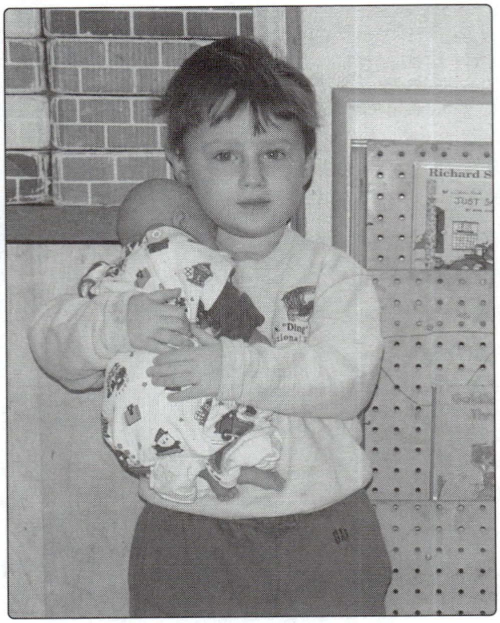

Developmentally appropriate practices influence all domains of the child's development, rather than focusing solely on the cognitive (courtesy of Play and Learn).

Developmentally Appropriate Practice

Developmentally appropriate practice (DAP) is a curriculum approach based on knowledge of the development of children's abilities and observation of individual differences, including abilities, interests, and culture. It can be applied in settings with children from infancy through middle childhood. The approach considers that all areas of development—physical, social, emotional, and cognitive—are important and that it is the early childhood practitioner's responsibility to plan an environment and support experiences to enhance all areas of development.

Curricula designed on developmentally appropriate practice contain a number of common characteristics. Typically, early childhood practitioners in a developmentally appropriate curriculum

- support the whole child, that is, all areas of the child's development;
- use observations of children's individual interests and developmental progress to plan the curriculum;
- promote children's active exploration and interactions with others;
- use learning materials and activities that are concrete and part of the lives of young children;
- provide for a wide range of developmental interests that meet the children's individual needs and skills;

- offer increased challenges as children's abilities and skills progress; and
- provide materials and allow time so children can choose activities (Bredekamp, 1987).

The **National Association of the Education for Young Children (NAEYC)** has defined specific standards of developmentally appropriate practice (Bredekamp, 1987). Guidelines are presented that describe how principles of developmental appropriateness can be applied in early childhood settings and set out the framework for developmentally appropriate curricula. You will learn more about these guidelines in Chapter 3, which considers voluntary standards for quality early childhood settings.

Montessori Schools

Montessori schools, following the philosophical approach of Maria Montessori (see Chapter 7), have proliferated in North America since the early 1960s. Montessori schools use unique methods and materials, and hire specially trained teachers. Sequences of prescribed tasks, using **didactic** Montessori materials, are presented to children in a designated order. Montessori teachers, or directresses, have received specialized training in an institution accredited for Montessori teacher training, which emphasizes observing children and presenting in **sequential steps** the directions for activities for which children are ready. The Montessori philosophy includes introducing children to varieties of **practical life skills,** such as washing dishes, sweeping floors, and watering plants. The curriculum also includes sensorial components, which involve providing materials to help children broaden and refine their sensory perceptions, and conceptual components, which means using concrete academic materials to introduce children to reading, writing, mathematics, and social studies.

There are particular terms that help explain the Montessori philosophy. Some of these include **absorbent mind,** Montessori's description of the ease with which young children learn unconsciously from the environment; didactic materials, those that teach children directly by making errors in their use obvious to the child for self-correction; and **sensitive periods,** the periods of learning during which children are particularly sensitive to particular stimuli. One abiding distinction of the Montessori philosophy is the respect for children and their abilities and accomplishments.

Many Montessori schools are for children ages two through six, although in some communities, Montessori education may continue through the elementary grades and beyond. In North America today there are Montessori schools that strictly follow the original techniques, such as the Association Montessori International (AMI), and others that follow practices adapted to North American culture and current thinking, such as the American Montessori Society (AMS).

High/Scope

Another current model in early childhood education and care programs is the **High/Scope** curriculum. This curriculum has developed under the leadership of David Weikart, who was a founder of the Perry Project, one of the best-known

2000
1950
1900
1850
1800
1750
1700
1650

1848
Interest in kindergartens is growing in Germany. Forty-four kindergartens are opened this year.

1850
Roman Catholic nuns open Canada's first crèche in Montreal.

intervention programs of the 1960s (see Timeline, p. 167). The High/Scope curriculum is based on Jean Piaget's constructivist theories of child development. "The curriculum rests on the fundamental premise that children are active learners, who learn best from activities that they themselves plan, carry out and reflect on" (Epstein, 1993, p. 30). As in most good early childhood programs, children learn in a variety of learning centres with plenty of appropriate materials. Teachers help children actively plan what they will do each day, after which the children carry out their plan during work time for self-selected activities and then review what they have done (**"plan, do, review"**). Early childhood practitioners join in children's play, asking questions to extend their thinking skills. The High/Scope curriculum identifies five ingredients of active learning for young children:

- materials for the child to explore;
- manipulation of materials by the child;
- choices by the child about what to do with the materials;
- language from the child; and
- support from the adult (Epstein, 1993).

THEORY INTO PRACTICE

High/Scope Preschool Key Experiences

Creative Representation

- Recognizing objects by sight, sound, touch, taste, and smell
- Imitating actions and sounds
- Relating to models, pictures, and photographs to real places and things
- Pretending and role-playing
- Making models out of clay, blocks, and other materials
- Drawing and painting

Language and Literacy

- Talking with others about personally meaningful experiences
- Describing objects, events, and relations
- Having fun with language: listening to stories and poems, making up stories and rhymes
- Writing in various ways: drawing, scribbling, letter-like forms, invented spelling, conventional forms
- Reading in various ways: reading story books, signs, symbols, one's own writing
- Dictating stories

Initiative and Social Relations

- Making and expressing choices, plans, and decisions
- Solving problems encountered in play
- Taking care of one's own needs

High/Scope Preschool Key Experiences (cont'd)

- Expressing feelings in words
- Participating in group routines
- Being sensitive to the feelings, interests, and needs of others
- Building relationships with children and adults
- Creating and experiencing collaborative play
- Dealing with social conflict

Movement

- Moving in nonlocomotor ways (anchored movement: bending, twisting, rocking, swinging one's arms)
- Moving in locomotor ways (nonanchored movement: running, jumping, hopping, skipping, marching, climbing)
- Moving with objects
- Expressing creativity in movement
- Describing movement
- Acting upon movement directions
- Feeling and expressing steady beat
- Moving in sequences to common beat

Music

- Moving to music
- Exploring and identifying sounds
- Exploring one's singing voice
- Developing melody
- Singing songs
- Playing simple musical instruments

Classification

- Exploring and describing similarities, differences, and the attributes of things
- Distinguishing and describing shapes
- Sorting and matching
- Using and describing something in several ways
- Holding more than one attribute in mind at a time
- Distinguishing between "some" and "all"
- Describing which characteristics something does not possess or which class it does not belong to

(cont'd)

High/Scope Preschool Key Experiences (cont'd)

Seriation

- Comparing attributes (longer/shorter, bigger/smaller)
- Arranging several things one after another in a series or pattern and describing the relationships (big/bigger/biggest, red/blue/red/blue)
- Fitting one ordered set of objects to another through trial and error (small cup/small saucer, medium cup/medium saucer, big cup/big saucer)

Number

- Comparing the numbers of things in two sets to determine "more," "fewer," "same amount"
- Arranging two sets of objects in one-to-one correspondence
- Counting objects
- Space—filling and emptying
- Fitting things together and taking them apart
- Changing the shape and arrangement of objects (wrapping, twisting, stretching, stacking, enclosing)
- Observing people, things, and places from different spatial viewpoints
- Experiencing and describing positions, directions, and distances in the play space, building, and neighbourhood
- Interpreting spatial relations in drawings, pictures, and photographs

Time

- Starting and stopping an action on signal
- Experiencing and describing rates of movement
- Experiencing and comparing time intervals
- Anticipating, remembering, and describing sequences of events

Source: Reprinted with permission of High/Scope Education Foundation.

Early childhood practitioners may study the High/Scope philosophy in short workshops and training institutes. An important part of the philosophy involves using small groups to help children focus on the following **key experiences**, a set of eight concepts based on Piaget's ideas of preschoolers' cognitive characteristics and learning potential:

- active learning initiated by children and involving sensory and manipulative materials
- use of language
- opportunities to represent their own ideas and experiences

- opportunities for classification
- use of seriation
- use of number concepts
- experience with spatial relationships
- opportunities to understand time concepts

Reggio Emilia

After World War II, preschools opened in the northern Italian town of Reggio Emilia. Today the municipal government supports thirty-five **Reggio Emilia** schools, serving children aged three to six and infants up to age three. About 47 percent of the town's preschool population and 33 percent of its infants attend Reggio Emilia schools (Gestwicki, 1995). This city-run early child development program has captured the attention and imagination of early childhood practitioners in Canada and around the world for several reasons. First, it enjoys a high degree of community support and is viewed as an essential part of a cohesive, healthy, and productive community. Second, the schools' physical beauty and attention to detail are evident to all who visit the program. Third, the program philosophy and curriculum builds on Lev Vygotsky's concept of the social construction of knowledge and skillfully integrates other theoretical concepts, including Jean Piaget's theory of cognitive development, John Dewey's concept of progressive education, Hugh Gardner's theory of multiple intelligences, and Urie Bronfenbrenner's ecological environment theory (Berk & Winsler, 1995).

Loris Malaguzzi was the founder and leader of Reggio Emilia's approach and programs. He based his system of early childhood education on a few key principles:

- child-centred programs, which respect children and emphasize the reciprocal adult-child relationship;
- an "environment as teacher" approach, which organizes space to promote relationships, creates aesthetically pleasing surroundings, promotes choices and activities, stimulates all areas of learning, and reflects children's ideas, values, and culture;
- a curriculum centred on children's interests, reflected in projects that are undertaken in considerable depth and detail;
- a spirit of collaboration between early childhood practitioners and young children in facilitating intellectual discovery through social process; and
- the participation of families as an integral part of the educational experience (Gestwicki, 1995).

Emergent curriculum follows and extends children's own interests (courtesy of Yes I Can! Nursery School).

Many early childhood practitioners are turning to an **emergent curriculum** model that allows them to incorporate some of the principles found in Reggio Emilia programs. However, Reggio Emilia is not a curriculum model that can be transported from its roots in northern Italy to Canadian settings. It is a way of thinking and interpreting the immediate surroundings (natural environment and social community) and following the lead of children and their families to create a unique early childhood setting. Emergent curriculum is an approach that encourages early childhood practitioners to really respond to their immediate surroundings—physical place and people—and guide children's natural curiosity about their environment to encourage learning.

Emergent curriculum first appeared in the 1970s, and Elizabeth Jones coined the term in the introduction to a curriculum book (Jones & Nimmo, 1994). Many North American early childhood practitioners are finding it a useful framework that integrates knowledge about child development and an approach to planning that allows the child's interests to lead (Goulet, 2001).

THEORY INTO PRACTICE

Processes of Emergent Curriculum

The Early Childhood Education diploma program and the eight lab centres at George Brown College have adopted an emergent curriculum model. Marie Goulet, ECE faculty member, prepared the following handout for ECE students, which describes how early childhood practitioners can support children's learning and respect their interests and explorations.

Provision

Provide materials and space for children to explore and play. Consider the needs of individuals and group development. Environments should be rich enough for each child to find her interests with opportunities to learn.

Sustain

Learning requires repeated practice to ensure integration and mastery. Provide materials and support to maintain practice at the same level of difficulty (horizontal curriculum). Strategies include: open-ended questions, broadcasting, joint-attention, imitation, parallel play, adding new materials that offer the same practice.

Enrich

Learning involves achievable challenges to extend and elaborate children's skills (vertical curriculum). Extensions involve problem-solving, increased complexity, new combinations, increasing logical thinking. Strategies include: adding materials with open-ended questions—(what can you do with this?), asking predictive questions (what would happen if ...), asking problem-solving questions (how can you ...), asking relationship questions (which one rolled further? Which ones belong together?)

Documentation and Representation

Document children's explorations and thinking by collecting and displaying the materials that track the growth of children's play. Represent what happened with children's work, photographs, videotapes, audiotapes. Documentation includes descriptions of children's actions and language that are connected to the children's purpose, development, person meaning and identity.

Source: Goulet, Marie (2001). Reprinted with permission.

Problem-Based Play

Play is the activity of early childhood and early child development programs. An understanding of play and its role as an essential part of early child development programs and curricula will be a central component of your studies in Early Childhood Education. Play is difficult to define because it is viewed somewhat differently in different cultural contexts and different early child development curricula contexts.

The Ontario Early Years Study recommended environments for young children that "offer children an array of opportunities to explore, discover and create ... Play-based, problem-solving with other children and an adult is an early learning strategy" (McCain & Mustard, pp. 159–60). Problem-based play is activity that captures and engages children in their quest to understand the world.

A bucket of blocks is an endless opportunity to arrange and rearrange (courtesy of Eglinton Public School).

Curriculum Issues

The early child development curriculum models described in the previous section are examples of the kinds of approaches you are likely to find in early childhood education and care programs in your communities. You can see that there are both similarities and differences among the models. Most early childhood practitioners tend to find themselves comfortable in one approach more than another. Many practices—what kinds of materials to make available to children, observing children's interactions with each other, or displaying representations that children have created—are common in a number of curriculum approaches.

THEORY INTO PRACTICE

Examples of Problem-Based Play

- A bucket of blocks offers endless opportunities to arrange and rearrange, much like a scientist or mathematician arranging and rearranging ideas to find a solution to a problem and build the capacity for flexible thinking.
- Storytelling and picture books draw children into fantasy and narratives and motivate reading and writing skills.
- Guided play with board games using number lines, dots on a die and small markers (to place on the number line) builds the understanding of numbers that underlies more complex mathematics.
- Make-believe or pretend play with costumes and props is practice in symbolic thought and problem-solving. It extends children's creativity and encourages innovation and adaptability.
- Art-making with paints or crayons and paper mirrors the child's developmental progress.

The same curriculum issues often challenge early childhood practitioners working in different settings and with different curriculum approaches. What is the best way to organize children's groups—same age or mixed-age groups? Should early childhood education and care programs try to ensure that children are ready to adapt to school settings, or should the schools adapt to children? How can early childhood programs support children who have developmental challenges or identified special needs? How can early childhood settings respect and support all young children (boys and girls) and their families from diverse cultural, linguistic, and religious backgrounds? Does electronic technology have a place in today's early child development programs?

1856
Margarethe Schurz, a student of Froebel's, begins a small German-language kindergarten for six children in her home in Watertown, Wisconsin.

Mixed-Age Groupings

Children in early child development programs are not always organized into groups of the same age and stage of development. In home-based programs it is more common to find children from different ages together in the same setting—similar to a family with siblings of various ages.

Those of you who have listened to grandparents and great-grandparents reminisce about attending one-room schoolhouses will understand that **mixed-age groupings** in education are not a new phenomenon. But the concept is receiving renewed attention in early education programs and has been used successfully not only in Canada but also in England, Sweden, and Italy. Rethinking the concepts of mixed-age groupings may assist current efforts in Canada to meet standards of individualization in developmentally appropriate practice.

A mixed-age grouping combines children across at least two, preferably three, chronological years. Usually, the grouping remains together for much of the time, with the oldest children moving on each year and new children joining as the youngest in the group. This means that children are in a group for at least two and sometimes three years and remain with the same early childhood practitioner over this time. In regulated group centres, multi-age groupings "refers to the placement of children of different ages together in activity and learning areas for substantial portions of the daily schedule" (Bernhard, Pollard, Eggers-Pierola & Morin, 2000, p. 80). Such an arrangement is more like a typical family or neighbourhood setting than like a classroom that educates children in "litters" of only age-mates.

Such a system has some clear social advantages. One benefit is that, over extended time, children's relationships with one another, and relationships between children, teachers, and parents can develop and provide security and deeper knowledge. Another advantage is that older children have opportunities to exhibit leadership skills with younger children and to assume responsibility. Older children in mixed-age groupings seem to increase their own regulation of their behaviour, perhaps taking seriously their roles as models for the younger children. Their prosocial, caring skills also increase. Younger children benefit by participating in the more complex forms of play developed by the older children and by imitating their behaviours. Cooperative behaviours increase for all children (Katz, Evangelou, & Hartman, 1991).

There are cognitive benefits for children in these groupings, as well. Rigid curricula with age-graded expectations must necessarily be relaxed in mixed-age settings. Children are allowed to develop and learn at their own rates without fear of failure and with less competition. Children's unique needs are more easily identified when teachers are not considering group goals; the curriculum is more likely to be matched to children's needs and skill-learning levels. Children whose knowledge is similar but different stimulate one another's mental growth and thinking. Therefore, **cooperative learning** and peer tutoring situations generally abound in mixed-age settings.

Early childhood practitioners are key in determining how effectively mixed-age groupings actually work to benefit children. Simply putting children of different ages together does not guarantee these benefits; early childhood practitioners need to carefully structure the environment, plan the curriculum, and play particular roles. As teachers learn about children's interests, abilities, learning styles, and choices for play partners, they can support individual growth. They plan activities that are child-initiated, so that children can find their own place on the developmental ladder. By freeing themselves from the role of director, teachers can step in to facilitate as children need their interaction. They may consciously create opportunities for children to work together, make suggestions about how children can help one another, or frame thought-provoking questions that suggest ways of working with others who are younger or older. Children, with adult guidance, can learn much about democratic practices by jointly solving the problems that naturally arise in a multi-age grouping, such as what to do when the younger ones don't want to listen to the longer books that the older ones love to have read chapter by chapter.

Early childhood practitioners often resist the move to mixed-age groupings at first, fearing that it will be too difficult to plan for great differences and that individual age-stage behaviours won't mix. However, early childhood practitioners usually change their opinions after some experience. They discover that they are teaching more responsively and appropriately. They benefit by learning about individual differences in new ways and by questioning their stereotypes and prejudices about behaviours that they may have assumed stemmed from age levels. They discover that the teaching and learning in their classrooms are shared among all participants. Perhaps this innovation will be in your professional future.

Adult and Child Focus

Early child development curriculum is often oriented to parents or other caregivers and to children. In some instances, parents or other caregivers attend the program with their children some or all of the time. In other early childhood settings that provide nonparental care, parents or caregivers do not participate directly in program activities with their children (except during drop-off and pick-up times or at special events). In either situation, the curriculum can be directed to both parents or other caregivers and children.

A two-pronged approach that is both parent- and child-oriented recognizes the power of early learning within the family and seeks to support parents' and other caregivers' participation in the children's early development.

2000 · 1950 · 1900 · 1850 · 1800 · 1750 · 1700 · 1650

1857
Canada's first English crèche, the Public Nursery, opens in Toronto. It later becomes an orphanage.

1858
Salle d'asile Saint Joseph provides care for young children in Montreal.

THEORY INTO PRACTICE

A Place to Grow

Sylvia was pregnant when she emigrated from China to Canada with her two year old daughter. Her husband was to follow later. One of the people who lived in the same apartment complex told Sylvia about the Parenting and Family Literacy Centre, run by the Toronto District School Board, located in the local school. [Parenting and Family Literacy Centres offer activities to support children's early learning and provide parenting support, education, and courses. They have both a child and adult curriculum focus.]

They found the center to be a warm and welcoming place, with parents and other caregivers, like grandparents, and young children engaged in activities that help young minds develop through play. There was music and storytelling and games and snacks. At the Centre, Sylvia's little girl found other children who spoke Cantonese, and Sylvia found toys and books in their language to take home. Both of them started to learn English. The center showed Sylvia how to make toys from ordinary household objects. She learned the importance of reading to her toddler. They made friends.

Source: McCain & Mustard. (1999), p. 105. Reprinted with permission.

Readiness for School Learning

Children's coping and competence when they enter formal schooling is related to their academic success—how well they do in school and whether they complete high school and pursue other training and education. There is growing interest in ensuring that all children enter school "ready to learn" as a way of improving academic achievement and social well-being. There is also concern that Canadians' literacy abilities and mathematical achievement fall behind many other countries (Willms, 2000). If the early years set the foundation for later academic achievement (as well as social competence), it makes sense to consider how well prepared children are for school environments. This does not mean that young children should learn to read earlier or that prereading, prewriting, or number skills are the only criteria for school **readiness**.

Gillian Doherty, an early childhood research expert, points to five components of readiness for school learning, based on research findings:

- physical well-being and appropriate motor development;
- emotional health and a positive approach to new experiences;
- age-appropriate social knowledge and competence;
- age-appropriate language skills; and
- age-appropriate general knowledge and cognitive skills (1997, p. 18).

The desire to improve children's readiness for school learning sometimes leads to an emphasis on **early academics** or the direct instruction of reading and number skills during the early years. Since the modern era of early childhood education began in the 1960s, one issue has generated a good deal of debate: how much academic content and method is appropriate for children in their early years? Although many professional early childhood practitioners understand and support developmentally appropriate practices for young children, some are less grounded in developmental knowledge. They are more easily influenced by administrators, parents, and policymakers who push for academic experiences for young children, which look a good deal like those presented to older children.

A number of reasons underlie this concern for accelerating young children's acquisition of academic skills.

- Many Canadians believe that the public education system is failing many young children, particularly those from socially and economically disadvantaged backgrounds, who were beginning school unprepared for the rigours of academic learning. Among the many programs that proliferated during the years following this initial concern were some that stressed academics in a highly structured, adult-controlled system.
- The school system operates kindergarten programs for five-year-old children in Canada. In many cases, other early-learning programs for preschool children are established within the school system. School systems are used to dealing with older children's development and learning and, thus, often based these new programs for the youngest ones on a dribbled-down version of what the older ones were doing, rather than recognizing that what is good for older children may be quite different for younger children. Based largely on the work of behavioural learning theorists, these programs for preschools and kindergartens often emphasize helping children learn the specific skills that they would need for later learning. The methods that early childhood practitioners employed in these academically oriented programs look suspiciously like those used in classrooms for much older children, where teachers tended to take on the primary role of instructing, drilling, assigning, directing, and testing for retained knowledge.
- Essentially, proponents of early academics have moved the academic curriculum to begin in earlier grades. What was once taught in Grade One is now frequently expected to be mastered in kindergarten, which means that the preschool years must be spent learning the subskills for reading and math that would bring success at that later stage. In many cases, administrators of school programs were pressured by policymakers to produce evidence of learning, which is more easily measured and quantified in the kinds of academic knowledge that can be drilled and taught by rote. Therefore, educational systems moved to earlier academics to attempt to answer some of the system's own questions about how best to achieve learning results and how to define its role in serving very young children.
- Today many families, unfamiliar with knowledge of early child development programs, are most comfortable in finding a program that resembles their own educational experiences. "What are you teaching them?" they ask. "Do they just play here?" As consumers who want to be assured that they are receiving their full money's worth, they, too, want to see measurable results, which, unfortunately for developmentally appropriate practice, may be more concrete and obvious when a preschooler has a worksheet to take home than when he has spent a productive morning building in the block area. "This assumption has enabled parents to accept the commonly held view that today's children are ready for learning earlier than in past generations" (Gestwicki, 1995, p. 21).

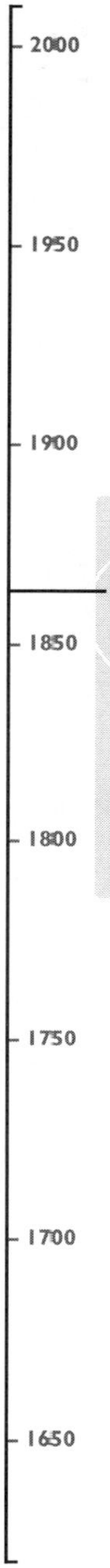

1860
There are 15 child care centres (crèches and salles d'asile) in Canada providing child care on a daily basis.

Making marks—with crayons, pencils and paints—is a step along the way to learning to print (courtesy of Eglinton Public School).

- Parents who are concerned about giving their children the "competitive edge" believe that getting children started off on the right track when they are very young should logically enable their later acceptance into the right school or program. This leaves parents vulnerable to accepting the influence of both professionals and entrepreneurs who promote ideas for early teaching.

Where is the real danger for young children in being caught between the opposite poles of this debate on developmentally appropriate practices and formally structured academic experiences? By ignoring the active nature of the young child's learning style, children's healthy development may be jeopardized.

Infants and young children do not just sit twiddling their thumbs, waiting for adults to teach them to read and do math. They expend a vast amount of time and effort in exploring and understanding their immediate world. Healthy education supports and encourages this spontaneous learning. Early instruction miseducates, not because it attempts to teach, but because it attempts to teach the wrong things at the wrong time. When we ignore what the child has to learn and instead impose what we want to teach, we put infants and young children at risk for no purpose (Elkind, 1988, p. 25).

What sorts of risks do many experts see in prematurely exposing children to academic learning? There are several.

- Children's self-esteem will suffer if they are unable to succeed at the tasks that seem so important to adults. Many children, especially young children, just cannot learn in the ways that academic instruction requires, no matter how hard they try. They cannot sit still, comprehend, or use the material they have learned by rote, or even see its importance. They feel incompetent, as they realize that they have failed to meet adults' standards, without understanding that what they have failed at is alien to their nature and learning style. When these kinds of learning are not meaningful to children, they feel no sense of self-worth, even when they do master some learning.
- Another component of healthy development that is thought to be at risk in academic preschool programs is self-control (Greenberg, 1990). Learning to make good choices is an important part of self-discipline. In a teacher-directed and teacher-managed environment, young children have few opportunities to make choices. The longitudinal studies of the preschool programs in Ypsilanti, Michigan, suggest more positive social and emotional adjustment of adolescents and adults who participated as children in preschool programs that allowed them to choose and initiate their activities (Schweinhart, Weikart, & Larner, 1986).
- Formal instruction often puts excessive and inappropriate demands on young children, which can result in an overload of stress. A number of studies confirm that there is increased stress on children in developmentally inappropriate learning situations (Burts, Charlesworth, & Fleege, 1992).
- Lastly, it seems likely that the long-term effects of rushing children prematurely into formal academics may be less than positive. Frequently, children's attitudes

and dispositions to learning are negatively affected by the stress and circumstances of those first learning experiences. Evidence also suggests that children whose introduction to academic content and methods has been delayed past preschool and kindergarten have had at least equally positive results in later learning, rather than being handicapped when compared with peers whose academics began earlier (Burts et al., 1992). This seems to support the notion that waiting until children are developmentally ready for later kinds of school learning is beneficial.

Early childhood professionals are not unanimous in this position. There are those who believe that early direct instruction enhances children's later academic success, and that the earlier this direct instruction is started, the better (Gersten & Keating, 1987, in Jensen & Chevalier, 1990). It will be important for you to immerse yourself in understanding the research and statements of both viewpoints. Only by forming a personal philosophy about appropriate practices for children will early childhood practitioners be able to adequately defend their position to others.

You will need to be prepared to offer convincing and solid rationales for structuring environments and learning experiences in which children initiate their own active learning through play. These environments and experiences are supported by facilitators who recognize that teaching based on active learning provides opportunities to develop *all* domains of the child's growth, and is more likely to support readiness for all learning and success in school.

Also early childhood practitioners must recognize that literacy and numeracy are important skill areas. Individual Canadians need these skills to be successful in almost any workplace and to participate fully in a democratic society. These are basic skills to survival in our society as we enter the twenty-first century. Therefore, early childhood practitioners can understand that families are often anxious that their children acquire the skills necessary for learning to read and compute numbers when they enter formal schooling. We should be concerned too. However, direct instruction in isolated skills such as letter recognition or matching things that are the same and different will not be nearly as effective as other curriculum strategies.

Early childhood practitioners can provide environments that do prepare children for learning to read and understand mathematics. Early child development environments can be structured in a number of ways that support emergent literacy and emergent numeracy during the early years and then early literacy and numeracy abilities as they develop.

Emergent literacy is the development of children's ability to use print forms of language and includes making marks, scribbling, reciting stories from memory, and printing letters. Writing and reading become meaningful, and children understand their purpose even if they cannot read and write themselves. Literacy develops relationship to oral language, but is a separate ability. Literacy is the symbolic representation of oral language.

We can support emergent literacy by

- understanding that literacy develops in close relationship to oral language;
- speaking with children about things that interest them;
- engaging in reading out loud and storytelling;
- developing a sense of narrative;

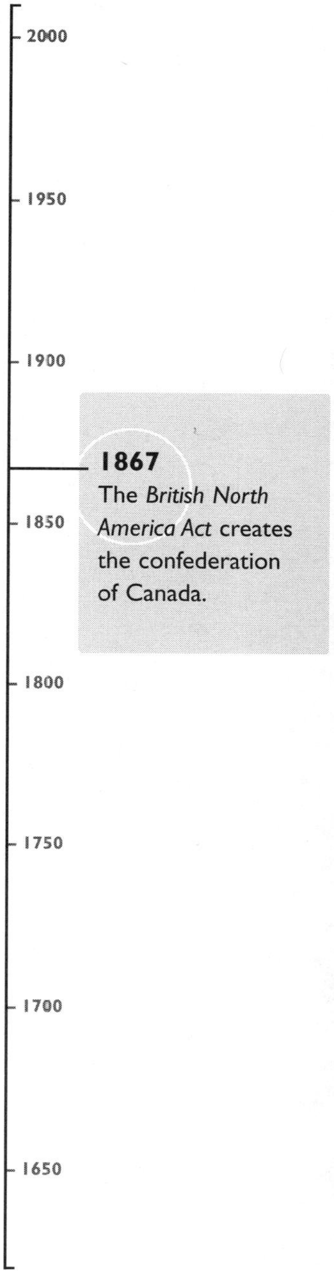

1871
In Ontario, the *School Act* makes schooling compulsory for children aged 7 to 12 for four months of the year. Schools in some boards were open six days a week and eleven months of the year.

- valuing and modeling adult literacy use;
- enriching our environment with books and print;
- encouraging play reading and writing; and
- offering activities that develop visual discrimination, fine motor control, and phonological awareness.

Emergent numeracy is the understanding of numbers and what they represent. It involves both an understanding of quantity (big and little, two comes after three) and quality (more and less, a little bit, and a lot). Understanding of quantity (and numbers) begins very early as children explore objects and their properties.

Monitoring Readiness for School Learning

The concept of readiness for school and learning creates the need to assess how well children are doing when they enter school. If the focus is on moving academic learning earlier into early childhood, the assessment is accomplished by testing young children. If being ready for school means being ready for the academic demands of the primary classroom, children are expected to enter school already knowing what formerly was not taught until Grade One. Now, as schools move to measure the quality of their instruction, it is increasingly important to them that children have achieved particular standards of learning at specific points.

Since **standardized tests** tend to focus on narrowly cognitive tasks, young children's overall development and strengths are not given equal importance in making decisions about readiness. And since young children do not have the requisite skills for successfully taking tests—sitting still, following directions, writing, or making specific marks—the test results are often inadequate measures of what they truly know and who they are. In setting out particular expectations for what children should already know before they enter or progress in school, the concept of readiness really becomes a method of gatekeeping, excluding those children who cannot be expected to adjust to the demands of school (Willer & Bredekamp, 1990).

If the focus on readiness takes into account physical, social, emotional, and cognitive areas of development (as outlined at the beginning of this section) we need a different kind of measurement to monitor how well children are doing during the early years. Such a measurement will also tell us how well early child development programs and communities are doing in supporting young children and their families.

The measurement of low birth weight is another universal marker that serves a similar function. It is a measure of maternal health and reflects the conditions surrounding pregnancies. It also points to problems in the physical and social environment and is a point of reference to monitor to see if changes to prenatal supports or social conditions are making a difference. One point after birth when it is possible to get a measure of the whole population of children is when they enter the school system. A readiness to learn or school readiness measure should be a tool that communities can use to assess how well they are responding to their children's needs. It also is a measure to monitor the impact of initiatives designed to support children during the early years. Readiness to learn can be used to provide special programming to children when they enter the school system, but is not intended for labelling children.

There is considerable interest across Canada to implement such a tool. The Canadian Centre for Studies of Children at Risk in Hamilton, Ontario, has developed and tested a tool that is a broad measure of children's development or readiness to learn at school entry.

Called the Early Development Instrument (EDI) (Janus & Offord, 2000), it is based on the concept of readiness to learn that includes all areas of development. The EDI measures children's ability to be cooperative and attend to the teacher, to benefit from educational activities, and get along with other children. The instrument consists of five domains (or scales): physical health and well-being, social knowledge and competence, emotional maturity, language and cognitive development, and general knowledge and communication skills. Kindergarten teachers complete the EDI for each child, but the results are reported at the population level for individual communities and schools. This means that all of the scores for the children in a kindergarten class are grouped together to give a picture on the overall readiness for school of that group of children living in that community.

Our recent understanding of how the brain develops from conception through the first five years of life emphasizes the importance of a measure of that development during this critical period. The key evidence from studies in the neuroscience and developmental psychology show that high-quality nurturance during this period leads to competence and coping skills. The ability to learn is enhanced and the risks of antisocial behaviour are reduced. The EDI is a tool that allows us to monitor how well children are doing when they enter school.

Inclusion in Early Childhood Education and Care Programs

There likely will be at least one young child with special developmental or learning needs enrolled in the first ECEC program you work in as a student or a graduate. Although the **inclusion** (also called integration, and formerly known as **mainstreaming**) of children began in the mid-1970s, only recently have we seen these programs become common in many communities.

Benefits have been identified in these programs for all the children involved, both those whose development follows typical patterns and those who require special supports and adaptations. Research shows that children with disabilities in nonsegregated settings are able to learn and practice many new skills through imitation of their peers. Children with disabilities demonstrate higher levels of social play and initiate more appropriate social interactions with peers than do children in **special education** preschool classes (Diamond, Hestenes, & O'Connor, 1994). The gains that children in integrated classrooms make in language, cognitive, and motor-skill development are comparable to those of children in special education classrooms. Therefore, they are able to take advantage of the activities that promote development while gaining an advantage in levels of play. More realistic expectations are placed on children with special needs who attend the same programs as children with typical abilities (Chandler, 1994). Children with special needs will be perceived as "less different" if they are not

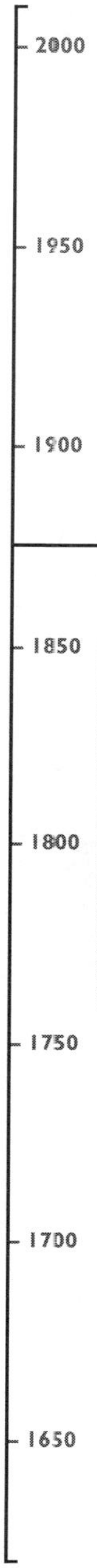

1873
First public kindergarten in the United States is established in the St. Louis school system, after a letter-writing campaign by Elizabeth Peabody. The first teacher is Susan Blow.

1878
The American Froebel Union is established by Elizabeth Peabody. It is the first professional association for those concerned with the education of young children.

excluded from environments with other children and, thus, may be more easily accepted by family, peers, and community.

Studies also show that children without disabilities benefit when children with special needs are included in preschool classrooms. In addition to making developmental gains equivalent to those that children make in noninclusive classrooms (Diamond et al., 1994), they may gain a more realistic view of people with disabilities and develop positive attitudes instead of prejudices toward others who are different from themselves. In inclusive settings, children are able to develop responsive, helpful behaviours toward others and become sensitive to the needs of others.

Parents of children with and without special needs generally support the concept of inclusion but express concerns about whether the needs of all children within the classroom will be met (Galant & Hanline, 1993). However, after experience in such classrooms, both groups of parents become far more positive about the impact on their children. For parents of children with special needs, their child's integration may help them feel less isolated within the community and more hopeful about their child's future.

What has inclusion been like for early childhood practitioners who have been trained to work with children whose development follows typical patterns? Initially, early childhood practitioners react with caution, fear, or negative responses, wondering whether the placement is wise for the child and fair to all the children. But in two studies (Giangreco and Kontos, reported in Diamond et al., 1994), early childhood practitioners noted that the children had become part of the group, without the disability being identified as their most important characteristic. Early childhood practitioners who worked in inclusive classrooms reported they had become more confident and flexible in their teaching, reflecting more on the needs of all the children in their class.

There can be barriers to successful inclusion. One is that there are philosophical and methodological differences between early childhood special education and regular early childhood education. The individualized teaching plans of early special education emphasize skill acquisition, structured use of instructional time, a strong behavioural orientation, and more teacher direction than is considered good practice in regular early childhood education. These differences create various issues of turf, with many special educators feeling that they can provide the best education for children with disabilities. However, it has been noted that "cooperation, collaboration, and mutual respect between regular and special education early childhood teachers and therapists was an important component of successful integrated programs" (Diamond et al., 1994, p. 71). When early childhood practitioners have the support of an intervention team that respects their expertise and their educational approaches, most believe that they can meet the needs of the children with disabilities in their classrooms.

It is most helpful to successful inclusion if specialists provide intervention work within naturally occurring situations in the setting, rather than disrupting the curriculum and routines. So the speech therapist comes in to present a group activity to *all* the children or to interact conversationally or in a game with small groups during free play time, rather than coming in to remove the young child with special language needs for an individual therapy session. So, too, the physical therapist encourages all the children to try rolling on her giant ball, realizing that the participation of children without physical limitations may encourage the child with cerebral palsy to take part. Besides preventing disruption of the program's routine, including all children avoids sending the message to other children that the child is not really a member of their group. The early childhood practitioner is more likely to feel supported when the members of the special intervention team share the common framework and goals.

Effective inclusion programs use naturalistic teaching strategies within the context of daily routines. "Activity-based intervention" strategies involve the kinds of teaching we discussed earlier in our considerations of developmentally appropriate practice. When using these strategies, the early childhood practitioner plans appropriate active learning experiences by taking into account goals for individual children. Thus, a child who is using a variety of fasteners on the dress-up clothes is working on a fine-motor goal, and one who is invited to use the puppets is working on a language goal. Because each child's goals are addressed within the context of daily activities, early childhood practitioners have the responsibility of targeting goals, planning materials and activities to further those goals, and regularly assessing progress toward those goals. The planning cycle is part of good practice in every ECEC setting, no matter what the ability of each participating child. In integrated settings, early childhood practitioners, often with the assistance of other experts, plan modifications in the environment or activity to facilitate involvement of children with special needs.

As early childhood practitioners understand fully the implications of individualizing programs and curricula, they will see that the inclusion of children with special needs in early childhood settings is simply an extension of this developmentally appropriate practice of individualization. They will be supported in their efforts to help the children with special needs by the active, ongoing involvement of parents, special education teachers and interventionists, and administrators. There has been much learning about inclusion in the past ten years. As you enter the profession, your task is to continue learning how to provide education that respects the uniqueness of all children.

Antibias Curriculum

When you first work with young children in programs, you may find yourself working with children and families who look very much like you, having grown up in similar cultural and community settings, and with the same racial heritage, religious beliefs, and historical traditions. However, given the diversity of the Canadian population, it is more probable that your early child development program will include families with varying structures, including two parents of different genders, two of the same gender, single parents of either gender, and families headed by grandparents, adoptive parents, or foster parents. There may be families of varying income and educational levels. There may be homeless families, families in which children are abused or neglected, and others in which children are well cared for. Families may represent one of any number of the religious, racial, and ethnic groups that live in Canada today. You may also work with children who are learning English as a second language. And, as we have just discussed, you may also work with children whose physical or mental abilities are delayed. But differences can make people uncomfortable, particularly when those

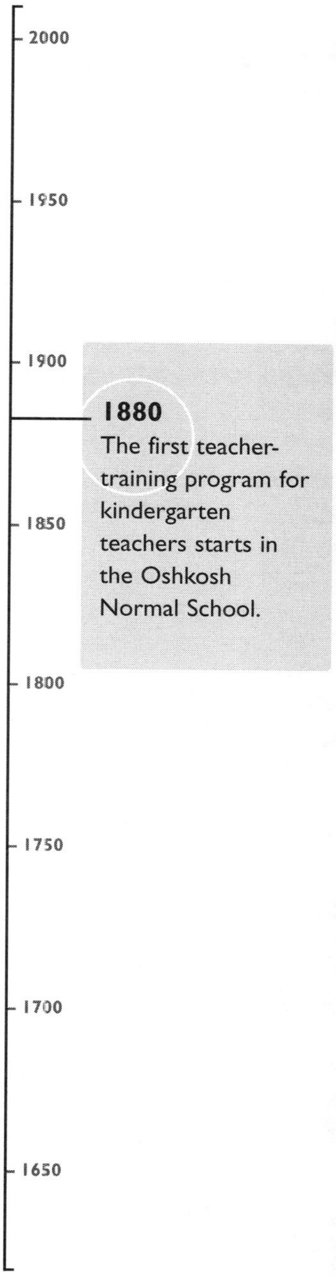

differences seem threatening, as they do when no one talks about the reason for the differences, and about the similarities that exist along with the differences. This feeling of discomfort is exactly the rationale for discussions in the early education community about the need for an **antibias** curriculum in educating young children.

Very young children are busily constructing their own self-identity, as well as attitudes about that identity, by observing the ways they are different from and similar to other people and by absorbing the others' verbal and nonverbal messages about the differences. Young children construct self-identity and attitudes through the interaction of three factors:

- experience with their bodies. Children are exploring questions about what their bodies look like and how they are different from those of others.
- experience with their social environments. Children need to know about messages they get, either explicitly or implicitly, about the facts and differences they observe. They need to know about society's evaluation of those differences.
- cognitive functioning. Young children's methods of thinking produce some confused or illogical conclusions, such as the assumption that genital identification or colour might be open to change, or disability contagious.

1882
Ontario establishes a board of health that promotes public health in the schools.

In this process of identity, children need adults to help them explore some of their questions and curiosity about differences. We know that even very young children can perceive and uncritically absorb the negative messages about diversity in the world, and that these messages can be both powerful and harmful in producing bias. Prejudice and bias are destructive forces, and they can hurt all children.

On one hand, struggling against bias that declares a person inferior because of gender, race, ethnicity, or disability sucks energy from and undercuts a child's full development. On the other hand, learning to believe they are superior because they are white, or male, or able-bodied, dehumanizes and distorts reality for growing children (Derman-Sparks, 1989, p. ix).

Children need adults who can help them develop positive attitudes about their own identity and that of others. They also need adults who can help them deal with the reality of the world, as well as show them that they can participate in changing ideas from the past to improve their future. The goals of the antibias curriculum are to enable all children "to construct a knowledgeable, confident self-identity; to develop comfortable, empathetic, and just interaction with diversity; and to develop critical thinking and the skills for standing up for oneself and others in the face of injustice" (Derman-Sparks, 1989, p. ix).

Those who see this work as an integral part of what classroom teachers do are quick to explain that the antibias curriculum is an aim and an approach that permeates the existing environment and curriculum, rather than something that is added on. It is looking at everything early childhood practitioners do and say and at the children's interaction and classroom life "through an antibias lens." The attitude of respect for all people that lies behind the antibias attitude is consistent with the sensitivity to individual children and families. Furthermore, cultural diversity and values are a declared principle of developmentally appropriate practice. But the thinking behind the antibias curriculum argues that it is important to go beyond respect and create an environment in which children can actively explore questions about disabilities, gender, and race, in order to understand differences, appreciate similarities, and recognize and confront biased ideas and behaviours.

MEET THE EARLY CHILDHOOD WORKFORCE

Zeenat Janmohamed, ECE Faculty, George Brown College, Toronto

Zeenat graduated from the Early Childhood Education diploma program at Seneca College in 1985. Since 1995, Zeenat has taught ECE students at George Brown College.

"Racism, sexism and class bias is economically, politically, educationally and institutionally produced. Teachers [college faculty] can help [ECE] students overcome these social barriers by engaging them in exploring different ways of resisting oppression. As an educator in a community college environment, I encourage students to challenge their understanding of the norm. In class discussions when I raise various forms of discrimination or prejudice, students often respond by talking about the importance of helping children "get along" and learning to accept all people regardless of difference ... I challenge their common perceptions of people living on the street, new refugees, or gays and lesbians ... In order for [ECE] students to articulate and understand systemic barriers, it is the responsibility of educators to create opportunities for transformation ....

When [ECE] students are asked how they integrate anti-bias approaches in field placement, countless numbers have discussed the Spanish music they shared, or how they made fruit salad with 'exotic' fruit. In an effort to be inclusive and to encourage a pluralist approach to education, we have lost the opportunity to challenge the dominant culture ... Clearly a more strategic process in the discourse of anti-racism education needs to develop so that students with early childhood diplomas do not limit their experience to simulated or mock celebrations of Diwali and Hannukah. Instead of focusing on festivals, there needs to be a greater emphasis on the diverse child rearing practices which go beyond the dominant Western understanding of child development."

Source: Janmohamed (2001). Unpublished paper, pp. 20–27.

1883
Concerns about preschool-age children being left unsupervised are brought to the Toronto School Board by its chief inspector, James L. Hughes.

G. Stanley Hall publishes "The Content of Children's Minds" in the *Princeton Review.*

The Toronto public school system may establish kindergartens on an optional basis.

In other reading, you will likely come across the term "**multicultural** curriculum." This refers to teaching children about other people's cultures, in the hope that they will learn to respect one another and not develop prejudice. This positive intent is obviously similar to the antibias approach, but those who practise the antibias curriculum suggest that the multicultural approach too frequently deteriorates into a "tourist curriculum" (Derman-Sparks, 1989, p. 7). This negative outcome results from presenting other cultures as exotic and foreign, with different food, holiday celebrations, clothing, and so on, rather than dealing with people's real-life experiences. "Children 'visit' non-White cultures, and then 'go home' to the daily classroom, which reflects only the dominant culture" (Derman-Sparks, 1989, p. 7). The danger here is that such experiences fail to communicate real understanding, and stereotypes about differences may actually be reinforced. In addition, the multicultural curriculum usually focuses on other countries—for example, China or Mexico—rather than on the diverse people in

1886 Chicago kindergarten college is established.

our own country whom children actually come in contact with. The antibias approach tries to avoid some of the dangers of this tourist approach. It retains some of the positive ideas from the multicultural curriculum and, at the same time, includes more than just cultural diversity. An antibias curriculum addresses ability, age, and gender differences, as well as the problem of stereotypical or biased behaviour in children's interactions.

There are those who feel that it is beyond the capacity of young children to absorb these concepts. Yet others feel that young children are themselves interested in questions about diversity, and that they may exhibit fearful or prejudiced behaviour when they are not encouraged to understand differences in a positive way.

How do early childhood practitioners prepare themselves to deal with antibias issues with young children? A first and ongoing step is to confront their own attitudes and discomforts, to become aware of how their own identities and attitudes were formed by early experiences. This is a process that may bring pain as well as awareness, but it helps early childhood practitioners identify their prejudices, which need to be uncovered before they will be able to do antibias work with children.

Early childhood practitioners then examine their environments, materials, and curriculum activities to find and eliminate stereotypical messages, messages of omission, and to answer the question "What am I currently doing in the curriculum about gender/race/culture/different physical abilities/stereotyping and discriminatory behaviour?" They observe the children and discover the issues in the community that affect their lives, as well as learning the family and community resources for specific antibias activities. Early childhood practitioners watch for opportunities when children's spoken and unspoken questions show that they need help to understand differences or to be supported in confronting unfair treatment.

Such examinations help early childhood practitioners discover the steps they will take to slowly make their classrooms into environments where diversity is accepted, and where children know they can explore these important issues with supportive adults. Early childhood practitioners and parents work together to explore how to handle sensitive and emotional issues, such as how to celebrate holidays, which holidays to celebrate, or whether holidays should be celebrated at all in early education programs. These are deeply felt concerns, and it is through dialogue that adults support one another to help children develop healthy self-identity and attitudes of acceptance of diversity. As Canada's diversity grows, children will be empowered and enriched to interact comfortably with those whose lives, appearance, and experiences are unlike their own. Issues raised in the discussion of multicultural and antibias education free all early childhood practitioners to find new and positive answers. Involve yourself in these dialogues in later reading, course work, and experiences. We will all benefit.

Technology in Early Education

There was a time, not long ago, when there was almost universal horror at the mention of computers and young children in the same sentence. The image of children sitting in front of computer screens suggested a passive learning mode.

THEORY INTO PRACTICE

Activities to Raise Early Childhood Practitioner Consciousness of Biases

Sample activities for early childhood practitioner consciousness-raising:

- Talk about your earliest memories related to your understanding of your ethnic/racial identity, your gender identity, your family's social class, and your physical abilities and limitations. What was good or difficult as you learned about these aspects of your identity?
- Discuss how you agree or disagree with your parents' views on race, gender, social class, ethnicity, and physical abilities. If you disagree, how did your own ideas develop?
- Take a popular European fairy tale (such as *Snow White* or *The Three Bears*) and rewrite it, switching the characters' genders.
- Describe an incidence of racism or sexism that you have encountered in your daily life. What actions did you take? What would you like to do differently?

Sources: Adapted from Derman-Sparks (1989) and Kilbride (1997).

Most felt that this passive mode conflicted with the idea that children construct their understandings of the world by manipulating objects and actively interacting with others within their environments. It was felt that the computer screen could offer only two-dimensional symbolic representation, instead of the sensory experience and social interaction needed for real early childhood learning.

Should preschool children learn to use computers? (courtesy of Eglinton Public School)

Early childhood practitioners are currently attempting to come to terms with the idea that **technology** can be an effective and interesting additional choice in the early childhood education and care (ECEC) curriculum. Indeed, sometimes it is their own "technophobia" that causes early childhood practitioners to dismiss computers so abruptly. In fact, research and observation of children using computers suggests that computers offer children another way to play, enhancing social, emotional, and cognitive development. To ensure that computers in early childhood classrooms support ideas of developmentally appropriate practice and nurture overall child development, it is helpful to keep certain assumptions about computers in mind:

- Computer use is a social activity.
- Computer use is a child-initiated and child-directed activity.

- Computer software is selected so that it allows children to explore, experiment, and solve problems.
- Computers are one of many materials in a classroom (adapted from Davidson & Wright, 1994).

In fact, while adults are struggling to keep up and feel comfortable with technology, young children accept it as a normal part of their environment. Observers note that they are quite comfortable with computers, exhibiting curiosity, rather than fear, when given a new software program. Much recent research indicates that children's learning on computers is generally positive. Children have increased opportunities for social interaction, as many prefer to work with one or two partners, rather than alone, and will more often request help from peers than from the teacher.

1887
The Montreal Day Nursery program opens.

Ontario becomes the first jurisdiction in the world to officially recognize kindergarten as part of the elementary school system.

There are also high levels of spoken communication and cooperation at the computer. "They interact more frequently and in different ways than they interact when engaged with traditional activities, such as puzzles or blocks. Whereas in sociodramatic play there tends to be a 'leader' of communication, children share leadership roles on the computer" (Clements, 1994, p. 43). Computer use has also been shown to enhance children's belief in their own competency and their enthusiasm for academic subjects such as reading (Orabuchi, in Clements, 1994).

The benefits of computers depend on the kind of software children use and its developmental appropriateness for young children's thinking abilities. Much research has also been done recently on this aspect of technology (see lists and discussion in Wright & Shade, 1994, and Hohmann, Carmody, & McCabe-Branz, 1995). The following criteria for appropriate computer programs suggest overall principles of developmentally appropriate practice:

- Children can use the program independently, without asking for help, no matter what their reading ability.
- Children can control the pace and direction of the program.
- Children have opportunities to explore a number of concepts on different levels.
- Children receive prompt feedback and feel successful.
- There is multisensory capacity for children's natural learning style.
- The program is enjoyable, fun to use, and takes imagination to explore (adapted from Narodick, 1992, in Bowman & Beyer, 1994).

Another caution regarding the use of technology is that inequities must not be repeated. Girls and boys from all ethnocultural and linguistic backgrounds must be encouraged to use computer software and develop their technological skills. Therefore, the context of the software cannot contain any gender, racial, or cultural biases. All children must feel comfortable and included in the images and actions created by the software. Technology may also be very useful for young children with special needs (Behrmann & Lahn, 1994) by offering alternative communication systems to some children with communication disorders. Children who have difficulty with fine motor coordination (needed for printing and writing) may benefit from using a keyboard instead.

It appears probable that technology will be used increasingly in early childhood programs. You will have opportunities during your program of professional preparation to explore these issues in depth and to discuss how much computers can offer in active learning environments.

SUMMARY

Early childhood education and care programs include a variety of curriculum approaches. Curriculum issues currently under discussion—such as academics and the related issues of "readiness" and achievement testing, inclusion, the antibias curriculum, mixed-age groupings, and technology in early childhood settings—directly affect the kinds of programs offered to young children. Other issues will present themselves as you continue your studies and enter the dynamic and developing field of early childhood care and education.

REVIEW QUESTIONS

1. Describe different types of approaches to the early child development curriculum.
2. Describe some of the issues related to early academics, readiness, and achievement testing.
3. Discuss what is meant by "inclusion," and list the benefits for the children and adults who are part of integrated early education classrooms.
4. Describe what is meant by an "antibias curriculum" and the rationale for it. How does an antibias approach differ from a multicultural approach?
5. Identify what is meant by "mixed-age groupings" in early education and some of the advantages for children and adults.
6. Describe current thinking about the appropriateness of using technology in the classroom in early education.

STUDY ACTIVITIES

1. Visit an early child development program that is implementing a specific curriculum approach.
2. Go to www.ece.nelson.com and search for information about Reggio Emilia.
3. Visit an early child development setting in your community that includes young children with special learning or developmental needs. Discuss with the early childhood practitioners the advantages and disadvantages for the children and adults involved. Describe any obvious modifications that have been made in the physical environment to meet the needs of children, and explain why these alterations were made. List any other specialists involved in planning and therapy for the children.

4. Go to the library and find several books that an early childhood practitioner could use to help raise children's awareness and positive acceptance of diversity in various forms. Share these books with fellow classmates. There is a helpful list in Derman-Sparks's book to get you started.
5. Learn if your community has any early childhood programs or mixed-age groupings in the school system. If so, try to visit a class to observe the children's interactions and the differences in learning environment and curriculum that distinguish the programs.
6. Try to observe young children using computers—maybe in the children's section of your local library or large bookstore. How does the software encourage creativity, problem-solving and logical thinking, social interaction, and independence? What questions about technology does this observation suggest to you for further study?
7. Find out how many languages are spoken by students in your class. Brainstorm with each other about how this resource could be used in your community.

KEY TERMS

absorbent mind: Phrase used by Montessori to describe the active, natural learning style of children in their first years.

antibias curriculum: Philosophical approach developed by Louise Derman-Sparks according to which classroom practices and materials are to be used that foster each child's (1) construction of a knowledgeable, confident self-identity; (2) comfortable, empathetic interactions with diverse people; (3) critical thinking about bias; and (4) ability to stand up for him- or herself and others in the face of bias.

cooperative learning: Learning in which environments and activities are structured so that children can work together.

curriculum: Every learning experience that happens in an early childhood setting, including planned and spontaneous activities and interactions.

developmentally appropriate practice (DAP): NAEYC-defined standards that base teaching practices on age-level standards for children's abilities and on observations for individual differences, including abilities, interests, and cultures.

didactic: Usually refers to materials that teach because they are inherently self-correcting. Montessori developed many didactic materials.

early academics: The introduction of academic content and teaching methods to children before the primary grades.

emergent curriculum: Early childhood practitioners follow children's leads by observing their interests and needs and then planning the learning environment.

emergent literacy: Awareness of the meaning of print that develops before early reading and writing.

emergent numeracy: Awareness of the meaning of numbers that precedes ability to use numbers in simple computations.

High/Scope: Curriculum based on Piagetian cognitive principles, developed after the Perry study model.

inclusion: Placement of individuals with special needs in classrooms with typically developing individuals, and in which special services are provided within the classroom setting.

key experiences: Eight important cognitive components of the High/Scope curriculum.

mainstreaming: This term is now less frequently used than "inclusion."

mixed-age groupings: Arrangements that group children together across several ages and frequently keep them together for two or more years, rather than separating them by chronological age.

Montessori schools: Schools based on the philosophy of Maria Montessori that, as such, emphasize sensory learning, practical life skills, and didactic materials.

multicultural: Curriculum that teaches awareness of the diversity of cultures and of cultural experience.

National Association for the Education of Young Children (NAEYC): Largest early childhood professional organization in the United States, established in 1926. Current membership over 90 000. Source of position statements on developmentally appropriate practice, code of ethics, etc.

philosophy: One's beliefs and attitude. Related to early education, one's ideas about how children learn, and how teachers teach.

plan, do, review: Basic methods of the High/Scope cognitive curriculum, in which children make activity choices with teachers' assistance, and then carry out the activity and report on it.

practical life skills: Skills used in daily life, such as washing dishes and sweeping floors. Part of the Montessori curriculum.

readiness: State of being ready. Used in early childhood education, the term indicates a child's ability to learn particular tasks. Readiness tests are frequently used to determine readiness for kindergarten and/or Grade One academic learning.

Reggio Emilia: Early childhood programs in Reggio Emilia, Italy that are world famous for their child-centred and extensive project approach to learning.

sensitive period: Term used by Montessori to indicate time of readiness for particular learning.

sequential steps: Predictable patterns of development and learning.

special education: Branch of professional study that centres on techniques for working with children with special needs.

standardized tests: Assessment techniques using results accumulated for large groups of children, producing standards or norms for evaluating children's successful learning.

technology: Refers to the knowledge that drives today's machines, particularly computers.

SUGGESTED READINGS

Early Academics, Readiness, and Testing

Bredekamp, Sue, & Rosegrant, Teresa (Eds.). (1992). *Reaching potentials: Appropriate curriculum and assessment for young children.* Washington, DC: NAEYC.

Charlesworth, Rosalind. (1989, March). Behind before they start? Deciding how to deal with the risk of kindergarten failure. *Young Children*, 44 (3), 5–13.

Kamii, Constance. (1990). *Achievement testing in the early grades: The games grownups play.* Washington, DC: NAEYC.

Meisels, Samuel. (1987, January). Uses and abuses of developmental screening and school readiness testing. *Young Children,* 42 (2), 4–6, 68–73.

Puckett, Margaret B., & Black, Janet K. (1994). *Authentic assessment of the young child: Celebrating development and learning.* New York: Merrill.

Inclusion

Canning, M. & Lyon, P. (1990). "Young children with special needs." In I. Doxey (Ed.), *Child care and education: Canadian dimensions.* Scarborough, ON: ITP Nelson, 254–68.

Irwin, S., Lero, D., & Brophy, K. (2000). *A matter of urgency: Including children with special needs in child care in Canada.* Wreck Cove, NS: Breton Books.

Rose, Deborah F., & Smith, Barbara J. (1993, May). Preschool mainstreaming: Attitude barriers and strategies for addressing them. *Young Children,* 48 (4), 59–62.

Spodek, Bernard, & Saracho, Olivia N. (1994). *Dealing with individual differences in the early childhood classroom.* New York: Longman.

The SpeciaLink Newsletters from SpeciaLink: The National Child Care Inclusion Network P.O. Box 775, Sydney, NS B1P 6G9.

Widerstrom, A. H. (1986, December). Educating young handicapped children: What can early childhood education contribute? *Childhood Education,* 63, 78–83.

Wolery, Mark, Holcombe, Ariane, Venn, Martha L., Brookfield, Jeffri, Huffman, Kay, Schroeder, Carol, Martin, Catherine G., & Fleming, Lucy A. (1993, November). Mainstreaming in early childhood programs: Current status and relevant issues. *Young Children,* 49 (1), 78–84.

Antibias Curriculum

Bernhard, J., Lefebvre, M., Chud, G., & Lange, R. (1995). *Path to equity: Cultural, linguistic, and racial diversity in Canadian early childhood education.* Toronto: York Lanes Press, Inc.

Chud, G. (1993). Anti-bias education: An approach for today and tomorrow. *Interaction* (Summer), 18–20.

Clark, Leilani, DeWolf, Sheridan, & Clark, Carl. (1992, July). Teaching teachers to avoid having culturally assaultive classrooms. *Young Children,* 47 (5), 42–49.

Gonzalez-Menza, J. (1993). *Multicultural issues in child care.* Mountainview, CA: Mayfield Publishing Co.

Jones, Elizabeth, & Derman-Sparks, Louise. (1992, January). Meeting the challenge of diversity. *Young Children,* 47 (2), 12–17.

Kilbride, K. (1997). *Include me too! Human diversity in early childhood education.* Toronto: Harcourt Brace.

Sheldon, Amy. (1990, January). Kings are royaler than queens: Language and socialization. *Young Children,* 45 (2), 4–9.

Contact: Early Childhood Multicultural Services, Third Floor, 210 West Broadway, Vancouver, BC V5Y 3W2.

Mixed-Age Grouping

Stone, Sandra. (1994/1995, Winter). Strategies for teaching children in multiage classrooms. *Childhood Education,* 71 (2), 102–05.

Theilheimer, Rachel. (1993, July). Something for everyone: Benefits of mixed-age grouping for children, parents, and teachers. *Young Children,* 48 (5), 82–87.

Technology

Burns, Susan M., Goin, Laura, & Donlon, Jan Tribble. (1990, January). A computer in my room. *Young Children,* 45 (2), 62–67.

Clements, Douglas H., Nastasi, Bonnie K., & Swaminathan, Sudha. (1993, January). Young children and computers: Crossroads and directions from research. *Young Children,* 48 (2), 56–64.

Clements, Douglas H., & Swaminathan, Sudha. (1995). Technology and school change: New lamps for old? *Childhood Education, Annual Theme Issue,* 71 (5), 275–81.

Wright, June L., & Shade, Daniel D. (Eds.). (1994). *Young children: Active learners in a technological age.* Washington, DC: NAEYC.

CHAPTER THREE

Quality in Early Child Development Programs

OBJECTIVES

After studying this chapter, students will be able to

- discuss several specific components of quality in early childhood education and care programs;
- describe specific program and policy decisions that exemplify each component;
- identify what is meant by the Canadian Child Care Federation's (CCCF) standards on quality child care and what is included; and
- identify several components that are not found in quality programs.

Quality in early childhood settings takes many forms and has unique faces. Yet there are specific descriptions that can be applied no matter what the age group or population served or the structure of the program. It may seem a daunting task to describe and recognize factors that denote quality in early childhood programs. Indeed, it is a task that researchers, scholars, and commissions have laboured over for years, and their combined efforts still precipitate spirited debate. To some extent, the difficulty is that quality is often defined subjectively. If you have visited sites of early child development programs, you have likely been attracted to (or dismayed by) some space in the room or some activity that resonated with meaning for you, often for subconscious reasons.

But standards for quality must go beyond subjective opinion, to find their basis in concrete and observable phenomena that can be discussed and explained. In this chapter, we will explore some of the specific components of quality in early education and describe how these components may manifest themselves. You will be asked to actively consider each component as we move through the discussion, to help you as you begin to construct your own standards of quality in early education. In later course work, you will learn more about how early childhood practitioners create the environments that allow excellent education to occur; for now, it is important that you begin to evaluate child care and learning environments.

Thinking about Quality

High-quality early child care and education can be defined from many perspectives, including those of children, parents, early childhood practitioners, researchers, employers, and the community. When parents are asked to describe what they are looking for in a program for their children, they often list health and safety as their primary concerns. In a recent U.S. study of quality in family child care and relative care, both mothers and caregivers identified health and safety as prime concerns, as well as communication with the family child care provider and the provider's warmth and attention to the child (Galinsky et al., 1994). Early childhood practitioners, on the other hand, are likely to define a high-quality program as one that

- "supports and assists the child's physical, emotional, social, language, and intellectual development; and
- "supports and complements the family in its child-rearing role" (CCCF, 1991, in Doherty-Derkowski, 1995, p. 4).

Katz (1993) directs us to consider five different but interrelated perspectives as we consider quality in an early childhood program. The perspectives she describes are those of experts, children participating in early childhood settings, parents who are using early child development services for their children, early childhood practitioners and other caregivers who work directly with young children to provide early child development programs, and the community or society at large.

- The expert perspective is a *top-down* perspective that examines measurable and quantifiable characteristics, such as adult-child ratios and training of early childhood practitioners, which set the stage for excellent early childhood education to occur. Many of these things are defined and regulation requirements discussed in this and later chapters. It is important to recognize the correlation between these factors and quality.
- The children's perspective is a *bottom-up* perspective that describes the quality of life experienced by children within the program. This perspective is obtained when adults take children's perspective; propose whether program practices would help children feel individually welcomed and securely accepted within a program; and find engaging, challenging, and absorbing learning experiences. As we think about stories of children in various early child development programs, we draw conclusions about how those settings would affect children's lives.
- The viewpoint of staff members (early childhood practitioners and others) is the *inside* perspective. Job satisfaction and retention of qualified personnel over time has a major impact on quality. As we consider the components of quality, think about how early childhood practitioners would perceive their work if these components were either present or absent.
- The parents' perspective, or the *outside-inside* perspective, considers the quality and impact of relationships between early childhood practitioners and families. The elements crucial from the perspective of parents are mutual respect and relationships that allow early childhood practitioners to create caring and learning environments that are individually responsive to children.
- The community or society perspective is the *outside* perspective, which describes the relationship between the early childhood program and its community and the larger society beyond. As programs respond to community needs and expectations, and as social decisions and support mesh, good things can happen for children and families.

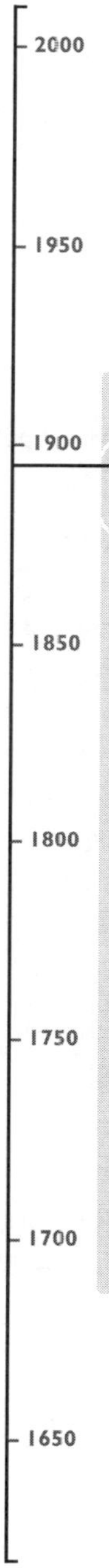

1889
Hester How allows students to bring siblings to school to reduce student absenteeism.

Fourteen kindergartens were in operation in Toronto schools, with a total enrolment of 981 children.

A kindergarten association is established. It later became the Toronto Early Childhood Association.

THEORY INTO PRACTICE

Child Care Sector Study on Quality

In Canada, the Child Care Sector Study (Beach et al., 1998) used the early childhood practitioners' definition quoted above to identify the key elements of quality in child care settings and the factors necessary to ensure quality.

Elements of Quality

The elements of quality include both the ways in which child care environments are organized and the nature of interactions supported by that environment—in other words, the structure and process elements. A high-quality child care environment incorporates the following principles:

- Responsive, stable caregiver-child interactions that are sensitive to children's needs, provide support and encouragement, and set consistent behavioural expectations are of primary importance in both home- and centre-based child care.
- Caregiver knowledge of child development (usually indicated by postsecondary education credentials) supports positive adult-child interactions in both family child care and child care centres.
- The caregiver-child ratio and group size, which allows the caregiver to be aware of and involved, are important to adult-child interactions in child care centres.
- Low **staff turnover** creates a stable and consistent caregiver-child relationship. High turnover reduces stability.
- Daily routines, including planned activities and free play, with health and safety provisions, support children's learning and protect their well-being in family child care homes and child care centres.
- A child-appropriate physical environment allows children to take part in learning opportunities and encourages increased positive interactions with caregivers.
- A positive organizational climate, which includes opportunities for caregivers to be involved in decision-making, creates a coherent administrative framework to support caregivers in centres. In family child care, support and contact with other caregivers is part of a positive organizational climate.
- In family child care settings, caregiver "intentionality" is associated with better outcomes for children. Intentionality is the conscious commitment to caring for children and the provision of nurturing, stimulating child care.

Ensuring Quality

The elements of quality and the quality of the care provided are affected by factors outside the immediate program setting. The following principles ensure quality:

- In centre-based programs, adult-child interactions are influenced by the caregiver's job satisfaction level, which, in part, is determined by the work environment, including wages and benefits, general working conditions, and administration. Low job satisfaction is associated with higher staff turnover, which reduces the stability of relationships between caregivers and children. Compensation levels and other aspects of the work environment are determined by the level of available funding.
- High-quality child care is more likely to be found in nonprofit than in commercial child care settings. However, there are both low-quality nonprofit centres and high-quality commercial centres.

Child Care Sector Study on Quality (cont'd)

- The level of government regulation, particularly for caregiver education, caregiver-child ratios, and group size, does influence the quality of care in centre-based settings.
- Quality may be encouraged through professional standards of practice for individual caregivers and for child care settings.

Source: Adapted from Beach et al. (1998), pp. 32–34.

Early childhood practitioners need to be aware of the complex **interrelationships** of all participants when considering quality. There are differences in what is important in quality early child development programs and questions of when one view should be given more weight than another. The values and beliefs that shape perceptions of quality vary in different cultural contexts.

Gillian Doherty (2000) has recently reviewed the "reality of multiple perspectives" on quality and suggests that the level of quality in a particular early child development program depends on who is making the judgment. Doherty states that it is important to recognize that there are different perspectives and to move away from a judgment of quality that is defined by experts only, without the input of other stakeholders. She does conclude that "there appears to be agreement that there are some values that are so critical to the well-being of children that they should be a core part of any definition of quality" (2000, p. 4). These values are

Sensitivity to children's interests, worries, and passions is central to definition of quality early childhood education and care environments (courtesy of Eglinton Public School).

- safe care;
- healthful care;
- individualized care that promotes equal opportunity;
- care that provides developmentally appropriate stimulation;
- care that is characterized by positive interaction with adults;
- care that encourages individual emotional growth; and
- care that promotes positive relationships with other children.

Doherty cautions us to keep in mind that there are many ways to carry out these universal values. Our responses can and should vary to reflect our own cultural realities and those of the children and families we work with in early child development settings. "Understanding and ensuring quality involves a continuous process of reconciling the perspectives of different stakeholders within the broader context of universal values. It is not a prescriptive exercise" (Doherty, 2000 p. 5).

Be aware of your own perspective on what is essential in a high-quality early child development program as you consider the research on this subject. Your perspective may be challenged or reinforced. Your ideas about quality will most likely change as your knowledge and skills broaden. Pay attention and be aware of the changes.

A Bird's-Eye View of Quality

1891 The Crèche (later called the Victoria Day Care Centre) opens in Toronto.

Let's look at several ECEC settings to see if our quick glance can yield first clues about quality.

Robin's family child care home. Robin Ferguson, a thirty-two-year-old mother of two school-age children and a three year old, cares for her own toddler along with another three year old, one two year old, and eight-month-old twins in her comfortable home in a small subdivision. She has cared for children at home since her older two were babies.

This morning finds the three year olds busy playing in the family room that adjoins the large kitchen. One of them is playing with baby dolls in an area with child-sized kitchen furniture, and one is stringing beads and spools. The child with the beads wanders into the kitchen, where Robin is feeding breakfast to one of the babies, while the other infant sits near her on the floor, fitting cups into each other. Robin helps the bead-stringer tie the beads into a necklace. When he tries to slip the necklace onto the baby on the floor, who protests weakly, Robin redirects him in a conversation admiring the beads, and then changes the topic to the bird feeder on the deck that they filled the day before. The baby on the floor is still unhappy, and Robin gently talks with her as she continues to feed the baby's brother. When this does not soothe her, Robin puts out some Cheerios for the baby she had been feeding in the highchair, rolls the highchair to the doorway, where she can keep an eye on that baby, and scoops up the crying baby. She takes her to the changing table in the family room, commenting to the child playing with the doll that she has to change the baby, and wonders if she has to change her baby, too. She keeps her eyes on the baby's face as she changes the diaper, although her soft conversation includes the older child, as well as the other older child, who has wandered into the room. At Robin's suggestion, he comes over to join her as she sings to the baby, who is now grinning cheerfully.

Sam Cooper's K/1 classroom. Sam Cooper teaches in a class that includes five to seven year olds in a public elementary school. During the few minutes that we are in his large, attractive classroom, we see a number of groups, containing three to five children, busy in several areas in the classroom. Three children are talking to one another as they puzzle over a computer game. Four boys are sitting at a round table, busy with notebooks, crayons, pencils, and markers. One boy explains to us that they are writing in their journals. Sam arrives and responds to one child's comment about his work with a specific and enthusiastic response that indicates his familiarity with the child's story. He encourages another child to share his idea with another, and he suggests to a third child that his friend might help him figure out the spelling for the word he asked about.

Sam's glance searches the room, and he moves over to two girls who are in heated discussion about balancing objects on the scale. Sam does not interrupt, listening seriously to each contribution, nodding thoughtfully and supportively, then asking a question. As the girls seem to move into cooperative activity, Sam looks around at other busy children. He notices a child standing by the bookshelf, moves over to discuss his selection, and speaks with him quietly, finally leaving him after giving him a pat and a smile. A child and an adult enter the classroom. Sam moves over to greet them both, and he engages in relaxed conversation with the parent, who seems equally relaxed with Sam. Sam gives the children a prearranged signal to gather at the meeting carpet; the parent has been invited in advance to conduct a mini-lesson about American Sign Language, since the children are interested right now in different methods of communication.

Maia Chen's toddlers. Maia Chen is one of two early childhood practitioners for a group of ten children aged fourteen months to twenty-four months in a church-sponsored child care centre in a large city. We find them playing outdoors. Maia has added several large, sturdy cartons to their usual playground equipment of climber, low slide, large sand area, and three small swings. Four of the toddlers have climbed into the biggest box: Maia sits near them, commenting on their activity. "You climbed right in the big box, didn't you, Justin? Sarah's in the box with you. Oh my, here comes DuJawn, he's climbing in with you. D.H. thinks there's room too—maybe if you scoot over a bit. Look at that—four children in the box. And jumping, too. It's a jumping box." The children laugh with excitement, and Maia joins in. A shriek from the sandbox gets her immediate attention. "Oh Derek, that looks uncomfortable. I think Jennifer's trying to tell you that you're crowding her," she says as one toddler attempts to sit where another child already sits. "How about we find a space for you to dig over here? There—now you both can make big holes. Let me see—oh, that hole *is* getting big." Meanwhile, Maia's colleague is pushing three children on the swing, and another child climbs repeatedly up the steps and slides down the slide, never far from Maia's gaze and encouraging smile. Maia reaches into her pocket and pulls out a notebook, in which she jots down two words of a brief observation that she will expand later, and turns to the children who are now pushing the boxes around the playground.

Three early childhood programs, three early childhood practitioners, and their children, all serving to illustrate some of the components of excellence. Some fundamental concepts lie behind the practice of each of these early childhood practitioners. Think about it: What has each educator done that seems excellent to you?

1892
La Presse in Montreal proposes child care centres for working mothers.

The East-End Crèche opens in Toronto.

Defining Quality for Early Childhood Practitioners

Standards for quality early child development programs have been defined by several professional organizations. These are important resources with which to become familiar as you learn more about early education.

Later in this chapter we will turn to the national statement on standards for quality early child development programs published by the CCCF in 2000. We will also look at the **position statement** published by the NAEYC in 1987, since this is the most widely acknowledged standard for quality in accrediting early childhood programs in the United States and has influenced Canadian perspectives on quality. But, first, we will examine the components of quality in the vignettes just described. These same components will be found in any excellent early child development program and are supported by considerable research and consensus among early childhood practitioners.

Dynamic components. These are the child's daily experiences that are based on the relationships children develop with adults and other children. Dynamic components define the nature of the interactions and activities that early childhood practitioners, children, and their families engage in each day.

Framework components. These are the structural elements that define the social and physical environment in ECEC settings. These elements are important because they also influence the quality of the dynamic components. The actual practice of early childhood practitioners can benefit or suffer depending on the framework—the number of children, the kind of physical space, the types of play materials and activities, and the level of knowledge and understanding early childhood practitioners have of child development.

Context components. These are elements outside the immediate ECEC environment that influence both the framework and the dynamic components of quality. Organizational climate, funding that determines wages and benefits, regulations, sponsorship, and auspice create the structure of the social and physical environment. In turn, they also set the stage for how early childhood practitioners relate and connect with children and their families.

In quality programs, these components are interrelated and, sometimes, indefinable as separate entities. But we shall try to identify them individually. As we go through this chapter, keep your pencil and notebook handy. You will be asked to think about your own knowledge and experience with the components of

THEORY INTO PRACTICE

You Bet I Care!

Caring and Learning Environments

A recent Canadian study of child care centres and regulated family child care confirms the findings from earlier research in Canada and the United States (Goelman et al, 2000). The key factor is the caregiver and his sensitivity to be responsive to the needs of the children and their families.

Higher levels of sensitivity were associated with

- higher levels of remuneration;
- higher levels of early child development-specific education;
- better benefits;
- higher staff levels of satisfaction with colleagues and the work environment;
- use of site as a student-teacher field placement;
- subsidized rent and/or utilities (that allowed for higher salaries);
- favourable staff/child ratios; and
- nonprofit operation.

quality. Take your time to see what you already know about good early childhood programs.

Dynamics of Quality

The central element in the dynamics of early child development programs is the early childhood practitioner and what she believes, knows, and is, as a person. This is an important idea to state at the outset, because all of the dimensions in early childhood settings radiate from this basic premise. No matter how good the administration, the education system or philosophy, or the community support, everything of crucial importance in the learning and caring environment comes from the decisions and actions of the early childhood practitioner. Some of these components are based in attitudes, values, and ideas, and some are based in areas of knowledge. But *all* translate into specific actions the early childhood practitioner takes that have far-reaching implications for everyone involved. What an astonishing, humbling thought for all of us drawn to consider caring for and educating young children.

The basic components, which we will discuss separately, include the three R's: **respect** for the uniqueness of individuals, **responsiveness** to children's and family's needs, and **reciprocity** of learning interaction.

1893
Model day nursery at Chicago World's Fair cares for 10 000 children of sight-seeing parents.

Respect

To respect someone is to honour and to show consideration and esteem for that person. All of these phrases suggest that respect is a genuine regard for and sensitivity to self and others. How does respect manifest itself in an early child development setting?

Respect for children and families. There are several key considerations in respecting children. The principles of developmentally appropriate practice, as stated by the Canadian Child Care Federation (CCCF) (1991), the National Association for the Education of Young Children (NAEYC) (see Bredekamp, 1990), the **Association of Childhood Education International (ACEI)** (Moyer et al., 1987, and Isenberg & Quisenberry, 1988), as well as by other early childhood organizations, emphasize the uniqueness of each child as a crucial determinant in early education and care. The different interests, abilities, talents, and styles of learning, the range of developmental differences within children, as well as the pace with which children move through development, are particular for each child and deserve the early childhood practitioner's sensitivity and response.

High-quality programs for young children show this respect for individuality in every decision that is made about children's care and curriculum, from considering what to do about naptime for toddlers to choosing materials for the art table so that preschoolers may find something to whet a particular interest or to guarantee certain success. This is a good time for you to note any practices you are aware of in early care and education settings that demonstrate respect for children's individuality or, conversely, any program decisions that fail to demonstrate this component and that treat all children as if they all had the same needs and interests.

1896

John Dewey begins an experimental lab school at the University of Chicago for four and five year olds, called subprimary rather than kindergarten.

In New York, Susan Blow is at Teachers College representing the Froebelian point of view, and Patty Smith Hill represents the newer developmental approach.

The need for respect demands that early childhood practitioners first acquire knowledge of the children that goes far beyond merely learning characteristics of age-level norms. Such constantly growing knowledge demands that educators continually observe and communicate, as well as avoid the hasty conclusion that they "know the children." Early childhood practitioners who truly respect children's individuality find themselves always in a suspenseful state of not quite knowing, and always being open to new perceptions and evidence of growth and change.

Respect for children also implies recognizing children as capable and interesting people, who are also filled with the potential and the desire to learn, grow, and develop. Early childhood practitioners listen carefully to children, recognizing that their insights are valid and that their questions are important. They talk with children individually, about things that are of interest and consequence to both child and adult. These are genuine conversations that follow the same rules as those of adult communication; they are not just "pop quizzes" or questions such as "What colour is the ball?" or "How many bears do I have in my hand?" Early childhood practitioners talk *with* children, not at them.

Children are enjoyed as people, rather than seen as being slightly entertaining because of their less-developed abilities and understanding of the world (*not* as in "Did you hear that? Isn't that cute—he called the lobster a monster.") Early childhood practitioners accept children's feelings and social interactions with seriousness, but also recognize children's need for guidance and assistance in moving toward greater **self-regulation** (*not* as in, "It's nothing to get upset about. He probably doesn't even understand much about moving to another province.") Early childhood practitioners see this need for help and guidance not as a defect, but as necessary assistance in the progression of learning (*not* as in "I'm sick and tired of hearing toddlers yell.") Respectful guidance of young children means that educators use techniques and communication that safeguard positive feelings of **self-esteem.** Guidance methods that humiliate or cause feelings of shame and doubt are not seen in excellent programs.

Adults with this kind of respect for childhood are not in a hurry to move children to more advanced learning or behaviours that they are not yet ready for. Instead, they recognize that childhood has an importance of its own, not just as a preparation stage for a later time. This is an important question to consider: How do adults convey to children that their childlike abilities and characteristics are appreciated, enjoyed, and accepted?

Early childhood practitioners also respect children's individuality when they help children learn to recognize and comfortably acknowledge the **diversity** of race, culture, religion, socioeconomic experiences, gender, and physical ability that exists within Canadian communities and in early childhood education and care settings. Play materials, visitors, activities, and conversations all help children accept their **self-identity** and that of others. Each child identifies with a specific family and its background. As early childhood practitioners interact with children's parents, welcoming them into communication and collaboration, children perceive that their own family culture is respected. When parents are drawn into partnership, they are able to help early childhood practitioners truly understand their children and to guide early childhood practitioners in creating a curriculum that is responsive to their needs and goals for their children. Without such dialogue, early childhood practitioners are in constant danger of overstepping the boundaries of appropriateness in the multicultural world in which we live. Respect for individuality includes this acknowledgment of family needs and preferences and the active attempt to work with families. Genuine respect comes when parents and early childhood practitioners recognize that each has a role in children's lives that is very different, but complementary, rather than cause for antagonism. What are some of the practices in early childhood programs that convey respect for different cultures and that reach out for collaboration with families?

In essence, showing respect means that programs and early childhood practitioners are **child-sensitive**—that is, they notice that children are unique, acknowledge this as important, and use this knowledge as a significant basis in planning the total program. "Respect" may be a more descriptive term than the frequently used "child-centred," which may seem rather one-sided or totally indulgent toward children. Respect demands responsiveness.

Each child has unique interests (courtesy of Yes I Can! Nursery School).

Respect for self. Early childhood practitioners who have respect for self recognize the right and personal obligation to be authentic to the values and principles that drive their lives. This begins as early childhood practitioners consider the philosophy and practices of potential employers, to determine whether these are compatible with their own deeply held personal convictions. Nothing more quickly destroys the joy of personal or professional life than working in a setting that calls for dissonance with self. Sometimes early childhood practitioners feel they can make compromises that would allow them to keep their job while balancing with their principles. It is true that there are, as yet, few perfect worlds, and that early childhood practitioners will likely have to live with some situations that are less than ideal. Nevertheless, to be able to maintain congruent feelings of personal and professional integrity, early childhood practitioners will need to identify those ideals and issues for which there is no room for compromise.

Such insights sometimes come slowly, after experience, as with the early childhood practitioner who left her first job after several months, saying ruefully, "I didn't realize how strongly I felt about teamwork. I now know that being part of a supportive community of adults and children is necessary for my ability to grow and see myself as a real contributor in the small world of the child care centre. I will never again take a job that doesn't emphasize that feeling of community." Some early childhood practitioners might want to be able to express their religious beliefs in their daily work; they will need to find programs that permit them to do so. The same is true for early childhood practitioners who believe so strongly in being able to espouse particular lifestyles that they would be uncomfortable working with other adults who deem them unacceptable. Early childhood practitioners must identify their defining principles.

Beyond being aware of who they are, early childhood practitioners also need to value the worth of their ideas and abilities. Humility is an important characteristic

1899
John Dewey publishes *The School and Society*, outlining his philosophy of progressivism—that education is a means of social reform.

for professionals. But this is not a self-effacing humility; rather, it is a confident recognition of one's strengths and capabilities, coupled with an understanding of the need to join with others for maximum effect. In other words, early childhood practitioners do not have to be all-knowing or all-answering to any question, but can draw on the strengths of others, both adults and children. Early childhood practitioners who value their worth are confident in knowing that they have a contribution to make in early care and education. They know that their passions and ideas are worth sharing, that their presence is valuable. Such confidence leads toward excellence.

Responsiveness

"Good quality child care provides emotional security, frequent communication and encouragement to play and explore. It begins with a nurturing, stimulating relationship between the child and the caregiver, and extends to positive relationships and appropriate activities with other children" (Human Resources Development Canada, 1994, p. 11).

Responsive adults are able to read children's verbal and nonverbal signals, and they react promptly to children's needs and requests. Responsiveness is a sensitivity to a child's emotional state and mood, as well as attention to his physical needs. Children's trust emerges when adults are emotionally and physically available and reliable in meeting their needs. Infants and young children are more likely to form a secure attachment to responsive early childhood practitioners than they are to less-responsive caregivers.

Research findings in both centre- and home-based programs reveal the benefits of responsive early childhood practitioners. In studies of children in child care centres, responsive adults are associated with

- children who have more positive social skills with peers (see Rubenstein & Howes, 1979, 1983; Tzelepis et al., 1983; Whitebook et al., 1990, reported in Doherty-Derkowski, 1995); and
- children with higher cognitive and language skills (see Carew, 1980; Melhuish et al., 1990a, 1990b; Rubenstein & Howe, 1983; Whitebook et al., 1990).

Research studies find the same connections in home-based child care settings. Responsive adults bring out better social skills and higher levels of intellectual skills in children (Clarke-Stewart, 1987; Galinsky et al., 1994, reported in Doherty-Derkowski, 1995).

Doherty-Derkowski (1995, p. 26) states that the concept of responsiveness combines

- age-appropriateness,
- appropriateness for this particular child at this time, and appropriateness in this particular situation.

Effective early childhood programs recognize that learning takes place in the context of responsive relationships. Much time and effort go into building and sustaining significant relationships. These relationships form complex webs that extend in many directions: from early childhood practitioner to child; child to child; early childhood practitioner to parent; early childhood practitioner to early childhood practitioner; parent to parent; and, most important, parent to child. The ultimate statement on the importance of relationships in early education comes from the late Loris Malaguzzi (1993, p. 9), of the Reggio Emilia programs in Italy:

> Although (from our experience in Reggio Emilia) we know how strongly children represent the centre of our educational system, we continue to be convinced that without attention to the central importance of children and families, our view of children is incomplete; therefore, our proposition is to consider a triad at the centre of education—children, early childhood practitioners, and families. To think of a dyad of only an early childhood practitioner and a child is to create an artificial world that does not reflect reality.
>
> ... We strive to create an amiable school where children, early childhood practitioners, and families feel a sense of well-being; therefore, the organization of the school, contents, functions, procedures, motivations, and interests, is designed to bring together the three central protagonists—children, early childhood practitioners, and parents—and to intensify the interrelationships among them.

"Intensify the interrelationships"—what a strong image. Those who offer good early childhood care and education make procedural and environmental decisions based on their potential for encouraging relationships to flourish. The current practice of assigning **primary caregivers** for the youngest children is based on this concept, as are mixed-age groupings (where children remain together and with the same early childhood practitioners over two or three years), which many schools and programs are exploring for continuity. How different this is from making decisions based on efficiency; for example, if an early childhood practitioner had a feeding-table device that allowed her to feed six infants simultaneously, time might be saved, but the quality of personal and physical interaction that promotes emotional attachment would likely suffer. But imagine the benefit to interrelationships if the time-honoured custom of encouraging kindergarten and primary children to "do your own work," by working in silence at separate desks, were changed to having children face one another at tables and encouraging interaction and cooperative learning. Or consider the effects of changing the traditional policy of having the fewest staff members available at the beginning and end of the day so as to permit time for early childhood practitioners to talk with parents and with one another. What other specific decisions in the procedures, functions, and environment of early childhood settings can you imagine that could enhance the quality of the relationships within the programs?

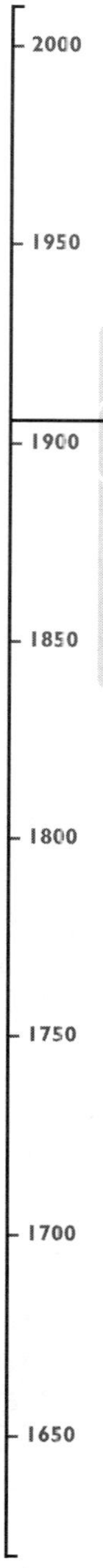

1903
In Ontario, legislation is passed allowing municipalities to purchase land for public parks.

Reciprocity of Learning

If early childhood practitioners believe that learning takes place in the context of many complex adult-child relationships and does not result solely from one-way pronouncements and instruction, learning becomes a reciprocal relationship, in which every participant within the system shares responsibility. Most early childhood practitioners find this concept exciting, as it removes from them the burden of being the sole resource for learning.

Children construct their own increasing understanding of the world by interacting with materials and people in their environments. High-quality programs understand that individuals construct their own reality, with all participants playing both teaching and learning roles. An early childhood practitioner aptly stated this idea to Margaret Yonemura (1986, p. 50) when she said, "'We all have some learning to do from each other,' expressing her underlying view that we are all resources with practical knowledge ... the 'all' was inclusive of all children and adults."

The image that conveys the concept of a *nonreciprocal* learning environment is that of the liquid in a pitcher being poured into an empty cup. The early childhood practitioner is the pitcher, the cup is the child, and the action is all one way, with the child taking in knowledge from the early childhood practitioner and giving nothing back, and with no one else involved in the transaction. Instead, the image that could exemplify a reciprocal learning environment, where learning is multi-directional rather than following a single line, is that of a game in which many balls are being thrown at random around a group. A ball leaves one hand, flying across the room to another hand, crossed by the flight of another that goes from a different hand to still another. Another ball suddenly surprises the first thrower, and so on. The possibilities of connections and combinations is limitless. This is the

THEORY INTO PRACTICE

Responsiveness in Infant Early Childhood Settings

The caregiver enters the room. Sean looks up, eyebrows relaxed, lower eyelids raised, wrinkles in the outer corners of the eyes, cheeks raised, mouth open, and lips up and drawn back. The caregiver recognizes this joyful welcome and responds with her own joyful expression and a warm hug. This type of give and take characterizes the "dance" known as the responsive relationship.

Responsive relationships between caregivers and infants are the foundation upon which infants grow and flourish. Quality infant care provides these relationships. It supports the parent-child relationship, as well as offering security that promotes infant development within the centre. To do this, the caregiver must establish security through responsive relationships. She must manage stimulation to reduce stress. Knowledge of the role of temperament is also required to assist the caregiver's understanding of individual differences among infants. In addition, the caregiver must also be aware of her own needs and how these needs can impact upon her relationship with the infant. When the caregiver reads the infant's signals and replies, follows the child's lead and knows the child's rhythms, the child learns that she is understood. The child's security and self-esteem are enhanced.

Responsive relationships are ensured when the caregiver obtains information from parents; observes the infant; demonstrates respect for the infant's feelings and behaviour; is emotionally available to the infant and develops creative verbal and nonverbal responses to the infant. Responsive caregivers must also manage stimulation, finding the right timing, kind and amount of stimulation. In order to manage stimulation, a caregiver must be sensitive to individual variations in state or level of awareness.

Source: Excerpted from "Building responsive relationships in infant care," Marie Goulet (April 1995) pp. 9–13. Reprinted with permission.

richness of a reciprocal learning environment, in which all participants can teach and learn from one another through interaction.

What would a reciprocal learning environment be like from an early childhood practitioner's point of view? One of the best descriptions comes from Vivian Paley (1990, p. 136). In her work she continually marvels at how she learns from the children themselves how to present ideas that they can comprehend. Her mistakes with using adult logic are gently changed as she listens carefully to the children and has new insights into the world they have constructed. As Paley also points out, the children learn from one another in ways she cannot teach them. Here she is reflecting on a child's words:

"Samantha's explanation, 'You're really a helicopter really but you're pretending a baby,' could never be used by an early childhood practitioner. The statement can only be made by another child, because it must stay within the child's context of reality... In this single exchange between two young children there are important implications for classroom teaching at all levels. Children are able to teach one another best if they are permitted to interact socially and playfully throughout the day." Read some of Paley's work, such as *Mollie Is Three* (1986), *Bad Guys Don't Have Birthdays* (1988), *The Boy Who Would Be a Helicopter* (1990), and grow into your own understanding of how good early childhood practitioners learn from children. Janet Gonzalez-Mena also reminds us, in *Multicultural Issues in Child Care* (1993), how early childhood practitioners can change their limited perspectives as they come to learn reciprocally by listening to parents who have very different views about raising children.

This early childhood practitioner is listening to the child's explanation about the "water pieces falling down" (courtesy of George Brown College).

One of the most important, and often undeveloped, reciprocities in early childhood programs is the interaction that should exist between early childhood practitioners and colleagues. Professional growth and learning does not take place in a vacuum. Opportunities for supportive dialogue and shared reflections, and for joint observation and goal-setting provide early childhood practitioners with stimulation and challenges. Excellent programs provide encouragement and environmental supports for forming real collegial systems. Reciprocal learning is possible only when early childhood practitioners show attitudes of respect for all, and when they recognize the importance of relationships and nurture them.

Think about this concept of reciprocity in good learning environments. What evidence of reciprocity have you encountered in your own learning experiences?

Framework of Quality

The framework of quality is constructed from basic building blocks, including an underlying knowledge of child development and early education practices, the number of children for whom each early childhood practitioner is responsible, the interrelationship and integration of learning experiences, and environments that are prepared for active learning.

1906
Montreal begins the first regular and systematic medical inspection of school pupils in Canada.

1906–08
Playground associations are established in Hamilton, London, Ottawa, and other centres in Ontario.

Underlying Developmental Understanding

When early childhood practitioners make decisions based on facts and **theories** from child development knowledge and research, they are using standards of developmental appropriateness. The question is always this: Is our practice supported by what we know to be true about children in general, and these children in particular? The CCCF (1991, p. 4) states that "all practices [in a quality early childhood education and care setting] that take place are based on sound child development theories and current research." While recognizing that the quality of an early childhood program is affected by many factors, the NAEYC states that a major determinant of program quality is the "extent to which knowledge of child development is applied in program practices" (Bredekamp, 1987, p. 1).

Early childhood practitioners in quality programs work from a base of knowledge of child development that has taught them much about children's abilities and interests at various stages. Therefore, they are able to structure learning experiences in which children can find their own developmental level, guaranteeing success without fear of failure. Recognizing that children, even within a particular chronological grouping, may be far apart in their actual development, they provide materials that children can use with increasing levels of sophistication and skill. Children's freedom to make choices, coupled with support and challenge from the early childhood practitioner, helps them find the activities they are ready for.

An appropriate curriculum in a good early childhood education and care program is created through the interaction of children and all the involved adults, including parents, and through early childhood practitioners' careful observations of children and knowledge of developmental tasks and goals. That is, early childhood practitioners are acutely aware of the direction they will help children move toward because they recognize the sequence of development in all domains. Children's interests direct the actual activities, themes, and projects.

Because early childhood practitioners recognize that young children's learning styles are active and hands-on, the strategies they use to support active learning may look quite different from what many think of as typical teaching behaviours. Early childhood practitioners in good programs for young children will not be found instructing or lecturing from the front of the room; indeed, you would have a difficult time finding a front to the room! They interact with individuals or small groups busy at work or play. Embedded in their interaction may be challenges or suggestions of new directions to take in exploration, questions that may stimulate more activity or extend thinking, and comments that reinforce or provide information. This is subtle and supportive teaching, adapted to each situation. Occasionally, you will see early childhood practitioners working with the whole group, but there is much less direct instruction even in this form, as children and adults interact. As they listen to children's responses and ideas, early childhood practitioners are continually assessing their progress and seeing what new learning they can facilitate. Teaching styles and strategies are shaped by developmental knowledge of how young children learn. How does this description of early childhood practitioners' strategies match your image of what you would do as an early childhood practitioner of young children?

Related postsecondary education increases the likelihood that all caregivers are early childhood practitioners and have an understanding of development to guide their work with young children. The Canadian Child Care Federation (CCCF) National Statement on Quality in Child Care (1991, p. 3) states that caregivers of young children in centre-based settings should "have experience and formal postsecondary accredited early childhood education and care. Their training minimally includes the study of child development and developmentally appropriate practice for the early years and includes supervised practicum experiences to ensure the transfer of knowledge."

The CCCF National Statement on Quality Child Care (1991, p. 3) also considers the training of caregivers who work with young children in home-based early childhood education and care settings. "In terms of learned skills for working with young children, it is important that a provider [in family child care] has an understanding of the developmental needs of children. This information can be acquired through formal child related courses of study and through a variety of life experiences."

The final report of the Child Care Sector Study (Beach et al., 1998, pp. 140–41) recommends that provincial and territorial governments "establish training and education requirements for all caregivers in all regulated settings and an implementation plan with targets and timetables. Requirements should include a two-year postsecondary ECE or equivalent for centre-based caregivers."

Research findings consistently back up these recommendations for postsecondary ECE to prepare early childhood practitioners. In Canada, staff with postsecondary ECE qualifications are more likely to be associated with high-quality early child development settings and better child development outcomes than those without these qualifications (Goelman & Pence, 1987; Doherty & Stuart, 1996; Lyon & Canning, 1995).

Gillian Doherty (1996, p. 44) tells us that the findings are quite understandable. Education in child development and care

- assists the adult to understand children's developmental stages and needs. This, in turn, increases the likelihood that the adult will provide activities that are both stimulating and appropriate for the child's developmental level, and will not impose unrealistic expectations;
- helps to compensate for the fact that initially the caregiver cannot know the child's developmental level and needs as well as the parent. An understanding of typical child development enables the caregiver to make "educated guesses" about what is appropriate and desirable for the child; and
- assists the adult to understand and manage the more complex group dynamics and processes among unrelated children who may not have the same history of familiarity and compromise as do brothers and sisters.

Do you think your own studies in a postsecondary early child development program will affect your understanding of young children's development? Take a moment to reflect on what you are learning about children's development and how this may be changing some of your behaviours.

Earlier we discussed the dynamic components of quality in early child development programs—respect, responsiveness, and reciprocity. Your capacity to

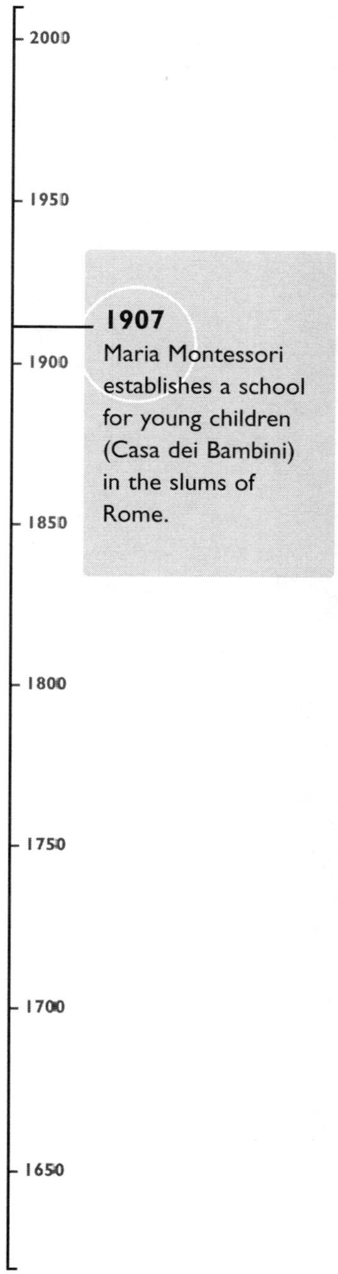

respect, respond, and reciprocate with young children is clearly the crux of the issue of quality. Of course, your own values, attitudes, and other personal characteristics have an important role in your capacity. But so does your underlying understanding of development. As you become more knowledgeable about the principles of child development and developmentally appropriate practice, your capacity to respect, respond, and reciprocate with all children increases.

College and university early child development programs recognize the importance of both theoretical and practical knowledge in preparing early childhood practitioners. Theoretical knowledge encompasses child development principles and learning theory. Practical knowledge focuses on varied strategies and when to use them in guiding children, making decisions that take the immediate situation into account, and following appropriate "rules of thumb" in day-to-day routines (Vander Ven, 1994). We will examine the curricula offered in Canadian college and university programs further in Chapter 5.

1908
London School Clinic is established by the McMillan sisters in response to concern that British children were not reaching school age healthy enough to learn.

Number of Children

It is common sense that the total number of children a caregiver is responsible for will affect the quality of care and education children receive. Can you imagine the quality of care or child development outcomes in situations where one caregiver is responsible for six infants? Basic safety is missing in this situation, let alone an environment of respect, responsiveness, and reciprocity.

Optimal child-adult ratios in early childhood settings allow early childhood practitioners to interact frequently with each child, respond promptly to children's needs, and observe individual children and the group dynamics. The overall size of the group is another important component. If the group size becomes too large, even when the number of children each early childhood practitioner is responsible for is small, quality deteriorates. The adults must pay more attention to overall group organization and schedules, taking away from their individualized attention and flexibility in following the lead of children's activity. In large groups, children's play and daily routines must fit into a group schedule and pattern. It becomes much more difficult to respond to individual children without creating chaos in the group.

Research studies help us to understand the optimal child-adult ratios and maximum group sizes for different ages of children. The CCCF has used this information to prepare a chart that illustrates this information for different age groups, based on the 1985 research of the National Academy of Early Childhood Programs, a U.S. organization (see Table 3.1).

These are the optimal adult-child ratios and maximum group sizes for centre-based ECEC programs.

Where a multi-age grouping exists, the adult-child ratio and group size requirements must be based on the age of the majority of children in the group. When infants are included, the ratio and group size for infants must be maintained.

Think about your experiences in early childhood settings during field placements or while working. Can you remember times when you were responsible for more children than is recommended in Table 3.1? What was it like? Were you able to engage in conversations? Could you respond to children's nonverbal signals?

TABLE 3.1

Recommended Group Size and Child-Caregiver Ratios in Child Care Centres

Age	*Group Size*												
	6	8	10	12	14	16	18	20	22	24	26	28	30
0–12 months	1:3												
12–24 months	1:3	1:4											
2 years		1:4	1:5	1:6									
3 years			1:5	1:6	1:7								
4 & 5 years						1:8	1:9						
5 & 6 years						1:8	1:9						
6–9 years								1:10	1:11	1:12			
10– years										1:12	1:13	1:14	1:15

Notes: Ratios are determined with the adult having full-time responsibility for the care of children. If the caregiver must carry out other administrative duties or support services, an additional adult is required to maintain the designated ratios.

Source: Canadian Child Care Federation (1991).

Did you observe their activity and the dynamics of the group? Or were you mainly focused on "crowd control" and on the safety of children? Was your communication mainly directed at the whole group, rather than at individual children?

Integration of Learning Experiences

High-quality programs consider the **whole child,** meaning that there is equal attention paid to needs and growth in all domains of development: **physical, cognitive, emotional, social,** and **moral.** Physical competence, including both **gross motor** and **fine motor** skills, and emotional, social, and intellectual development, are all recognized as important areas of growth that occur simultaneously in young children. "Simultaneously" does not mean at an equal level; rather, it means that all of this learning is occurring in the child at the same time. Furthermore, what is happening in one aspect of development has an effect on what is happening in other aspects. For example, success in learning within the cognitive curriculum of the early elementary years can be predicated on the child's comfort in social situations and ability to respond to the directions and guidance of a new adult. Infants' language development parallels their acquisition of motor abilities to explore the world firsthand. Toddlers' frustration and temper outbursts may be directly related to their limited vocabulary for expression, and so on.

Considering the whole child also means taking into account the child's family. The world at a school or centre makes up only a part, and a small part, of a child's life experiences. Family life is included in the classroom, and the child is recognized at all times as a family member and participant.

Those who sponsor good programs know that they cannot expect children to learn in a fragmented way, with learning or subject matter broken up into individual, isolated lessons. Imagine the nonsense of planning an infant's day to

2000
1950
1909
Depots are set up in Hamilton, Ontario, to provide "pure" milk for bottle-fed infants.

The West End Crèche opens in Toronto.
1900
1850
1800
1750
1700
1650

include a time to practise the physical skills of crawling on the mat, followed by a short period of language instruction, with an experience in emotional closeness and bonding to come next! Instead, early childhood practitioners sit near babies crawling on the mat, while smiling, talking, encouraging, and interacting. Development in all domains is being nurtured simultaneously—learning is integrated.

In the same way, a good curriculum for older children consists of an **integrated curriculum** of whole activities, rather than separate subject lessons. Instead of providing a language period, followed by math and then science, teachers have children participate in a cooking activity that allows children to learn all of those concepts in a meaningful activity, as well as develop and use the skills involved.

Integrated learning for children in high-quality programs does not happen by chance. Early childhood practitioners who observe children continually come to know their abilities, their strengths, and the areas of development that need particular support. They assess individual development in the light of their knowledge of the predictable sequence of developmental abilities, so that they have for each child a sense of where the child is now and what the next steps will be. These **assessments** are always enhanced by the early childhood practitioner's dialogues with parents, so that objectives are based on the best available information. In good programs for young children, early childhood practitioners have a clear sense of direction to guide them in planning the most appropriate learning experiences for each child. The learning may look spontaneous and involve choices and action on the child's part, but it is part of a careful overall plan for each child.

Another factor that contributes to whole rather than fragmented experiences for children and families is the wholeness of the early childhood program itself. Such wholeness is achieved only when a clearly articulated common philosophy connects all participants to an understood framework that is translated into daily actions. Defining common beliefs is crucial to full participation for both staff and families, so that adults knowingly commit themselves to the stated fundamental beliefs that will bind them. Quality programs have clearly stated and understood philosophies to guide their decisionmaking.

Physical Environment for Active and Cooperative Learning

The physical **environment** is a powerful influence on our behaviour. Think about how you change your behaviour in different spaces. Large, open spaces encourage big movements. Cozy, small spaces encourage quieter activity. The physical environment gives us messages about what kinds of behaviours are expected.

It follows that excellent programs for young children have environments structured so that early childhood practitioners are able to continually observe children. Can you think of environmental arrangements, curriculum methods, and early childhood practitioner functioning that allow early childhood practitioners to continually observe the children in their care? Specific practices include encouraging individual exploration or small-group interaction, rather than whole-group activities led by early childhood practitioners; providing **open-ended** materials that do not require adult instruction or assistance; and defining observation and record-keeping as important tasks for early childhood practitioners come to mind. What else can you add?

Programs designed for children to learn by interacting with materials and people structure the time and space environment for active learning. Classroom environments reflect early childhood practitioners' beliefs about children and how they learn, about the form and content of the curriculum, and about the importance of social connections. When early childhood practitioners see environment as an important component of quality, they create physical arrangements for opportunities that encourage "encounter, communication, and relationships" (Gandini, 1993, p. 6).

Because early childhood practitioners know that children learn through direct manipulation, play, and work with varieties of materials, furniture is used to display the available choices in a logical and organized way, encouraging children's choices and productive use. The environment is organized into logically separated **interest centres,** so that children may decide to work or play alone, with a few other children, or with a larger group. The coherence and unspoken messages of invitation, challenge, and order allow children to feel a measure of control over their learning activities and methods. In quality learning environments, children have opportunities to make choices about which activities they are interested in, about how they will structure their learning activity, and whom they will play and work with, and where.

Environment does convey messages. It can say, "This is a place where you can decide some things that you are going to do today," or "The early childhood practitioner will always be the one in charge of telling you what to do." It can say, "Work with a friend if you like," or "Stay in your seat and work by yourself." It can say, "Here are some things that will allow you to succeed," or "This stuff is really hard and the early childhood practitioner thinks you won't be able to learn it without her." It can say, "They care about your work here," or "What you do isn't really very important." It can say, "What you do, think, and communicate with your play is significant," or "After you play then we'll do the important things." It can say, "Your rights to work undisturbed and undistracted will be protected," or "This is not a peaceful place." It can say, "Childhood is valued and appreciated as a separate and distinct time of life," or "Childhood is merely a superficially cute stage." It can say, "You can do as much as possible for yourself here," or "You'll need to depend on grownups to do things for you." It can say, "There are some limits on things that aren't safe or acceptable," or "You'll have to make mistakes and then we'll stop you." It can say, "You belong here as a member of this classroom community," or "Your individuality is not respected." Think about it—in what ways might the design and placement of classroom furniture and materials, the decoration on the walls, or the classroom schedule and routines reflect these messages? What other messages have you been aware of in physical environments? Programs that understand children's active involvement in discovery learning support this involvement with the way they arrange time and space.

High-quality programs for young children emphasize materials and experiences, knowing that children learn through doing, not through simply being told

2000
1950
1900
1850
1800
1750
1700
1650

1910
University of Manitoba introduces a child development component to the Faculty of Home Economics.

The Ottawa and Toronto school boards establish "forest schools" to maximize the time "sickly" children spend outdoors.

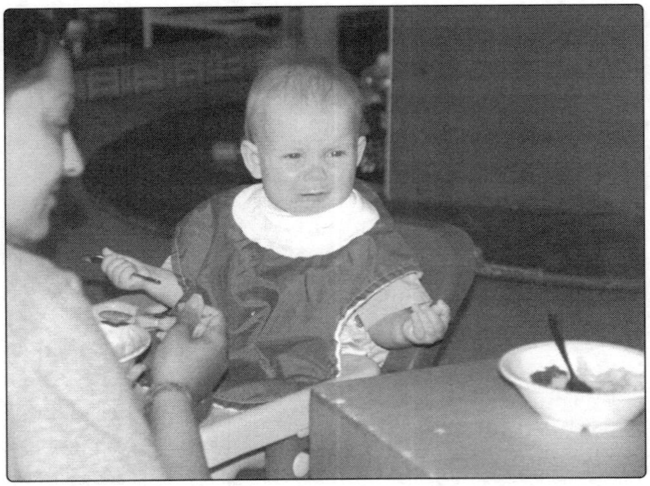

Daily routines should be an integral part of the daily curriculum. They are learning opportunities that need to respect and to respond to individual children's needs, abilities, and desires (courtesy of Yes I Can! Nursery School).

about concepts. Excellent early childhood education and care programs include enticing materials that invite touching. This variety of "loose parts," to use Nicholson's term for open-ended materials (1974), allows children to explore, invent, combine, create, and communicate. Materials are in good repair, are safe, and match the children's manipulative abilities. As well, materials have been carefully selected for the learning environment to correspond with the early childhood practitioners' goals and children's interests. Think of some of the materials that have been part of any good early childhood program you know. What were some open-ended items that children used in interesting ways?

Large blocks of uninterrupted time are needed for children to become meaningfully involved in their play. The daily schedule will be organized to permit participation, practice, and repetition as children's repertoire of skills and knowledge grows, rather than fragmenting the day into defined times for separate activities, with many confusing **transitions.** When programs recognize that children's self-esteem and confidence as learners result from meaningful, personal involvement in activities, the daily plan permits such involvement.

Quality programs for young children recognize the importance of environmental decisions and create environments that support interaction and activity. The physical environment can greatly enhance early childhood practitioners' abilities to respect, respond, and reciprocate.

Context of Quality

The framework of an early child development program sets the stage for the dynamics that occur among children and early childhood practitioners each day. The context of quality describes the elements outside the immediate early childhood setting that influence both the framework elements and the children's day-to-day experiences. Each early child development program creates a distinctive work environment for early childhood practitioners, which will influence her ability to provide responsive, respectful, reciprocal, and reliable experiences for children. The work environment, in turn, is influenced by funding, government regulation, nonregulatory measures, and the sponsorship of early child development programs.

Think about your field placements. How did the work environment vary from one setting to another? Did the roles of the director or supervisor differ between settings? What kind of remuneration did early childhood practitioners receive? Was there a difference in the level of government funding compared to the fees paid by parents? Who was ultimately responsible for the care and education of the children? Did any of these arrangements and practices make a difference to how early childhood practitioners worked with young children?

The context of quality in early child development programs is a complex and sometimes confusing array of details that seem to have little to do with the work of providing quality nurturing and stimulation to young children. But they are important, and they either contribute or present barriers to your capacity as an early childhood practitioner. We will consider the various elements of context of quality, including the work environment, funding, regulation and voluntary standards as well as their impact on quality.

Work Environment

A number of interrelated elements make up the work environment of a child care centre, nursery school, home-based setting, or kindergarten classroom. These elements include the style of administration and supervision, organizational climate, compensation, working conditions, and staff turnover rates, all of which influence early childhood practitioners' interactions with young children and their ability to organize the physical environment and learning experiences.

Administrative style. In centre-based early childhood education and care programs, the role of the director or supervisor includes program and professional leadership; information, financial, and personnel management; parent and community relationships; and legal compliance. All of these functions underlie the early child development program. Like a building's foundation, the elements are often invisible if they are effective. But ineffective administration and management, like cracks in a building's foundation, have noticeable negative effects.

Program and professional leadership. Child care centre directors, nursery school supervisors, and public school principals share common ground in the influence their leadership has on the early childhood practitioners and teachers they supervise.

One U.S. study of the director's administrative style in 400 child care centres (Helburn, 1995, reported in Doherty, in press) found that a director's involvement in both curriculum planning and in the professional community of early childhood education was associated with higher levels of overall quality. A recent study of the child care sector reports that the director in a centre-based early childhood education and care program "is a gatekeeper to program quality who establishes standards of practice and expectations for others to follow" (Beach et al., 1998, p. 99). As well, research identifies school principals who provide academic leadership to teaching staff as a critical component (King & Peart, 1990; Sylva, 1994).

In other words, a competent leader demonstrates and encourages practices that support the dynamics and framework elements of quality, increasing the likelihood that early childhood practitioners who are working directly with the children will do likewise.

Information, financial, and personnel management. Directors and supervisors in centre-based programs manage information systems, monitor finances, and supervise early childhood practitioners and other staff members. This role involves a myriad of details and technical skills. It also involves the ability to set clear job expectations, monitor staff performance, and provide constructive feedback. The director's success in carrying out these responsibilities determines the effectiveness

2000 – 1950 – 1900 – 1850 – 1800 – 1750 – 1700 – 1650

1911
Rachel and Margaret McMillan establish the Deptford School, an open-air nursery school in London, inventing the term "nursery school."

Ottawa Day Nursery opens.

The Vancouver Infants Centre opens at the infants hospital.

First Montessori school in the United States is established.

of the centre's organization. A well-run centre pays staff members on time and makes accurate deductions; keeps children's records updated and within easy access in case of emergencies; and provides clear, written expectations of job responsibilities.

Parent and community relationships. The director is an important link in establishing partnerships with parents and with other groups in the community.

Legal compliance. All early child development centre-based programs that provide nonparental care are subject to numerous regulations and requirements. The director needs to ensure that these obligations are met and that the necessary records are available for verification. A good information system will go a long way toward fulfilling these obligations.

1913
Caroline Pratt, frustrated with the Froebelian-based teacher preparation at Teachers College, begins an experimental school for five year olds in New York City.

The administrative style also shapes the work environment in family child care settings. In family child care, an early childhood practitioner sets his own expectations for program standards, manages information and finances, develops parent and community relationships, and ensures compliance with all legal requirements. His ability to carry out these responsibilities sets the administrative style and influences the quality of the family child care program.

In in-home child care settings, the working relationship between the early childhood practitioner and the employer is the basis of the administrative style. Both verbal and written communication are key components of the working relationship. The early childhood practitioner can enhance the administrative style by ensuring that written employment agreements outline both parties' responsibilities and expectations. Because in-home child care takes place in the child's family home, clear communication practices help to maintain personal boundaries. Communication includes the ongoing exchange of information, as well as the discussion of problems and the exploration of solutions. The important exchange of information between family members and an early childhood practitioner is helped by written daily logs, opportunities for daily verbal reports and transitions, regular parent-nanny conferences, and job performance reviews (Bassett, 1998). Early childhood practitioners can work with employers to establish these kinds of practices. If an employment agency is involved, it is likely to provide administrative support in these areas.

Compensation and working conditions. The income, benefits, and working conditions early childhood practitioners receive in early child development settings are important components of the work environment. These context components do have an impact on job satisfaction and turnover, which influence the dynamics of quality. We will discuss these issues further in Chapters 6 and 9.

Funding

It is fairly simple to determine the cost of a child's participation, whether in a home- or centre-based setting. The biggest expense (usually 80 to 90 percent) is the cost of early childhood practitioners who provide the services; other expenses include the costs of the physical environment, food, supplies, and play materials. To calculate the cost per child, the total expenses are divided by the number of children enrolled in the program. In Canada, funding for the costs of early child programs comes mostly from two sources: parents who pay the costs or fees and public money from governments, which may fund early child development programs or subsidize the parent's fees.

The impact of funding on the quality of early child development programs is also easy to determine. Quality early child development programs depend on skilled early childhood practitioners. Reasonable compensation levels help to attract and retain competent early childhood practitioners in early child development programs. To offer reasonable compensation rates, early child development programs require higher levels of funding, either from higher parent fees or from higher levels of government funding. But higher funding reduces the affordability of child care, both to parents and to governments, making it less accessible to children and their families. We will discuss more about the tensions between quality, compensation, and accessibility in Chapter 9. Now we will look more closely at the costs to parents and to governments, as well as at the types of funding that governments provide in Canada.

The cost of ECEC programs to parents. Parents are responsible for most of the costs of child care centres, nursery schools, regulated and unregulated family child care, and in-home child care. Some parents receive subsidies, which are intended to help low-income families meet the costs of regulated child care. Parents do not pay for the cost of kindergarten programs within the school system, most early intervention services, and early childhood services offered within family resource programs.

The cost of child care centres, nursery schools, and regulated family child care varies from one provincial/territorial jurisdiction to another. Early childhood practitioners and other caregivers receive different levels of compensation, and there are differences in the types of direct funding to these programs by provincial/territorial governments. Also, there are significant variations in the amount of fee subsidies available to parents.

The cost of ECEC programs to government. Governments at all levels contribute to the cost of early child development programs in different ways. Some governments directly fund the programs themselves to cover partial or full operating costs. Public funding to programs also provides fee assistance to low-income families. Other government funding goes to parents to cover the costs of child care, usually as part of employment assistance and training programs. Kindergarten programs operating within the education system are publicly funded as part of the school system and do not charge fees to parents. Finally, governments also fund child care expenses through the income tax system in the form of the Child Care Expense Deduction. Governments' contributions to the funding of early child development programs are complex and subject to frequent change.

The level of public funding influences the compensation levels of early childhood practitioners across all early childhood settings, which, in turn, affects the education qualification levels and turnover rates of early childhood practitioners. Consistent, qualified early childhood practitioners are more likely to provide respectful, responsive, reciprocal, and reliable daily experiences for young children in early childhood settings. Early child development programs that receive higher levels of funding are likely to offer better compensation to early childhood practitioners.

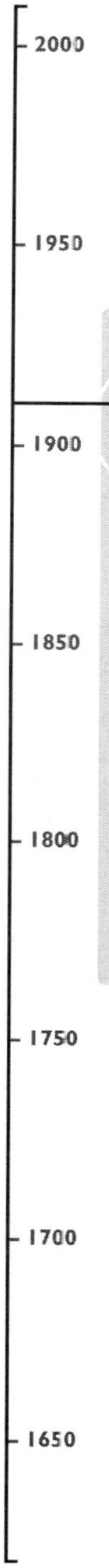

1914
Caroline Pratt designs the first set of unit blocks.

The school nurse is moved into the central position in school medical programs.

Jost Mission Day Nursery opens in Halifax.

Regulation

Of the variables that can be regulated (that is, those with set standards), three that affect positive child outcomes in early childhood programs include adult educational qualifications, group size, and adult-child ratios (*Young Children,* 1993, January). These factors determine the quality of relationships within the program.

In Canada, these factors are regulated by provincial/territorial legislation for regulated child care settings. There is considerable variation from one jurisdiction to the next, and no single jurisdiction has regulations for educational qualifications, adult-child ratios, and maximum group size that meet the criteria recommended by the CCCF and the NAEYC.

Kindergarten programs operate within the legislative framework for education in each province and territory. Kindergarten teachers in public school programs

THEORY INTO PRACTICE

Public Funding for ECEC Programs in Canada

Government Funding for Regulated Child Care Programs

Federal Government

As of April 1, 1996, the federal government consolidated its social spending into the Canada Health and Social transfer (CHST). The CHST combines the federal transfer of funds for postsecondary education, health programs (formerly the Established Programs Financing), and social welfare, which includes funding for child care programs (formerly the Canada Assistance Program), into one block fund. The CHST funding carries few conditions on how the provinces and territories may spend the money. The provinces and territories are no longer required to put up 50 percent funding to access federal dollars for child care programs. Nor are they required to spend any amount of the CHST on child care programs.

In addition to the CHST, the federal government contributes indirectly to regulated child care programs through the Child Care Tax Deduction.

Provincial/Territorial Government

Each province and territory has established fee subsidies to low-income families with specific criteria for eligibility. The target fee subsidies are usually intended for use only in regulated child care settings. In British Columbia and the Northwest Territories, provincial fee subsidies may be used in unregulated family child care settings (Friendly, 1994).

All provinces and territories offer a number of recurring or operating grants to regulated child care programs. Operating grants are public funding that is paid directly to the child care program to support a portion of its overall operating expenses. Some of these grants are aimed at raising staff salaries or other program expenses, whereas others are applied generally to program budgets

Through the federal/provincial/territorial Early Child Development Agreement, the federal government is providing $2.2 billion to provincial and territorial governments for new early child development initiatives, including extending and expanding early childhood education and care programs.

Source: Childcare Resource & Research Unit, 2000 and 2001.

TABLE 3.2

Staffing Requirements for Child Care Centres, By Province/Territory, 1999

Province/Territory	*Staff-Child Ratios* *Age Group*[a]	*Ratio*	*Staff Education/Experience Required*
Newfoundland	0–2		Centre supervisors in preschool and school age need 1-year ECE certificate or related degrees and 1 year or working experience in a licensed centre, or 2-year ECE diploma with no experience required
	2–3	1:6	
Prince Edward Island	0–2	1:3	Centre supervisors and 1 full-time staff must have 1 - or 2-year early childhood development diploma or university child study degree; centre supervisors with 1-year certificates must have at least 3 years' experience; those with 2-year diploma must have at least 2 years' experience.
	2–3	1:5	
	3–5	1:10	
	5–6	1:12	
	7+	1:15	Certification required: 30 hours of in-service training every 3 years for all staff.
Nova Scotia	0–5	1:7[b]	Centre supervisor and 2/3 of staff must have 1- or 2-year ECE certificate or diploma, or 2 years experience, plus 2 course and 35-hour workshop on child development and curriculum. All staff must have first-aid training.
	17 mo–5	1:12	
	5–12	1:15	
New Brunswick	0–2	1:3	Until 2001, no ECE training, education, or experience required. One-quarter of all program staff and the supervisor will be required to have at least an ECE certificate. Staff must be at least 16 years old. Staff 16–19 years old are required to be supervised by a staff member who is at least 19 years old. All staff must have first-aid training.
	2–3	1:5	
	3–4	1:7	
	4–5	1:10	
	5–6	1:12	
	6–12	1:15	
Quebec	0–18 mo	1:5	One-third of staff in licensed centre must have a DEC (2-year ECE diploma equivalent) or AEC (1-year ECE certificate equivalent) and 3 years' experience. No staff training requirements for school-age care.
	18 mo–5	1:8	
	6–12	1:15	
Ontario	0–18 mo	3:10	Centre supervisor must have 2-year ECE diploma plus 2 years experience. One staff member within each group of children must have 2-year ECE diploma or equivalent.
	18 mo–2.5	1:5	
	2.5–5	1:8	
	5–6	1:12	
	6–10	1:15	
Manitoba	3–12 mo	1:3	Qualification levels: Child Care Worker (CCW) III: 2-year diploma and advanced certificate, or approved 4-year degree; CCWII: approved 2-year diploma program or a Child Day Care Competency-Based Assessment Program; CCWI: Grade 12 or 1 postsecondary accredited child care course; CCA (child care assistant): less than Grade 12.
	1–2	1:4	
	2–3	1:5	
	3–4	1:8	
	5–6	1:10	
	4–5	1:9	
	6–12	1:15	

(cont'd)

TABLE 3.2

Staffing Requirements for Child Care Centres, By Province/Territory, 1999 (cont'd)

Province/Territory	*Staff-Child Ratios* *Age Group*[a]	*Ratio*	*Staff Education/Experience Required*
Manitoba (cont'd)	nursery school		
	3 mo–2	1:4	Centre supervisors must be qualified as CCWIII, plus 1-year experience, and 2/3 of centre's preschool staff must be CCWII or III. School-age and nursery school directors can be CCWII. All staff must have first-aid training.
	2–6	1:10	
Saskatchewan	6–18 mo	1:3	Centre-based supervisors must have 1-year certificate in child care, or equivalent. All staff must have a minimum of 130 hours of child care orientation course or equivalent. Staff working with children with special needs must have additional training.
	18–30 mo	1:5	
	30 mo–6	1:10	
	6–12	1:15	
Alberta	0–12 mo	1:3	Three qualification levels: Level 1 (50-hour orientation program); Level 2 (1-year certificate); and Level 3 (2-year diploma). Program directors required to have 2-year diploma in early childhood education. All centre staff must have at least Level 1, and 1 in 4 staff must have Level 2. No training required for school-age care.
	13–18 mo	1:4	
	19–35 mo	1:6	
	3–4.5	1:8	
	4.5–6	1:10	
	nursery schools		
	0–18 mo	1:6	
	19–35 mo	1:10	
	3–5	1:12	
	5–6	1:15	
British Columbia	0–3	1:4	Legislation indicates 3 qualification levels: early childhood educator (1-year basic ECE program and 500 hours of supervised work experience); infant/toddler educator (basic ECE plus post-basic infant/toddler care and education); and special needs educator (basic ECE plus post-basic in special needs). Each preschool group requires 1 early childhood educator plus assistants. Each group (5–8) infants/toddlers requires 1 early childhood educator and 1 infant/toddler educator. In designated special needs child care facilities, one special needs educator is required for each group of children with additional early childhood educators and assistants to maintain a 1:4 ratio. No training requirements for school-age children.
	3–5	1:8	
	5–6	1:10	
	7–12	1:15	
	preschool	1:10	
	specialized	1:4	
Yukon	0–18 mo	1:4	By 1997: 50% staff must have CCI 60-hour introduction course. By 1999: another 30% must have CCWII (1-year program) or equivalent. By 2000: another 20% must have CCWIII (2-year program).
	18 mo–2	1:6	
	3–6	1:8	
	6–12	1:12	

(cont'd)

TABLE 3.2

Staffing Requirements for Child Care Centres, By Province/Territory, 1999 (cont'd)

Province/Territory	Staff-Child Ratios: Age Group[a]	Ratio	Staff Education/Experience Required
Northwest Territories	0–12 mo	1:3	No early childhood education qualification requirements. First-aid certificate required.
	13–24 mo	1:4	
	25–35 mo	1:6	
	3	1:8	
	4	1:9	
	5–11	1:10	

[a]Age given in years, unless noted otherwise.

[b] There are no specific ratios for children 0–17 months, but the recommended ratio is 1:4.

Sources: Data from Beach et al. (1998) and Childcare Resource and Research Unit (1997).

are required to meet provincial/territorial qualification requirements for education and certification. In all jurisdictions outside the Northwest Territories, requirements include an undergraduate university degree and specialized teacher preparation education. In many instances, these requirements are combined in a bachelor of education degree, sometimes with specialization in primary education required. In the Northwest Territories, an undergraduate degree is not required, although this is under review, and classroom assistants and Aboriginal-language specialists are qualified to teach kindergarten under the supervision of a certified teacher. Kindergarten teachers are not required to also have ECE credentials or experience, but primary teacher preparation programs and employees often prefer that teachers have this background.

Only three provinces and territories stipulate adult-child ratios and maximum group size for public kindergarten programs. In Yukon, the maximum group size is 23 and the ratio is 1 adult for every 23 children; in New Brunswick the maximum group size is 20 and the ratio is 1 to 20; and in Quebec the ratio for junior kindergarten is 1 to 17 with a maximum of 34 children per group, while senior kindergarten has a maximum group size of 20 and a 1 to 20 ratio.

Voluntary Standards

Government regulations really establish minimum standards only to support the framework and dynamic elements of quality, They can influence important structural elements of quality, in particular requirements for early childhood practitioners with ECE postsecondary qualifications, maximum group sizes, and the maximum number of children for every adult. But there are limitations: requirements vary from one jurisdiction to another, monitoring and enforcement cannot ensure full compliance, many of the crucial dynamic elements of quality cannot be regulated, and many early child development settings operate outside the system of regulation.

1915
Maria Montessori has an exhibition classroom at the World's Fair in San Francisco.

Prenatal care is introduced through baby clinics in Regina, Saskatchewan.

Voluntary standards offer another mechanism to support and ensure the presence of dynamic, framework, and context elements of quality. The standards are established outside legislation and regulations. In most instances, voluntary standards for early child development programs are established by professional organizations outside the government.

National Standards for Quality Child Care

The CCCF (2000) developed standards for both family- and group-based settings. The statement incorporates major research findings and input from many individuals across Canada. It is meant to be applied in diverse types of services and to be a living document that increases public understanding of and support for quality early child development programs as an essential part of the well-being of Canada's children.

The CCCF statement on program standards identifies what needs to be in place in early childhood education and care settings in ten areas:

- *Conceptual framework.* Program philosophy and a statement of goals and objectives serve as a basis for decisionmaking, daily practice, and program evaluation.
- *Program policies and procedural guidelines.* Program policies and procedural guidelines are in place to enable the conceptual framework to be translated into practice and to support practice consistent with the conceptual framework.
- *Program facility.* The physical facility protects the health and safety of children and adults in the child care program and enables and supports developmentally appropriate programming.
- *Supplies and equipment.* Supplies and equipment of the child care program protect the health and safety of both the children and the adults in the setting and enable and support developmentally appropriate and culturally sensitive programming.
- *Human resources.* The number and skills of persons working in the program meet program needs.
- *Partnerships with children's families.* Partnerships with children's families are encouraged through policies and procedures that enable and support cooperation and collaboration between program staff and families.
- *Recordkeeping and information system.* A reliable system of record and information storage and retrieval is in place to monitor resources and ensure a purposeful program.
- *Administration.* Administrative policies and procedural guidelines support the program and, consequently, service delivery.
- *Program evaluation.* Mechanisms for program evaluation and monitoring are in place and routinely practised.
- *Governance Standards.* The board of directors or owner sets the overall direction of the centre and is legally responsible for what it does.

The CCCF statement (p. 2) on quality begins with the following advice:

> This is your national statement on quality child care. Take ownership of it. Take it back to your work as a functional support for your efforts to improve quality

> child care. Take it back to your day care. Discuss it at staff meetings. Incorporate it into your parent policies. Teach it to your students. Do research on quality care. Share the information with parents. Invite governments to study it in the development or revision of standards. Use it as a tool in your lobby efforts. It is really important that you realize that we mean this to be a living document, a document that will continue to change, that will continue to be integrated into future editions.

Take time now to consider the indicators of quality for home- and centre-based early childhood education and care programs. Identify how you would apply these broad indicators to ensure high-quality early childhood education and care programs, and then check your applications against the statement in the Appendix. Can you find the dynamic, framework, and contextual components of quality in this statement? Are there any contradictions or missing pieces?

Accreditation. All of the components of quality programs that have been discussed in this chapter are discussed in detail in the NAEYC's position statement *Developmentally Appropriate Practice in Early Childhood Programs Serving Children from Birth through Age 8* (Bredekamp, 1987). That statement represents a consensus of the thinking of many early childhood professionals about appropriate and inappropriate practices in programs that span the time from infancy through the early elementary years. In the opening position statement, NAEYC affirms that a "quality early childhood program provides a safe and nurturing environment that promotes the physical, social, emotional, and cognitive development of young children while responding to the needs of families" (Bredekamp, 1987, p. 1). Guidelines are presented in the position statement that describe how principles of developmental appropriateness can be applied to four components of early childhood programs: curriculum, adult-child interactions, relations between family and program, and developmental evaluation of children. The statements describe the specifics of integrated components for appropriate and inappropriate practice for infants, toddlers, three year olds, four and five year olds, and children from five to eight years.

In 1985, the NAEYC established and administered a national, voluntary **accreditation** system for ECEC programs in the United States. The accreditation process is based on ten factors: the physical environment, health and safety, nutrition and food service, administration, staff qualifications and development, interactions among staff and children, staff-parent interaction, curriculum, staffing, and evaluation, with specific criteria defined in each factor (Bredekamp, 1990). The process itself involves self-study, guided by the NAEYC publication Accreditation Criteria and Procedures (Bredekamp, 1991).

Administrators, parents, and staff work on constructive improvements to create high-quality programs; excellence is then verified by visits and assessment from representatives of the NAEYC. The position statement on DAP (1987) made some of the desired outcomes regarding self-study and validation visits more explicit. The National Academy of Early Childhood Education Programs is a body within the NAEYC that administers the voluntary accreditation system for child care programs.

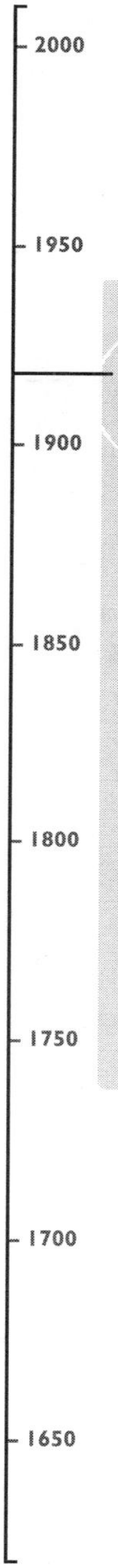

1916
Women in Manitoba, Alberta, and Saskatchewan gain the right to vote in public elections.

Lucy Sprague Mitchell and Harriet Johnson begin the bureau of Educational Experiments in New York City. The Bureau is later to become the Bank Street School of Education.

1916–20 Mother's allowances/pensions are introduced in four provinces (British Columbia, Alberta, Manitoba, and Ontario). This provides financial assistance to poor, single mothers and their young children.

Evaluating Quality

Evaluating the quality of ECEC programs is important to early childhood practitioners, to parents, and to policymakers, who might ask the following questions:

- What is the level of quality offered by an individual ECEC program?
- What type of ECEC service offers the best experience for my young child?
- What is the most effective use of public funding to promote quality ECEC programs?
- How can the impact of changes in staff qualifications, funding, or daily routines be monitored?

But quality can be looked at from different perspectives, as we saw at the beginning of this chapter. We have also considered the various dynamic, framework, and context elements of quality in early childhood settings. Which of these elements is most critical in evaluating the quality of early child development services?

The first step in answering these questions is to consider which aspect of quality to measure. Policymakers and researchers use three different possibilities:

- program ingredients or characteristics, which are mostly the framework and context quality elements;
- child or parent outcomes, which include the achievements, behaviours, and characteristics of children or parents; and
- program outcomes, which attempt to measure what children actually experience, or the dynamics of quality.

Evaluation Tools

Researchers and early childhood practitioners have developed tools to evaluate early child development programs. The Early Childhood Environment Rating Scale (ECERS) (Harms & Clifford, 1980) provides a scale with which to review preschool centre-based child care programs. The scale focuses on the physical environment and looks at the use of space, play materials, and learning experiences, as well as at adult-child interactions. There are thirty-seven items on the scale, with a continuum of possible performance. ECERS is the most frequently used measure of quality for research studies and is also useful as a tool to assist individual program development. In addition to the scale for preschool ECERS settings, there are comparable tools for infant and toddler settings, school-age settings, and family child care settings.

Not Found in Quality

Before we finish this discussion about what quality early childhood programs look like, it may be useful to identify what you will *not* see in a developmentally appropriate program.

Institutionalization

When decisions in programs are made to fit adult needs or preconceptions, children are often expected to behave and learn in ways that fit with the requirements of the institution, rather than in ways that nurture their growth and development. What do we mean by this? As you search your memory for early childhood education experiences, you may recall occasions when children have had to do things that were not necessarily good for them, their learning, or their self-esteem, but were deemed necessary for the good of the institution. Examples of this might include keeping exhausted toddlers awake so they can eat their lunch at the time convenient for the kitchen staff; insisting that first graders eat a silent lunch, so they can finish in twenty minutes and allow the next group to use the cafeteria; demanding that three-year-olds all sleep on their tummies at naptime, so they won't look around the room and prolong time for falling asleep; and insisting that two-year-olds give up all pacifiers.

Sometimes inappropriate practices exist because "that's the way we've always done it," and no one is applying the test of developmental appropriateness to it. Sometimes inappropriate practices exist because the adults in charge are thinking about adult convenience, efficiency, time, or budgets. Sometimes they exist because the adults lack child development knowledge or awareness of current thinking about appropriate practices. And sometimes they exist because adults believe that children need to endure negatives to strengthen their character. Whatever the reason, practices that do not nurture development and learning through emphasis on support and acceptance of individuals are bad for us all and dehumanize society.

THEORY INTO PRACTICE

What Is Quality Child Care About?

It's about warmth and caring and interesting things to do. It's about high self-esteem and genuine concern about the quality of everyone's day. It's about playing games and singing songs and playing house and holding and laughing and having a nice time. It's about everyone being accepted and respected without reservation—and telling each other this in lots of ways. It's about overlooking transgressions so we can get on with things that really count. It's about children and adults spending the day together and looking forward to spending tomorrow together, too.

What Is Poor Quality Child Care About?

It's about criticism and harsh voices and stern faces and frowns. It's about battles of will between children and adults. It's about threats and time-outs for *everything*. It's about too many rules and bribery and adults who always stand up and no one can sit in their laps. It's about not much that's interesting going on and waiting for time to go home and wishing you didn't have to come back tomorrow.

Source: Allen (1991), p. 18. Used with permission.

1917
Women with property are permitted to hold office in Saskatchewan. Women in British Columbia and Ontario gain the right to vote in provincial elections.

Caroline Pratt's Play School (later to become City and Country School) is growing. Lucy Sprague is one of its teachers.

Failure

If early childhood schools and programs stand ready to adapt to children's individual needs and achievements, there will always be opportunities for children to find a comfortable learning level and style and, thus, to succeed. But when programs apply their own arbitrarily drawn standards for success to children at various ages, there are too many occasions when children will have to fail, since arbitrary standards allow no room for individual timetables. For example, readiness testing before gaining admission to kindergarten is going to exclude some children who have failed to meet the school system's standard for readiness. Proof of having completed toilet training before moving on to the next preschool classroom will negatively characterize those children who are not yet ready. Not yet being ready to move from two to one nap per day may mean that some young toddlers will be denied moving on to the wonderful stimulation of the next class. Standardized testing at the end of Grade Two will find those children deficient who are learning at a less-standardized rate. As long as schools and programs apply firm and arbitrary standards, with no room for individual needs or developmental patterns, children will fail, and they and their parents will be burdened by a negative evaluation that may have a lasting impact on future learning and development (Kamii, 1990).

Indifference

Quality programs for young children depend on adults who have visions of wonderful worlds to support children and families, and who are unwilling to accept mediocre facilities, policies, or curricula as the way it has to be. When early childhood practitioners and caregivers are not knowledgeable enough to be able to know the mediocre from the excellent, when they become overwhelmed by the discrepancies between what they know should be and what is, or when they become indifferent to lack of quality, too many less-than-wonderful situations for children will exist. Indifference is an enemy to good early childhood education.

This is a good time for you to consider other conditions that you believe you should not find in good early childhood education and care programs for young children. Write them down for later discussion. Recognizing what is opposite to excellence in early childhood education and care will help you define your personal standards.

Partners in Quality

The CCCF and its provincial/affiliate organizations completed a research and development project, called Partners in Quality, on how all the partners or stakeholders in early childhood education and care can work together to improve quality. The project's activities include a questionnaire survey of views on quality, including the views of parents and family members, early childhood practitioners and other caregivers, early childhood education and care program administrators, child care organizations, governments, training institutions, and communities.

The project distributed copies of surveys across Canada in 1997 and collected thousands of responses from early childhood practitioners and other stakeholders. The results have been analyzed and used to develop a national framework of resources for enhancing quality including manuals and worksheets to promote community and leadership development. ECE students are partners in quality and have distinctive, if changing, perspectives on quality. The questions in the survey are the basis for a useful tool to consider the similarities and differences between your perspectives and those of your colleagues. The Theory into Practice below provides a questionnaire for ECE students. Compare your responses with your classmates'.

THEORY INTO PRACTICE

Partners in Quality Survey for ECE Students

- Complete the following sentence: I consider a child care/early childhood setting as being high quality when it has the following characteristics:
- The protection of the child's health and safety is a basic requirement for any child care/early childhood setting. List the three other most important requirements that you would seek in such a setting for your own child.
- What are the two most important ways in which an early childhood practitioner or an early childhood setting can support families?
- What kinds of involvement with families do you feel is desirable (e.g., the type and frequency of information sharing)?
- What do you believe are the main barriers to family involvement?
- How do you think children should be involved in establishing and implementing daily activities in an early childhood setting? (Identify the age range of the children you have in mind.)
- Here is a list of mechanisms for promoting quality. Rank their importance as very important, important, somewhat important, not too important or not applicable.
  - government regulations
  - government monitoring of its regulations
  - recurring government funding, e.g., an annual grant
  - Early Childhood Education training before starting to provide child care/early childhood education
  - in-service professional development
  - parent/family involvement
  - community involvement
  - staff mutual support groups
  - trade unions
  - professional associations
  - program accreditation
  - program research
  - other (please specify)

(cont'd)

Partners in Quality Survey for ECE Students (cont'd)

- Referring to the mechanisms identified in the question above, please rank the five you believe are the most important for promoting quality.
- Do you believe that any of the mechanisms listed above are barriers to promoting quality? If yes, which mechanisms? How?
- In what ways do you, as a student early childhood practitioner, try to involve families?
- How do you address situations in which you and the family have different beliefs about or approaches to childrearing?
- Indicate if the programs where you have completed or are currently completing a field placement have regular contact or links with any of the community services listed below. Be specific about the type of link e.g., "information-sharing about specific children"; "joint in-service training programs"; "sit on the same community services planning committee."
  – elementary school
  – family resource program
  – public child welfare agency (child and family services)
  – public health nurse or unit
  – municipal parks and recreation department
  – other
- What informal or formal approaches have you found effective in assessing the quality of your daily practice?
- What type of observation notes, program plans, journal and/or record-keeping do you use in your field placement?
- What opportunities to you have for dialogue with colleagues while you are in field placement?
- What changes or additions would you like to make to your field placement program to improve its quality? What is holding you back?
- Have you any additional comments about quality or what is required to provide quality?

Source: Adapted from Canadian Child Care Federation (1997).

SUMMARY

The Canadian Child Care Federation and other organizations have described standards for developmentally appropriate practice. These standards include specific criteria for curriculum, adult-child interaction, relationships with families, and physical early childhood environments for children from birth to twelve years and their families. In this chapter we examined the dynamic, framework, and context components of quality early childhood settings. These components include respect for the individuals involved, responsiveness to children's needs, reciprocity of learning for all participants, developmental knowledge, optimal numbers of children, integrated learning experiences, physical environments to facilitate active learning, organizational climate, funding, regulation, and nonregulatory mechanisms. Components not found in quality programs include institutionalization, failure, and indifference.

REVIEW QUESTIONS

1. Describe several components to be found in quality early childhood programs. Why are these elements necessary?
2. For each component described above, discuss several practices that might be included in program structure or function.
3. Discuss the CCCF statement on quality child care for both family child care and child care centres.
4. Identify what is meant by the NAEYC position statement on developmentally appropriate practice.
5. Identify several components not found in quality programs. Explain why these practices should be avoided.

STUDY ACTIVITIES

1. Complete the Partners in Quality questionnaire on pages 95–96 and compare your answers with others in your class. Perhaps you can collect students' answers and analyze what elements of quality your class of ECE students rates as the most important.
2. In small groups, discuss with classmates the statements and experiences you reflected on and recorded in your notebook throughout the chapter. Identify experiences that seem to support components discussed in the chapter. Identify experiences that seem at variance with the components. Identify additional ideas discussed by participants.
3. Write a personal statement of your belief or philosophy of quality early childhood care and education, based on your thinking and reading to this point.
4. Prepare a statement to deliver at a school board meeting that discusses whether or not to exclude five year olds who fail to achieve a specific score on a readiness test given before admission.
5. Prepare answers to the following questions from parents:
   a. Why do the children spend most of their time playing in your classroom?
   b. Why are some of the two year olds beginning toilet training, and some not?
   c. Why doesn't my four year old bring home artwork every day, as my neighbour's child does?
   d. Why do you call it teaching when you're not doing reading or math lessons or anything like that?
6. Read the CCCF national statement on quality child care. What key words do you find in the statement?
7. Go to www.ece.nelson.com. What can you find out about different perspectives on quality?

KEY TERMS

accreditation: System of voluntary evaluation of excellence in early childhood centres, administered by the National Academy for Early Childhood Education and established by the NAEYC.

assessment: Evaluation of abilities, skills, and knowledge when referring to persons, or components when referring to environments.

Association of Childhood Education International (ACEI): Oldest early childhood professional organization in the United States. Originally associated with the kindergarten movement.

child-sensitive: (Also called child-centred.) Programs whose practices are responsive to knowledge and observations about children.

cognitive development: Related to mental functions of thinking, perceiving, and learning.

diversity: Variety of differences that exist in a classroom, community, culture, or country. Diversity may refer to ability, gender, age, race, culture, and so on.

emotional development: Related to feelings and expression of them.

environment: Everything that surrounds the children that, therefore, affects their lives. Includes physical arrangements of time and space, materials and activities within the environment, and the people available for interaction.

fine motor: Related to the smaller muscles of the body and limbs, such as those in the fingers, toes, face, sphincters.

gross motor: Related to the whole body and to the larger muscles such as those in the legs, arms, trunk.

integrated curriculum: Centres on activities that address many aspects of development and knowledge, rather than separating curriculum into numerous skill areas.

interest centres: Also called learning centres. Classroom areas arranged for particular activities chosen by children. Examples are art, block, and book centres.

interrelationships: Connections between one aspect of development and another, and between people.

moral development: Related to acquiring a sense of right and wrong behaviour and to the ability to control one's actions according to these internalized standards.

open-ended: Activities, materials, or communications that permit various responses and reactions, rather than one fixed correct response.

physical development: Related to growth and coordination.

position statement: Statement of philosophy of a professional organization, used to guide the practice of professionals.

primary caregiver: Person assigned responsibility for a small subgroup of children within a larger group.

reciprocity: Mutual give-and-take.

respect: Recognition of, and sensitivity to, self and others.

responsiveness: Being sensitive to a child's emotional and physical needs, and taking actions to meet those needs.

self-esteem: Sense of self-worth, or value placed on the image of self.

self-identity: Image of self, constructed by feedback from others and cognitive understanding of gender, race, ability, and the cultural messages about these components.

self-regulation: Ability to regulate one's response to challenges and stresses, related behaviour, attention, and memory.

social development: Steps in learning appropriate social interaction with peers and adults.

staff turnover: Rate at which practitioners leave their place of employment and seek other work.

theory: Set of ideas, principles, or explanations that explain phenomena.

transition: Period of change. Daily classroom schedule contains several, e.g., when children change classrooms.

whole child: Theory that supports recognizing the separate and interrelated aspects of the individual, including the domains of the physical, cognitive, language, emotional, and social development.

SUGGESTED READINGS

Balaban, Nancy. (1990, March). Statement to the Montgomery County Council. *Young Children,* 45 (3), 12–16.

———. (1995, March). Reaffirming a national commitment to children. *Young Children,* 50 (3), 61–63.

Bredekamp, S., & Rosegrant, T. (Eds.). (1991). *Reaching potentials: Appropriate curriculum and assessment for young children.* Washington, DC: NAEYC.

Christie, J. F., & Wardel, F. (1992, March). How much time is needed for play? *Young Children,* 47 (3), 28–32.

Canadian Child Care Federation. (2000). *Partners in Quality: Tools for administrators in child care settings.* Ottawa: CCCF.

Doherty, G. (2001). Regulations as a strategy for promoting quality in child care settings. In *Research Connections Canada: Supporting Children and Families,* 6.

Greenberg, Polly. (1990, January). Why not academic preschool? (Part 1) *Young Children,* 45 (2), 70–80.

Greenman, Jim. (1994, November). Institutionalized childhoods: Reconsidering our part in the lives of children. *Child Care Information Exchange,* 63–67.

Honig, Alice S. (1989, May). Quality infant/toddler caregiving: Are there magic recipes? *Young Children,* 44 (4), 4–10.

Kamii, Constance. (1985, September). Leading primary education toward excellence. *Young Children,* 40 (6), 3–9.

Katz, Lilian G. (1994, November). What should young children be learning? *Child Care Information Exchange,* 23–25.

Kelman, Anna. (1990, March). Choices for children. *Young Children,* 45 (3), 42–45.

Moss, P., & Pense, A. (Eds.). (1994). *Valuing quality in early childhood services.* New York: Teachers College Press.

NASBE. (1991). *Caring communities: Supporting young children and families.* The Report of the National Task Force on School Readiness. Alexandria, VA: Author. NAESP. (1990). *Early childhood education and the elementary school principal: Standards for quality programs for young children.* Alexandria, VA: Author.

Sava, Samuel G. (1987, March). Development, not academics. *Young Children,* 42 (3), 15.

Section Two

The Early Childhood Workforce

In Section One we examined the organization of early child development programs in Canada. We also discussed quality in early childhood settings. From this framework, it is time to look at the roles of early childhood practitioners, who make up the early childhood workforce. Chapter 4 will highlight the responsibilities of those who are part of the early childhood workforce. Chapter 5 will examine early childhood practitioners' motivations for entering the field and paths of early child development. Chapter 6 will look at the work environment and the opportunities and challenges that lie ahead for the early childhood workforce.

Be prepared to think honestly about yourself in relation to the early childhood workforce. Could this be your life?

CHAPTER FOUR Early Childhood Practitioners

CHAPTER FIVE Becoming an Early Childhood Practitioner

CHAPTER SIX The Work Environment

From *The Good Preschool Teacher*

For these six women, teaching involves a search for meaning in the world. Teaching has become for each a life project, a calling, a vocation that is an organizing center of all other activities. Teaching is past and future as well as present, it is background as well as foreground, it is depth as well as surface. Teaching is pain and humor, joy and anger, dreariness and epiphany. For these six, teaching is world building, it is architecture and design, it is purpose and moral enterprise. Teaching is a way of being in the world that breaks through the boundaries of the traditional job and in the process redefines all life and teaching itself.

"Teaching as identity" is the clearest theme to emerge in this inquiry, and "teaching as identity" is the frame through which each portrait makes sense. In these portraits, there is no clear line delineating the person and the teacher. Teaching is not simply what one does, it is who one is. Teaching is a life, a way of being in the world, an intentional circle for those six outstanding teachers.

There are, of course, teachers who are narrower in their concerns and more clearly bounded in their jobs than those outstanding teachers are. And yet, teaching is the kind of activity that calls out strongly for an investing of oneself. For many, perhaps most teachers, the sense of calling exists. It may be only a flicker of memory or a feeling dulled by years of bureaucratic maneuvering, endless demands, and excruciating complexity; it may exist now only as a shadowy palimpsest, that little erasure that leaves tracks on the page. But somewhere along the way, teaching called out to teachers as a chance to love children, to make a difference in their lives, to remake the world. Teachers somewhere, sometime felt called to teach (William Ayers. [1989]. The Good Preschool Teacher [pp. 130–31]).

CHAPTER FOUR
Early Childhood Practitioners

OBJECTIVES

After studying this chapter, students will be able to

- define the terms "early childhood workforce" and "early childhood practitioners";
- describe the profiles of early childhood practitioners who work with young children in early child development programs and in home settings;
- identify several distinct roles played by early childhood practitioners; and
- discuss how the key roles of early childhood practitioners support children's optimal development.

Early Childhood Practitioners

Early childhood practitioners work with young children and their families (courtesy of Yes I Can! Nursery School).

Over 300 000 individuals, mostly women, work with young children in early child development programs and home care settings across Canada. This chapter will focus on those people who are trained and educated as early childhood practitioners and who may work in child care centres, nursery school programs, home-based child care, family resource programs, and kindergartens.

You are preparing to enter the early childhood workforce. As an early childhood practitioner, you will have several roles in caring for and educating young children. You will ensure that children are safe and that their basic physical needs are met. You will provide opportunities for learning by creating the framework for good programs and curricula. You will make connections with the larger community that supports children and their families. But your most important role will be getting to know, understanding, and relating to each child and her family.

This image goes far beyond the dictionary definition of the word "teacher." The role of an early childhood practitioner in any early child development setting is more complex than that of a traditional teacher. The term "caregiver" implies too custodial a function and, to many, overlooks the importance of helping children to develop knowledge and skills. As the search for appropriate terminology continues, the term "early childhood workforce" seems most appropriate to describe the sector of individuals who work with young children and their families in many different settings. "Early childhood practitioner" describes those within the early childhood workforce who have specific training and education in child development and in early child development practices.

The Early Childhood Workforce Defined

The 300 000 individuals who now work with young children and their families in home-based and group settings have different job titles, earn different salaries and benefits, and spend each day in different kinds of environments. But much of what they do each day is similar.

Table 4.1 introduces you to some of the similarities and differences among those who work with young children in Canada. Early childhood practitioners, as well as those who do not have specific training or education, are included in each group.

Within each of the groups of individuals who work with young children and their families in Table 4.1, there are early childhood practitioners who have specific postsecondary qualifications in early child development, usually an early childhood education (ECE) credential. In regulated child care centres, 70 percent of all staff working with children have a one-year, two-year, three-year, or higher credential in ECE (Doherty et al, 2000).

Early childhood practitioners may work in any of the early child development settings described in Chapter 1. In Canada, we find early childhood practitioners in child care centres, nursery schools, kindergarten classrooms, early intervention

TABLE 4.1
Working with Young Children

	Total (N)	*Postsecondary Education*	*Female (%)*	*Average Income ($)*
Child Care Centre[a] (including school age child care) & Nursery School Staff	48 000	86.3%	96.3	22 717
Unregulated Family Child Providers	155 000	30%	98.4	n/a
Regulated Family Child Providers	15 000	35%	98.0	n/a
In-home Caregivers[b] (including nannies)	115 000	22%	98.8	12 220
Family Resource Program Staff[c]	4 000–5 000	n/a		n/a
Early Intervention Staff	26 000+	n/a		n/a
Kindergarten Teachers	21 000	95%		

[a] Includes school-age child care

[b] Includes nannies

[c] Includes family-child drop-in programs, family literacy programs, toy-lending libraries, etc.

Sources: Adapted from Beach, Bertrand, & Cleveland (1998); Doherty, Lero, Goelman, LeGrange, & Tougas (2000).

services, family literacy programs, family resource programs, family child care, and in-home care. Early childhood practitioners, working in a variety of early child development settings, make up the early childhood workforce.

A Diverse Workforce

"Demonstrating diversity in the workforce to children through the visibility of men and women ... alongside the visibility of people of colour, is an important goal for enhancing the quality of childcare" (Cameron et al, 1999, in Moss, 2000).

Is Canada's early childhood workforce a workforce that is representative of the population at large? Look around your class, and consider your field placements.

The first observation that you are likely to make is that the early childhood workforce in Canada is predominantly female—about 97 percent. Early childhood workforces in other countries—Sweden, United States, Australia, Britain, New Zealand, Spain, France, Denmark, Italy, etc.—are also predominately female (Moss, 2000). In the next section, we will look at some of the challenges facing men in early childhood settings. The care, education, and development of young children remain women's responsibilities. The early childhood workforce is probably the most gendered workforce in North America, Europe, Australia, or New Zealand (Moss, 2000).

However, early childhood practitioners and ECE classrooms do represent Canada's cultural, linguistic, religious, and ethnic diversity, which is an enormous strength of the early childhood workforce. In urban centres, immigrants, newcomers, and visible minorities make up a large proportion of the population.

Men in the Early Childhood Workforce

Few men work in early childhood education and care programs (courtesy of Play and Learn).

If you are a male entering the early childhood field, there are additional issues of respect for you to consider. At this time, men constitute only a small part of all early childhood practitioners in Canada, probably about 3 percent (Beach et al., 1998). Professional attitudes and practices may have been defined too narrowly as a result of the dominance of females in the field. But it is likely that men avoid the field for specific reasons related to status, economic conditions, and bias.

Most cultures convey to most men the idea that they have to compete in the job field as a measure of their worth. We have already discussed the relatively low status accorded to early childhood teachers, and many men are reluctant to enter a field that is accorded so little recognition. Many men also report that family, friends, and even academic counsellors strongly discourage their working with young children, as they feel it is far beneath their talents. When considering long-term career opportunities, as most men have been socialized to do, the early childhood field appears to have relatively little opportunity for advancement.

1918

The Canadian National Committee for Mental Hygiene is established and transfers ideas on child study from the United States.

Women gain full federal franchise. Women in Nova Scotia gain the right to vote in provincial elections.

The poor salaries in early childhood education and care programs probably encourage more men than women to look for other employment opportunities. Even today, many men still consider careers in the light of being a major contributor to the financial support of a family, and it is difficult to imagine bearing the costs of raising a family on the limited salaries of most early childhood practitioners. The majority of men who work in the early childhood field are married, with at least one child (Robinson, 1988), and many find they cannot afford to work for long in the field. Men certainly have more options than women do for employment even today, and would more likely opt for those that pay substantially more than early childhood education. And yet this contributes to a vicious cycle: it seems probable that if more men entered the field, the low salaries accorded to a mostly female profession would increase.

Although in the decades since the women's movement, more men play a more active role in parenting, indicating their abilities and interest in nurturing young children, there is still widespread gender bias. This bias translates into both overt action and covert attitudes that prevent males from feeling acceptance in early childhood centres and schools, to say nothing of within ECE postsecondary education programs. Early childhood education is associated with the role of mothering, and men, who obviously cannot be mothers, are seen as less capable of caring for young children. The teaching role is associated with a variety of characteristics that are generally classified as female traits, such as patience, gentle nurturing, and emotional sensitivity, as opposed to aggressiveness and emotional control. Men who are willing to take on these roles are unfairly subjected to conjecture and suspicion about their masculinity.

A few widely publicized cases of sexual abuse in the past decade have added fuel to this discriminatory fire, to the point where men are often subjected to humiliating questions and restrictions. There have been instances when directors have refused to allow male students in rooms where diapers would be changed or have refused employment to well-qualified candidates on the basis that "the parents would be uncomfortable." Men do report bias directed toward them by directors, parents, and even female co-workers. Small wonder that few choose to become uncomfortable minorities within the early childhood field.

And this is a pity. It would appear that men in early childhood programs have much to offer. Seifert (1988) refers to a "compensation hypothesis," meaning that male early childhood practitioners could perhaps compensate for the lack of male involvement in the lives of many young children, as well as offer children of both sexes a model of a caring, nurturing male. He also speaks of a "social equity hypothesis," meaning that the presence of men in these programs would help society in general, and children in particular, to discover the many options available to men and women. Little research has been done on male early childhood practitioners; however, it is clear that their presence in the workforce could add a stronger voice for the emergence of a true profession, in which colleagues are not restricted by current stereotypes of "women's work."

As a male, are you willing to recognize the stereotypes and counter them with your own personality and ideals? As a female, are you able to support males as true colleagues, who can work with you for the nurturing and care of young children, as well as for the growth of the profession?

Early Childhood Practitioner Roles

"We need reflective and well-informed practitioners, who do not assume there is one best practice which suits all, but who are able to recognize, explore and discuss the arc of human possibilities" (Penn, 1999, p. 4).

When we refer to early childhood practitioner roles, we are referring to the particular functions and behaviours that early childhood practitioners are expected to perform and exhibit. If you were to ask experienced early childhood practitioners what they do on any given day, they might respond with a list of nouns associated with particular actions. That list might include nurse, diplomat, housekeeper, artist, musician, judge, cook, friend, bookkeeper, entertainer, and instructor. They might add some specific skills that have been handy: repairing toys, detecting guilt or sources of strange odours, restoring physical order from chaos, unstopping toilets, and determining fair solutions to playground conflicts.

As we discussed in Chapter 3, early childhood practitioners are critical to the provision of high-quality early child development. Above all else, early childhood practitioners must recognize their roles and responsibilities in supporting optimal child development and respecting family relationships. A recent review of research studies in neuroscience, social sciences, education, health, and child development identified four broad determinants of optimal child development: protection, relationships, opportunities, and communities (Guy, 1997). These determinants provide a framework for identifying the early childhood practitioners' key roles in supporting children's optimal development in early child development settings.

Providing Protection

Early childhood practitioners meet children's physical needs for safety, nutrition, health care, and hygiene. **Caregiving** means ensuring that children eat healthy foods, get enough rest, and are in safe, secure physical environments.

Early childhood practitioners working with younger children will find that they spend a great deal of time changing diapers and cleansing, feeding children or helping them learn to feed themselves, serving food and cleaning up afterward, helping with hand-washing and face-wiping, and changing clothes after spills or accidents. Through such commonplace daily acts, adults provide children with gentleness and demonstrate skills that children can eventually acquire themselves. Physical care is a core part of the daily routine and is not something to rush through. Daily routines are opportunities to provide responsive care and learning moments.

Safe and healthy daily routines protect children (courtesy of George Brown College).

Providing protection also means protecting children from harm. Safety must be a priority for all early childhood practitioners. You have an obligation to be aware of situations that put children at risk of harm. You also have an obligation to take action if you suspect any form of child maltreatment.

1919

Women in New Brunswick gain the right to vote in provincial elections.

Margaret McMillan publishes *The Nursery School*. (Her sister Rachel died in 1917.)

Harriet Johnson establishes the Nursery School of the Bureau of Educational Experiments.

Establishing Relationships

With the youngest children, the warm physical contact and responsiveness that accompany the providing of physical needs is an interrelated and inseparable part of establishing relationships. The gentle stroking and soft crooning that accompanies the cleansing bath and the warmly enfolding arms that hold the baby being fed the bottle are both **nurturing** to overall development. Young children's healthy social/emotional development (and their cognitive development) depends on their involvement in warm, supportive relationships with caring adults. The most important relationships, of course, are with their parents and other family members. But early childhood practitioners involved in their care also hold key roles in healthy personality development.

Often, early childhood practitioners are the first people outside the family with whom children have caring relationships. The need for warm, responsive relationships does not disappear with toddlerhood; preschoolers and school-age children also look for affection and responsiveness from early childhood practitioners.

Relationships are established through sustained positive interactions between children and early childhood practitioners. In positive interactions, adults are responsive to children's verbal and nonverbal communication and encourage two-way or turn-taking conversation. Some children enjoy a lot of physical contact, whereas others do not. Early childhood practitioners can demonstrate warmth and emotional responsiveness to children in many ways: in gentle smiles and eye contact; in a warm tone of voice; in personal attention and shared moments. The exact manner of interaction that builds relationships is individual. Children know when they are genuinely cared for, even though the message may come in different ways.

For babies and toddlers, positive interaction may include being "smoothers of jangled feelings ..., comforters, ... facilitators of parent-child separations" (Balaban, 1992, pp. 69–70), as well as rockers, singers, and tummy-kissers. For preschoolers, care may include touching a child on the arm when passing, offering moments of quiet conversation, and bestowing a special hug to say good morning. For school-age children, it may be offering a joke that the child can appreciate, teasing gently about private secrets, and giving a personal wink or a thumbs-up sign. All of these interactions tell children that early childhood practitioners know and like them as unique individuals.

Another part of the early childhood practitioner's role in creating positive relationships is setting limits and guiding behaviour. Young children slowly understand which behaviour is acceptable in their homes and in other environments. Children gradually internalize this understanding to regulate their own behaviour. In the meantime, they need adults who understand how difficult a process this is, and who can guide them positively and effectively as they learn to live within limits.

Young children need to have adults around them who are using their power with warmth, support, encouragement, and good explanations of the limits they must impose. As early childhood practitioners base their guidance decisions on their developmental understandings of how children think, learn, and develop impulse control, they are more likely to guide as firm and friendly adults. Within the context of caring relationships, children come to trust that adults will help them regulate their behaviour.

Early childhood practitioners use both direct and indirect guidance as part of their relationships with young children.

- **Indirect guidance** refers to the behind-the-scenes arrangements that early childhood practitioners make in the environment that not only prevent problems from occurring but also help children learn appropriate behaviour. Indirect guidance actually reduces the number of conflicts or problem situations that arise, making the atmosphere more positive and reducing the need for direct adult intervention.
- **Direct guidance** includes either verbal or nonverbal messages that early childhood practitioners send to children about appropriate behaviours. These messages include explanations of what the children's limits are, the reasons for them, and what changes in behaviour are necessary. Early childhood practitioners learn to communicate clearly, in terms that children can understand. They are careful to use words and techniques that teach without shaming or demeaning children. They are less concerned with merely stopping the undesirable behaviour or disciplining the child, and more concerned about what the child can learn about behaviour. Children need to discover that their behaviour affects the way others respond to them and that they cannot hurt other people or ignore their rights. They need to learn that some ways of expressing feelings are unacceptable, especially when they infringe on the rights of others. They need to learn that they can discuss differences and disagreements and do not have to rely on physical force to solve problems. Children need to learn that adults will keep them safe while helping all the children in the group learn to live within limits.

Building positive relationships with children means both directly and indirectly guiding children's behaviour, based on an understanding of children's development. Positive relationships do not include harsh or punitive measures or the withdrawal of warmth and affection. Besides undermining both the adult-child relationship and the child's self-esteem, these actions are ineffective in either changing behaviours or supporting children's abilities to regulate their own behaviour.

Early childhood practitioners set up the environment to offer learning opportunities for each child (courtesy of Yes I Can! Nursery School).

Designing Learning Opportunities

Children whose physical and emotional needs are met are primed to learn from the world around them. They are active and ready receptors for stimulation from new experiences, which provide them with the learning they need for later competencies.

As resource persons who help children as they construct their own continually shifting knowledge of the world, early childhood practitioners facilitate children's active learning. They provide time, space, materials, and support for children's active explorations to promote development and learning. They choose to make specific opportunities available based on their knowledge of individual children, child development, and family and ethnocultural goals and needs.

1920
Uniform franchise is established through the *Dominion Election Act*, making permanent the right of women to be elected to Parliament.

In designing learning opportunities that contribute to optimal development, early childhood practitioners

- are continually learning more about educational theory and practice in general and, in particular, about the children and families with whom they work. They seek out new information and experiences of others, gather information about their own group of children through observations, and set specific goals for supporting children's development and learning.
- organize and create the environment for children's active learning. They decide on the aesthetics of the environment, creating a look that invites children to participate. They set the order and organization of the children's environment, placing objects where they will attract notice and can be used most effectively. They consider what materials, toys, and objects will enhance the environment for particular children.
- match their teaching strategies to what is required to best assist a child's learning. The term "scaffolding" describes the process of helping children reach new levels of understanding (Berk & Winsler, 1995). Russian psychologist Lev Vygotsky used this term to describe the kind of assistance that adults give children in their learning. "If the children have gone from [point] A to B and are getting very close to C, sometimes to reach C, the child needs to borrow assistance from the adult at that very special moment" (Filippini, quoted in Edwards, 1994, p. 153). In helping a child take the next step, the adult functions rather like a scaffold that allows the child to reach further than would be possible unassisted.

Bringing children to learning opportunities is part of the supportive relationship between an early childhood practitioner and a child; the child learns through active involvement, not through passively receiving information.

Early childhood practitioners open up learning opportunities for young children when they expect children to become confident and competent. These expectations encourage a young child's hopefulness.

> An amazing quality of the human brain is to create an image of the future. To make an internalization of a better place, a better way, a better life, a better world. This capacity is called hope ... Some of the most influential people in any person's life may be someone they have never met. They have used that person to create an inner image to aspire to, to idealize, to idolize. Role models, mentors, heroes—all can provide critical formative experiences for children (Perry, 1996, pp. 8–9).

Connecting with Communities

Early childhood practitioners are part of broader communities—neighbourhood, faith, ethnocultural, school, professional, and workplace communities. They can connect the families they work with to the resources and supports available in various communities. They can connect children to other services to meet their health or developmental needs or to programs that offer specialized opportunities for recreation, sports, and cultural activities.

Across North America, there is a trend in child and family services, including early child development programs, to improve the coordination and integration of a continuum of services at the local level (Benner, in press). Early child development programs have an important place within this continuum, but early childhood practitioners will need to be actively involved in planning, delivering, and evaluating these services. Child care centres, nursery schools, and family child care homes cannot best serve children and families if they remain outside this continuum, which includes players from education, health, and social services.

In Canada, there are currently a number of exciting initiatives that provide a holistic view of community. They have incorporated early child development into a continuum of services and programs that are founded on principles of community decisionmaking and autonomy. These initiatives include South East Grey Community Outreach and the Better Beginnings, Better Futures projects discussed in Chapter 1.

Children live in families and families live in communities (courtesy of George Brown College).

What Early Childhood Practitioners Do

A number of initiatives have identified the key skills and competencies that early childhood practitioners working with young children in early childhood settings need. The recent Child Care Sector Study (Beach et al., 1998) describes the skills and competencies early childhood practitioners need to work with young children, whether in centre or home-based environments.

An early childhood practitioner will

- ensure that the physical environment and daily practices of caregiving promote the health, safety, and well-being of children in care;
- establish a working partnership with parents that supports their responsibilities to their children;
- develop and maintain a responsive relationship with each child and with the children as a group;
- plan and provide daily learning opportunities, routines, and activities that promote positive child development;
- observe and think about children's activity and behaviour;
- act in a manner consistent with principles of fairness, equity, and diversity to support the development and learning of individual children within the context of family, culture, and society; and
- work in partnership with other community members to support the well-being of families (Beach et al., 1998, p. 4).

What Is This Early Childhood Practitioner Doing?

Before we leave this discussion of roles, it will be useful for you to read the following story of a family child care provider and identify the various roles she is playing in her morning of interaction with her group of children. Although not all of the roles just discussed are represented in this account, it will be useful to see

2000
1950
1900
1850
1800
1750
1700
1650

1921
The first Canadian maternity protection legislation is introduced in British Columbia.

Patty Smith Hill begins a lab nursery school at Columbia Teachers College.

the integration of the roles in a typical scenario of good practice, and, not incidentally, representations of the excellent practice we discussed in the preceding chapter.

A Puzzle, a Picnic, and a Vision: Family Day Care at Its Best

Not long ago I was fortunate enough to be present during an extraordinary drama between a family day care provider and the children in her care.

It was late morning. A three-year-old and a four-year-old were racing small, cast-iron cars along a homemade highway running from the arm of a comfortable chair to the living-room wall. A willing toddler retrieved the cars that had met their demise on the moulding. A mobile eight-month-old infant followed in his wake.

In the kitchen, oblivious to the noise and excitement, Joel, a four-year-old boy, sat at the table, intently working on a hundred-piece puzzle.

After some sixty cars had wrecked themselves on the wall, the older children tired of the game. Hungry and ready for lunch, they advanced toward the kitchen table.

Absorbed in the puzzle, Joel was unaware of the danger headed his way; but the caregiver could see what was about to happen. Scooping up the infant, she placed herself between the group and the kitchen. "Joel is working at the table," she said. "I'll go and see how he's doing. You wait here."

She stood over the boy a moment before she said, "The other children are hungry. Are you ready to stop for lunch?"

Joel looked up, but only for an instant. "I'm not," he said.

Passing me on her way to the children, the provider explained, "He worked on a different puzzle for over an hour yesterday. When he was finished, he picked out this one to do today."

To the children she said, "Joel is still working on the puzzle." She gave out the information evenly, the way someone might read a telephone number from the phone book.

"So?" asked the four-year-old.

"Let's take a minute and think about what we should do." The caregiver released the infant and sat down on the floor with the children. "Let's see. We could wait and give Joel time to finish what he's doing."

"We're hungry now!"

"Well, then that won't work. What else could we do?"

The infant shrieked, but no one else said anything.

"I can think of things," the caregiver said in the kind of voice that lets children know something special is about to happen.

"Like what?"

"Well, we could fast. We could skip lunch for today and see how we felt." The provider dangled this as bait, but the preschoolers knew enough not to bite.

"All right, we could make Joel stop. We could tell him it isn't fair for him to keep working when we're hungry. We could mess up his puzzle because he's kept us waiting."

Unaware of the momentousness of this suggestion, the infant and the toddler were making their way back to the cars. Transfixed, the older children held their breath.

"We shouldn't do that," the four-year-old finally said.

"I agree, Sara. I'm glad you said that." The caregiver smiled at Sara and then, with a flourish, pulled the rabbit out of the hat: "We could have a picnic, instead, in the living room."

"A picnic! A picnic! We want the picnic!"

Looking down from his observation post in his highchair, the infant watched as the caregiver, the toddler, and the two preschool children spread a blanket, a plastic tablecloth, sandwiches, milk, and fruit on the rug.

When everyone was seated, Joel left the table and joined the group.

"How come he gets to have a picnic with us?" the three-year-old asked.

"Why?" the caregiver returned the question.

"Because he wasn't finished with the puzzle."

"Is it finished now?" she asked Joel. She couldn't see the tabletop from the floor, where she was kneeling.

"Not yet."

"You'll be able to work on it again later, after the picnic, if you want to," she said.

"Yeah. Because we're having a picnic and you didn't have to move it!" the three-year-old exclaimed.

This whole drama took no more than five minutes, and yet, so inspired was this caregiver's nurturing, the moment lives with me still. She had a vision of what a child—a person—could become. She nurtured the children's highest qualities—the ability to listen to their own conscience, the ability to talk about their feelings, the ability to correct their mistakes and go on. She was looking for more than good behaviour from them; she had a commitment to the human spirit.

She knew intuitively how to guide them. "We could mess up Joel's puzzle," she said in a voice so beguiling it caused my heart to skip a beat. Only a moment before, the children had been on their way to do just that. But the caregiver knew what she was doing. The children didn't take off for the kitchen; they paused and considered. "We shouldn't do that," Sara said. And this time Sara knew that she was right.

The same skill is evident in the provider's conversation with Joel. She could easily have decided whether or not to let him continue with the puzzle. But more is at issue here than lunch. It is important for Joel to learn to recognize his feelings and speak on his own behalf. And so the caregiver invites him to consider the problem. His answer reveals her nurturing—he knows that he matters.

As I left this home, I thought of how fortunate these children were to have this adult for their caregiver. Her vision was profound, her instincts were true, and her touch was light. Child care can't get much better than that (Baker, Amy C. [1992]. A puzzle, a picnic and a vision: Family day care at its best. *Young Children,* 47 (5), 36–38).

Consider the effectiveness of the various roles this teacher played. These are the roles you too will play as you become an early childhood professional.

2000
1950
1900
1850
1800
1750
1700
1650

1922
Women in Prince Edward Island gain the right to vote and to hold elected office.

1923
Arnold Gesell writes *The Preschool Child.*

SUMMARY

Early childhood practitioners take on numerous roles and responsibilities in various early child development settings. Motivations to enter the early childhood workforce include enjoying children, wanting to make a difference to families, desiring variety and challenges, and seeing a demand for early childhood practitioners in the workforce. Competent early childhood practitioners bring together personal characteristics and dispositions with knowledge and experience.

REVIEW QUESTIONS

1. Identify several of the roles of early childhood practitioners discussed in this chapter.
2. Discuss briefly your understanding of the rationale for each role, including specifics about what that role might look like in particular settings.
3. Discuss the interrelationships among the various roles.

STUDY ACTIVITIES

1. Identify the early childhood practitioner roles that you find in the following examples. Remember that the various roles are interrelated, so you may find more than one role.
   a. A family child care provider writes a newsletter describing some of the children's recent learning activities. In the letter, she includes a copy of her philosophy of discipline, and invites parents to comment on it. She also solicits donations of old kitchen utensils.
   b. An early childhood practitioner of infants displays family photos, under clear plastic paper, above the baseboards in the room.
   c. A preschool early childhood practitioner prepares for her fall conferences with families by organizing the anecdotal observations the two teachers in the room made on each child and noting questions that she wants to ask each parent.
   d. A kindergarten teacher clips newspaper articles on pending local legislation that will affect adult-child ratios in child care centres, and posts them on her parent bulletin board, along with the addresses of local representatives.
   e. A Head Start early childhood practitioner arranges for her director to observe a child in her group who frequently hurts other children.
   f. An early childhood practitioner of toddlers creates a cozy area in her room with piles of soft pillows and stuffed animals. She frequently sits here with children.
   g. A kindergarten teacher adds several new containers of objects to his counting centre.
   h. An early childhood practitioner of infants rocks a fussy baby, then places him on a mat with some colourful objects just out of his reach. He's been crawling for a week now.

2. Go back to the scenarios that began this chapter. Now what roles do you find?
3. Complete the following statements:
   When I first thought about becoming an early childhood practitioner, I thought that educating meant …
   Now I'm beginning to think that educating also means …
   One early childhood practitioner role that appeals to me a great deal is …
   One early childhood practitioner role that concerns me at this point is …
4. Observe in an early childhood education and care setting for a morning. Take detailed objective notes on one scenario. Later, label each early childhood practitioner role and define its purpose.

KEY TERMS

caregiving: Physical nurturing and protection of young children.
direct guidance: Teaching children appropriate behaviour through direct methods, including verbal and nonverbal messages. An alternative term, discipline, is used infrequently because of its negative connotations.
indirect guidance: Arrangements teachers make in the time and space environment that influence children's behaviour by both preventing problems and creating a positive learning environment.
nurturing: Providing care and fulfilling needs to promote development.

SUGGESTED READINGS

Benjamin, Ann C. (1994). Observations in early childhood classrooms: Advice from the field. *Young Children,* 49 (6), 14–20.

Bergen, Doris. (1994, Summer). Developing the art and science of team teaching. *Childhood Education,* 70 (4), 242–243.

Bredekamp, Sue. (1993, November). Reflections on Reggio Emilia. *Young Children,* 49 (1), 3–16.

Duckworth, Eleanor. (1987). *The having of wonderful ideas and other essays on teaching and learning.* New York: Teachers College Press.

Feeney, Stephanie, & Chun, Robyn. (1985, November). Effective teachers of young children. *Young Children,* 41 (1), 47–52.

Ferguson, E. E. (1995). *Child care ... becoming visible.* Halifax: Child Care Connection, NS.

Klass, Carol S. (1987, March). Childrearing interactions within developmental home- or center-based early education. *Young Children,* 42 (3), 9–13, 67–70.

Nelson, Bryan G., & Sheppard, Bruce (Eds.). (1992). *Men in child care and early education: A handbook for administrators and educators.* Minneapolis MN: Men in Child Care Project.

Powell, Douglas R. (1986, September). Effects of program models and teaching practices. *Young Children,* 41 (6), 60–66.

Rogers, Dwight L., Waller, Cathleen Boggs, and Perrin, Marilyn Sheerer. (1987, May). Learning more about what makes a good teacher good through collaborative research in the classroom. *Young Children,* 42 (4), 34–39.

Saracho, Olivia N. (1988). Cognitive style and early childhood practice. In Bernard Spodek, Olivia N. Saracho, & Donald L. Petus (Eds.), *Professionalism and the early childhood practitioner.* New York: Teachers College Press.

Spodek, Bernard, Saracho, Olivia N., & Petus, Donald L. (1988). Professionalism, semi-professionalism, and craftsmanship. In Bernard Spodek, Olivia N. Saracho, & Donald L. Petus (Eds.), *Professionalism and the early childhood practitioner.* New York: Teachers College Press.

Stephens, Karen. (1994, January). Bringing light to the darkness: A tribute to teachers. *Young Children,* 49 (2), 44–46.

CHAPTER FIVE

Becoming an Early Childhood Practitioner

OBJECTIVES

After studying this chapter, students will be able to

- identify and discuss some motivations for entering the early childhood workforce;
- discuss the importance of self-knowledge in becoming an early childhood practitioner;
- identify stages of development as a student and as a practising early childhood practitioner;
- discuss why different types of educational preparation for early childhood practitioners exist; and
- identify and describe the Canadian Child Care Federation's Training Guidelines.

Why do individuals choose to be early childhood practitioners and make a career in early child development? We will look at the motivations of several people who decided to work with young children. We will also explore the personal characteristics and attitudes that are important for members of the early childhood workforce.

The enjoyment of being with children and guiding their development of new skills and abilities is strong motivation for many early childhood practitioners (courtesy of Play and Learn).

Why Become an Early Childhood Practitioner?

In Canada each year a few thousand women and men enter the field of early child development through college ECE programs. But, the truth is, many do not end up working in early childhood settings, and many others who begin a career in child care centres, kindergartens, or family child care programs leave their work for other employment. Why do they come or decide to leave? Clearly, educating and caring for young children demands much of those who accept the task. So the education and care of young children is not an endeavour that can be entered into casually or accidentally—at least, not if the early childhood practitioner is to remain satisfied and productive and stay in the field.

Your participation in an ECE postsecondary education program represents your own intention to examine issues of early child development with some seriousness. As you read about early childhood practitioners who decided to work with young children, think about your own motivations and expectations of what working with young children as a career can offer you.

2000
1950
1924
1900
1850
1800
1750
1700
1650

1924
Lawrence Frank, administering the Rockefeller Foundation grant, is the first to begin using a multidisciplinary approach to studying children at Columbia University.

St. George's School for Child Study (later the Institute of Child Study) at the University of Toronto and McGill Day Nursery are founded through a grant from the Laura Spellman Rockefeller Foundation.

Motivators toward a given profession are both large and small. Some are related to human characteristics, including the emotions and meaning for the individual, and others are related to structural aspects, meaning the way the profession itself is organized. We will consider the human motivators first—enjoyment of children and making a difference to children and their families—and then look at the motivators related to the structure of the sector.

Enjoyment of Children

For most early childhood practitioners, the fact that they genuinely enjoy being with young children is the main reason they consider a career in early child education care, and it is definitely the reason they stay.

> "What other job can you have that lets you play with kids, and watch them playing? They're great—I mean really fun. Amazing things come out of their mouths. They know so much, and they're learning so much. And never, I mean never, boring. Kids are so real."—Connie
>
> "I'd never really paid much attention to children until I had my own. Then I was blown away. There's these little people, and every single day they find something new. And the world is really brand new for them, and their eyes are so wide, taking it all in. And I get to look at it all again too, because I'm with them. It's like getting a second chance at the world."—David
>
> "I just like being with kids. They're honest and funny and smart, and they love your singing, even if nobody else does. I can just let myself go and be silly right along with them. Nothing else is so much fun."—Laura
>
> "Hutch was a year old and my first day with him I was hooked ... Hutch was just a joy to be with ... I remember one time early on, it just happened that I had Hutch and Eric on the same day because their mothers asked me, and I was fascinated by the interaction between them. It wasn't some kind of underdeveloped, superficial play that you might read about. I was amazed at how full and engaged it was."—JoAnne

A deep and genuine enjoyment of children and a desire to be with them is a primary motivator for early childhood practitioners. They recognize that children are interesting and valuable in their own right, and they don't approach them primarily as persons in need of change or instruction. This enjoyment goes beyond the sentimental view that is frequently phrased as "I just love children, they're so sweet." Early childhood practitioners can enjoy young children while still being quite realistic about their developmental characteristics. These characteristics may include stubbornness, illogical thinking, uncontrolled outbursts, and a tendency not to care about other people's needs or rights. They recognize that children are not always sweet—they can scream, whine, and kick, and are frequently smelly and sticky!

Making a Difference to Children and Families

Another primary motivation for most early childhood practitioners is the tangible evidence that they are making a difference in the lives of children and their families. Every early childhood practitioner knows of some specific impact she has had on children, and that impact has a lasting effect on these children's lives. Research proves it. **Longitudinal studies** of children show that participating in early education contributed to their later school success and social adjustment.

One review of the impact of early childhood education and care programs on later development shows that the results of thirteen longitudinal studies point in the same direction (Tremblay & Craig, 1995). Children who participated in early childhood programs tended to complete higher levels of education, have fewer social problems, and more easily find employment than their counterparts who did not to participate in early education programs.

A particularly well-known longitudinal study in the United States shows that early education programs have a dramatic impact in the lives of children who are at risk. Lazer and Darlington (1982) demonstrate that early education programs enable children to progress satisfactorily through later grades without failure or the need for special assistance. The Perry Preschool Project of Ypsilanti, Michigan, has followed participants of a compensatory preschool program for twenty-seven years (Schweinhart and Weikart, 1993), as well as contemporaries who did not have the benefit of an early education program. There were noteworthy differences not only in school success but also in social factors such as involvement with crime, marriage, and job and salary levels. The researchers found specific positive correlations between participation in early childhood learning experiences and later positive accomplishments. *"It was the development of specific personal and social dispositions that enabled a high-quality early childhood program to significantly influence participants' adult performance"* [emphasis added] (Schweinhart and Weikart, 1993, pp. 11–12).

In France, all children may attend preschool programs, which are part of the public school system, and most children attend at least two years of full-time preschool education before entering formal schooling in Grade One. Researchers and educators have found that children from all socioeconomic and family backgrounds are likely to be more successful in elementary school if they have at least two years' preschool experience (Bergmann, 1996).

A Canadian survey (Canadian Child Care Advocacy Association and Canadian Child Care Federation, 1992) identifies enjoyment in working with children as the primary source of satisfaction for staff working in child care centres. Family child care providers echoed this sentiment in the recent Child Care Sector Study (Beach et al., 1998) and in a survey of regulated family child care (Goss, 1998).

As a child, how did a particular adult make a difference in your life? As an adult, have you been able to touch a child's life, with lasting impact? How was that important to you?

Of course, children do not come as independent beings to early child care and education settings—they are part of families. Early childhood practitioners reach beyond the confines of their settings to support the lives of families, acknowledging that strong and stable families mean healthy communities. Parents of the youngest children need information and emotional support to best nurture their children's development. They need the security of knowing that they can leave their children in the hands of trusted and competent early childhood practitioners while they go to work. They need connections with community resources that will help them perform all the tasks of parenting. They need to trust early childhood practitioners to help their children become interested in learning and in school.

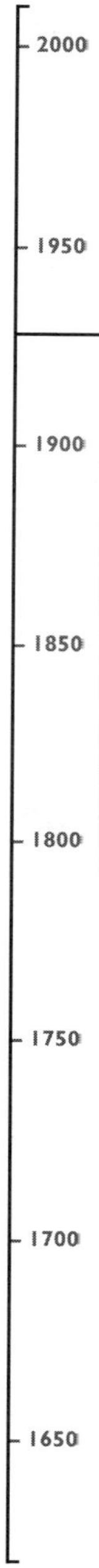

1925
Women over twenty-five gain the right to vote in Newfoundland.

Sigmund Freud's theory of personality development, stressing the importance of the early years and relationships, is published.

1926

Jean Piaget publishes *The Language and Thought of the Child*, outlining the four periods of the growth of intelligence.

Canada's first nursery schools open at the St. George School for Child Study in Toronto and at McGill University in Montreal.

The Canadian Council on Child Welfare sponsors prenatal and preschool letters to mothers eager to find out the most up-to-date child-rearing advice.

They need to be welcomed into the centre, school, or home to contribute to their children's lives away from the family home. They need to feel respected by other adults and by society for their primary importance in their children's growth and development.

Early childhood practitioners are on the frontlines, between parents and the society at large. They are often the first people outside the family who have prolonged contact with children and parents. Their support and relationships may be crucial determinants in families' optimum functioning. They have the power to help parents understand and accept their children and to support parent-child relationships.

Variety and Challenge

Some people are attracted to work that is never predictable, in which challenges and questions arise continually, and each day is likely to be different from the day before. Life in early child care and education settings with children from birth through elementary school is just that—and then some.

As you read the descriptions of quality programs in Chapter 3, you probably realized that the active, hands-on learning that is required means that early childhood practitioners live in a world of movement, noise, choice, and activity. In such an environment, there are an infinite number of tasks that call for attention simultaneously. A typical scenario may be four children splashing at the water table, while another needs assistance in the bathroom and yet another is crying quietly in the book area, when the phone rings and a parent walks in the door. Working with young children demands energy, knowledge, patience, and creativity, as well as the capacity to be flexible in the face of diverse demands.

Much of the variety in an early childhood practitioner's day comes from the children and their families. With rapidly developing children, change is the operative word. Today's crawlers are tomorrow's walkers; abilities to use a spoon, catch a ball, or climb stairs are acquired in quick spurts; and the use of language often seems to explode overnight. The diversity of **inherited** potential and environmental experience offered by each family means that a roomful of three-year-old children will have stunningly different interests, needs, and ideas.

Early childhood practitioners have to develop the skills and intuition to adapt to a great many unknowns in the course of a day. Sometimes the small events indicate larger happenings: the six-month-old baby seems unusually fussy, until you find a new tooth about to pop through; the six year old cries when he's forgotten his library book, then reveals that he left it in his Dad's car when Dad picked him up from Mom's house after the weekend visit.

You learn to make adjustments to children's individual styles and idiosyncrasies, as well as to new skills and abilities. The variety and challenges of working with unique children and families should motivate, rather than frustrate, you. Early childhood practitioners who are stimulated and excited by this part of the work environment are continually challenged to learn new things and develop new skills.

Early childhood practitioners encounter beings who are constantly changing and becoming, and they themselves are challenged to continue to grow. The sector is not clearly defined; rather, it is dynamic and offers a number of paths to follow.

Demand for Early Childhood Practitioners

One motivator for people considering a particular occupation or field is the availability of job opportunities; that is, the need for employees in the sector. The good news about early child care and education is that there appears to be no end to the demand for competent early childhood practitioners. Whether you are drawn to work with school-age children or with younger ones, there are many opportunities in every community across Canada.

The demand for early childhood practitioners to work with young children and their families is growing (courtesy of George Brown College).

Growing as an Early Childhood Practitioner

"A good teacher is first and foremost a person, and this fact is the most important and determining thing about him" (Combs, Blume, Newman, & Wass, 1974, p. 6). If you think back on the good educators you have known, you will likely confirm this statement. No doubt each educator stands out in your mind as a unique person, with a particular personality, values, beliefs, methods, and techniques.

Self-Knowledge

A good early childhood practitioner is the sum of the experiences, thoughts, and learning that have occurred in her life, all of which interact with the environment in which she works with young children. Since the self is the tool with which early childhood practitioners fulfill their caring and educating roles, both new and experienced members of the early childhood workforce benefit from an understanding of themselves—both a self-awareness and a self-acceptance.

Early childhood practitioners who examine their personal lives, including the experiences that have helped shape their values and the ways they live and relate with others, take part in an ongoing, important process. A word that is frequently used to describe educators who take the time and effort to analyze their actions and evaluate their experiences is *reflective* (Kochendorfer, 1994). An effective early childhood practitioner extends this concept to reflect on her whole life.

Take a moment to reflect on your own life. What early influences made you the person you are today? Who were the important people in your life? What picture comes to mind as a place you liked to be? What were you like when you were ten? Who gave you your name, and why?

What strong childhood memories do you have? How do these memories influence you as you think about working with children? Do you see an influence from your family on your decision to care for and educate young children? How have you been influenced by colleagues or mentors? What outstanding early childhood practitioners or caregivers do you remember? What were their central teaching and caring ideas and styles?

When did you decide to work with young children? What was it about early childhood education that attracted you? Were there any chance factors that led you to consider early education? How did you feel after you had made the decision? What are your goals for children?

What is important to you beyond early education? What do you do about it? What concerns you most about the state of the world and about the state of children and families? What have you read recently that was important to you? What do you see yourself doing five or ten years from now?

2000
1950
1928
John Watson publishes *Psychological Care of Infant and Child*, which suggests that parents not hug or kiss their children, or let them sit on laps.
1900
1850
1929
Women are deemed "persons" and can therefore be appointed to the Senate.
1800
1750
1700
1650

Disposition, Knowledge, and Experiences

It is impossible to list all the criteria and characteristics of the perfect early childhood practitioner. But we do know that good early childhood practitioners are a combination of particular dispositions, knowledge, and experiences that contribute to their abilities to care for and educate young children.

A disposition is an attitude or a tendency to behave in a certain way (Katz, 1995; Carter & Curtis, 1994). It is closely linked to personal characteristics that typify how an individual thinks, relates to others, and deals with daily life. Some characteristics that we expect from adults who will spend their days working with young children include patience, energy, enthusiasm, and a sense of humour.

The recent Canadian Child Care Sector Study on quality (Beach et al., 1998) asked early childhood practitioners and other caregivers working in child care settings what they thought were the most important characteristics for those who work with children. Patience topped the list, followed by respect for children and the ability to communicate.

There are several other dispositions that are important qualities for early childhood practitioners:

Commitment to self-discovery. Good early childhood practitioners recognize that the preparation for their work is an ongoing, never-ending process. Just as they recognize that their work with children is a process of facilitating development over time, rather than achieving specific, easily observable and quick goals, so, too, early childhood practitioners acknowledge that they continue to create their teaching, their understanding of what they do and how they do it, and their personal perspective. They continue to question, to puzzle, to set goals, to remain open to new learning. In so doing, they model for children the active process of becoming human beings. "The good early childhood practitioner is not a finished product, but a human process" (Combs et al., 1974, p. 144).

Being committed to self-discovery allows early childhood practitioners to grow. Early childhood practitioners must recognize and welcome their constant state of growth and change, and realize that they do not yet know what lies ahead, and what directions they will take. This uncertainty should not be regarded as daunting, but intriguing with its wealth of possibilities and unknowns. Good early childhood practitioners are curious about themselves, as well as about much else in their worlds. In addition to curiosity, early childhood practitioners need an ability to engage in self-criticism, to uncover the new challenges and demands that call for growth.

Feelings of adequacy. Competent early childhood practitioners accept themselves and feel comfortable with who they are. Basic feelings of adequacy allow early childhood practitioners to perform their daily tasks with the confidence and maturity to focus on the job at hand, rather than constantly evaluating their ability to perform or others' responses to them. Early childhood practitioners need to be able to transcend their own needs and pay attention to those of others. Working with young children is not about meeting the early childhood practitioners' needs for attention and affection. This is not to say that there are no emotional rewards for early childhood practitioners from their interaction with children and their families; rather, these emotional rewards are not sought to adjust early childhood practitioners' feelings of inadequacy. The early childhood practitioner must be focused on the needs of others, not self-indulgent of personal needs.

Flexible thinking. Each day, early childhood practitioners are required to make hundreds of decisions, both large and small. They have to select responses and methods that are appropriate to complex and unique circumstances. Often they are faced with situations that they have not encountered before and with behaviours that require them to adapt existing ideas and techniques to fit complicated circumstances. They must create their own answers to immediate questions and move beyond the theories to find appropriate answers to concrete conditions and feelings for children and adults.

The child care environment demands that early childhood practitioners depend on their own problem-solving abilities, rather than resorting to a "cook" approach that predetermines what the solutions are. Early childhood practitioners work from a solid knowledge base but are able to fit the knowledge of children, philosophy, and goals to the immediate conditions and discover the best course of action.

Autonomy. Early childhood practitioners who have developed autonomy are able to function in the flexible manner described above. They do not heavily depend on others' opinions to guide their actions or decisions. They understand their own goals and motivations and feel comfortable functioning independently. To be able to make choices, early childhood practitioners must have a healthy sense of autonomy, which allows them to accept ultimate responsibility for their work and their decisions. Being autonomous does not necessarily mean being solitary, however. Early childhood practitioners can assume responsibility and leadership, while still being able to work as a member of a team.

Empathy. Early childhood practitioners require the ability to identify with others, including the children with whom they work, their parents, and the other team members. Empathy is based on understanding the complexity of human development, and on respect and concern for others' personalities and feelings. Empathy grows as early childhood practitioners are able to focus on others' needs and feelings, an ability that arises from the early childhood practitioner's maturity and willingness to transcend issues of self. This allows early childhood practitioners to be really tuned in to children and adults, to recognize and accept their differences. Early childhood practitioners gain this knowledge of others by observing and making connections to allow others to reveal themselves.

Empathy prevents early childhood practitioners from exhibiting biases or prejudices; they are able to get past narrow thinking, as they take another's perspective, rather than shying away from others who are different.

2000
1950
1900
1850
1800
1750
1700
1650

1930

Over 1500 registered public health nurses are at work in Canada organizing and conducting well-baby clinics, visiting families at home to inspect school-age children, and sponsoring Little Mothers Leagues and Junior Red Cross branches.

McGill University Nursery School closes.

Fisher Price Toy Company is established.

1931
Mother Craft Society is introduced to Canada-trained well-baby nurses.

A sense of the "big picture." Early childhood practitioners need to have a philosophy that connects their work with young children to the important issues in their lives, so as not to get bogged down in the minutiae of early care. This larger philosophy is likely what Almy (1975) referred to in her list of important personal characteristics of early childhood teachers as "world-mindedness," an interest and a view that extend beyond the family child care or group care setting and connect the early childhood practitioner to webs of deeper meaning and social relevance. When early childhood practitioners understand the values and themes that are important in their lives, they find ways of embodying these ideas in their early childhood settings and practices.

Knowledge of child development and early education practices and experience are essential to competent early childhood practitioners. Research findings are clear that early childhood practitioners are more likely to be competent when they have postsecondary education, particularly in child development and early childhood education (Helburn et al., 1995; Goelman & Pence, 1987; Doherty & Stuart, 1996; Lyon & Canning, 1995). As noted in Chapter 3, educational qualifications are a key factor of quality. Experience also builds practical knowledge and helps early childhood practitioners refine their skills and their practice. "Teaching is an eminently practical activity, best learned in the exercise of it and in the thoughtful reflection that must accompany that" (Ayers, 1993, p. 12).

Stages of Early Childhood Practitioner Development

As you are beginning your training and career in early child development, it may be interesting for you to learn that some study has been done on the various sequences of professional development and growth of early childhood practitioners. Recognizing the common emotions and experiences and their needed supports at the various stages may be helpful. Katz (1995) discusses four distinct phases, with unique developmental tasks and training needs. The stages are generally linked to experience acquired over time.

Stage I: Survival. During this period, which may last at least a year, early childhood practitioners are mainly concerned with surviving—getting through the day or week in one piece, doing the work, and being accepted without doing something dreadful. Many early childhood practitioners feel inadequate and unprepared as they face the realities of a classroom of energetic, developing children. Early childhood practitioners at this stage need direct help with specific skills, as well as encouragement, reassurance, comfort, support, and understanding, as they develop basic concepts of what young children are like and what to expect of them.

Stage II: Consolidation. After some time, early childhood practitioners believe they will survive the immediate crises and stay in the profession. They are now ready to consolidate their overall achievements and to concentrate on learning specific new skills. They begin to focus on individual children and situations that are troublesome and to look for answers to questions for children whose behaviour departs from the norm. Early childhood practitioners at this stage benefit from discussion with more experienced colleagues about possible alternatives for action and resources.

THEORY INTO PRACTICE

Another View of Dispositions

Seven dispositions characterize master early childhood practitioners and are the desired outcomes of training and education:

- delighting in and being curious about children;
- valuing children's play;
- expecting continuous change and challenge;
- being willing to take risks and make mistakes;
- providing time for regular personal reflection and self-examination;
- seeking collaboration and peer support; and
- acting as an advocate for children and families (Carter & Curtis, 1994).

Source: Adapted from Front Line Profile, IDEAS, Volume 4, Number 2, December 1997, p. 19.

Stage III: Renewal. Frequently, after several years of teaching, early childhood practitioners get tired of doing the same old things and the same activities for successive groups of children. They become interested in learning about new developments in the field. It is often useful for early childhood practitioners at this stage to meet colleagues from other programs, to attend local and regional conferences, to read more widely, and to set new learning goals for themselves. Supports for all of these educational endeavours are helpful.

Stage IV: Maturity. After several years of teaching, many early childhood practitioners have reached a comfortable level of confidence in their own abilities. The questions they ask are deeper, more philosophical, and more abstract. Early childhood practitioners at this stage search for insight, perspective, and realism. These early childhood practitioners are ready to interact with other educators as they work on problem areas on many different levels. (See more about these stages in Katz, 1995.)

Stages of Student Early Childhood Practitioner Development

You may also be interested to know that someone has organized some typical patterns that student early childhood practitioners often experience. It might be comforting to know that you are not alone when some of these feelings and situations occur during your first practicum and teaching experiences. If you read through them all, you will see that there *is* light at the end of the tunnel, even though on some days, that seems doubtful.

1932
Subsidies begin to decrease and costs to provide services increase; parent fees rise to twenty-five cents a day.

Phase I: Anxiety/euphoria

- Easily identifies with children—because it is important to be liked by them.
- Starts making friends—an important early step in developing the authority relationship necessary in teaching.

Phase II: Confusion/clarity

- Thinks, "I don't know anything about planning curriculum."
- Asks, "What are rules?"
- Asks, "How should routines proceed?"
- Wonders, "When should I intercede between children?"
- Sometimes avoids situations out of fear or not knowing whether it is the early childhood practitioner's role.

Phase III: Competence/adequacy

- Feels triumph in guiding children.
- Comes up with a good idea during planning.
- Gets a hug.
- Needs strong reinforcement from supervisor. Focuses on the positive.
- Even though progress has been made, still feels everyone else seems to "know it all."
- Feels there's so much to learn.
- Wonders, "Will I ever be that good?"
- Still finds it difficult to be a controlling figure or disciplinarian.

Phase IV: Criticism/new awareness

- Says, "If it were my classroom, I would do it differently."
- Begins to notice imperfections.
- Starts to find fault with practicum of early childhood practitioner.
- Starts to question as part of the growth process.

Phase V: More confidence/greater inadequacy

- No longer questions—has acquired a sense of stability.
- Knows she'll make it.
- Experiences success more frequently.
- Wants more responsibility.

Phase VI: Loss/relief

- Has to depart from children with whom she's become close.
- Puts standards of performance in proper perspective.
- Finds returning to classes somewhat difficult (Caruso, 1977).

You will notice that individual variations during the phases may range from experiencing success and confidence to discovering that this may not be the field that best matches student capabilities. Many student early childhood practitioners find that keeping a **journal** during their experiences helps them express their feelings and keep track of their learning experiences (Surbeck, 1994).

> Growing early childhood practitioners is different from training them. Oddly, we more often think about growing plants than about growing people. People, especially the young, are to be domesticated—trained as dogs and horses are—to make them reliable, responsible members of society ... An alternative to domestication is liberation (Freire, 1970). Early childhood practitioners, like other people need some of both (Jones, 1993, p. xi).

ECE Postsecondary Education

As you prepare to join the early childhood workforce, you will want to know what educational training and professional experience you need to work with young children in early childhood education settings. As you can imagine from what you have already learned about early education, the answers to this question are not simple. Quite different educational requirements are necessary for the various kinds of settings.

ECE students are guided by early childhood practitioners in their field placements (courtesy of George Brown College).

Diversity in Early Childhood Practitioner Preparation

Cheryl W. is currently a student at a community college, where she will earn a diploma in ECE. She plans to work in a child care program with preschool children. The centre where she hopes to teach has two early childhood practitioners with four-year degrees in ECE and teacher certification credentials; one who has a bachelor's degree with a major in abnormal psychology; two who have earned a certificate in ECE offered through continuing education at a community college; and four who have no related training.

Christy S. is a university student, completing a bachelor of education degree, with a major in primary education. When she graduates, she will be eligible to apply for certification as a school teacher. She hopes to be a kindergarten teacher. After high school, Christy took a year off school and worked in a YMCA kindergarten child care program as an assistant. During that year, she participated in professional development activities, such as a recreation leadership course, creative movement workshops, and conflict resolution sessions, which were offered by her employer.

Charmaine is a family child care provider and a student in a family child care certificate program offered through continuing education at the local community college. She has five children under six years attending her family child care program and looks forward to using some of the new ideas she is learning about in her classes. Charmaine has been providing family child care for seven years.

Why are there such differences in the educational paths these women took, when many of them have similar career goals? What do these differences in education and training mean to their performance in ECE programs and, ultimately, to the question of quality in programs for young children? Today you will find such differences in every program for children, in every province or territory and location you visit. What do you already know about the provincial/territorial requirements where you hope to work?

One reason for the multiplicity of educational paths is that there are also several typical routes by which people become early childhood professionals: the traditional route, the parent route, and the "serendipitous" route (Bredekamp, 1992). When early childhood practitioners come into the profession via different routes, the profession must respond with different kinds of training and education.

- *Preservice training* and education refers to professional education acquired before entering the early childhood workforce.
- *In-service training* and education refers to professional education and development acquired, while employed in an ECEC setting.

THEORY INTO PRACTICE

This was written in 1920 by Jessie Van Stanton, who would later become director of the Bank Street Nursery School, after one year of teaching at the City and Country School. It was an assignment to write an exposition about the ideal teacher. She says she wrote it while "oppressed by a sense of dreadful inadequacy, convinced that I could never learn to be a teacher." This summary recognizes that there is no ideal for an early childhood practitioner.

The Ideal Teacher and How She Grows

The teacher of young children should have a strong physique and a strong well-balanced nervous system. She should be plump and round and have curly short hair. Her cheeks should be rosy and her teeth pearly white. She should have a pleasing personality and a quiet firm manner. She should be poised and of a high moral character. She should have sentiment but not sentimentality. She should be gentle but not sloppy, strong but not impetuous when bitten or scratched.

She should have a fair education. By this I mean she should take a doctor's degree in psychology and medicine. Sociology as a background is advisable.

She should be an experienced carpenter, mason, and plumber and a thoroughly trained musician and poet. At least five years' practical experience in each of these branches is essential.

She should be a close observer and a judge of character and should be able to deal with young and old. She should be able to hypnotize the parents of her young pupils and to cause them to change life-long habits of thinking in two mothers' meetings.

Tested at birth and found to have an I.Q. of 150, she was taken from her parents and brought up in totally disinfected surroundings. She was given a cold bath and a globule of gland each morning and her health was carefully watched.

From early childhood she was given every sort of tool and taught to practise close observation. She spent every morning seeing and sawing. In the afternoons, her time was spent on music, hearing, howling, and handling. She spent her evenings practising manners, masonry, mechanics, mesmerism, and musing on metamorphosis. Thus she acquired early in life the inestimably useful habit of using every second to its highest capacity. *Now, at 63 she is ready!*

But, added to all the virtues and attainments for a teacher, the ideal director should have a spine of steel—to stand the long hours and the tremendous demands made upon her. Her spine must bend easily, however, so that she can crawl under radiators with the cleaning man to dig out the dirt—or put away blocks for two-year-olds on shelves close to the floor. She should have feet of iron so that she can go up and downstairs tirelessly all day long from kitchen to roof to classroom to office.

Now she has studied. Now she has taught. *Now, at 83—she is ready to direct!*

Source: Stanton (1990). Used with permission of New York State AEYC. Originally published in *New York Nursery Education News* (Winter 1954).

Regulation and Certification

Why do early childhood practitioners take different educational paths to achieve similar goals? One reason is found in the differences among the provincial/territorial licensing requirements in regulated child care and nursery school settings. (Look back at Chapter 3 to review the various qualification requirements.)

- Preservice professional education is not required for all positions in licensed child care or nursery school programs in any province or territory.
- There are few educational requirements for regulated family child care in any jurisdiction, although there are some requirements for early childhood practitioners in these settings to participate in professional development activities.
- In some jurisdictions there are no educational qualifications for staff in regulated child care programs.

There are also variations in the requirements for government approval of educational qualifications. As we discussed in Chapter 3, in some provincial jurisdictions, governments issue licenses or certificates or authorize specific classifications to early childhood practitioners.

- In British Columbia, the government issues a license to practice to early childhood practitioners who have the necessary educational qualifications and hours of work experience.
- In Alberta, Prince Edward Island, and Manitoba, early childhood practitioners must submit their education qualifications to the provincial government, which will then assign the appropriate classification for regulated centre-based programs.
- In Newfoundland, all qualified ECE staff in regulated child care centres are required (by the provincial child care legislation) to have their certification from the provincial organization, the Association of Early Childhood Educators, Newfoundland and Labrador.

Each province and territory also has different requirements for early childhood practitioners who are kindergarten and primary or early elementary teachers within the school system, and those who are working in other types of early child development programs. In all provinces and territories, teachers in the school system are required to have a university degree and teacher education qualifications. There are a few professional education programs in Canada that offer joint ECE and teacher education programs. Early childhood practitioners who complete these programs are qualified for either system.

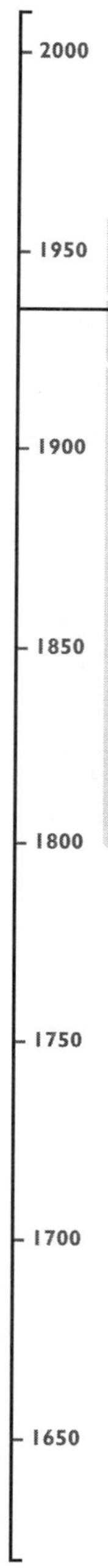

1933
Commission des assurances sociale du Québec proposes establishment of public-funded child care.

About twenty day nurseries exist in Canada, serving approximately 2500 children.

Equivalency

The recognition of out-of-province educational qualifications to meet requirements for credentials is another aspect of regulation and certification. Each jurisdiction (with the exception of New Brunswick, the Northwest Territories, and Newfoundland, where there are no educational qualification requirements) sets up its own process for determining if educational qualifications from outside that province/territory are comparable or equivalent to those recognized within its boundaries.

1934
The Dionne quintuplets are born in Callander, Ontario.

It is important for you to be familiar with the requirements in your province or territory. If you think you may wish to move to a different part of Canada, you will want to know how your credentials will apply. You may also want to know how the credits you are now earning could be applied to further education.

College and University ECE Programs

In Canada, postsecondary education, like regulated child care, is the responsibility of the provinces and territories. The federal government does transfer funding to provincial and territorial governments for postsecondary education, and to support research, and participates in initiatives to coordinate education across Canada. But it is the provincial and territorial governments that establish postsecondary institutions and policies for their governance. So, like the delivery of regulated child care and other early childhood education and care programs, each jurisdiction sets up its own system. This adds to the diversity of early childhood workforce across Canada.

Most of the preparatory training and education of early childhood practitioners is offered through two types of postsecondary institutions—colleges and universities. There are more than 120 postsecondary institutions offering ECEC certificate, diploma, and degree programs across Canada (ACCC, 1997).

The length of programs varies from two-semester certificate programs to three- or four-year degree programs. Degree programs are usually offered in universities, although there are a few colleges that now have degree-granting status. Some programs are full-time, and others are available part-time through continuing education or distance education programs.

Provinces and territories establish broad guidelines and procedures for both college and university programs. Universities are autonomous institutions, with their own boards of governors, which establish their own curriculum content and methods of teaching and assessment within the broad provincial/territorial policy framework. Colleges may be autonomous, or they may be directly operated by the provincial/territorial government. College programs usually receive more direction from provincial/territorial governments than do university programs.

Universities provide degree programs leading to certification for teaching kindergarten and elementary grades in the school system. Teacher education programs are driven by provincial/territorial policies and curriculum guidelines with little consideration of the ECEC sector. This is similar to the situation in the United States (Spodek, 1994).

ECE Certificate and Diploma Programs

Canadian colleges and Quebec's CEGEP programs offer fifty ECE certificate and seventy-six ECE diploma programs (Beach et al., 1998) and graduate approximately 4000 early childhood practitioners each year. The programs prepare early childhood practitioners to work with young children in early childhood education and care settings, particularly child care centres and nursery schools. Specific curriculum content in ECE certificate and diploma programs is shaped by the particular institution, by provincial/territorial postsecondary education requirements,

MAKING IT HAPPEN

Red River Community College, Winnipeg, Manitoba

Red River College is a postsecondary institution located in Winnipeg, Manitoba, which offers a two-year ECE diploma program. Graduates from this program are eligible for classification as Early Childhood Educator II by the Manitoba government, and are thus qualified staff members in Manitoba's licensed child care centres and nursery schools.

The ECE diploma program was first offered at Red River College in 1971. Since that time, the program has continuously evolved to remain current and to meet the needs of its community. For example, when the provincial government increased its requirements for educational qualifications for program staff in regulated child care centres, Red River College moved management courses to a post-diploma program and increased the focus on child development, curriculum, and professionalism in the ECE diploma program. When the federal government established the Aboriginal Child Care initiative, Red River College developed many partnerships with First Nations communities for community-based training in rural and remote communities.

Since 1986, the ECE program at Red River College has delivered the program following a competency-based model. The curriculum is organized around core competencies for entry-level early childhood practitioners as defined by experts in the field. Each of the courses or modules has a learning package or "Com Pac," which faculty members review regularly and update to include current research and best practices. The individual learning packages are clear about what the student needs to learn and how she can demonstrate the learning. Faculty members produce an instructor's guide, accompanying readings, and assignments for each module. In addition to course work, ECE students complete six field placements, the equivalent of about nineteen weeks.

The detailed, thoroughly developed curriculum has made the program very portable, providing flexibility for various delivery modes and communities. As well, with the course content organized in learning packages, there is considerable flexibility for students to switch from part-time or distance education to the full-time program with full recognition of their course credits. It has also allowed for a consistent recognition of credits through Prior Learning Assessment.

Currently, the ECE program is offered on-site on a full-time basis as well as through continuing education, distance education, and Prior Learning Assessment. The program is also piloting an ECE Workplace program in an attempt to relieve the shortage of trained ECEs in Manitoba. In this program, students are experienced employees who continue to work three days a week and attend classes two days a week. Prior Learning Assessment, workplace practicum, Gap training, and regular classes are all components of this accelerated program.

The program has also been delivered full-time in ten locations in First Nations communities outside Winnipeg, including several remote fly-in communities. These projects enabled many students to graduate who would never have been able to attend classes in Winnipeg. The competency-based approach, with learning objectives clearly defined, allows for considerable community and cultural relevance. There are many paths to competency, and under the supervision of an experienced off-site coordinator, instructors and students can adapt learning tasks and assignments to be relevant to their communities. Inspired by both the shortage of culturally relevant resources and the fine work being done by aboriginal ECE students at Red River College, the program was successful in obtaining funding to develop a video series *Our Children, Our Ways: Early Childhood Education in First Nations and Inuit Communities*. This series is now used across Canada and internationally with diverse groups.

(cont'd)

Red River Community College, Winnipeg, Manitoba (cont'd)

The program has also collaborated with other agencies to develop and deliver full-time programs that include some ECE content as well as other curriculum content, tailored to meet the needs of a specific group.

In partnership with community agencies the program has developed and delivered a Family Daycare Training Project, a Child Care Assistant Training Project, a Family Support Worker Training project, and a Childminders program, which combines part-time ECE and English and a Second Language content for immigrant women.

With a shortage of trained ECEs in Manitoba, graduates from Red River College enjoy a healthy employment record. In 2000/01, 100 percent reported employment in areas related to their studies. Also, Red River College graduates may transfer their ECE diploma credits to university studies. The college has several articulation agreements with universities enabling graduates to obtain credits toward degrees in the areas of education, developmental studies, family studies, and human services.

Most ECE graduates go on to join the early childhood workforce and find employment in a child care centre or nursery school. Many return to Red River College to participate in one of its continuing education post-diploma programs (Studies in Aboriginal Child Care, Studies in Special Needs Child Care, or Infant Care). With experience and this additional professional education, a number of graduates move on to positions as directors of child care centres or in early childhood intervention services. The ECE graduates from Red River College meet the requirements for staff qualifications in most jurisdictions across Canada.

In Canada, Red River College is an acknowledged leader in several areas, including

- implementation of a competency-based approach to ECE programs;
- articulation agreements with universities, allowing ECE grads to articulate to degree programs;
- very high student accessibility through alternative but equivalent delivery of ECE programs;
- community-based training programs in partnership with urban aboriginal and First Nations communities; and
- the production of multimedia resources for ECE training.

In 1999 the program embarked on a substantial program review process that included the input of community, graduates, and faculty. The main outcomes of this review were the reorganization of content into forty-hour packages, delivering all practicum as block placements and more emphasis on emergent curriculum and creating high-quality play environments for young children.

Other innovative initiatives include the proposed development of a core curriculum resource that would translate the most recent research findings related to neurological development in young children to a flexible, learner-friendly core multimedia resource. As well, the program has just received funding to develop a video series and documentary for broadcast profiling best practices in Family Resource Programs across Canada.

The program has also recently participated in two international projects, in Vietnam and the Philippines, which involved staff from Red River College going to those countries and Vietnamese and Phillipine staff spending time at Red River College.

The ECE program at Red River College continues to evolve to maintain currency and quality and to meet the needs of its community.

Source: Based on interview with Joan Kunderman, Chair, Community Services, Red River College (2001).

by provincial/territorial requirements for regulated child care, and by the early childhood workforce and its organizations in that jurisdiction.

In spite of differing institutional and provincial/territorial requirements, the recent Child Care Sector Study (Beach et al., 1998) found remarkable similarities across all Canadian ECE certificate and diploma programs.

ECE students at Red River College in Winnipeg, Manitoba.

- All programs include course content related to child development, teaching/caregiving practices, and behaviour guidance.
- Most (about 90 percent) include course content in health, safety and nutrition, observation skills, interpersonal communication, and the foundations or history of early education theory.
- Ninety percent focus their program's course content on centre-based settings, whereas only 10 percent include specific curriculum content related to other ECEC settings, such as family child care.
- All certificate and diploma ECE programs include field placement requirements.
- ECE certificate and diploma programs contain general education content, in addition to the professional education related to ECEC.

The study also identified a number of gaps in ECE certificate and diploma programs, including the need for a more family-centred approach, for more content related to infant/toddler and school-age care, for the inclusion of children with special needs, and for greater attention to cultural, linguistic, and racial diversity.

Guidelines for ECE Programs

To prepare early childhood practitioners, postsecondary education programs must expose students to general knowledge and specific competencies associated with all groups within the early childhood span—infants, toddlers, preschoolers, and primary-aged children. Students must learn about children with needs beyond those of the typically developing child. They must also have course work and practica that particularize their knowledge about some of these age groups.

In addition to this general knowledge, some professionals in particular roles, such as a director of an early childhood program or a family child care provider, require specialized knowledge. Some roles also will require greater depth of knowledge for teachers to be effective, such as learning more about special education or language delays. And because of the complexity of their roles, early childhood practitioners must reach beyond child development and early childhood education and learn the communication skills that will allow them to work effectively with colleagues on the teaching team and on more extended professional teams, with parents and others in families, and with the community beyond. At each successive level of preparation, from the introductory courses and certificate programs through to graduate degree levels, professionals become increasingly able to apply, analyze, and refine the core knowledge to improve practices. It is recognized that higher levels of general education are linked to quality in programs for children, as professionals are able to apply their specialized knowledge. General knowledge also helps early childhood teachers create learning experiences for children that draw on the broad content of general studies in the arts and sciences.

The Canadian Child Care Federation (CCCF) has responded to the need to improve the preparation of early childhood practitioners and to ensure greater consistency of ECE postsecondary education programs. In 1989, the CCCF established a national training committee to study educational issues and concerns. Over the next several years, the committee facilitated a series of national discussions about ECE postsecondary programs. These discussions led to the 1993 CCCF conference, where there was clear support to move forward to develop national guidelines. After considerable consultation with interested individuals, organizations, educational institutions, and government departments across the country, the committee prepared *National Guidelines for Training in Child Development*. The guidelines are built on common values and beliefs and identify the essential components of the ECE postsecondary education necessary to prepare entry-level early childhood practitioners. The guidelines include items related to curriculum

THEORY INTO PRACTICE

National Guidelines in ECE Training

Program Content

- Studies in self-awareness and communication are essential for the learning and application of knowledge, skills, and attitudes.
- Knowledge of human development and an appreciation of the role of the family serve as the foundation for promoting each child's optimal development.
- Specialized studies in early child development form the framework for supporting professional theory and practice.

Learning Environment

- Faculty in ECE programs play a key role in facilitating a student-centred learning process.
- Administrators have a responsibility to support an optimal learning environment.
- Students have a responsibility to contribute to an optimal learning environment.
- Advisory bodies, as representatives of the early childhood community, ensure the link between training programs and the broader early childhood education and care field.
- Governments and policymakers have responsibility to provide the financial support and the regulatory framework.
- All stakeholders share responsibility for ensuring the delivery of quality program content in a quality learning environment.

Source: Canadian Childcare Federation (1997). Used with permission.

content and the learning environment. They can be applied to full-time, part-time, and distance education programs. In addition to the guidelines, the committee developed a companion document, *Towards Excellence in ECCE Training Programs: A Self-Assessment Guide,* which provides a tool for educational institutions to review their ECEC programs.

To date, several community colleges have reviewed their ECE programs using the guidelines and assessment guide. The guidelines provide a common reference point for ECE postsecondary programs offered by institutions in various provinces and territories and could be the basis to develop closer coordination of programs, which would enable early childhood practitioners to more easily transfer their ECE credits.

NAEYC has just revised and expanded its guidelines for associate and baccalaureate professional preparation programs (1995). Those changes reflect a clear emphasis on preparing practitioners for working in environments that fully include children with disabilities and special learning and developmental needs with typically developing children. In addition, the revised guidelines include a greater depth of preparation for working with families and a more family-centred approach, as well as more in-depth preparation for working collaboratively with other adults. A final area of change in the guidelines is increased emphasis on individualizing curriculum and assessment to reflect cultural and linguistic diversity (Bredekamp, 1995).

Other Professional ECEC Preparation Programs

ECE university degree programs. There are about a dozen ECE degree programs offered at Canadian universities, which can extend early childhood practitioners' career opportunities. Some of these university programs are offered as bachelor of arts programs; others are bachelor of child studies or child and youth studies. The

MAKING IT HAPPEN

Certificate in Family Supports with Specialization in Family Resource Programs

In September 1996, the Continuing Education Division of Ryerson Polytechnic University initiated a certificate program related to family resource programs. The program is offered both on-site at Ryerson and through distance education. It includes eight university-level credit courses. The program is designed for individuals working in family resource programs or related community services, some of whom are early childhood practitioners with ECEC qualifications.

The courses reflect an inclusive and dynamic approach to changing families and communities and promote family resource programs as a support to all families. The certificate program encourages participants to expand existing knowledge and skills and allows them to transfer credits to other areas of university undergraduate study.

Source: Canadian Childcare Federation (1997). Used with permission.

ECE degree credentials extend early childhood practitioners' careers. A few programs have developed in collaboration with ECE college-level programs or offer concurrent ECE degree/diploma programs.

Admission requirements to ECE degree programs vary. A few require an ECE diploma or other postsecondary education, and others may give advanced standing for college-level ECE credentials.

The curriculum of ECE degree programs often focuses on working with children with special needs and on early intervention strategies in specialized and regular early childhood education and care settings. Many programs offer students specialization options. Some ECE degree programs emphasize preparation for further graduate study.

1940
Women in Quebec gain the right to vote in the provincial elections, completing the enfranchisement of women in Canada.

1941
4.5 percent of married women in Canada work outside the home.

Family child care certificate programs. Many early childhood practitioners who have completed ECE certificate or diploma programs operate family child care programs in their own homes. Other family child caregivers enroll in ECE certificate or diploma programs through continuing education or distance education. Only British Columbia and Ontario provide certificate programs specifically designed for family child care.

Specialized post-certificate and post-diploma programs. A number of postsecondary education institutions in British Columbia, Manitoba, and Ontario offer post-certificate or post-diploma programs specializing in child care management and administration, children with special needs, infants and toddlers, and school-age children.

Teacher education programs. Teachers in the school system in Canada complete a university degree and specific teacher education programs to be eligible for provincial/territorial teaching credentials. Teacher education programs usually lead to a bachelor of education degree. B.Ed. degrees may be earned as a first university degree or as a second degree (Canadian Education Association, 1993).

A first degree B.Ed. program takes four to five years to complete and is open to students with a secondary school diploma. It is often offered as a conjoint, concurrent, or integrated program and includes study both in professional training for elementary or secondary education and in an academic discipline area. Students then graduate with both a B.Ed. and another degree such as a B.A. or a B.Sc.

Students with ECE certificates or diplomas are usually eligible for admission to these programs, even if they do not have the required secondary school diploma. In a few instances, there are articulation agreements between a B.Ed. program in a university and an ECE program in a college that allow for advanced standing into the second year of the teacher education program.

Other teacher education programs are organized as a second degree (a B.Ed.), after the completion of another university degree program, including a degree in early childhood education or child studies.

Mentoring

During the past several years, a new pattern has been added to opportunities for professional growth and learning in many early childhood communities. It is possible that you will have an opportunity to benefit from a mentoring program, either as a student or as a professional. In mentoring programs, experienced and effective early childhood practitioners are given specialized training, to help beginning early childhood practitioners gain skills and become more effective practitioners. Students or beginning early childhood practitioners, called protégés, mentees, or apprentices, are paired with the **mentor** early childhood practitioner, generally in the mentor's early child development program. Here, during experiential learning, mentors act as backup support, offering feedback that allows the beginner to move to a higher level of skilled performance.

Obviously, both participants gain from such arrangements. Experienced early childhood practitioners receive recognition, advanced education, and increased salaries when they act as mentors. They can remain in their early child development settings with children, using their developed skills and expertise, and yet feel that they have advanced in their own professional careers, taking on new positions of leadership in their centres and communities. They are more likely to remain in the field, with the new status and enhanced salaries, as well as with the new interest in leadership and professional development. Their own practice is likely to improve, as the mentors increase their reflection on their teaching to share with the protégé. Protégés are also more likely to remain in the field over time, having been given the opportunity to develop the skills needed to prevent the overwhelming frustration of the unskilled. Having formed a relationship with an experienced early childhood practitioner, they see that early childhood is a

MAKING IT HAPPEN

Mentoring Programs in Canada

Recent initiatives in Canada studied the effectiveness of a mentoring approach in ECE field placements:

- Pollard (1996) examines both mentor and supervisory roles involved in the relationship between an early childhood education student and the child care centre staff member who is a cooperating teacher in field placement situations. The study concludes that early childhood education students in field placement benefit from a mentor-mentee relationship with the sponsoring/cooperating teacher.
- Singleton (1997) tried out a mentoring model with ECE students completing field placement at the lab centres at the College of the North Atlantic in St. John's, Newfoundland. The mentors planned scheduled conferences and journal writing for students to reflect on their practice experiences. The study reports benefits for both the mentors and the students because the mentoring process led both to examine, reflect on, and improve their childhood care practices.
- Child Care Connections—NS developed the Child Care Substitute Youth Internship program (Ferguson, 1998) for fifteen youth in Halifax and Sydney. The interns spent two weeks in a child care centre and then had two weeks of course work, combined with sessions between the intern, mentor, and practice coordinator during the practice period to reflect on and evaluate their performance. The four-week cycle continued over six months. Each intern was matched with a mentor who was a skilled child care practitioner working in one of the child care settings. Mentors received training in communication, problem-solving, and teamwork to support their work with the interns. The mentors and interns both reported valuable learning and insights from their participation in the mentoring process.

viable profession over time. They have received the coaching, guidance, and counselling that help during the initial adjustments to any new experience. In lessening the probability of staff turnover and in developing more educated and skilled practitioners, mentoring programs are making important contributions to communities and early childhood programs. Therefore, the quality of services to children and families is directly affected by the development of such opportunities.

There are diverse ways to structure mentoring programs, according to community resources and needs. For examples of successful programs, see Whitebook, Hnatiuk, & Bellm, 1994.

SUMMARY

This chapter suggests that good early childhood practitioners are people who have certain dispositions and characteristics that allow them to grow as early childhood practitioners and as individuals. These attributes include having feelings of adequacy, the ability to think flexibly, a sense of autonomy, an ability to empathize, values and philosophy that enable teachers to take in the big picture, the ability to disclose oneself to others, and the acceptance of being a growing person. Knowledge and skill development will enhance teacher effectiveness. Growing early childhood practitioners—and student early childhood practitioners—experience common problems and feelings at different stages. This chapter has discussed the varying methods of preparing for professional roles in the early childhood field. There is little uniformity in current requirements, owing to the nature of provincial/territorial regulatory requirements for licensed early child development programs, the provision of postsecondary education in Canada, and the autonomy of most educational institutions. Currently, organizations in Canada and the United States are attempting to create a comprehensive framework for the preparation of early childhood practitioners. This framework includes consistent guidelines for postsecondary education programs, flexibility of entry point and preparation options, and an increase in the level of postsecondary education that is required for all early childhood practitioners.

After reading this chapter, you likely have a clearer perspective of where your own goals and training fit into the larger picture. It is important that you understand the options open to you and comprehend the purposes behind the design of your current program. A well-prepared early childhood professional can contribute most to these continuing discussions.

REVIEW QUESTIONS

1. Identify and discuss the motivators described in this chapter.
2. Which motivators have personal meaning for you?
3. Discuss your understanding of why self-knowledge is important to the development of an early childhood practitioner.
4. Describe several attributes that are important for early childhood practitioner growth and optimum functioning.
5. Describe several characteristics of the various stages of early childhood practitioner and student early childhood practitioner development.

6. Discuss the differences between licensing and certification.
7. Identify reasons for diversity in early childhood preparation requirements.
8. Discuss the concept of mentoring as it is currently being used in early education professional development.

STUDY ACTIVITIES

1. Interview early childhood practitioners about why they chose this field and their motivations for staying. Also ask them about the characteristics they feel workers in this field need, and about things they dislike about the field or their job.
2. Many of the following questions are for your consideration in your private notebook or journal. Some are appropriate to share with a partner.
   a. List ten words that you would use to describe yourself, to yourself only. List ten words that someone who knows you well would use to describe you. List ten words that an acquaintance would use to describe you. List ten words with which you would *like* to be described. Reflect on these lists. What do the similarities and differences mean?
   b. How have you changed in the last three years? What would you like to accomplish in the next three years? What thing that you don't know how to do right now would you like to learn in the future?
   c. What is your favourite thing to do alone? With others? What is something that is special to you that you would like to share with a child?
   d. Complete these sentences:
      I want to be an early childhood practitioner who ...
      I want to be an early childhood practitioner who believes ...
      I want to be an early childhood practitioner who feels ...
      I don't want to be an early childhood practitioner who ...
      I don't want to be an early childhood practitioner who believes ...
      I don't want to be an early childhood practitioner who feels ...
      I feel most competent when ...
      I feel most unsure of myself when ...
      It really bothers me when children ...
      I love it when children ...
   e. Ask your ECE faculty to describe their educational background.
   f. In what stage of early childhood practitioner or student early childhood practitioner development do you think you are at this time? What makes you think so? What would be most helpful to your growth as an early childhood practitioner right now?
   g. Analyze your early childhood program to see how the various courses provide the kinds of knowledge recommended by the CCCF statement on core knowledge for early childhood professionals, practical experience, and general knowledge.
   h. Check if there are any other ECE-related programs offered at your college. Is your ECE program eligible for advanced standing or special consideration for admission into an ECE degree program?

KEY TERMS

inherited: Characteristics directly transmitted to children from parents' genetic combinations.

journal: In early childhood education, notes made by student early childhood practitioner to document progress of children and teacher.

longitudinal studies: Research that follows the same individuals over a long time.

mentor: Experienced individual who supports someone with less experience.

SUGGESTED READINGS

Bowman, B. (1990). Self-reflection as an element of professionalism. *Teachers College Record,* 90 (3), 444–51.

Brand, Susan. F. (1990). Undergraduates and beginning preschool teachers working with young children: Educational and developmental issues. *Young Children,* 45 (2), 19–24.

Bredekamp, Sue. (1992, September). The early childhood profession coming together. *Young Children,* 47 (6), 36–39.

Ferguson, E. (1997). *Child care administrator credentialing: A work in progress.* Halifax: Child Care Connection—NS.

Greenberg, Polly. (1992, November). Making the most of it: Expanding professional horizons. *Young Children,* 48 (1), 58–61.

Greenough, Kathrin. (1993). Moving out of silence: The CDA process with Alaska native teachers. In Elizabeth Jones (Ed.), *Growing teachers: Partnerships in staff development.* Washington, DC: NAEYC.

Han, Eunhye. (1995). Reflection is essential in teacher education. *Childhood Education,* 71 (4), 228–30.

Heiss, Gayle. (1989, May). Preschool teaching priorities: Reflections of a former home morning care provider. *Young Children,* 44 (4), 31–36.

Humphrey, Susan. (1989, November). Becoming a better kindergarten teacher: The case of myself. *Young Children,* 45 (1), 16–22.

Johnson, Julienne, & McCracken, Janet B. (1994). *The early childhood career lattice: Perspectives on professional development.* Washington, DC: NAEYC.

Katz, Lilian. (1990, February). *On teaching.* Child Care Information Exchange, 3–4.

NAEYC. (1991, November). A vision for early childhood professional development. *Young Children,* 47 (1), 35–37.

Willer, Barbara, & Bredekamp, Sue. (1993, May). A "new" paradigm of early childhood professional development. *Young Children,* 48 (4), 63–66.

CHAPTER SIX

The Work Environment

OBJECTIVES

After studying this chapter, students will be able to

- discuss the characteristics of the work environment in early childhood settings;
- identify and describe what is meant by the term "career lattice," and discuss various career options in the early childhood field;
- identify and describe working conditions in early child development centres;
- describe several helpful supports for early childhood practitioners facing challenges; and
- prepare to apply for employment in an early child development setting.

In this chapter, we will juxtapose the positive aspects of a career in the early childhood workforce with the challenges faced by the early childhood workforce in Canada today. The early childhood workforce also offers opportunities for a variety of career directions, and options for skilled early childhood practitioners in Canada are expanding. The words and stories of early childhood practitioners who find fulfillment and satisfaction in their work suggest that the early childhood workforce can allow you to achieve your personal and professional goals. The recognition of our value in working with young children and supporting families is growing.

Nevertheless, there are challenges, and early childhood practitioners are often overworked, underpaid, and undervalued. But the field is expanding and becoming more visible, which expands the horizons for early childhood practitioners.

Some of the challenges must be accepted as things that cannot be changed. Others you can remedy through your own efforts and commitment as you become increasingly competent in the role of early childhood practitioner.

Career Directions

As you look ahead to your career, you may be looking toward thirty or more years. During that period, you may be involved in different kinds of work in the early education field. Today, as the field continues to grow and expand, there are numerous possibilities for those who choose to focus their life's work on early childhood.

As larger numbers of children live in families where both parents work, more early childhood professionals are needed to care for and teach these children. Families also need people to help them find care, and they need assistance paying for this care. In addition, our society has become more concerned with how this care affects young children and what quality child care really means.

These concerns have led to more jobs: we need more family and early intervention specialists to help families find the kind of care they need; more researchers to find out what produces quality child care programs; more trainers and consultants to ensure child care programs are meeting the quality requirements; and more administrators to help organize all these people and projects.

Careers in the early childhood field include those that serve children directly; those that serve families directly; those that organize services for children and families; those that provide information about children and families; and those that provide goods and services to businesses and governments and affect children and families. We will explore some of these careers here. Remember that within each type of position, there are many different job positions, and the particular responsibilities and requirements, as well as salary ranges, will vary with specific organizations.

Many early childhood practitioners work directly with children in a centre-based program (courtesy of Play and Learn).

Careers That Serve Children Directly

Working in centre-based programs. Child care centres or nursery schools will likely be a starting point for many of you. Almost every other related career requires that professionals have some direct experience with young children. Many of you may decide this is where you want to stay for many years; others may consider the experience an important foundation for other goals. From your reading of earlier chapters, you will know that there are many different kinds of centre-based programs to choose from, depending on your educational preparation. These may include public programs, such as kindergartens or primary grades in the school system; before- and after-school care programs; Head Start programs; child care centres that may be operated by churches, not-for-profit corporations, or for-profit organizations or owners; employer- or government-sponsored agencies; and part-day preschools. Some early childhood practitioners may operate out of a centre-based program but work as home visitors either in family child care or an early intervention service.

Early childhood practitioners in home-based programs. Early childhood practitioners may work with young children in their own home or in the child's home. In-home early childhood practitioners, sometimes referred to as nannies, may be hired by individual families. Family child care programs may be regulated or may operate outside provincial/territorial child care requirements beyond adhering to the maximum number of children. Many early childhood practitioners choose to work with young children in their own homes for reasons of convenience for their own families and preference for the types of relationships fostered in the home-like setting.

Resource teachers. Early childhood practitioners may decide to work with children with special needs and their families to ensure that they are able to fully participate in early childhood education and care programs. As the special needs of each child and the concerns of the family are identified, resource teachers and other service providers work with the family to create a plan to meet those needs.

This plan is called an Individualized Program Plan (IPP) or Individualized Family Services Plan (IFSP), depending on the age of the child. (You will learn much more about this terminology and this process in your later coursework on children with special needs.) Resource teachers work with other early childhood practitioners to implement the plan and include the child in the daily program. Resource teachers often help families access other support and remedial services as needed.

Child development specialist on health team. Professionals with a background in child development are often included on the assessment and education teams that provide services to children with special needs. Education specialists may interact with children during screening or assessment sessions, may help provide appropriate stimulation for children in small group therapy sessions, or may provide appropriate knowledge and guidance for family members. Some health departments include child development specialists on their teams to provide information about typical development of young children for parents. Other child development specialists may be part of the staff facilitating children's play in the play therapy room of hospital pediatric units. Some health agencies, such as the Red Cross, employ early childhood practitioners in their community education programs.

Recreation leaders. Some recreation facilities, such as city park and recreation programs and residential camps, employ early childhood practitioners to staff and administer their recreation programs for children. There are companies who sponsor exercise, dance, and fitness programs for children, and others that provide child care for parents who participate. Some library systems employ individuals with early childhood backgrounds to be storytellers and provide other services in children's rooms.

Careers That Serve Families Directly

Family support program staff. Family resource programs, parenting centres, and family literacy programs usually have a multidisciplinary team that includes early childhood practitioners. In these programs, early childhood practitioners will work directly with young children and their family members and other caregivers. They plan and carry out curriculum that is both adult- and child-focused and strives to actively involve parents and other caregivers in their children's early development and learning.

Family specialists. Family specialists include a broad spectrum of early childhood professionals who help families gain access to the services they need to care for their children. Many community agencies support families in the complex tasks of parenting and help find them the resources they need. A family specialist may provide information and education, refer families to services, help them gain access to funds to pay for services, or give direct support services. Family specialists may deliver services in agency offices, in child care programs, or in the family's home. Some have specialized expertise, such as child care referral counsellors, social workers, or family counsellors. Examples of programs that employ family specialists are child care resource and referral agencies, Head Start programs, community mental health and child abuse prevention agencies, and health departments.

1942
The *Dominion-Provincial Wartime Day Nurseries Agreement* is established. The federal government passes an Order-in-Council authorizing the Ministry of Labour to enter into cost-sharing agreements with any provincial government willing to establish day care services.

1942–45
Six community-based preschool child care centres are established in Montreal to support the war effort. Ontario establishes twenty-eight preschool and forty-two school-age day nursery programs.

Early intervention specialists design individualized programs to support the development of children with special needs (courtesy of Play and Learn).

Early intervention specialists. Early intervention is an interdisciplinary field that includes health, human services, and educational services for young children with special needs and their families. These specialists work directly with children and their families, in both homes and child care programs. They also work with other specialists who provide direct services to children and their families, such as child care providers, speech and language diagnosticians and therapists, physical and occupational therapists, medical personnel, and social workers. Early intervention specialists often develop and use Individual Program Plans or Individual Family Plans that are used by resource teachers (described in the previous section).

Early intervention specialists coordinate activities between the family and the other professionals working with the child, and they help plan and deliver services. These duties require specialists to use their knowledge of child development, assessment, and family needs. They must also be skillful in working with other professionals, in communicating with parents, and in applying techniques for working successfully with children.

Careers That Organize Services

Early childhood education and care program administrators. Program administrators usually have several years of early childhood education and care work experience, as well as specialized studies in administration and business matters. Administrators have responsibility for ensuring that programs offer developmentally appropriate experiences for children and meet all legal standards; helping teachers to grow and develop professionally; supporting the needs of families and involving them in their children's lives at the centre; and supervising the daily flow of all centre operations, including maintaining staff, collecting fees, meeting nutritional needs, and ordering equipment, materials, and supplies. Fiscal management is an additional challenge, and some may also have responsibilities for fundraising. Because the director is usually the one who handles crises, the job may include plumbing, first aid in emergencies, social work, and counselling on any given day. The differences in centre-based programs help determine what is expected of directors. Some very large centres also have an assistant director to help meet these responsibilities.

In large early childhood settings, program administrators may have much greater responsibility for arranging wider goals and priorities than for dealing with the day-to-day minute details of operating a program. For example, an upper-level manager may be the executive director of a child care resource and referral agency or a director of a multisite Head Start program. Program or project coordinators often manage a single specialty area, such as the education coordinator of a program, or a child care coordinator in a local department of social services.

Regulators. Regulators, or licensing specialists, ensure that early childhood programs comply with government requirements by visiting programs. In addition, many regulators provide directors and teachers with technical assistance and training to help them meet the standards. The knowledge required by regulators is comprehensive: they must know child development, appropriate programming and curriculum, effective guidance, and health and safety precautions. In addition, communication skills for working in the often difficult situation of monitoring others' practice are required. Most regulators are employed by provincial or local government agencies. Some monitor child care centres or homes; others monitor food programs or complaints of abuse or neglect.

Consultants. Early childhood consultants provide assistance and information to businesses, communities, and other organizations to help them develop child care programs or meet various standards. Usually, consultants work on-site to help the organization assess its current program, resources, and future plans. There is a growing need for consultants to help employers work out methods of providing for their employees' child care needs. In addition to knowledge of child development and child care program administration, consultants who work with the corporate community need specialized knowledge about market research and employee benefits.

Careers That Provide Information

Resource librarian/information officer. The expanding knowledge base about the science and practice of early child development creates new opportunities for early childhood practitioners. Resource centres, research institutes, early childhood organizations and governments are exploring ways to translate new knowledge and information to the early childhood workforce, parents, and communities. Early childhood practitioners understand the daily lives of front-line staff and families in early childhood settings and can translate what new policies and research findings may be relevant.

MEET THE EARLY CHILDHOOD WORKFORCE

Michelle Turiano

Michelle is a resource librarian at the Childcare Resource and Research Unit at the University of Toronto. (Many of the references and information boxes in this book have been supplied by the Childcare Resource and Resource Unit.) Michelle graduated from Ryerson University in 1995 with a B.A.A. in ECE.

While in high school, Michelle worked in an after-school program. She was drawn to the close, responsive relationships that the early childhood staff in the program established with young children and their families. Michelle recalls how one early childhood practitioner (who was her supervisor) made a huge difference in the lives of children who participated in the program. "I decided that I would like to do that," says Michelle. "The selling point of entering the field is the gratification that comes from relationships with children and their parents. At story time the children would cuddle up. They valued my presence and I valued their milestones. During my first field placement, I pushed myself hard to make

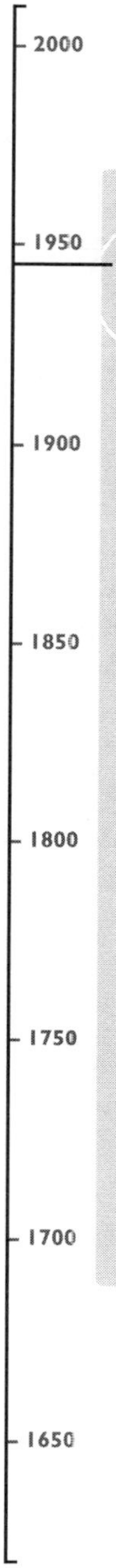

1943
British Columbia becomes the first province to license child care centres under welfare legislation.

Representatives from the government, the Children's Welfare Council, and the Institute of Child Studies attend a conference to establish guidelines for the wartime day nurseries in Ontario and Quebec.

Ontario establishes The Day Nurseries Branch, Canada's first provincial child care authority.

room for small talk with parents in the morning and at the end of the day. I found it rewarding to share the day with parents."

Michelle's experiences as a front-line practitioner are put to good use at the Childcare Resource and Research Unit. Michelle is responsible for maintaining the documents in the Unit's vast library, preparing new content for the Web site and gathering new information, reports, and documents from across Canada and from international sources. Michelle is able to sort through a vast amount of material and pull out what is really relevant to early child development policies and practices. Because she understands the day-to-day realities of early childhood education and care programs, Michelle is able to guide ECE students and practitioners to new research findings and policies. She also provides assistance to the Unit's various research projects, and finds that her practical experience brings a valuable perspective to the research work.

Michelle may work directly in an early childhood education and care program in the future. But for now, "The work here is challenging and another way to view the field. It is exciting to know about current policy developments, different reports, and cutting-edge research. I am able to bring this information to the field and to students entering the field."

1943
Junior kindergartens are introduced into Toronto Public Schools.

Marsh Report on Social Security in Canada recommends that maternity leave and benefits be extended to Canadian women.

Researchers. As programs and policies regarding children continue to expand, there is an ever-increasing need to understand all aspects of children's development and programming. Researchers generally focus attention on specific service aspects, such as child care, health services, early intervention programs, and nutrition. Much research focuses on children at various stages of development and on family variables and needs. Possible employers include colleges and universities, research institutes that may be affiliated with universities, government agencies, foundations, professional associations, and advocacy organizations. In addition to having a wide background in child and family development knowledge, many researchers need knowledge of observation and other data collection methods, as well as data analysis techniques. Many researchers begin their careers as part-time research assistants or interns.

Trainers and instructors. Those who work with practitioners have the important responsibility of helping those adults gain the knowledge and skills they need to work effectively with young children and their families. In addition to having a depth of knowledge in child development and all aspects of effective early childhood education and programming, they must understand adult development and learning as well as useful teaching strategies. Effective trainers and instructors have generally had considerable firsthand experience in early childhood settings with children. Many experienced early childhood professionals train others part-time in addition to working in the classroom with young children, by participating in workshops, taking part in local conferences, or teaching part-time at a local college. This is a satisfying way of making a larger contribution to the field, while still maintaining the connection to early childhood education and care programs.

You will recall from the discussion of mentoring that many early childhood practitioners work in their own programs with children while supporting students and other new practitioners in learning effective practice through mentoring. Others, such as regulators, family specialists, and early intervention specialists, also occasionally provide training as part of their jobs. Still others, such as college and university ECE faculty or high school child development teachers, teach other practitioners full-time. Some large child care programs with multiple centres employ their own educational specialists to work with their staff.

Authors and editors. There are a number of local and national publications devoted entirely to children's and family issues. These vary from the large, glossy magazines sold nationally on newsstands and by subscription to local publications sometimes distributed free of charge in doctors' offices and schools. All of these publications offer professionals opportunities to share their knowledge about children and families in print.

Careers That Provide Goods and Services

Merchandisers. There is a large consumer market for developmentally appropriate toys and materials, books, teaching materials, clothing, and other items used in the care and teaching of young children. Merchandisers may sell their products in stores or at local or national conferences. Many corporations prefer to have their items marketed by individuals with knowledge of child development.

Politicians and advocates. Although a background or experience in child development is not a prerequisite, gaining a voice in elected local, provincial/territorial, or federal government is one way individuals who care about children's and family issues can be effective. Legislators have a direct impact on creating the laws and policy that influence the lives of families and of childhood education programs. In our modern world of increasingly larger organizations set up to deal with the complexities of the legislative process, some people with early childhood interests may find there are opportunities to influence others by working as advocates with advocacy groups. Here, again, child development and other professional knowledge can make legislators and advocates far more effective.

Career Pathways

A conceptual framework of ECE postsecondary education and development must achieve a balance between inclusivity and exclusivity. It must fully embrace the diversity of roles and levels of preparation required for early childhood practitioners working with young children to provide high-quality services. It must also recognize that individuals enter the profession with diverse educational qualifications and experience, and promote a system that encourages ongoing professional development for individuals at all levels and in all roles. The framework must also set high standards for professional performance and distinguish the specialized skills and knowledge of the early childhood profession from those of other professions. (See Johnson & McCracken, 1994, p. 11.)

The symbol of a **career lattice** conveys the needed combination of diversity and development. The career lattice includes the idea of the multiple roles and levels (horizontal strands) and settings (vertical strands) within early childhood education, as well as diverse entry points and bridges between roles and settings in terms of professional preparation and responsibility (diagonals). It allows for the

1945
Attachment theory is discussed by Rene Spitz in *Hospitalism: An Inquiry into the Genesis of Psychiatric Conditions in Early Childhood.*

The federal Family Allowance program is introduced.

Quebec closes wartime centres. Ontario tries to but is stopped by the effective public campaign mounted by the Day Nursery and Day Care Parents Association.

possibility of upward movement with increasing responsibility and compensation within that role, as well as intersecting movement across the various roles. This seems a better image for a field as diverse as early childhood education than the more traditional ladder, which implies that there is just one clearly defined path to professional development and that one must step over others to advance.

The lattice, on the other hand, allows for the scenario of someone who is employed as a child care assistant without any preservice training. After two years of experience, she is given a promotion to lead early childhood practitioner and encouraged to enter a program at a community college, where she completes a two-year degree in early childhood education. Some time later, she seeks employment as a lead early childhood practitioner in a Head Start program, where she is encouraged to continue her education in a degree program. A number of years later, she applies for the assistant director position at her original child care centre.

The career lattice concept also allows for the scenario of a parent in a Head Start program who is hired as a bus driver/aide. He takes an introductory course in ECE at a local community college, then completes an ECE certificate through continuing education and, after a number of years, becomes a lead teacher in a Head Start program. While teaching in the program, he continues his education to obtain his teacher's certification, and he later becomes a kindergarten teacher in a local elementary school.

The lattice allows for the entry of people like these into the profession, and allows them to continue their professional growth and move into various roles. It allows entry, as well, for those who followed the more traditional routes of enrolling in a teacher certification program or degree program immediately after high school graduation, obtaining preservice credentials before gaining employment in the field, and then, perhaps, making several changes in the early childhood settings chosen for employment.

The lattice distinguishes the early childhood field from the early childhood profession; the field includes anyone engaged in providing early childhood services, whereas the profession includes those who have acquired some professional training and are on a professional path. It defines distinct categories on a continuum of professional development.

Early childhood practitioners work in many different types of settings and under many different working conditions (courtesy of George Brown College).

Working Conditions

Working conditions are the income, benefits, workloads, and characteristics of the social and physical environment. Provincial regulations and legislation about employment standards set the basic requirements for a minimum wage, mandatory vacations, and maximum numbers of hours of work per week.

Early childhood practitioners may be employed by parents, nonprofit organizations, local or provincial governments, businesses, or school boards, or they may be self-employed. The type of early child development setting affects the employment status of early childhood practitioners and the related working conditions.

The working conditions in each early child development setting create a climate for adults and children. If the salary and benefits are reasonable, job responsibilities and obligations are clear, health and safety are protected, and early childhood practitioners are valued, the working climate is positive. A positive work climate helps create a quality setting for young children and their families. (See Chapter 3 for further discussion about quality.)

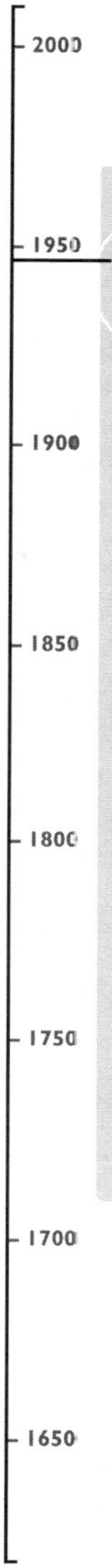

1946
The Common Sense Book of Baby and Child Care is published by Dr. Benjamin Spock.

Arnold Gesell, doing child development research at Yale, establishes the first descriptions of children at different chronological ages, using normative data.

The withdrawal of federal government funding leads to the closure of all forty-two school-age programs in Ontario.

Compensation

The good news is that income levels of early childhood practitioners are improving. For example, the average annual wage of an early childhood practitioner (who has primary responsibility for a group of children) in a regulated child care centre rose to an average annual salary of $22 717 in 1998, up from $18 000 in 1991 (Doherty et al, 2000).

But the still too-low income levels of early childhood practitioners are a major issue for the sector. Recent studies report that low remuneration remain a significant problem for those working in centre-based, family child care, and in-home settings, as well as for government officials, representatives from provincial/territorial and national sector organizations, and faculty members from colleges and universities (Beach et al, 1998; Doherty et al, 2000).

It is difficult to compare income levels for early childhood practitioners in different settings. Early childhood practitioners who work with children in centre-based programs or care for young children in the child's own home are usually employees. But early childhood practitioners in regulated or unregulated family child care settings are self-employed. Wages earned as an employee are subject to different deductions and taxation than is income earned through self-employment. Nevertheless, it is quite clear that both groups receive remuneration that fails to reflect the responsibilities and skills involved in working with young children.

Centre-based early childhood practitioners. A recent Child Care Sector Study reports that early childhood practitioners "with a college diploma or certificate working full-time and for the full year [in licensed child care programs] received less than 75 percent of the annual income of the average full-time, full-year female worker with the same education" (Beach et al., 1998, p. 77). Table 6.1 compares 1991 and 1998 child care wages in each province and territory.

You Bet I Care! (2000) is a major study of the wages and working conditions of staff employed in licensed child care centres across Canada in 1998. Higher education, more experience, unionization, and nonprofit status were factors found to be related to higher income levels, even within the same province or territory. *You Bet I Care!* also reported on benefits that staff in child care centres received. Only 25 percent of full-time staff reported pension benefits, and 26 percent did not have sick leave benefits. About half of the staff members in the survey reported that they did not have long-term disability benefits.

Early childhood practitioners in regulated family child care. A 1996 survey of caregivers in regulated family child care reported on the wages and working conditions for this group (Goss Gilroy, 1998). Most are self-employed, and their child

1946 (cont'd)
The American baby boom begins.

Ontario legislature passes the *Day Nurseries Act*. Federal funding for day nurseries ends.

1947
The Canadian baby boom begins and continues until 1964.

TABLE 6.1

Hourly Wages of Early Childhood Practitioners in Centre-Based Child Care in 1991 and 1998

Province/Territory	*1991* *Child care hourly wage*	*1998* *Child care hourly wage*
Newfoundland	6.03	6.76
Prince Edward Island	7.25	7.54
Nova Scotia	7.64	8.51
New Brunswick	6.19	7.12
Quebec	10.25	11.04
Ontario	11.51	13.48
Manitoba	9.29	9.49
Saskatchewan	7.52	10.47
Alberta	6.76	8.36
British Columbia	8.94	12.07
Yukon	9.58	11.71
Northwest Territories	n/a	

Sources: Doherty, Lero, Goelman, LaGrange, & Tougas (2000); Canadian Day Care Advocacy Association and Canadian Child Care Federation (1992).

care programs are set up as small businesses. They are able to deduct reasonable expenses from the child care fees they receive, either directly from parents or through government fee subsidy programs. In 1996 the average gross income (before eligible deductions) was $15 600; the average net income (after deductions) was $8400.

The income of early childhood practitioners in family child care is affected by the number of children in the settings. The maximum number of children is determined by the provincial/territorial regulations and varies from one region to another, but many providers care for fewer children than is allowed by government regulations.

The 1996 survey of regulated family child care found that providers have access to few employment benefits—only 7 percent had paid vacation time and only 2 percent had access to pension plans or sick leave (Goss Gilroy, 1998). Early childhood practitioners working in family child care are self-employed and, therefore, would have to purchase benefits on their own.

In the United States some of the highest-paid early childhood practitioners employed in child care centres earn less than $16 000 annually (Whitebook, Howes, & Phillips, 1993), and the average foregone annual wage—the additional money the worker could have earned in other female-dominated occupations—was $5238 for early childhood practitioners and $3582 for assistants (NAEYC, 1995).

Employment Contracts

A contract of employment outlines the terms or "rules" of an employment arrangement. It is an agreement between the employer (e.g., the early child development setting) and the employees (i.e., early childhood practitioners). In a unionized setting it is called a collective agreement and is negotiated by a process of collective bargaining.

A contract of employment usually includes

- job title;
- employment status;
- salary;
- benefits;
- work requirements; and
- employment procedures (hiring, termination).

Health and Safety

Your early childhood education studies emphasize the issues of health and safety of children in various early child development settings. Policies and practices ensure the physical environment is safe. There are also significant issues related to early childhood practitioners' health and safety in their work environments; early childhood practitioners are particularly vulnerable to infectious diseases, musculoskeletal disorders, and stress.

Getting sick is an occupational hazard in early childhood settings. Young children get colds, with runny noses and crusty eyes. They get gastrointestinal viruses and may vomit or have diarrhea. Early childhood practitioners who are exposed to these germs, and other common childhood diseases, risk becoming sick. Personal health care and preventive measures, as well as excellent hygiene practices, in the early childhood setting help to reduce the likelihood of illness. But, compared to most work environments, there is increased exposure to infectious diseases in work with young children.

Working with young children is physically demanding. Babies, children, furnishings, and equipment are carried or moved throughout the day's activities. Backs, knees, and joints can suffer.

Health and safety challenges are reduced by good working conditions that minimize the risks. As well, when illness or injury happens, sick leave and extended health care benefits can make a big difference.

Stress and Isolation

Early childhood practitioners are exposed to several sources of stress in their daily work environments, including the need for constant attention to keep children safe; frustrations connected to a lack of resources and too many demands from too many people; children's unpredictable interests, emotions, and behaviours; and emotional involvement with children and their families. Although the variety of situations and constant interactions with people are what attract and motivate many early childhood practitioners to work with young children, these same characteristics can be overwhelming.

Often in early child development settings, resources—human and physical—are in short supply. There may be too many children and too few adults to provide optimal experiences for young children. The quality of materials that the early

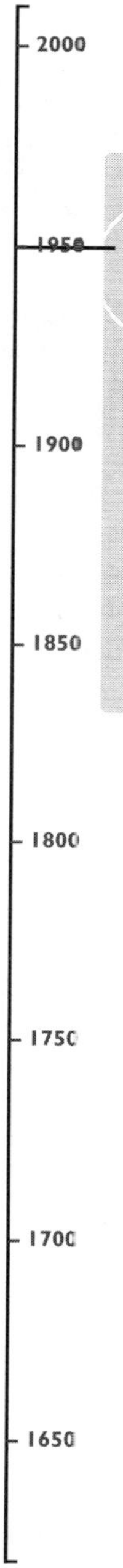

1950
Eric Erikson publishes *Childhood and Society*, outlining his theory of psychosocial development.

Nursery Education Association of Ontario (later AECEO) is started.

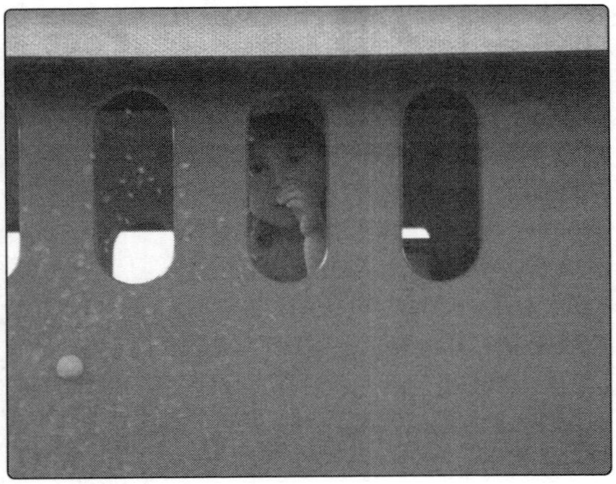

Working with young children can be isolating (courtesy of George Brown College).

childhood practitioner can provide for children is constrained by the program's budget. Early childhood practitioners know what they should do with children and what they would like to do, but they are often unable to do so because of situational constraints.

Beach et al. (p. 129) report that "the demands for, and expectations of child care are changing and often the caregiver bears the brunt of the inability of the sector to respond adequately to these changes. The sector is having difficulty responding to the needs of parents and offering high-quality care. Yet parents are becoming more aware of the importance of quality care and early experiences to their child's development, and are expecting more of caregivers." As resources become tighter and family expectations grow, early childhood practitioners feel the squeeze in trying to meet increased expectations with fewer materials, and with cutbacks in public health, social services, and public school programs.

The fact is that most of your time as an early childhood practitioner will be spent in the company of young children from infancy through to twelve years of age. Early childhood practitioners often work with other colleagues and assistants, but most of their conversations and contacts will be with the children. Even when parents and other family members are invited to spend time in the early childhood setting, the amount of time they spend there will be a small fraction of the workday. Though working with young children is rewarding and satisfying, it is also demanding.

The 1996 survey of providers in regulated family child care settings (Goss Gilroy, 1998) identified isolation and lack of contact with other adults as a work challenge. Almost half of the providers in the survey report dissatisfaction with the long and irregular hours, which interfered with their personal and family lives.

We will examine some of the characteristics of the work environment that often contribute to stress and isolation.

Variety and unpredictability. There is constant change in work with young children; no two days are the same. Although a moderate amount of variety is stimulating and exciting to most people, a constant barrage of unpredictable events may result in psychological overload. The need to adapt continually may be exhausting, both physically and emotionally, even while it is exhilarating.

Imagine the adjustments an early childhood practitioner needs to make when she discovers that her co-worker has had a car break down and is running late; that two of the children in her class have just been diagnosed as having chicken pox, and the other ten families need to be informed; that there is no red paint for the planned art activity this morning and the child who is painting at the easel is demanding a substitute for the blue paint provided; that the puzzle table that is usually quite attractive to her group of children draws only uninterested glances this morning; that the birthday party scheduled for lunchtime will have to be moved to morning snack time to accommodate working parents' schedules; and that the child who usually separates from her grandmother easily has chosen today to run after her with heartbroken screams. After order is restored, the early childhood practitioner may feel a boost in self-confidence as she realizes that she has coped with these and other impromptu decisions and situations, but, at the time she has undoubtedly experienced a good deal of the stress that comes with having to adapt quickly.

Let's analyze the reasons behind the stress in such a scenario. One reason is that early childhood practitioners are inextricably bound to a variety of other adults, both co-workers and the parents of the children. Their schedules, needs, and wishes, to say nothing of values, all have to be acknowledged and do have an effect on the early childhood practitioner's actions and schedule. A second reason is that early childhood practitioners have so many roles to play, and to play simultaneously. Concurrently, this early childhood practitioner is concerned with ensuring the health and safety of children, communicating with parents, planning and providing appropriate learning resources and a daily schedule, and providing emotional support. Small wonder that there is room for stress. Another reason is that young children are "predictably unpredictable" (Hyson, 1982, p. 26).

Fast-changing interests, emotions, and behaviour mean that teachers are never quite sure how children will react to activities or situations, as in the case of the unexpected **separation** distress. With emphasis in the early childhood curriculum on children's choice and independent exploratory play, early childhood practitioners regularly have to adapt to children's preferences in the use of materials. The unpredictable process of responding to children's choices means that early childhood practitioners face stress in trying to provide appropriate curriculum activities and materials, often without advance planning.

For many early childhood practitioners, this spectacular lack of limits—the variety, freedom, open-ended, and unpredictable nature of the early childhood setting—is attractive. For others, this same variety and freedom may create anxiety and stress. What about you? Are you the kind of person who enjoys seeing your capacities challenged, who feels energized when stretched into new territory and decisions? Or are you continually anxious when you move into unfamiliar territory?

Frustrations. Early childhood practitioners often have chosen their profession because of their high ideals and their desire to influence children's development in important ways. "They impose upon themselves the responsibility for unlocking each child's potential" (Hyson, 1982, p. 27). Each child comes with a particular background and cultural context and with unique abilities, interests, and needs. There are enormous differences in what each child needs from the teacher and in the teacher's ability to respond to each child and family. Good early childhood practitioners try their best to rise to each specific challenge and to find the resources to help when they are unable to do the job themselves. They spread themselves around as much as they can, often painfully aware that although there are one or perhaps two adults in a setting, there may be three, five, or even ten times as many children, all demanding and deserving of help. There is never enough time to do all that early childhood practitioners feel should be done, and it is frustrating to feel unable to meet that need.

Frustration also comes from the complexities of the early childhood practitioner's tasks. The children are the main focus, of course, but every early childhood practitioner would like to have more time to talk with parents, to share ideas with colleagues, to prepare new play materials, and to reach out to educate the community about young children's needs. The multiplicity of aims and goals may

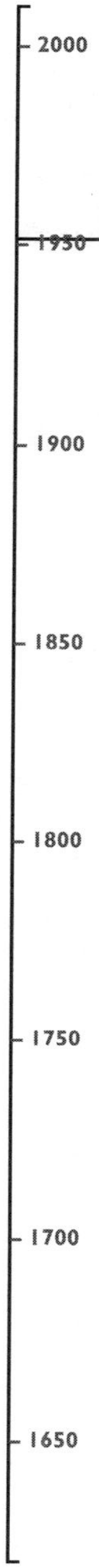

1951
11.2 percent of married women in Canada work outside the home.

be daunting. Time just does not permit a sense of completion in all these activities, and early childhood practitioners can feel frustrated in knowing that they could do their jobs better, "if only ..."

Frustration comes from not only being unable to accomplish everything one would like with children or families but also the fact that resources are often inadequate to support the work of early child development. In times of tight budgets, it is extremely rare to work in settings that provide all the materials and equipment that early childhood practitioners feel are essential to their task. Indeed, the programs that serve the needs of children and families often seem to be the first to feel the sting of budget cuts. On a day-to-day basis, limited resources mean constant efforts to stretch inadequate staff, materials, and supplies, in ways that are frustrating and sometimes downright humiliating. "I'm sorry," apologizes the infant early childhood practitioner, "We've run out of diapers so I'll have to ask you to bring some in for your child." "Only one cup of juice; we'll get seconds of water." "Just one dot of glue."

Frustration is a part of many occupations today that are stretched to do complex jobs with inadequate supports or resources. If early childhood practitioners are to avoid the **burnout** that comes with ever-mounting frustration, they will need to find the satisfaction that outweighs the frustrations in their work. What about your own response to frustration? Can you learn to live with the idea that your best efforts may never be quite enough to meet all needs? Understanding that there will be inevitable frustrations is important to your future well-being as a teacher. Your ability to focus on accomplishments instead of on uncompleted tasks is a factor of a positive outlook, related to the old question: Is this glass half-full or half-empty? What is your answer?

1952
John Bowlby publishes *Maternal Care and Mental Health.*

1953
Arnold Gesell publishes *Infant Development.*

1954
B.F. Skinner publishes *Science and Human Behaviour*, applying behaviourist theory to parenting and education.

Changing times. One challenge for early childhood practitioners is that their roles, which are always somewhat loosely defined, are even more nebulous as society's demands and needs for the care and education of young children continue to grow and multiply. They sometimes feel they are asked to become counsellors, psychologists, and social workers, as they tend to the stresses of both children and parents.

What about you? How do you feel about the idea that the changes in contemporary society expand general perceptions about what early childhood practitioners should be doing? Can you be realistic about your contribution to society without placing impossible demands upon yourself, or allowing others to impose them on you?

Attachment and loss. In many early childhood settings, children leave and move on to new relationships with other teachers at least once a year. Early childhood practitioners work by creating warm and nurturing relationships with the children in their care and with their families. Especially in child care situations with very young children, they are intimately involved with the daily care that leads to closeness and with the momentous events that bring satisfaction to both child and adult. Those mutual relationships often lead to strong attachments, which also fulfill some of the early childhood practitioners' emotional needs. Early childhood practitioners often discover that the severing of ties with children and families on a regular, cyclical basis is difficult.

Some early childhood practitioners protect themselves from the pain of loss by not forming meaningful attachments to the children under their care; however, they are then unable to be truly effective with children, as they have not created reciprocal relationships. Early childhood practitioners have to learn, instead, to recognize the temporary, though important, nature of their work with children. The inevitable sense of loss will be tempered by the knowledge of what they have accomplished through the relationship. Early childhood practitioners also must learn to have most of their emotional needs met outside classroom relationships, lest they become too dependent on those relationships.

> It was all connected, I realized, to a feeling of loss. Cathy and I were not only leaving the children, we were losing the ability to influence their lives. Who would these children talk to when they left us? Would their parents continue our work? Would their next caregivers love them as much as we did? There were no answers to these questions. In the end, we told ourselves that it didn't really matter. We had done our best: we had given the kids a good beginning. No one could take that away from them—or from us, either (Wollman, 1994, p. 269).

What about you? Are you able to fulfill most of your emotional needs through the relationships in your personal life, so that you will not become too dependent on the relationships with the children you teach? Can you balance the sadness of leaving relationships with children with the realization of your indelible effect on their lives?

Adult isolation. Many administrators complain that their staff congregate together on the playground. Rather than being a sign of indifference to the need for supervising children, this congregating probably speaks of early childhood practitioners' needs for adult communication. Infrequent staff meetings or quick breaks are often the only chance early childhood practitioners have to speak with other adults for any period. And the isolation of adults who work alone means not only that they are often hungry to talk and share ideas with other adults, but also that the total responsibility for the care of the children rests on them. It is not unusual to hear caregivers in home child care settings comment that they never even have anyone to leave the children with so they can use the bathroom or make a telephone call. Such continuous on-duty time and lack of privacy can be exhausting, both physically and emotionally.

What about you? How would you respond to spending most of your time in the company of young children, rather than with other adults? Would the feeling of complete responsibility for a group during the day regularly result in severe stress? If so, you will need to find ways to renew yourself. Has this been something you've been able to do in the past when you have borne a lot of responsibility? If not, you may want to consider how you will cope with isolation.

Process rather than product. Perhaps one of the most difficult parts of being an early childhood practitioner is that there is not necessarily a finished product to show for all the effort. What results there are may be intangible and virtually unobservable. Human development is a process of becoming. Parents get the rewards, as well as the headaches, of being able to watch the unfolding results of their efforts over long periods of time. Their child eventually becomes a responsible, caring adult in the best scenario, and parents are able to see the payoff of their efforts and worries. Early childhood practitioners, on the other hand, function for a relatively brief period in children's lives. Although they may see progress

2000
1950
1900
1850
1800
1750
1700
1650

1957
Sputnik is launched by the Soviet Union (U.S.S.R.), precipitating much discussion about the effectiveness of American and Canadian education.

1960s
Many boards of education in Ontario establish junior kindergartens.

There is a resurgence of interest in Montessori schools in Canada and the USA, following the post-Sputnik discussions about the need for more structured programs for early education.

during the time they work with children, they usually have no way of knowing what the final outcomes will be. The doctor knows when the medicine that was prescribed cures the symptoms; the TV repairperson knows when the correct wire has been connected to fix the problem; the painter sees the appearance of the house improve with a new coat of paint. Much of what early childhood practitioners do is based on hope and belief in the future benefits of their present efforts.

What about you? Are you the kind of person who needs immediate feedback and evidence that your work is productive? If so, teaching may be difficult for you, since so much of it is based on a concept of promise of future results. Can you happily live with the concept that your efforts may bear fruit long after you are around to see it?

Ideals versus realities. Few early childhood practitioners enter the profession without some images of what it will be like and without ideals that drive their efforts. Though you may have imagined working with always delightfully smiling and responsive children, your studies and practical experience will likely replace this image with a more realistic one before you have finished your program. But this is not to suggest that your ideals also need replacing. One of the most disillusioning things that early childhood practitioners often encounter early on is people and institutions that try to strip them of their ideals. Although it is certainly true that beginning early childhood practitioners have to adapt to the real situations in which they find themselves, this by no means implies that they must discard the ideals and goals that drive them. In fact, those ideals allow early childhood practitioners to continue to grow, develop, and demand the best of themselves, their colleagues, and their schools. "It is important to be both a dreamer and a doer, to hold onto ideals but also to struggle continually to enact those ideals in concrete situations" (Ayers, 1993, p. 131).

What about you? Can you maintain your ideals even when people try to convince you that the real world demands giving them up? Can you be realistic enough to accept the things that cannot be changed and to work with others to change things that can and should be changed?

Respect and Recognition

In Canada, the early childhood workforce struggles to be recognized as a valued, occupation and to be adequately compensated. Unfortunately, there is a societal lack of respect for those who educate and care for young children. It is widely viewed as work that is an extension of women's traditional roles as mothers and homemakers. "Caring—the looking after, responding to and supporting of others—has traditionally been carried out by women and is often viewed as women's natural role" (Beach et al., 1998, p. 123).

Surveys and consultations with early childhood practitioners working in child care centres, nursery schools, family child care, and in-home child care identify lack of respect as a significant problem and a barrier to recognition of the value of their work (Beach et al., 1998; Goss Gilroy, 1998; Canadian Child Care Advocacy Association and Canadian Child Care Federation, 1992). The lack of respect is related to low compensation levels throughout the sector. If Canadians valued early child development, public investment and pay levels would be higher, and early childhood practitioners would have more incentives to stay in the field and to continue expanding their knowledge and skills related to child development and early childhood education practices.

Some related terminology implies that the work is not very important. Who wants to be called a "baby sitter" when, to most of us, the term implies a temporary, and purely custodial, function? The provincial/territorial legislative bodies that set training requirements for regulated child care programs suggest that not everyone requires special preparation. Part of this attitude is no doubt a side effect of the sexist assumption in society that work that is primarily carried out by mothers and other women is not very valuable. It is an interesting paradox that our North American culture pays vehement verbal respect to the importance of children and the power of mothering, yet finds few concrete ways to translate this rhetorical tribute into real support (Modligiani, 1988). In Canada and the United States, teaching of the youngest children, even in the early grades in elementary schools, has always been primarily women's work and, thus, is linked with a sense that it is somehow inferior and requires little skill or knowledge.

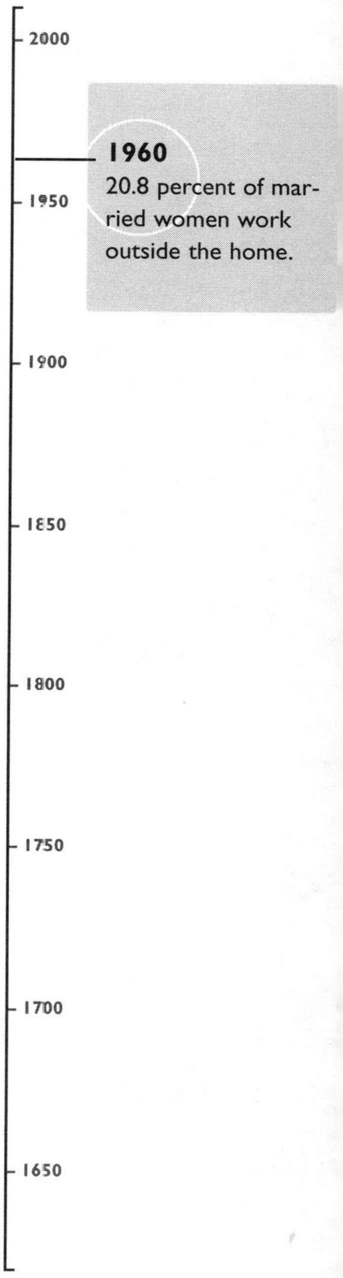

Facing Challenges

As you prepare to enter the early childhood workforce, it is important that you face the less attractive aspects of the sector. There is no question that the work is important and necessary and the field needs many new early childhood practitioners every year. But those who are entering the early childhood workforce will also face some challenges.

Think about how the people close to you responded when you told them you were considering early child development as a career. Write these responses in your journal.

When Christie S. announced to her friends and family that she would major in early childhood education and planned to be an early childhood practitioner, she was dismayed by their responses. Rather than congratulating her on her choice of doing meaningful, important work, as she saw it, her parents kept asking "Why?" To Christie, their question seemed to imply "Why do this work, when you obviously have a good deal of intelligence and could do something more important?" And "Why be a teacher of young children, when other work could be far more profitable for you?" Her father put it rather bluntly: "I'm afraid I'll still have to support you after you are finished with college." And one friend said, "You call that educating—what can you possibly teach those little kids?" Perhaps you have also encountered these responses; they are indeed discouraging, since they imply an absence of respect for the work that you have decided is meaningful and important to you. And, even more discouraging, these questions reflect some of the issues and attitudes that trouble the early childhood profession today.

When Christie turned to early childhood practitioners who were already employed in early childhood centres and schools for reassurance and support, she was dismayed to discover that here, too, were numerous rumbles of discontent. One spoke to her of his frustration with the gender bias he encountered; another complained of the stress of dealing with too many conflicting needs and unmet goals. Several stated that they would be moving on to work that allowed them to

1961 Pampers disposable diapers are introduced in a test market, along with an advertising campaign to promote their usage.

J. McVicker Hunt, in *Intelligence and Experience*, states that deficits in early stimulation can be corrected by early education.

earn salaries that would reflect their education. It took a number of conversations with an early childhood practitioner who had also experienced frustration and stress, but who still felt that the satisfactions of the work were strong enough to outweigh the negatives, before Christie was reassured that her decision was a good one. She continued with her plans to become an early childhood practitioner, knowing that there would be financial, physical, emotional, and social hurdles ahead. It is interesting to note that some of the advantages that were discussed in earlier chapters also contribute to the stress and frustration that early childhood practitioners experience. Almy suggests that those working with young children would need a tolerance for ambiguity (1975). This implies that the same characteristic of the work that makes it attractive also makes it difficult. Ambiguity suggests, as well, that tolerance for stress is a highly individual and subjective response; what is intolerable to one person may provide interest and excitement to another.

Solutions to many of the challenges are intricately connected with the challenges of quality, affordability, and accessibility in early child development programs. We will consider these challenges and how we can take action in Chapter 9.

Now, let's examine the central challenges that early childhood practitioners face today:

1. Early childhood practitioners are subject to many sources of stress and often work in isolated settings.
2. Financial compensation is low compared with other occupations, and benefits are scarce.
3. Early childhood practitioners face health and safety risks in their work environments.
4. Early childhood practitioners are often confronted by those who do not recognize their skills and the value of their work with young children.
5. Men who choose to work with young children face gender-bias challenges.

The first challenge listed is inevitable for those who choose to work in early childhood settings. This challenge is an unchangeable reality, which early childhood practitioners must come to terms with as part of the nature of the occupation. The next four areas of challenge can be changed and, indeed, are slowly changing, through the efforts of early childhood practitioners, families, communities, and Canadian society as a whole. In the final chapters of this book, we will examine those efforts, but here we will simply consider the challenges themselves.

There will be subtle and not-so-subtle pressures on you to give up the idea of caring for and educating young children, and to leave it to someone else to change the world, while you are urged to grow up and recognize harsh realities as unchangeable facts. You, and you alone, know how you feel as you contemplate these challenges and how you think about them as they will affect your future teaching. Think long and hard about the difficulties involved in early childhood education. There is a long line of early childhood practitioners, past and present, who hope that you, too, will decide this is worthy work and that you will take the challenge.

You are encouraged to take every opportunity to discuss these ideas with experienced early childhood practitioners, to hear about their real experiences and how they have made their choices to stay, and taken action to face the challenges. Let us consider some of the support available.

Knowledge

Opportunities to learn and develop skills will help early childhood practitioners deal with the pressure of decisionmaking and adaptation. The more you learn about current research and knowledge in early education, the more confident you will feel in the face of challenges and opposition. It is less frustrating when early childhood practitioners can focus on what they can do with children, rather than on what they cannot. Learning about current social conditions helps teachers see their positions within community efforts to support families. Understanding lifelong development will help teachers focus on the process of which they are a part. Continuing to learn and grow as a teacher allows teachers to face challenges with increasing confidence.

You are now embarking on a program of preservice education. But you also will continue to acquire knowledge after you have completed that initial degree. The challenges will keep coming; it is important that your knowledge continues also.

Support

Early childhood practitioners often find support from their colleagues (courtesy of Play and Learn).

Challenges seem less overwhelming when early childhood practitioners are not attempting to face them alone. Forming connections with colleagues allows early childhood practitioners to release feelings and get emotional support, to generate more ideas for solutions to problems both large and small, and to decrease feelings of isolation. Early childhood practitioners can obtain this collegial support from several sources. It is important to choose employment in a setting where collegiality is encouraged and facilitated by administrative support for teamwork and opportunities for collaboration. Membership in local professional organizations also can provide a sense of collective purpose and strength. A source of support often overlooked by many teachers is parents. Taking the time to form relationships and communicate with parents not only decreases many challenges but also increases insights and strength for meeting existing challenges.

Alliances

The challenges that can be changed—those that are primarily related to social attitudes and social policy decisions—require the work and effort of all early childhood professionals. The energy and sense of optimism that result from joining alliances and working together contribute to early childhood practitioners' sense of professional well-being, as well as producing concrete results. The progress that has been made in the past decade in gaining political support for children's issues proves the efficacy of joint efforts when trying to educate the public or garner legislative or corporate support. Meeting challenges requires memberships in alliances; it is virtually impossible to go it alone. We will come back to the question of alliances in Chapter 8.

Supervisor Support

In early childhood settings, supervisors can either add to or ease caregiver stress and frustration. It is important that early childhood practitioners open communication with their supervisors, who can be a positive source of support for individuals and for developing team opportunities. Early childhood practitioners need to avoid distancing themselves from their supervisors and to recognize that their supervisors have the same interests.

As one early childhood practitioner stated, "All of my good feelings [about the early childhood practitioner role] are to a large extent due to the fact that my supervisor is a person who goes out of her way for all of us. She trusts her staff in making decisions, is always searching for new ways of improving the program, and is open to ideas for these changes. Most important she listens ... Our parents respect her, because they know they can depend on her honesty and integrity. As a result, the atmosphere is friendly and warm and it's a good place to work" (Carson, 1978).

SUMMARY

An increasing number of career opportunities exist for the early childhood workforce. Those entering the workforce should investigate the possibilities in both education and employment that match their personal and professional goals.

Working conditions for early childhood practitioners are improving, but those who enter the early childhood workforce face some challenges including stress and isolation, compensation, and health and safety risks. Early childhood practitioners who expect to meet these challenges will benefit from knowledge, support systems, and supervisor support.

REVIEW QUESTIONS

1. Identify six of the challenges discussed in this chapter, including at least one that is being slowly changed by the efforts of early childhood professionals.
2. Describe at least three of the conditions that help early childhood practitioners face challenges.
3. Discuss what is meant by a "career lattice" and its significance for the early childhood workforce.

4. Discuss various career options within the early childhood field, under the following headings: working directly with children, working directly with families, organizing services for children and families, providing information about children and families, and providing goods and services affecting children and families.

STUDY ACTIVITIES

1. Go to www.ece.nelson.com. Look in the associations' Web sites to find information about employment opportunities for early childhood practitioners.
2. Think about any negative responses others had about your decision to become an early childhood practitioner (which you recorded in your journal). How do you now feel about these responses?
3. Reflect on the challenges of the field. Are there any that may seriously influence your wish to become an early childhood practitioner? Are there any that you feel will not be a particular problem for you?
4. Work with classmates to identify people in as many of the careers listed in this chapter as you can. Choose three careers that you would like to investigate further. If possible, make an appointment to discuss the work with someone currently involved in it.

KEY TERMS

burnout: Physical and emotional weariness caused by stress related to work.
career lattice: Diversity of educational backgrounds, entry points, and employment opportunities within the early childhood field.
separation: Process of learning to be apart from someone the child feels attached to. A developmental task of toddlerhood.
stress: Physical and emotional responses associated with coping with situations beyond one's control.

SUGGESTED READINGS

Carter, Gloria. J. (1992, September). How can the teaching intern deal with the disparity between how she is taught to teach and how she is expected to teach in "real world" primary grades? *Young Children,* 47 (6), 68–72.

Doherty, G., Lero, D., Goelman, H., LaGrange, A., & Tougas, J. (2000). *You bet I care! A Canada-wide study on: Wages, working conditions and practices in child care centres.* Guelph: Centre for Families, Work, and Well-being, University of Guelph, Ontario.

Jorde-Bloom, Paula. (1988, September). Teachers need "TLC" too. *Young Children,* 43 (6), 4–8.

Lay-Dopyera, Margaret, & Dopyera, John. (1990). *Becoming a teacher of young children* (4th ed.). New York: McGraw-Hill.

Section Three

The Early Childhood Workforce Comes of Age

Having considered personal motivations and challenges in becoming an early childhood practitioner, we now will consider the workforce at large. In this section we will consider the early childhood workforce: what it has been, what it is now, and the directions for the future. Chapter 7 looks back at the historical roots of work with young children. Chapter 8 discusses the early childhood workforce as a modern profession that creates philosophical bases for discussion of common ideals, practices, and ethics. Chapter 9 describes the role of advocacy in addressing the challenging issues facing the workforce and leaves students to embark on their personal journeys as early childhood practitioners and passionate early child development advocates. Again, you are challenged to see yourself taking a place within the ranks of this workforce.

CHAPTER SEVEN The Roots of Early Child Development Programs in Canada

CHAPTER EIGHT The Modern Profession

CHAPTER NINE Advocacy

Why Ontario Should Act Now

We know that the development of the brain in the early years of life, particularly the first three years, sets the base of competence and coping skills for the later stages of life. Improving the prospects for the next generation of Ontarians—with respect to school performance, health and quality of life, and success in the labour market—will improve the future for all of us.

The entrants to the workforce of 2025 will be born next year [2000]. From this generation will come a key factor in determining the wealth base of Ontario in 25 years. They will be Ontario's community leader and innovators in the next century. Brain development in the period from conception to six years sets a base for learning, behaviour and health over the life cycle. Ensuring that all our future citizens are able to develop their full potential has to be a high priority for everyone. It is crucial if we are to reverse "the real brain drain."

Investment by all sectors of society in the early years is as important as our investment in education to ensure Ontario has a highly competent and well-educated population, all necessary for a strong economy and a thriving democracy. Ontario has an opportunity to create a better future for our children and grandchildren—by building a strong base that exists today and engaging all sectors of society in establishing a new "system" for early child development and parenting. That system will provide the base for children's learning and development in the school system and postsecondary education system. Action now will put our children and our society on a firmer foundation for the future. This action is necessary, not only to keep a reasonable standard of living, but also because it is the right thing to do for our young children ...

We can take a major leap into the future, just as we did when we had the chance to provide safe water and immunize all children against diseases that had taken a terrible toll in infancy for centuries. When science provided us with the tools—inoculation against polio, smallpox, diphtheria and other scourges of childhood—we used them. We used them to protect individual children and society as a whole. We have new knowledge today. We must seize the opportunity to use that knowledge to benefit all children.

We believe the priorities and choices are clear. (Reprinted with permission from M. McCain & J. F. Mustard. [1999]. The Early Years, pp. 2–3.)

CHAPTER SEVEN
The Roots of Early Child Development Programs in Canada

OBJECTIVES

After studying this chapter, students will be able to

- discuss several reasons for examining the history of early child development programs;
- identify contributors to beliefs about early education and childrearing;
- recognize the role of psychology in current early childhood care and education practices;
- discuss social context in the history of early child development;
- differentiate between beginnings of child care systems, education systems, and recreation systems; and
- identify contributors to the field of early childhood care and education.

Many eyes glaze over with a section on history. Those who are beginning in a profession today may feel there is little relevance about learning in what people did in past centuries or decades. But the roots of early education beliefs and practices have clear connections to conditions, patterns, and beliefs that you will encounter in your immediate experience. A wise person once commented that unless people study history, they will be doomed to repeat the errors of the past. In the case of early education, we study history to understand how it has created our present and to better comprehend the work to be done in the future.

As you have been reading through the text, you have encountered the Timeline in the margin of some pages, documenting events, sociological trends, and contributions of people who have been of some importance in creating our present. This Timeline symbolizes the idea that our current concerns and immediate issues are played against the backdrop of ideas and events from earlier times. As we examine contemporary conditions in early care and education, some of the reasons these conditions exist may become clear when seen in the light of past history.

In this chapter, you will look at the historical roots of work with young children in centre-based settings and the patterns that have developed into the present. You will find out how our understanding of early childhood has evolved over the past few hundred years. We begin by examining two divergent philosophical approaches to how children develop and learn. We then move to an overview of this century's study of psychology and its influence on early childhood education and child-rearing practices.

We will also consider how cultural, technological, political, and economic contexts have shaped how we view children's growth and development during their early years.

You will then learn the story of Canada's earliest child care, kindergarten, and recreation programs, trace the development of nursery schools and related teacher preparation from the 1920s on, follow the development of day nurseries during World War II, and note the explosion of child care centres during the 1980s.

Reasons for Understanding the History of Early Education

There are important reasons to understand the roots of the multiple traditions that have created the complex world of early education. First, it is worthwhile to recognize that the ideas and passionate efforts of many people live on today in the daily practice of contemporary early educators. The significance of play in children's active learning, for example, is an idea that was formed before the defining of developmentally appropriate practice in the late twentieth century. The concept of uniqueness in learning styles and the need for early childhood programs to include families also have earlier roots, as we shall see. "The lack of historical perspective leads us to interpret too much of what passes for reform as new when, in fact, much of the reform had an earlier history" (Perrone, 1991, p. 120). There is a sense of continuity, of being in a long line of persons who have cared about young children, that is somewhat humbling, as well as uplifting, for today's professionals. So, on the Timeline and in this and the next two chapters, you will learn about people who have contributed to the theory and practice of what we do.

Second, it is crucial in appreciating the importance of early education to understand that real events in history and sociological trends shape a profession and its practitioners, in response to particular events and needs in society. So, as you have read the Timeline, you have noted statistics that suggest changes in family structure and resulting needs, as well as events that have focused national attention on children's education. These events, which comprise the social context of early child development in Canada, will be discussed in this chapter as well.

A baby having a bath before the days of running water.

Third, it is important for developing professionals to be able to articulate the theoretical basis for their practice—the *why* that undergirds the *what* of what they do. The contributions of the various thinkers and researchers in the field of early education have been noted on the Timeline and will be discussed in this chapter in connection with the educational trends they influenced.

Fourth, and last, is the reason that understanding of our diversity in structure, professional preparation, social status, and attitudes comes with recognizing the separate forms of early childhood education that developed in North America in the last decades of the nineteenth century and into the twentieth century. These forms are only recently coming together, and not without difficulty; each form has developed its own approach, practices, and culture. In fact, this last idea may be the most compelling reason to examine our historical roots—to help us see that the difficulties in getting professional consensus and public support result from differences within the structure of the profession, rather than from a lack of will or ability to work together. Once we understand the original separations, it may be easier to remove them. It is this last idea that will focus our discussion in this chapter, as we identify questions and trends that divided early education professionals and separated professionals from policymakers.

Philosophical Views of Early Childhood

The establishment of the first early child development programs extends back through the lines of philosophical thinking about children and education.

The English philosopher John Locke (1632–1704) wrote of the importance of early experiences and was one of the first to point out the individuality of children. He noted the importance of children's learning through play and emphasized natural education, rather than the harsh discipline more common in his time.

Another who wrote of the importance of natural methods of childrearing was the French philosopher Jean Jacques Rousseau (1712–78). In *Emile,* his fictional account of raising a child, he discussed the need to protect children from the evils of society and the importance of planning educational experiences that were directly related to children's interests.

Both Locke and Rousseau influenced the thinking of Johann Pestalozzi (1746–1827), who is known as the first early childhood educator, since he opened a school for young children in Yverdon, Switzerland, in 1801. He believed that children are capable of making their own discoveries, and he encouraged his teachers to respect each child's individuality. His school was influential in allowing others to observe his theories in practice; Robert Owen, who opened the first infant school, visited his school, and Friedrich Froebel taught there.

The Study and Science of Psychology

Sigmund Freud (1856–1939) was the founder of psychoanalytic theory. His work with adults with psychological problems led him to believe that childhood is the source of difficulties in personality formation. His work describing the stages of emotional development in childhood forever changed our thinking about the nature of the child. Freud's ideas were more readily adopted by the nursery school movement than by those in kindergarten education, partly because of the work of Susan Isaacs (1885–1948). In 1929, her book *The Nursery Years* interpreted Freudian theory for teachers and suggested how early education schools should apply this new insight about the unconscious to their work with children.

Erik Erikson (1902–94) further developed Freud's thought into a **psychosocial** theory of personality development, extending through the life span. He theorized that each stage of life offers a specific psychological struggle that contributes to a major aspect of personality. Erikson was one of the first to suggest that children develop in the context of their societies' expectations and prohibitions. His theory continues to influence our understanding of the environmental responses and supports that help children achieve healthy development.

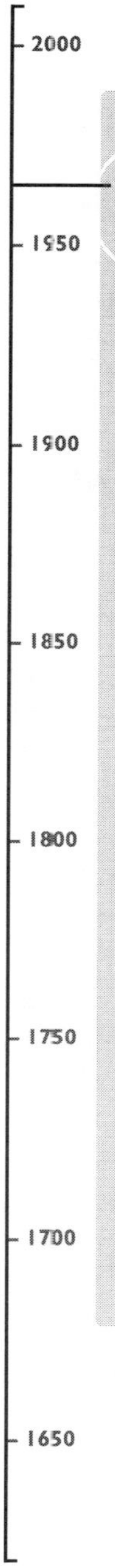

1962
Jean Piaget publishes *Play, Dreams, and Imitation in Children* in English.

Dr. Susan Gray establishes Early Training Project in Tennessee, which includes centre-based programs and home visitors.

David Weikart and associates establish The Perry Preschool Project in Ypsilanti, Michigan. This is the most famous of the early preschool programs established for early intervention because of the extensive longitudinal follow-up on the children in the program.

1963
IDS Harlem Project is established by Martin and Cynthia Deutsch in Harlem.

Jean Piaget's (1896–1980) substantial body of work has greatly influenced early childhood education today, with his focus on the cognitive development of children. His theory explained how children's understanding of the world continues to grow through adaptations made as a result of interaction between the individual and the physical and social environment. Piaget has influenced the way we believe that children think and learn, with implications for our educational settings that allow children to explore, experiment, and manipulate materials.

Lev Vygotsky (1896–1934), one of Piaget's contemporaries whose work has been translated into English more recently, also believed that children construct their own understandings of the world. He differed from Piaget by emphasizing that social interaction and experience with others is the major learning process for children. For Vygotsky, the development of language is a primary task of learning. Vygotsky's theory has important implications for early educators because it stresses the adult's role in assisting children with collaborative learning.

Lawrence Kohlberg (1927–87), influenced by the work of Piaget, developed his own theory of moral reasoning based on cognitive development. Both Piaget and Kohlberg believed that children form their own moral views based on their development of reasoning. Unlike Piaget, Kohlberg differentiated between the moral reasoning of children and adults. Kohlberg believed that children's moral development occurs in three fixed and invariable developmental levels of moral thinking through which all children pass, though perhaps at different rates.

B.F. Skinner (1904–90) was the most prominent theorist of **behavioural theory;** before him, Thorndike and Watson had been proponents of this theory. Behaviourists believe that children are shaped by external forces in the environment and that almost all behaviour can be learned through experience. Specific behaviours, according to learning theory, can be strengthened or weakened by the responses children are given following the behaviour.

A theorist who has received recent attention related to educational issues is Howard Gardner (1943–), with his theory of **multiple intelligences.** He identifies seven intelligences: linguistic, musical, logical-mathematical, spatial, bodily kinesthetic, interpersonal, and intrapersonal, rather than measuring intelligence from one perspective. This theory has interesting implications for the individualizing of learning experiences and expectations.

Developmental Health

Developmental health is a new conceptual framework for understanding human development. It integrates knowledge from the biological sciences with knowledge from developmental psychology and other sciences. Developmental health explains how our earliest social and physical experiences shape the brain's development and set the foundation for learning, behaviour, and health. Dan Keating and Clyde Hertzman, both Canadian scientists, and their colleagues at the Canadian Institute for Advanced Research, have marshaled evidence to support the developmental health theory (1999).

Social Context

Ideas about early childhood and psychological theories of child development and early learning do not occur in a vacuum; they are always products of the larger social context, which includes the structure of families, work, governments, and communities; the dominant political and economic systems; and the values and beliefs of a particular society.

Early Beginnings: Education and Social Welfare

Today's early childhood settings find their beginnings in the nineteenth century. In the early part of that century, Canada as a nation did not exist. Instead, a British colony struggled to bring together the worlds of French- and English-speaking people, or Upper and Lower Canada. The birth of Canada as a nation within the British Empire was a practical solution.

From the mid-nineteenth into the twentieth century, industrialization grew in Canada. Commercial and manufacturing development brought rapid growth in urban centres, such as Montreal and Toronto. In the early part of the nineteenth century, women and children, as well as men, were sought by employers to work outside the home in factories. However, by the middle of the 1800s, the supply of workers exceeded the jobs available. Children from working-class and poor families were now often idle, without supervision or direction. Immigration had increased to meet the demands of populating the new nation and filling its workforce. The resulting pressures of rapid population growth and industrialization brought problems for children and their families. Child advocates and charitable organization solutions looked for ways to organize and structure young children's time and activity.

Infant Schools

In the early nineteenth century, Robert Owen, a factory owner in Scotland, wanted to increase the number of workers available in the local community to work in his factory. His strong commitment to the education of young children led him to establish infant schools to educate very young children and to provide supplemental care while parents worked (Pence, 1990).

The infant schools were popular, and the early experiment in early childhood education and care tied to labour participation spread to North America (Pence, 1990; Prochner, 2000). Infant schools were introduced to Canada in the 1830s by factory owners in Halifax who wanted to attract women and older children to work in their factories (Pence, 1990). These infant schools and others located in Canadian urban centres were based on the models developed in Scotland. Particularly in North America, the philosophy of infant schools promoted the notion that these early education settings were beneficial to all children (Pence, 1990). But by middle of the nineteenth century, Victorian attitudes and beliefs

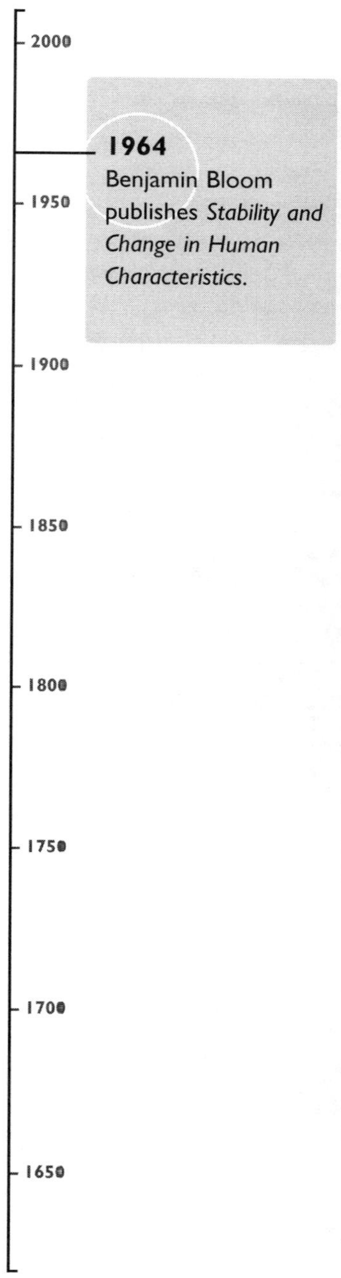

1965
Project Head Start is established in the US on May 18 by President Lyndon B. Johnson, as part of the War on Poverty, and administered under the Office of Economic Opportunity.

A summer program of Head Start is held in the United States, serving 536 108 children in programs that ran two to six weeks.

about the sanctity of motherhood brought about the demise of infant schools in Britain and the United States (Pence, 1990), although in Canada some infant schools continued to operate into the 1870s (Prochner, 2000).

Crèches, Day Nurseries, and Salles d'Asile

In the 1800s, work was the only alternative to starvation for many mothers who were poor, widowed, or deserted. Many women who had young children were employed as domestic servants or wet nurses.

Beginning in the mid-nineteenth century and continuing into the twentieth century in Montreal and Toronto and then other Canadian urban centres, **philanthropic organizations,** churches, and settlement houses founded institutions to care for the children of poor working mothers. These institutions were established to encourage the poor to help themselves through their own labour (Prochner, 2000). The early child care centres were called salles d'asile, crèches, nurseries, or day nurseries. Crèches, first established in France, were **custodial settings** in which the children of poor, working women were looked after. In Britain, these programs were called day nurseries. Salles d'asile in Quebec were primarily educational programs for young children from working poor and wealthier families.

Roman Catholic nuns opened Canada's first crèche in Montreal in 1850 (Schulz, 1978). In 1857, a group of wealthy Toronto women started the Public Nursery, which was the first English crèche (Prochner, 1996). Its initial focus was to provide infant day care but later became an orphanage for girls. Roman Catholic nuns in Montreal established numerous salles d'asile in the 1850s (Schultz, 1978).

Around the turn of the century, crèches or day nursery programs expanded across English-speaking Canada (Schultz, 1978). They provided care, sometimes into the school years, to young children of poor, working mothers or of mothers who were physically unable to care for their children. In 1887, the English-speaking Montreal Day Nursery programs opened, and were followed in the next twenty years by crèches in Toronto, Ottawa, and Halifax.

Canada's early crèche and day nursery programs were typically funded by private charities and sponsored by well-to-do women volunteers or by churches, missions, or settlement houses (organizations that provided services to new immigrants to Canada). Mothers were usually asked to pay a small fee for the care of their children. Government involvement in either funding or monitoring these early programs was minimal.

In Vancouver the development of early centres to care for children of working mothers was the one instance that did involve government participation (Schulz, 1978). In 1910, the Infants Hospital in Vancouver opened a crèche for infants and preschool children of working mothers, and the Associated Charities and the City of Vancouver jointly organized another crèche. In 1916, both programs were placed under the jurisdiction of the Health Department of British Columbia (Schulz, 1978). The salles d'asile received provincial grants from the Quebec government, as well as charitable donations and parent fees.

Children experienced custodial care at most crèches and day nurseries at the turn of the century. Physical space was often limited to a couple of rooms. The staff included untrained nannies and housekeepers, usually supervised by a matron. In

MAKING IT HAPPEN

The Crèche, Toronto

Hester How taught Grade Four in downtown Toronto in the 1880s at Elizabeth Street Public School. She became concerned that many of her students, particularly the girls, were frequently staying home to look after younger siblings. She allowed them to bring the young brothers and sisters to class. Once the problem became more visible and unmanageable, How convinced the school trustees to support the establishment of The Crèche in 1891. The Crèche was operated by a group of volunteer women as a charitable service. It cared for preschool children and for school-age children outside school hours.

Hester How in her Grade Four classroom with her students' young siblings in attendance (courtesy of Archives and Museum, Toronto Board of Education).

one report, a matron, a nanny, and a cook were responsible for forty-five infants and children (Schultz, 1978). For school-age children, care usually included a hot lunch and loose supervision by a housekeeper after school (Young, 1994). Hygienic practices and routines to encourage obedience and moral development kept children physically safe, clean, and behaved. This kind of care was often and Engelm… keted under … trademark DISTAR.

Canada Assistance Plan makes federal funds available for subsidized day care.

The *Ontario Day Nurseries Act* is amended to allow school-age children to participate in licensed group child care.

…ught to address the dangerous side effects of poverty and neglect faced by many the children living in urban poverty (Schultz, 1978)

The crèches and day nurseries did provide other types of services to children and their families, including health care; the distribution of free milk, food, and clothing; and social events. But perhaps the most notable service apart from the care of young children was their employment assistance.

The primary stated purpose of many crèches was

> to enable struggling and deserving women to help themselves, by taking care of their children by the day, or the week, and by so doing make it easier for the parent to earn the necessary means of support for her family" (ref. 16 in Schulz, 1978, p. 140).

In addition to the provision of care for the children of working mothers, many of the early crèches operated as employment agencies for the mothers themselves. The crèches that operated to provide employment for destitute young mothers were usually provided by charitable organizations, efforts of upper-class women, and private donations. Young women who found themselves alone with a child to support in urban settings used these services to care for their children while they worked, often as domestic servants.

> The majority of the women using the centres were domestic servants, and a chronic shortage of domestic help throughout the early part of the twentieth century made these agencies very useful to the same class of women who were members and managers of the centres. Annual reports consistently outline the number of days of care provided for children, and the number of days of employment provided for parents in the same paragraph (Schulz, 1978, p. 141).

The institutional records of the early crèches reveal examples of job training for young mothers. For instance, during World War I, the West End Crèche in Toronto set up a laundry so that the women could do soldiers' washing and offered night classes in how to do laundry (Schulz, 1978). By the end of the Depression, mothers with young children could usually find regular employment, so the employment agency aspect of these centres was discontinued (Schulz, 1978).

Public Schools

During the early nineteenth century, a variety of schools flourished in local communities throughout Upper and Lower Canada. They were supported by families, churches, and concerned public benefactors. Wealthy parents sent their children to private schools, and "ragged schools" were set up by individuals or charities to provide education and clothing for poor children.

Universal education was proposed as the answer to the ills facing a changing society. In order to be available to all children, it had to be free. School reformers proposed that common or parish schools, which had emerged in communities during the first half of the nineteenth century and had been partially subsidized by government grants in most provinces, be entirely sustained by a combination of provincial grants and assessment on local property. Egerton Ryerson, in his *Report on a System of Elementary Public Instruction for Upper Canada*, published in 1847, stated that "a system of general education amongst the people is the most effectual preventative of pauperism, and its natural companions, misery and crime."

The campaign for universal education was based on

- the realization that an educated population would stimulate economic growth and ensure social stability;
- the recognition of education as a means of achieving desired social and political attitudes in support of the status quo, that is, loyalty to Britain and British institutions;
- the promotion of education as an equalizer that could break down barriers between classes, and increased social harmony; and
- the view of schools as agents of social control, which could influence the development of loyal citizens.

THEORY INTO PRACTICE

The Nineteenth Century: The Century of Schooling

1816 *Introduction of local boards of school trustees*

- Loose governmental involvement in administration of local schools (schools are private and voluntary; some are aided by government grants or church assistance; and all are supported by student fees)

1841 Common School Act

- First legislation to provide a uniform school system for United Province of Canada

1843 Second Common School Act

- Increased centralization

1844–76 *Egerton Ryerson*

- Appointed superintendent for schools in Canada West

1846 Common School Act *(Canada West)*

- Standardization of curriculum, textbooks, and teachers

1850 *Great Charter of Common School Education*

- Recognition of principle of universal property assessment

1867 British North America Act

- Confederation of Canada
- Provisions made for Roman Catholic separate schools

1871 School Act

- Ontario government compels communities to provide free common schools; attendance is compulsory for children aged seven to twelve for at least four months each year. Other provinces follow this direction.

Sources: Adapted from Corbett, B. (1989); Mathien (2000); Young (1994).

The mid-century shift of education as a responsibility of families, churches, and communities to that of being public institutions free and distinct from social and denominational interests was dramatic. Schools were becoming part of an educational system, instead of programs created by communities. (Of course, the accommodation of both Protestant and Roman Catholic interests did result in two public systems of education in most parts of Canada.) Egerton Ryerson was the superintendent of education for Upper Canada (later Ontario) from 1844 to 1876. He, along with Alexander Forrester in Nova Scotia and John Jessop in British Columbia, worked to establish universal, publicly funded elementary education (Gaffield, 1985). The School acts from 1841 to 1871 in what is now Ontario set out the foundation for that province's school system and for more education acts in most other parts of the country.

The new educational system responded to industry's growing demand for skilled labour and for scientific and technological advances. During the second half of the nineteenth century, the curricula introduced manual training, as a means to prepare students to enter the workforce, and an emphasis on rote learning. School inspectors ensured standardization of curriculum, textbooks, and teachers.

Kindergarten

Friedrich Froebel (1782–1852) is known as the originator of the kindergarten, a German word meaning "children's garden." He believed that children could grow and flourish like plants in the right environment, developing internal impulses that would unfold naturally. Kindergarten was designed for children between the ages of three and six, and Froebel's curriculum emphasized language, numbers, forms, and eye-hand coordination to train children in ways that would establish courtesy, punctuality, neatness, and cleanliness, as well as respect for others.

The most notable elements of his tradition were the curriculum materials that he called gifts and occupations. The gifts were a series of ten concrete, manipulative materials, such as woollen balls, wooden shapes and cubes of various sizes, tablets, wooden sticks, and a variety of natural objects, all presented to children at defined intervals, with specific skills and symbolic concepts to learn. The occupations included sewing, perforating paper, weaving, working with clay, and cutting paper. As Froebel said, "What the child tries to represent, he begins to understand" (Olmsted, 1992, p. 4). Although this idea seems very familiar to us, it was considered quite revolutionary to educate children in a group outside the home and use play materials.

During the 1870s Froebel kindergartens first appeared in Canada as private institutions supported by parent fees or charitable donations. Egerton Ryerson's work in establishing the school system in Ontario did much to lay the groundwork for acceptance of kindergartens in schools in Ontario and other provinces (Corbett, 1989). He added Pestalozzi's object lessons to the curriculum and accepted Christianity (but not specific formal religions) as central to the school curriculum. By 1871 school attendance was compulsory for children from seven to twelve years for at least four months a year, but children often attended for considerably more time. Particularly in urban areas, children younger than six years were often in attendance.

The first Canadian public school kindergarten program opened in 1883 at Louisa Street School, in Toronto. James L. Hughes, the chief school inspector for the Toronto School Board at the time, had studied the writings of Pestalozzi and Froebel and was convinced of the value of education to support the whole child. Faced with concerns about growing numbers of preschool children left unattended while their mothers worked, Hughes (who supported Hester How's efforts to initiate The Crèche to care for young children left on their own) introduced Froebel kindergartens. The program, which responded to concerns about young children's environments, was also a vehicle for promoting curriculum innovation (Mathien, 1990).

The Froebel kindergarten at the Toronto Normal School, circa 1890 (courtesy of Ontario Archives).

Ada Maream (who later married James Hughes) was the first teacher, or **kindergartener** (the term used for a Froebel kindergarten teacher), in the Toronto public school kindergarten. She had studied Froebel education in the United States and opened private kindergartens, first in New Brunswick and then in Toronto in 1878 (Corbett, 1989). Her first public school kindergarten was a class of eighty children. Seven kindergarteners-in-training assisted her. Maream's professional education and her strong connections with American kindergarten activities provided the basis for her leadership in Ontario's and Canada's kindergarten programs as part of school systems.

In 1885 Ontario officially recognized kindergartens as part of the public school system. In 1887 the province began to provide grants to school boards to establish kindergarten programs for children three to seven years old. Kindergarten was introduced in several schools in Toronto and in towns and cities across the province. By 1900 there were 120 public school kindergartens across Ontario (Mathien, 1990). Quebec established kindergartens within its school system in 1892.

In 1885 a kindergarten course was introduced at the Toronto Normal School (a teacher-training facility), and in 1889 the Toronto Kindergarten Association for kindergarten teachers was established.

THEORY INTO PRACTICE

Kindergartens in the United States

Froebel's ideas and model of the kindergarten were imported to the United States primarily by German immigrants who had been trained in Froebelian principles in Germany. In 1856 Margarethe Schurz opened the first German-language kindergarten in her home in Wisconsin and was the person who explained the Froebelian principles to Elizabeth Peabody. Elizabeth Peabody (1804–94) opened the first English-language

(cont'd)

Kindergartens in the United States (cont'd)

kindergarten, in Boston in 1860, and is generally credited with gaining acceptance for the kindergarten movement in the United States. It was Peabody who influenced the superintendent of the St. Louis Schools to sponsor the first public kindergarten in the United States in 1873. At one of her lectures, Milton Bradley, the toy manufacturer, first learned of Froebelian methods, and he later began to manufacture the gifts and occupations. Kindergarten normal schools followed, including the Oshkosh Normal School in 1880, the Chicago Kindergarten College in 1886, and the Wheelock School in Boston in 1888. In 1978 Peabody founded the American Froebel Association for interested kindergartners.

The first public school kindergarten opened in St. Louis, Missouri, in 1873. But public school sponsorship of kindergartens came slowly, and, for a time during the late-nineteenth and early-twentieth centuries, many were sponsored by churches, charitable organizations, and settlement houses. Patty Smith Hill, writing in 1926, said, "When the kindergarten was introduced into this country more than a century ago, it survived as a philanthropy long before it was accepted as an organic member of the educational system" (Hill, 1987, p. 12). This was the time of heavy immigration, when slums were developing in major cities, accompanied by the social problems of crime, delinquency, and other difficulties associated with rapid increases in foreign population. Settlement houses were created in major cities to combat the problems and to help new citizens assimilate. The philanthropists adopted kindergartens in the hope that improving the lot of young children could improve society. The first president of the New York Kindergarten Association stated, "The kindergarten age marks our earliest opportunities to catch the little Russian, the little Italian ... and begin to make good American citizens of them. The children are brought into a new social order" (Youcha, 1995, p. 148).

The standard kindergarten schedule of morning classroom programs left the kindergarten teachers free to visit in the afternoons, essentially functioning as social welfare workers, "seeking work for the unemployed; space in hospitals for ill mothers, sisters, and brothers; physicians who would remove adenoids and tonsils; or dentists who would extract diseased teeth" (Hill, 1987, p. 13). Gradually, according to Hill's account, the philanthropists turned to the schools to ask them to accept the kindergarten "as a member in good and regular standing." They usually asked to be able to use an empty room in the school, and the philanthropic agencies still paid for the teacher's salary and the program expenses. Gradually, the boards of education were persuaded to take full responsibility for the kindergartens, and so the kindergartens became part of the public school systems.

But there were problems in their acceptance within the school systems. Kindergarten teachers had been trained in normal schools, which were separate from other teacher-training schools, and that used curricula, materials and methods—encouraging children to "talk, sing, dance, dramatize, model, paint, draw, build, and construct" that seemed quite foreign to teachers who used a curriculum "based upon the acquirement of the three Rs in their baldest and most barren form" (Hill, 1987, p. 14). The kindergarten method of handling behavioural problems was considered soft and sentimental, and it certainly did not prepare the children to enter Grade One. Hill comments that kindergarten children were not welcomed by Grade One teachers, through no fault of either the kindergarten or Grade One teachers. "It was due to the fact that children were supposed to pass on from one teacher to another with continuity of work when the two teachers were trained in diametrically opposed philosophies, curricula, and methods" (Hill, 1987, p. 14). (This comment certainly sound like it was written today, instead of eighty years ago!)

Moving into the Twentieth Century

The first two decades of the twentieth century witnessed the ongoing evolution of early child care and education. The early crèches and day nurseries continued with a few additions, but the majority of working women relied on family and neighbourhood arrangements, rather than on organized child care programs. In urban settings across Canada, organized playgrounds responded to the needs of school-age children for out-of-school hours, and kindergarten programs continued to expand. Interest in child study heightened attention to different approaches to early education, including Montessori and nursery schools. Early educators looked to progressive education and psychological theory for answers on how best to educate young children. Some crèches and day nursery programs began to employ teachers with kindergartener training or those who were knowledgeable about child development.

World War I reminded Canadians of human vulnerabilities and the destruction that comes from hate and violence. After the war people wanted to get on with the business of raising children and earning a living in a safe, secure world. They wanted to benefit from growing modern conveniences and health advances, including new understandings of mental hygiene (the term used for mental health) from Freud and others.

The suffrage movement, led by middle-class women, never took issue with the notion that a woman's place was in the home. The right of women to work and the right of children to high-quality child care were not demands of the movement. Their campaign for support to single, widowed, and deserted women did result in the introduction of Mother's Allowance (social assistance or welfare) in the 1920s. Once women had gained some support for their children without having to enter the workforce, the need for crèches and day nurseries actually decreased.

Across Canada, the 1930s were difficult economic times for children and families. The Depression dragged on, with high levels of unemployment and little in the way of social support networks. There were drops in both women's participation in the labour market and in the birth rate. Many of the crèches and their related employment agencies for women were forced to close.

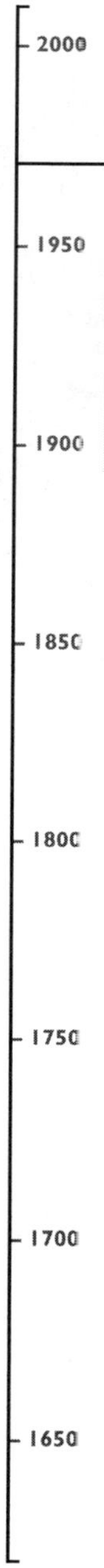

1967–68
Ontario establishes community college system and includes postsecondary Early Childhood Education diploma programs.

Children's Playgrounds

At the beginning of the nineteenth century, there was considerable interest in camping and fresh air experiences. Settlement houses acquired properties for family camping outside urban centres and made this a specific service available to supported children and their parents. The value of organized play programs for children in kindergarten was extended to older children. As well, concerns about unsupervised children on city streets and their enticement into criminal activity gave rise to the development of children's playgrounds, which offered supervised, outdoor play spaces for children outside school hours. Their development became

Children playing in a supervised Toronto playground circa 1910 (courtesy of Metropolitan Reference Library, Special Collections Centre).

connected to local governments and community organizations (Young, 1994). For example

- in 1900 inner-city settlement houses in Toronto operated summer playgrounds and provided noon-hour meals;
- in 1903 Ontario allowed municipalities to purchase land for public parks;
- in the 1900s playground associations were established in Toronto, Hamilton, London, Ottawa, and in other urban centres.

James Hughes viewed playgrounds as an extension of kindergartens and led the Toronto School Board's establishment of supervised playgrounds in eighteen of its schools; the Toronto municipal government took responsibility for programs in municipal parks (Young, 1994).

Public Health

In the eighteenth century, public health outreach programs played a role in helping to support mothers and their infants and young children. Immigration, urbanization, and industrialization brought larger numbers of people together, increasing the risk of infectious disease and mortality. Diseases that thrived in crowded, dirty living conditions could not be contained, and their transmission to upper-middle class environments was inevitable. Therefore, prevention of diseases attracted attention and interest from all socioeconomic groups.

Research evidence identified health prevention measures that could greatly reduce the spread of infectious disease and reduce mortality, particularly amongst infants and children (Sutherland, 1976). Canadian historian Neil Sutherland identifies three events in 1882 that shaped the direction of public health: "Louis Pasteur proved the effectiveness of immunization for anthrax in sheep, Robert Koch discovered the tuberculosis germ, and Ontario established a board of health" (1976, p. 40). The establishment of public health boards in other provinces followed. The emerging public health movement in Canada built upon the achievements of nineteenth-century improvements in sanitation and on the dramatic bacteriological discoveries of the 1880s and 1890s.

Public health initiatives brought prenatal and well-baby clinics, milk depots, and home visiting to communities throughout the province (Arnup, 1994; Sutherland, 1976). These measures reduced infant deaths and improved maternal health (Sutherland, 1976).

The public health nurse's responsibilities for well-baby clinics and home visits encompassed a parent education role. In many instances she became a friendly adviser who was able to suggest changes in caregiving practises and promote healthy habits (Sutherland, 1976). The approach to parent education was often patronizing and didactic (Kellerman & Kyle, 1999) but was typically directed at all families with young children, not just those living in poverty, consistent with the notion that everyone was vulnerable, and it was in everybody's best interest to avoid or prevent disease.

By 1914 the public health nurse's role in the schools was becoming central in school health programs (Sutherland, 1976). In addition to preventative health (such as routine inspections, maintenance of medical records, and teaching healthy habits, including nose-blowing and tooth-brushing drills), school nurses often led health clubs after school such as the Little Nurses League in Winnipeg or Little Mothers in Vancouver, Regina, Victoria, and Stratford (Sutherland, 1976). Such clubs provided after-school activities for school-aged children while promoting health prevention.

Public health initiatives in the early part of the twentieth century are the historical roots of many early child development programs for children and their families.

2000

1968
The University of Illinois establishes the ERIC clearinghouse on early childhood education.

Living and Learning, the Hall and Dennison report on the public school system, is released in Ontario and recommends child-centred learning environments.

1950

1900

1850

1800

1750

1700

1650

First Nursery Schools

Nursery schools evolved from British and European experiments at giving children from disadvantaged backgrounds an early head start (Prochner, 2000). Maria Montessori and Margaret and Rachel MacMillan established programs specifically designed to intervene in children's early lives and ameliorate the effects of poverty and poor housing conditions.

Maria Montessori. Maria Montessori (1870–1952) was the first Italian woman to earn a medical degree. From her observations of children, Montessori recognized the uniqueness of each child. Her phrase "the absorbent mind" captures her philosophy of children educating themselves, actively, in a prepared environment, using carefully selected didactic materials. From the opening of her *Casa dei Bambini* (Children's House) in the tenements of Rome in 1907 until the present, Montessori's philosophy has had an impact on early childhood education, although not universal acceptance. Her book, *The Montessori Method*, was translated into English and became available in North America. There were Montessori schools in Canada and the United States within the next decade, but the real popularity of Montessori education in North America did not happen until the resurgence of interest in early childhood education in the 1960s. Nevertheless, kindergartners and others working with young children were in the meantime exposed to her approach and did absorb some of her ideas and practices.

The lasting contributions of the Montessori philosophy include

- an attitude of respect for children;
- the idea that young children are essentially self-didactic, or learn through their own activities and adaptations;
- the concept that teachers learn through their interaction with children;
- the emphasis on a prepared, attractive environment, with child-sized furniture; and
- the focus on the quality of manipulative materials (Elkind, 1983).

Some contemporary educators suggest that much Montessori practice may not fall within the guidelines for developmentally appropriate practice due to

- its emphasis on work and the absence of fantasy;
- children's lack of freedom to experiment with materials, once they have been introduced to their appropriate use;

1969

The Report of the Royal Commission on the Status of Women recommends universal child care.

Sesame Street begins on public television on November 19.

- the discouragement of cooperative and collegial planning and conversation;
- the strictly sequenced series of activities; and
- tight teacher control of how space, time, and materials are used (Greenberg, 1990).

Because Montessori philosophy and training have been maintained so separately from other institutions within early education, Montessori and her schools maintain a unique place in our history.

McMillan Sisters. Meanwhile, in England, another form of early education was evolving for even younger children. A British government report in 1908 had pointed out that although most children were born healthy, about 80 percent arrived at school age in poor health. In response, that same year Rachel and Margaret McMillan established the London School Clinic for children under five years. Then in 1911, in Deptford, they established the Deptford School (later renamed the Rachel McMillan School) as an open-air **nursery school.** This was the first time that the term "nursery school" was used. The Deptford school emphasized healthy living for children, as well as nurturing for the whole child.

Throughout the early twentieth century, the McMillan sisters operated fresh air camps in Britain with a strong emphasis on hygiene, health, and physical activities to strengthen children's bodies (Prochner, 2000).

Other tenets of the nursery school philosophy were that schools should develop close links with children's families and communities and that teachers of young children should be well trained. Therefore the nursery school also functioned for many years as a teacher training-laboratory. Although Rachel McMillan died in 1917, Margaret continued her influential work in nursery education. She published *The Nursery School* in 1919, and her advocacy of government support for nursery schools led to the establishment of public nursery schools in England as part of the national education system.

Progressive Education and the Child Study Movement

The original Froebelian methods found in the early kindergarten programs began to be perceived as too rigid and abstract, centred as they were on prescribed methods and symbols with precisely ordered activities. During the 1920s new learning about children was emerging from university-based scientific research on child development, known as the **child study movement**. The "progressives" felt that the kindergarten curriculum should be based on scientific knowledge, rather than the mystic religious tenets of Froebelian philosophy.

The child study movement was brought about by two Americans, John Dewey and G. Stanley Hall, whose ideas about appropriate education for young children were influencing educators in both the United States and Canada. With his progressive philosophy, John Dewey (1859–1952) believed that education was a method of social reform, that information and knowledge would enable individuals to improve the quality of their lives, and that schools must represent life. He opposed the idealistic, religious philosophy of Froebel and his concept of "unfolding" and wanted to base the education of young children on scientific knowledge about their abilities. He was opposed to the traditional method of teaching children by **rote learning,** which is a form of learning by repetition, memory, and habit, rather than firsthand understanding. Dewey wanted the active involvement of the whole child and interaction, as described in *My Pedagogic Creed*, published in 1897. In 1896 he and his wife opened a laboratory at the University of Chicago for four and five year olds, which he called "subprimary," rather than a kindergarten, possibly to distinguish it from the Froebelian model. Dewey's influence on the growing early childhood education movement was profound, as was his influence on the whole scope of **progressive education** in North America, which promotes active involvement of the child in learning.

G. Stanley Hall (1844–1924) is credited with beginning the scientific study of children. His *Content of Children's Minds* (1883) focused on descriptions of children's concepts, and the educational implications of these findings for teachers. "Hall criticized Froebelian kindergarten theory as being superficial and fantastic—he considered that young children needed large, bold movements rather than the sedentary activities of gifts and occupations and asserted that free play could serve their developmental needs" (Spodek in Osborn, 1991, p. 76). Hall's influence on the movement away from the Froebelian kindergarten was eventually profound, although at a seminar in 1895, most of the kindergartners assembled to hear his child development research left infuriated.

At the same time, the mental hygiene movement was launched in the United States (Prochner, 2000). The movement shifted from treatment and prevention of mental illness to the promotion of mental health. Researchers and practitioners in mental health disciplines agreed that the roots of mental health were found in early childhood. Attention turned to the need to understand how children develop and what conditions are necessary to promote mental health.

The theories and practices of progressive education, the child study movement, and the mental hygiene movement converged to influence emerging kindergarten and nursery school programs in both Canada and the United States.

The Child Study Movement in Canada

In Canada, establishment of child study and nursery school programs was stimulated by the progressive education, child study, and mental hygiene movements in the United States and by the new social science—psychology. The experiments of the McMillan sisters in Britain and Maria Montessori in Italy did not find a direct translation in Canadian nursery schools that first appeared in the 1920s. However, they did offer powerful examples of how early child development programs could be targeted to disadvantaged children.

Because of the widespread interest in the development of young children generated by the child study movement, nursery schools were established in laboratory settings as part of child study programs. Funds were provided by the will of Laura Spelman Rockefeller in 1923 to establish child study centres in major research universities, such as Yale, the University of California, the University of Minnesota, the University of Iowa, and Columbia University.

The impetus to start a nursery school in Toronto came from this interest and from concerns about mental hygiene. Clare Hincks, a Toronto physician and a leader of the mental hygiene movement, worked closely with E. A. Bott, who headed the psychiatry department at the University of Toronto. They were

2000
1950
1900
1850
1800
1750
1700
1650

1970
White House Conference on Children identifies child care as one of the major problems facing the American public.

Toy libraries are introduced in Canada through community programs such as public libraries and community centres.

High/Scope Educational Research Foundation is incorporated.

successful in securing a grant from the Laura Spelman Rockerfeller Fund to establish the St. George's School for Child Study at the University of Toronto in 1925. It later became the Institute of Child Studies, a site of extensive child studies, as well as a working model of early childhood care and education practice.

McGill University also secured funding from the same source to establish the McGill University Day Care in Montreal, which operated from 1925 to 1930. In other university settings, child development studies were introduced as part of psychology and home economic departments.

William Blatz, a physician who graduated from the University of Toronto and then pursued a Ph.D. in psychology at the University of Chicago, is recognized as the founder of Canada's child study movement. In 1926 he opened a nursery school at St. George's School for Child Study. He created a program of guidance and education for children from two to five years of age that influenced practices and regulations in early childhood care and education across Canada for decades. From the beginning, parent education, early educator training, and child development research were also major components of the St. George's School for Child Study, in addition to the nursery school.

In the first year of operation, four adults developed the program for eighteen children at the school. Blatz's recent studies in psychology had introduced him to the work of Watson's behaviourism, Dewey's progressive education, and psychoanalytical ideas about mental hygiene. Because he did not have direct experience with either kindergarten or nursery school settings, he worked with the school staff to develop the program. Della Dingle, who had studied at the College of Home Economics at Cornell University in New York, was the first nursery school director.

Children follow lunch routine at St. George's Nursery School, circa 1930 (courtesy of Thomas Fisher Rare Book Room, University of Toronto).

Blatz and his team did not have a cohesive understanding of young children or a theoretical framework to guide their practice. Instead, they were eager to explore new ideas about child development and took a "let's watch the children and find out" approach (Raymond, 1991). Blatz did encourage a smooth-running routine that promoted the habit of regularity in the lives of young children. The nursery school established such practices as staggered enrollment and a staff timetable that covered every situation in a nursery school day. Within the parameters of daily routines (such as eating, washing hands, using the toilet, and napping), children were encouraged to explore environments designed for open-ended play activity. Over the years, the research of Blatz and others explored environmental influences on child development and the importance of adult-child relationships in developing emotional security.

The focus of the St. George's School for Child Study was not just child-oriented; parent education was championed from the beginning. Blatz began an infant-mother drop-in program and promoted parent training in his frequent lectures broadcast on CBC radio (Raymond 1991; Doyle & Young, 2000). Parents whose children attended the nursery school program at St. George's School for Child Study participated in parent education classes (Wright, 2000) and were encourage to read Blatz's publications for parents.

A dramatic event in the 1930s propelled William Blatz and his approach to child studies and parent education into a very public international spotlight. In 1934 the Dionne quintuplets were born to a French-Canadian couple living in rural northern Ontario. The babies were made wards of the province and cared for in a controlled environment (Wright, 2000). Blatz and his colleagues were given to opportunity to study the children for over two years. In retrospect, it was a horrific situation for the young children (Berton, 1977), but it did offer a unique opportunity to study environmental influences and development among five children who were thought to have identical genetic inheritance (Wright, 2000).

In 1938 the St. George's School for Child Study became the Institute of Child Study, and within a year began to offer courses to kindergarten teachers. During the next two decades, it remained a leader in the development of early childhood care and education programs.

Other nursery schools followed the lead of the St. George's Nursery School and began to open in the late 1920s. They joined private kindergarten programs (operated outside the public school system) in offering educational programs to young children. For example, by 1928, there were several private nursery school/kindergarten programs operating in Calgary. Most were half-day programs for middle-class children from age two to five years, which relied on parent fees. Many were unable to survive the economic depression of the 1930s.

Working Women: Day Nurseries to Early Childhood Education and Care

The past sixty years have repeatedly brought early childhood care and education to the forefront of social policy, education, and employment initiatives. The types of programs vary, but their relationship to women's employment in the labour force has remained constant. Periods of strong demand for the participation of women in the workforce resulted in government support of full-time early childhood care and education. However, during periods of women's decreased participation in the labour force, the emphasis shifted to part-time early childhood programs that focused on children's development and education needs.

World War II brought an acute labour shortage in Canada, and the government began to recruit women to work in industry. The government first recruited single women, then married women, and finally married women with children. After the war, attention turned to maintaining a prosperous economy in peaceful times, and social policies were introduced to ensure basic protection for families from the economic devastation of the Depression era. The consumption of goods and services, combined with a boom in babies, created the conditions for economic expansion and the expansion of government social programs. By the 1960s and 1970s, women were receiving higher levels of education and had greater employment opportunities. At the same time, lone-parent families, the need for more

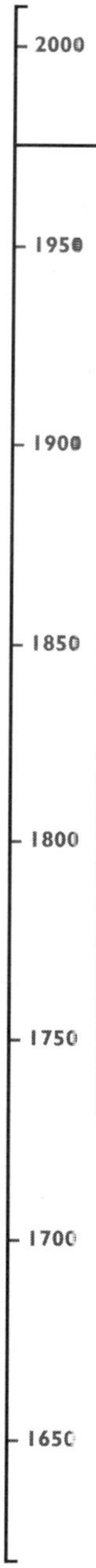

1970–73
Federal funding through Local Initiatives Projects for child care centres and advocacy groups encourages expansion of community-based child care as well as day care advocacy activities across Canada.

1970s
The introduction of legislated standards and limited funding for each province signals the beginning of the modern era of child care.

1971

Unemployment Insurance Act provides fifteen weeks of maternity benefits after twenty or more weeks of full employment.

The Mothercraft infant caregiver training program begins in Canada.

Health and Welfare Canada begins to collect statistical information on Canadian day cares.

There are 17 391 regulated day care spaces in Canada.

than one income to sustain family standards of living, and the consciousness-raising efforts of the women's movement led many women with young children into the workforce.

War-Time Day Nurseries

As women were recruited for work in war-related industries, the image of the perfect Canadian woman changed from being the foundation of hearth and home to the industrious beauty on the assembly line. The recruitment drive made it necessary to consider alternative care for children. In 1942 the federal government passed the *Dominion-Provincial Wartime Day Nurseries Agreement,* authorizing the Ministry of Labour to enter into cost-sharing agreements with any provincial government willing to establish day nursery services.

Quebec and Ontario were the only provinces to take advantage of the scheme—the other provinces maintained that the need did not exist. The federal government stipulated that at least 75 percent of day nursery spaces had to be given to children whose mothers worked in essential war industries. In both provinces, day nursery operating standards were established. In Ontario, the St. George's School of Child Study provided a model for a rapidly expanding system and established short training courses for staff.

By the end of the war, there were twenty-eight nurseries for preschoolers and forty-two programs for school-age children in Ontario; in Quebec there were five community-based centres. The Ontario centres were located in Toronto, Ottawa, Hamilton, and Windsor. The cost was $1.05 per day, shared roughly equally between parents and the federal and provincial governments. There was no provision for infant care.

After the war, governments undertook a campaign to get women back to hearth and home and introduced Family Allowance ("baby bonus") benefits. In Quebec, the government closed all five centres, despite enrollment to capacity and long waiting lists. In Ontario, all three governments (federal, provincial, and municipal) tried to close the centres but were met with strong resistance through the Day Nursery and Day Care Parents Association. Federal funding ceased, but the Ontario government then passed the *Day Nurseries Act,* which included provisions for licensing day nursery and nursery school programs and 50 percent provincial cost-sharing of fee subsidies to municipalities. The regulations were based on routines and practices in place at the Institute of Child Study (formerly called St. George's) nursery school. Sixteen of the twenty-eight preschool centres survived, but all forty-two school-age programs closed due to lack of funding and political support (Schulz, 1978).

Baby Boom Time

The campaign for women to remain with *kuchen* and *kinder* continued throughout the 1950s and early 1960s. To the extent that day nurseries existed in Canada, they continued to operate mostly as charitable institutions for women in need. However, with increased births and growing numbers of young children, nursery

THEORY INTO PRACTICE

Establishing a Day Care Centre

Procedure in establishing and operating Day Care Centre is gradually becoming crystallized. The School Board investigates the need by a prepared questionnaire sent to the parents. Upon receipt of twenty or more applications a plan and budget are set out to meet the specific requirements of each school setup. These are submitted by the board to the Provincial Advisory Committee, and from thence to the Department of Labour, Dominion Government. Upon approval they are returned to the Board for administration. Officers of the Wartime Day Nursery Branch of the Department of Public Welfare act to advise and assist in establishing each centre.

Fitting Day Care into a school organization is a challenge in ingenuity. Even its most obvious aspects of time, place and personnel have required considerable forethought. Adequate care necessitates a programme from 7:00 in the morning until 6:00 or 6:30 in the evening. Space in crowded school buildings is difficult to spare. Assembly halls, gymnasiums, music rooms and even classrooms are variously used. In a few cases the ideal has been reached—namely, provision of a room of their own. Here, surroundings can be made attractive and conducive to easy, enjoyable but busy living. A sense of belonging and possessing can be achieved, and thus, more basic objectives of the programme approached. A demonstration has been attempted in one Toronto school. The room—an empty classroom—is decorated in soft green; gay curtains relieve its classroom severity; coloured tables and chairs built to suit recreational needs break the monotony; a reading corner, furnished with comfortable chairs is conducive to relaxation.

Serving lunch at school has presented on first glance insurmountable difficulties. In a majority of centres, it has been possible, through community co-operation to arrange meal service in a nearby church or other community building. The Board remains in charge of this service, providing adequate facilities where these are lacking. The child's well being is built up and maintained by a nutrition programme planned through the Provincial Department. Staff from the Nutrition Department of the University of Toronto give directions to this programme.

For staff to carry out the Day Care Programme, authorities have turned in the first instance to those whose work they know and value—namely the teachers. These have given their services regularly and unstintingly in overtime hours. It is recognized, however, by those participating, that time is required for planning a Day Care Programme, time not possible to the busy teacher. In a few centres a regular staff has been provided whose one responsibility it is to plan for the needs of a particular group of children.

Source: Millichamp (1944, November), p. 248.

schools offering part-time enrichment programs for children (and breaks for full-time mothers) were popular throughout Canada.

Parent cooperative preschool programs, which had surfaced in Canada as part of the nursery school movement, expanded during this period. In 1950 Vaughan Road Nursery School was established in Toronto. Daisy Dotsch was the teacher in charge, and she became a leader in the development of preschool philosophy in Canada (Stevenson, 1990).

1971 (cont'd)
The Child Care Expense Deduction is introduced through the income tax system in Canada.

French immersion kindergartens are introduced into Toronto public schools.

The First National Day Care Conference is organized by the Canadian Council on Social Development and helps to focus on child care as a national issue.

The Nursery Education Association of Ontario (which later became the Association for Early Childhood Education, Ontario) started up in 1950. An organization for nursery school and day care teachers, it offered extension courses and began a voluntary system of certification for early childhood education staff.

The preschool curriculum during the 1950s reflected an emphasis on the personality development of young children. Blatz's theory of security guided the focus on the interaction between adults and children in preschool settings. The development of children's emotionally healthy relationships dominated preschool programs, and skill development was a less dominant goal (Millichamp, 1974).

Times A' Changing

The 1960s and 1970s saw enormous changes in Canadian society. A growing awareness of poverty and the importance of the early years led to increased attention to early childhood education research and curriculum methodology. Education reform became an important issue throughout school systems and universities. Federal legislation was introduced to provide assistance to low-income Canadians, including provision for day care fee subsidies. The booming economy created thousands of new jobs in the service sector, and the participation of women began to increase rapidly. In 1967 fewer than 20 percent of women with children younger than six years were in the labour force; by the end of the 1970s almost half of all women with children under six were in the labour force.

To meet the need for affordable child care, the federal government made day care a cost-shareable welfare service under the terms of the Canada Assistance Plan in 1966. By 1977 all provinces had passed legislation enabling their governments to take advantage of the federal funding.

Several provincial/territorial governments introduced legislation to regulate day care and nursery school programs and improve quality. As a result of funding and demand, the number of licensed day care centres began to rise.

During the 1960s and into the 1970s preschool education began to emphasize cognitive development from the cognitive development theory of Jean Piaget and the direct instruction theories, which come from a behaviourist orientation. The makeup of childhood care and education in Canada at this time was influenced by campus programs, government employment programs, and compensatory education initiatives. University campuses across Canada established day care centres for infants and toddlers, often as outgrowths of students' or women's organizations. Typically, these programs involved parents as participants, set up collective decisionmaking structures, and attempted to provide an environment free of gender stereotypes. They often operated at odds with the prevailing licensing and professional standards and guidelines. Many of the campus child care centres are still in existence. More importantly, they changed ECE postsecondary education programs and government regulations by illustrating support for full-day child care for very young children as a viable choice for families, not just for those in social need.

The federal government launched a number of employment programs such as the Local Initiatives Programs (LIP) during the late 1960s and early 1970s. In the winter of 1972–73, 215 children's programs, mostly day care centres, were created. LIP provided these programs with funds for workers to organize day care centres for neighbourhood children. Many LIP-initiated child care programs set up community/parent boards of directors and sought fee subsidies from provincial governments to continue operation after the LIP grants ran out.

Throughout the 1970s, day care activists across the country continued to campaign for more day care and higher-quality day care, pointing to the enormous gap between demand and supply. In 1973 there were only 28 000 regulated day care spaces in Canada. Although this rose to over 125 000 spaces by 1982, it in no way kept pace with need.

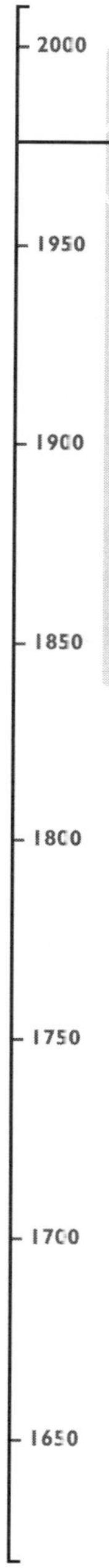

1972
Home Start programs are established in the United States to provide Head Start's comprehensive services to children and families at home.

First National Canadian Conference on day care.

Compensatory Preschool Programs

During the 1960s and 1970s child development research on the importance of early experiences on later abilities and the growing movements to end racial segregation and poverty in the United States pointed to early childhood care and education solutions.

Child advocates, policymakers, and researchers pursued early child development initiatives with high hopes that enriched early childhood experiences could change the life course of disadvantaged children, especially African-American children. Head Start, the Perry Preschool Project, and *Sesame Street* are three initiatives that attempted to improve disadvantaged children's developmental outcomes and ameliorate the effects of poverty, social marginalization, and racism.

1. Head Start was established in 1965 as part of President Lyndon Johnson's War on Poverty to expose young children (ages three to five years) living in disadvantaged environments to experiences that would minimize the effects of poverty and racial discrimination. The purpose of the program was to improve children's health and well-being; develop physical skills, social opportunities, and cognitive functioning; and involve parents in their children's educational experiences. In addition, the program offered medical, dental, and nutritional screening and services for children and social services for their families.
2. The Perry Preschool Project was a carefully designed experimental longitudinal study on the effects of compensatory preschool education program for children three and four years old in the 1960s. Preschool teachers worked closely with child development and educational experts to develop a cognitive-developmental program based on Piagetian theory of cognitive development. The findings dramatically endorse the financial and social benefits of compensatory preschool education as an intervention for disadvantaged, marginalized children (Schweinhart & Weikart, 1993). The follow-up studies found that children who attended the half-day program were more successful in school, less likely to be involved in the criminal justice system, and less likely to be on social assistance. The cost-benefit analysis of the results illustrated a seven-dollar savings for every dollar spent on the targeted program delivered to at-risk children and their families.
3. *Sesame Street*, the best-known children's educational show in North America, began in 1969. The Children's Television Workshop produced the daily program to foster intellectual and social development. The initial program proposal stated that *Sesame Street* would respond to "the national demand that we

give the disadvantaged a fair chance in the beginning" (Liebart & Sprafkin, 1988, p. 219). The creators drew on both ideology and research to design a show that would bring stimulation and opportunities for learning into disadvantaged children's homes through the television to compensate for resources presumed to be available to affluent families. Both earlier assessments (Liebert & Sprafkin, 1988) and more recent studies (Wright & Hudson, 1995) report positive benefits for low-income children who regularly view the show.

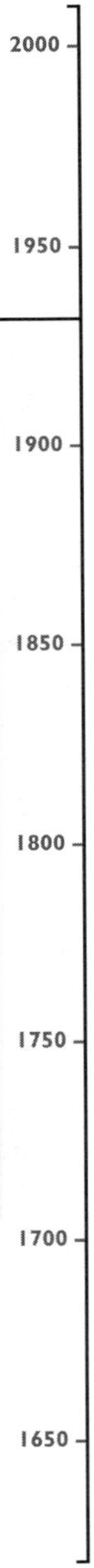

1973
West Side Family Place, a parent-child resource program, opens in Vancouver.

In the United States, the Children's Defense Fund of the Washington Research Project is established by Marian Wright Edelman, later to become the Children's Defense Fund.

There are only 28 000 regulated day care spaces in Canada.

A number of smaller-scale programs based on Head Start and the Perry Preschool Project emerged in Canada during the 1970s. Several of the programs funded by the federal government's youth employment program (Local Initiatives Program) launched preschool compensatory programs for poor children. These programs often adapted the cognitively oriented curriculum of the Perry Preschool Program and a Head Start approach that combined enriched early childhood education experiences with additional family supports.

In 1974, the Moncton Headstart Inc. began as a free day care for a few children whose parents could not afford outside care and were having difficulties in their parenting role (Bradshaw, 1997). It quickly adapted the Perry Preschool Program's curriculum and included programs to support families. Parent participation in the children's program and in parent sessions was required. Adjunct programs that meet families' basic physical needs (food, shelter, and safety) and adult education activities were also offered.

The University of Western Ontario Pre-school Project began in 1973 as an experiment to assess the impact of compensatory preschool education on low-income children (Howe, Jacobs, & Fiorentino, 2000). Mary J. Wright, who headed the project, developed a constructivist type of curriculum with an emphasis on cognitive development (similar to the High Scope curriculum) as well as a focus on the development of social competence and emotional control (derived from her work with William Blatz at the Institute for Child Studies). Children from economically disadvantaged families attended the University of Western Ontario Preschool for one or two years alongside children from middle-class families. Low-income children who attended the program for two years demonstrated greater cognitive and self-management skills than low-income children who attended for only one year. Follow-up assessments after Grade Three found that 82 percent of the low-income children who attended for either one or two years were at grade level compared to only 58 percent of a low-income control group of children who had not attended any preschool program. "Overall, the findings of the UWO Preschool Project supported the idea that compensatory education can have long-term beneficial outcomes for low–socioeconomic status children, particularly if the preschool experience is two years in length. A word of caution is in order: the study was based on a very small number of subjects, and it is difficult to generalize the findings" (Howe, Jacobs, & Fiorentino, 2000, p. 226).

The Child Care Decade

Action Day Care was established in Toronto in 1979 to advocate for free, high-quality, nonprofit child care for all families, with demands for a universal child care system. It grew out of a decade of rapid expansion of both regulated child care

THEORY INTO PRACTICE

The Federal Government and Child Care in the 1980s

Throughout the 1980s, a series of significant activities emphasized the federal government's role in the provision of child care for Canadians. Advocacy demanded that the federal government adopt a new policy role.

1982 The second Canadian Conference on Day Care brought advocates, early childhood experts, and policymakers from across Canada together for the first time in ten years. The tumultuous conference grabbed media attention and had several concrete outcomes. It passed resolutions calling for the enactment of national legislation to create a universally accessible, high-quality, nonprofit child care system and a mandate to form a broad-based national child care advocacy organization.

1983 The Canadian Day Care Advocacy Association (later called the Canadian Child Care Advocacy Association) was established. Each province and territory elected representatives to advocate for the development of a universal, publicly funded child care system.

1984 The Liberal government established the Task Force on Child Care to report to the federal government on the development of a system of quality child care in Canada.

1985 The Task Force on Child Care released its report, which was widely reported in the media. The report called for the development of publicly funded child care and paid parental leave to be implemented over a fifteen-year period. The background study and research conducted for the report provided the most thorough and comprehensive look at child care issues to date. At the same time as this report's release, the Special Committee on Child Care, established by the Conservative government, began a series of public hearings across the country, travelling to more than thirty locations in Canada, from Newfoundland to Vancouver.

1986 The Special Committee on Child Care made its report, which recommended increased tax credits for families, continuation of subsidies to low-income families through the Canada Assistance Plan, small operating grants for profit and nonprofit programs, enhancement of maternity leave to six months, and small capital grants to child care programs. It did not reflect the one thousand public submissions and was greeted with an unfavourable response from child care advocates and their supporters.

1987 At the end of 1987 the federal government announced its National Child Care strategy, which included new child care tax deductions for parents, a federal fund for special projects and research, and an intention to introduce a national Child Care Act. The act would allow $3.2 billion to be matched by the provinces over seven years; provinces would make decisions about how the funds would be spent.

1988 The federal government introduced the *Canada Child Care Act*, which met with widespread, persistent opposition. The act died on the Order Table on October 1, when the federal Conservative government called an election.

Source: Adapted from an information sheet prepared by Childcare Resource & Research Unit (1988). Used with permission.

programs and the labour force participation of women with young children. Action Day Care developed the neighbourhood hub model to deliver comprehensive child care programs and, together with other day care and early childhood advocates, labour groups, women's organizations, parents, and other concerned citizens, the new agenda for child care spread across the country. During the 1980s, provincial organizations with similar demands sprang up.

The 1980s saw rapid growth in other early child development programs. Several provinces introduced regulations for family care, and kindergarten within the school system expanded. Family support programs, such as family resource programs, toy-lending libraries, and parent-child drop-in programs expanded in many parts of the country (Beach & Bertrand, 2000). Some were aimed at high-risk families and focused on enhancing parenting skills. Others offered informal child care providers with opportunities to get together with each other, perhaps take part in some training activities and group play activities for the children. And some were provided to more advantaged at-home parents.

As you can see, kindergarten, nursery school, child care programs, and family support programs in both Canada and the United States came from very different traditions. Understanding the traditions, and the reasons behind the very different professional preparation patterns and organizations, may help today's early childhood practitioners find new ways to find common ground (see also Granucci, 1990).

SUMMARY

It is important to understand the heritage of early childhood education and care, to recognize the continuity with earlier practitioners and theorists, to realize how real events and trends influence the response of the profession, and to answer questions that separate early child development programs today. These questions include: Who cares for and educates the children? Who supports parents and other caregivers? What role will the federal and provincial/territorial governments play in early childhood education and care? Why are care and education seen as separate activities?

REVIEW QUESTIONS

1. List several reasons for studying the history of early education.
2. Identify the different occasions when the federal government has taken a role in supporting early care and education.
3. Explain why early care and education are thought of as two separate entities.
4. Discuss the separate developments of kindergartens and nursery schools in Europe and in North America, identifying key names in each tradition.
5. Order the following names chronologically, stating for each his or her importance in early childhood education and care: William Blatz; Erik Erikson; Sigmund Freud; Howard Gardner; Harriet Johnson; Lawrence Kohlberg; Ada Maream Hughes; Maria Montessori; Jean Piaget; Caroline Pratt; B. F. Skinner; Daniel Keating.

STUDY ACTIVITIES

1. Read one of the articles or books listed in the Suggested Readings or References section to learn more about one of the individuals in this chapter.
2. Write up the history of a nursery school or child care centre in your community.
3. Look for some older ECE texts in your college's library. What differences and similarities about early childhood education and care programs and practices do you see in comparison to today's texts?

KEY TERMS

behavioural theory: Also called "learning theory." A psychological theory developed in the United States, according to which behaviour is learned and can be modified by changing the environmental responses of reward and punishment.
child study movement: University-based scientific research on child development that began to be widespread in the 1920s and that facilitated the development of early childhood education.
custodial care: Looking after the basic (primarily physical) needs of children and protecting them from danger.
developmental health: Explains how our earliest social and physical experiences shape the brain's development and set the foundation for learning, behaviour, and health.
kindergarteners: Term used for the first Froebelian-trained kindergarten teachers in the nineteenth and early twentieth century.
multiple intelligences: Theory, originated by Howard Gardner, that suggests intelligence may be organized into seven different kinds of abilities.
nursery schools: Programs modelled after the MacMillan philosophy of physical, emotional, and social development, usually for three and four year olds and usually operated on a part-day basis.
parent cooperative: Nursery schools that involve parents in the classroom and in the administration of the school.
philanthropic organization: Organization established for the provision of social services to better the lives of its clients.
progressive education: Educational philosophy that, based on the tenets of John Dewey, promotes active involvement of the individual in learning.
psychosocial theory: Erik Erikson's theory of personality development, in which the individual must resolve conflict with the environment, including other persons.
rote learning: Learning by repetition, memory, and habit, rather than from first-hand understanding.

SUGGESTED READINGS

Baylor, R. (1965). *Elizabeth Peabody: Kindergarten pioneer.* Philadelphia: University of Pennsylvania Press.
Bradburn, E. (1989). *Margaret MacMillan: Portrait of a pioneer.* London: Routledge.
Froebel, Friedrich. (1896). *The education of man.* New York: Appleton.

Gardner, Howard. (1985). *Frames of mind: The theory of multiple intelligence.* New York: Basic Books.

Gesell, Arnold. (1940). The significance of the nursery school. *Childhood Education,* 1 (1), 11–20.

Greenberg, Polly. (1990, September). Head Start—Part of a multi-pronged anti-poverty effort for children and their families ... Before the beginning: A participant's view. *Young Children,* 45 (6), 41–52.

Isaacs, Susan. (1968). *The nursery years.* New York: Schocken Books.

Johnson, Harriet. (1928). *Children in the nursery school.* New York: John Day.

Kohlberg, Lawrence, & Lickona, Thomas. (1986). *The stages of ethical development: From childhood through old age.* New York: Harper Books.

MacMillan, Margaret. (1919). *The nursery school.* New York: E. P. Dutton.

Montessori, Maria. (1967). *The Montessori method.* (Anne E. George, Trans.) Cambridge, MA.

Paciorek, Karen, & Munro, Joyce. (Eds.). (1996). *Sources: Notable selections in early childhood education.* Guilford, CT: Dushkin Publishing Group. (Selections from writings by many of the early theorists and practitioners, such as Abigail Eliot, Katherine Read, Elizabeth Peabody, Patty Smith Hill, Friedrich Froebel, Lucy Sprague Mitchell, Maria Montessori, Harriet Johnson, John Dewey, Susan Blow, Margaret MacMillan, Arnold Gesell, Robert Owen, and G. Stanley Hall.)

Prentice, S. (2001). *Changing child care: Five decades of child care advocacy and policy in Canada.* Halifax: Fernwood Publishing.

Prochner, L. & Howe, N. (2000). *Early childhood care and education in Canada.* Vancouver: UBC Press.

Reeves, Carolyn, Howard, Esther, & Grace, Cathy. (1990, Fall). A model preschool: London's Rachel MacMillan Nursery School. *Dimensions,* 19 (1), 10–13.

Strong-Boag, V. (1982). "Intruders in the nursery: Childcare professionals reshape the years one to five, 1920–1940." In J. Parr (ed.), *Childhood and family in Canadian history*. Toronto: McClelland & Stewart.

Weber, E. (1984) *Ideas influencing early childhood education: A theoretical analysis.* New York: Teachers College Press.

Zinsser, C. (1988). The best day care there ever was. In *Early Childhood Education 88/89.* Guilford, CT: Dushkin Publishing Group.

CHAPTER EIGHT
The Modern Profession

OBJECTIVES

After studying this chapter, students will be able to

- identify eight characteristics of a profession and discuss how the early childhood workforce compares to each of the eight;
- identify Canadian and U.S. professional and advocacy organizations and some of their contributions to the field;
- explain the process and importance of the Canadian Child Care Federation's National Standards of Practice and identify major components of the statement; and
- describe the importance of a code of ethics for early childhood practitioners and identify Canadian examples.

As we consider the modern emergence of the early childhood workforce, it is both surprising and exciting to realize that you are entering this field only about thirty years into the modern era. As Chapter 7 explained, that era began with the rapid growth of day care centres in Canada and the establishment of Head Start programs and other compensatory programs in both Canada and the United States, which signalled a new period of interest in the importance of early childhood education. These early childhood education and care programs were also a reflection of the social currents that demanded equality for racial minorities and women and that emphasized the need to expand the field.

Thirty years is a relatively brief period when the development of a profession is considered. As we shall see, it took some time before early childhood leaders in Canada and the United States questioned how caring for and educating young children and supporting their families compared with other professions and how unity could strengthen the efforts of the field. This chapter will describe the discussions that guided some of the developments. Only in the past decade have dialogues, position statements, and other pronouncements by professional groups given early childhood practitioners substantive evidence that there are indeed unifying ideas and issues that help the profession continue to evolve.

The concept of a profession suggests that there is already unity and consensus. However, as you have already realized, early childhood practitioners have likely come to their work via different entry points and varied training, traditions, and preparation. And the work they do is itself very different. This diversity is atypical of many other professions. This chapter will continue to illustrate this diversity, as it traces the various steps that have been taken to define cohesive knowledge and practices in an evolving profession.

Is the Early Childhood Workforce a Profession?

1974
Margaret Birch, a minister in the Ontario government, proposes cutbacks to regulated child care programs in Ontario. Day Care Reform Action Alliance is established in Toronto to fight back.

Sarah C. works as an assistant in a classroom for toddlers. She took the job because it was advertised in her local newspaper, and she thought it might be fun for a while. She has no training beyond the orientation given by her director for new staff, and she has no plans to get anything beyond the sixteen hours of workshops she is required to attend each year by child care licensing requirements. Eventually, she'd like to study interior design.

Tom B. is a Head Start teacher. He took the job of assistant several years ago, after volunteering when his daughter attended a Head Start program. He liked the teaching and decided to stay and to take advantage of the continuing education ECE certificate program training offered. He has recently earned the ECE credential and is considering beginning work on a four-year degree, so that he will be certified to teach in elementary schools. This would offer him a better salary and benefits.

Mary A. is the lead teacher in a mixed-age preschool classroom. She completed an ECE degree at her local college, which she attended immediately after graduating from high school.

Rachel F. is a family child care provider. Originally trained as a data entry operator, she began her centre so that she could be at home while her own children were young. Although she is currently active in her local family child care provider organization and has completed a short training program for home care providers, she plans eventually to return to data entry.

Martha H. has almost completed a master's degree in early childhood education. After five years in preschool classrooms, she has just accepted a promotion that includes doing some administration as well as staff training at her centre.

Towanya B. is a teacher's aide in a Grade Two classroom. Her major duties include individual tutoring in reading and checking the children's work. She is a high school graduate with one semester of college general education courses.

Susanne S. teaches a K–2 mixed-age grouping in a private school.

Are these early childhood practitioners all early childhood **professionals**? What criteria did you use to decide whether they were professionals? Keep these criteria in mind as we continue our discussion of professionalism.

These individuals demonstrate the complexity of discussing early childhood education as a profession. Working in different strands of the field, with varying educational and training backgrounds and quite separate goals, they illustrate the breadth that must be included in any discussion about early childhood education as a profession. Throughout this book, we have been using the term "profession" to refer to those who are involved in teaching and caring in the early years. This is, in fact, the terminology being used by leaders in the field during the current discussions. For example, Sue Bredekamp, Director of Professional Development for NAEYC, clarifies her use of the term "early childhood professional": "The term early childhood professional or early childhood practitioner is used here to refer to individuals who are responsible for the care and education of children, birth through age eight, in centres, homes, and schools, and others who support that delivery of service" (Bredekamp, 1992, p. 52).

The origin of the word "professional" takes us back to "medieval times when an individual took *vows* in order to be received in a religious community. It was the act of openly declaring or publicly *professing* a belief or faith" (Giles & Proudfoot, 1994, p. 333). Over time, "professional" came to mean a particular occupation that required specific knowledge and high standards of practice (Giles & Proudfoot, 1994).

Whether or not early childhood education is truly a profession is something that has been debated in recent years. Sandra Griffin (1994), a Canadian early childhood practitioner, academic, and researcher, points out that if there is to be a child care system in Canada, the early childhood workforce must build a profession as the infrastructure to develop and maintain the quality of that system.

However, Milly Almy (1988) has stated flatly that, although the early childhood practitioner role requires professional attitudes and behaviours, it does not meet the standards for a profession. "With its shaky knowledge base, its ambiguous clientele, and its lack of a code of ethics, early childhood education qualifies only as an occupation or, at best, a semiprofession" (Almy, 1988, p. 50). (Note: We shall see that the lack of a code of ethics has been remedied since her statement.) Almy describes early childhood practitioners as double specialists; these double specialties include teaching young children, assessing their development, and working with children and with adults; and thinking concretely in practice, and formally, in theory. This certainly sounds like demanding work for something she claims is not a profession.

Ade (1982) agreed that the unique characteristics associated with professional status, including "specialized knowledge, a desirable service, and an assurance of quality, dependability, and effectiveness" (p. 25), were lacking in early childhood care, but that certain changes in the field could help move it closer to professionalization. The changes he identified are

- to require a greater familiarity with the field's knowledge base, thus increasing the specialized knowledge needed to practice and extending the length of the training period;
- to identify and establish uniform criteria for admitting new members, including entry criteria for training and content of training;
- to develop the kind of practitioner licensing system that would ensure meeting of criteria and exclusion of those who do not meet the criteria;
- to gain internal control of the licensing system to allow members to have input for requirements and thus greater self-regulation; and
- to obtain stronger voices by joining with parents and other decisionmakers to determine needs and the provision of appropriate services to meet those needs.

Spodek, Saracho, and Peters (1988) suggest that "professional" is used in a variety of ways, including individuals who are paid to do certain work, without implication of level of skill and those "with a high degree of skill and competence" (p. 7), but more often implying one of the learned professions, which require a high degree of training and usually involve mental rather than physical work. The fields that require less preparation and therefore often have lower levels of status, including teaching, social work, and counselling, might be called semiprofessions, having some but not all of the attributes of true professions.

1975
Two parent-child resource programs—the Parent Preschool Resource Centre in Ottawa and the Children's Storefront in Toronto—are established.

The National Association of Toy Libraries is established.

Today's early childhood practitioners are creating a new kind of profession (courtesy of Eglinton Public School).

Criteria of a Profession

Lilian Katz (1995) suggests that eight criteria must be met before an occupation can be classified as a profession, and that those in the early childhood community must work to gain consensus and take needed steps in the following areas.

Social necessity. The work of a profession is essential to the well-being of a society, and society would be weakened if the profession did not function. There are few in the early childhood field who do not believe in the absolute importance of the nurturance and development of children in the first eight years of life. Increasingly, public attention and support have been drawn to this truth by the advocacy efforts of practitioners. The longitudinal studies on effects of early education (Lazar, Hubbell, Murray, Rosthe, & Royce, 1977) offer empirical evidence about these previously more subjective ideas. But the lack of public respect and fiscal support for early childhood programs suggests that society does not fully understand the absolute necessity of these supports for children and families.

Altruism. A profession is said to be altruistic when it is directed toward service instead of profits and is performed unselfishly, with an emphasis on social goals. From the beginning, early childhood education has been grounded in principles of social improvement of the lives of children and families. Early childhood practitioners certainly rank highly in this characteristic of professionalism, since their salaries could be said to be truly sacrificial, and their concerns are with children and families.

Autonomy. Professionals are said to be autonomous when clients do not dictate what services are to be delivered, or how they are to be delivered. Early childhood practitioners are in the somewhat complicated position of defining children, parents, and society as clients and are challenged by trying to respond to various opposing ideas about goals and methods for practice. For example, what does the teacher do when parental demands suggest a curriculum that teachers feel is not in children's developmental interests? Or when the community defines family support systems that seem to usurp parental roles? Early childhood practitioners are often caught in the middle and are forced to respond to ideas that limit their autonomy. As they are often under the direction of school boards that argue for standards that teachers feel may not be in children's best interests, or of licensing requirements dictated by laypeople in the legislature, early childhood teachers are a long way from having the autonomy that would allow them to decide independently on optimal educational directions.

Distance from client. Since the practice of a profession requires applying knowledge to particular situations involving individuals, the relationship between professionals and those served is expected to be distinguished by emotional distance that would prevent clouded judgment. In this tradition, for example, doctors are not expected to treat members of their own families. Many practitioners in early education question this aspect of professionalism, noting that it is more important to meet children's needs for closeness and affection than to create professional distance. Early childhood teachers struggle with the idea of creating caring relationships with children and families while maintaining some distance that is helpful to all parties. Katz notes that maintaining an "optimum distance" allows teachers to be compassionate and caring and to exercise professional judgment, while protecting teachers from the dangers of emotional overinvolvement and burnout.

Code of ethics. A profession requires conformity to standards that are defined by a **code of ethics.** This code is adopted by all members of a profession to ensure that there will be uniformly high standards of acceptable conduct. In addition, there is a professional body that institutes procedures for ensuring that members do not violate this code. The NAEYC, after working through various stages to gain consensus, adopted and published a code of ethics in 1989. However, many practitioners are likely unaware of the code's existence, and there are no means of enforcing or disciplining those who violate its precepts. Nevertheless, the discussions that have ensued among professionals during the development of the code and since are a part of the process of defining the issues that unite the profession. A little later in this chapter, we will examine the code and its meaning for early childhood practitioners.

Standards of practice. Professions adopt standards of practice to ensure that professionals apply uniform procedures and principles in response to typical situations using their best professional judgment, and that no professional's behaviour will fall below the standards. One area still lacking in professionalization of the early childhood field is the identification of typical situations, as Lilian Katz points out in her classic discussion of how a professional early childhood teacher would teach in the situation where children are in a conflict over a tricycle (Katz, 1984). But a step toward standards of practice was taken when the NAEYC adopted the position statements on developmentally appropriate practice in 1987, and on curriculum and assessment in 1992. The discussions, debate, and the refining of these position statements will continue to move practitioners toward a profession with clear standards, as well as toward a fuller understanding of the nature of their work.

The Canadian Child Care Federation has prepared standards of practice for practitioners in child care settings (Doherty, 2000). The standards are applicable to early childhood practitioners working in regulated child care, preschool and nursery school programs, family child care, and family support programs. The standards are designed to identify specific skills and abilities necessary to perform required tasks in a competent manner.

Early childhood practitioners are guided by standards of practice (courtesy of Eglinton Public School).

Prolonged training. A major characteristic of a profession is that entrants are required to undergo extensive training, with requirements for entry that screen out individuals. The training is specialized and contains a common core of knowledge. The training is also expected to be difficult. By stretching cognitive abilities, it is more extensive than daily practice requires, and it is offered by accredited training institutions. In addition, regular continuing education is systematically required of the profession's practitioners.

1976
Lawrence Kohlberg describes his view of the states of moral reasoning in *Moral Stages and Moralization.*

As you will recall from our examples at the beginning of this chapter, there are enormous variations in the amounts and types of preservice and in-service training required of early childhood practitioners, ranging from almost nothing at all to advanced degrees. Further, because of the shortages of caregivers, trained or otherwise, there are virtually no entry requirements for practitioners, other than age and minimal literacy, in many settings. In too many cases, quality of programs is compromised by the lack of training and the acceptance of unprepared and unqualified workers. This is a complex dilemma that relates to some of the challenges regarding wages and status of the work that we discussed in Chapter 6. Yet, as we saw in that chapter, the career lattice plan attempts to respond to the varying entry points and training options that currently exist in the field, while facilitating professionalization for all who are employed in programs for young children.

Specialized knowledge. There is fairly universal agreement that a profession is work that involves specialized knowledge and skills that are based on a systematic body of principles. The knowledge is abstract, relevant to practice, and expressed in very technical terminology; it is also exclusive, known only to those who have been trained in the profession. In as much as developmentally appropriate practice is an attempt to base decisions about best practices for young children on child development research and current knowledge of learning, this offers a body of specialized knowledge to trained early childhood teachers. When, however, untrained teachers base their actions on goodwill and what comes naturally, they are likely using the best intuitions about childrearing, which parents and laypeople have used for generations.

"The problem for childcare workers is that the care of normal preschoolers is very familiar to everyone," and especially to their parent-clients. Thus for early childhood education, the main struggle with clients is to be acknowledged as "professional ... to make the status leap from babysitter to educator" (Joffe, 1977, p. 22). It is not surprising, therefore, that when these teachers base their actions on intuition, they do not garner the professional respect that they deserve.

In considering these eight characteristics of being a professional, you can see that the idea of a semiprofession, or of not quite achieving full professional status, has some validity. Yet the impressive progress that has been made in the past decade or so provides hope, and this progress has brought early childhood education closer to becoming a true profession.

However, saying that you are a professional doesn't make it so; others have to perceive you as professional. Vander Ven (1988) suggests that professionalization refers to the public recognition of and demand for a specialized service that can be provided only by people prepared to do it. It will require the joint efforts of us all to make this kind of professionalism a reality. And it is this movement toward professionalism that will overcome some of the challenges discussed in Chapter 6 and in Chapter 9.

Spodek et al. (1988) define seven questions that need to be the focus of efforts toward professionalism.

- How do we establish standards of quality for practitioners in the field?
- What standards of early childhood professionalism are reasonable in our field?
- How should entry to the field be determined?

- Should the field of early childhood education become more inclusive or more exclusive?
- How should standards be applied, and by whom?
- How should gender and economic issues be dealt with?
- How do we define professional knowledge and values?

Silin (1985) injects a cautionary note into these discussions about professionalism, noting that "early childhood education has a unique history that needs to be taken into account when exploring these issues" (p. 42). He suggests that our authority as professionals is not only based on knowledge of child development but also on moral, ethical, and aesthetic notions of what is right and true, knowledge that is not highly valued in our society. Therefore, too much emphasis on a search for professional status through "pedagogic reforms or technical knowledge" (p. 45) might deflect attention from relevant questions about the need for socio-economic and philosophical changes in society at large.

While the discussion of professionalism is ongoing, it is worth noting that even some leaders in the field recognize these kinds of concerns, as well as the benefits inherent in the idea of professionalizing. Specific concerns include

- the possibility that professionals will become separated from contact with children, as happens when increased qualifications and status are associated with positions that remove professionals from the classroom;
- parents' concern that increasing training and professionalism of early childhood teachers will mean increasing institutionalization and uncaring environments for children;
- a wish to avoid the hierarchical systems that result when distinctions are made between nonprofessionals, paraprofessionals, and more highly trained professionals;
- the increased cost of services that comes with improved compensation and that is necessary for raising the level of professionalism; and
- the possibility of excluding individuals who typically are denied access to higher education because of economic factors, yet who are needed in early childhood education to represent the cultural diversity of our society.

But the advantages of enhancing professionalism are unmistakable. The benefits include

- better care and education for children, resulting from a shared knowledge base;
- consistency across the settings and developmental continuity between programs;
- ethical behaviour;
- shared meanings and expanded knowledge; and
- improved compensation (Willer and Bredekamp, 1993).

Willer and Bredekamp (1993) express the problem of professionalism in a field that includes a wide variety of configurations of experience, training, and entry into the field. Drawing particular lines to show acceptable criteria for entry into the field will "always be too high for some and too low for others, and as the field changes over time, it will likely become obsolete" (p. 64).

It may be more useful to focus on the process of "professionalizing" and on the concept of "professionalism," rather than on the "professional." Defining early

2000 | 1950 | 1900 | 1850 | 1800 | 1750 | 1700 | 1650

1977
All provinces passed legislation enabling their governments to take advantage of the federal funding.

1978
The world's first tube baby, "Baby Louise," is born in England.

1979
Longitudinal study done by Lazar et al. reports that there are significant, lasting effects of the Head Start programs.

Action Day Care is established in Toronto. It is an organization of child care activists—parents, caregivers, and other supporters—who are calling for universal child care.

The United Nations officially names the "Year of the Child."

2000 · 1950 · 1900 · 1850 · 1800 · 1750 · 1700 · 1650

childhood practitioners as professional assumes that everyone who works with young children is involved in a process of professional development. "Professionalism" is the belief in the vision and ideals of a profession that guide daily practice. "Professionalization" describes the kinds of activities and actions that an early childhood practitioner or the early childhood workforce as a whole use to achieve the goal of a profession.

As a beginning early childhood practitioner, it is important that you recognize that there is work to be done in the profession as a whole but also begin to think of yourself as a member of a profession that has made—and continues to make—major contributions to Canadian families and society. This recognition will allow you to begin with a sense of self-respect and will enable you to represent the educated and articulate practitioners who work to gain public recognition of the value of early childhood education. Your program of training will help you see how your own competence and knowledge are developing as you progress through courses and practicums. You will see your own movement on the continuum toward increasingly professional behaviour and attitudes. As time goes by, you can work more effectively with children and families, truly representing the best of professional practice, and also contribute to this growing profession.

Professionalization of the Early Childhood Workforce in Canada

The professionalization of the Canadian early childhood workforce includes the following components:

- core knowledge;
- codes of ethics;
- standards of practice;
- credentialing;
- self-governance; and
- professional development.

In this chapter we will look at the six components and their current status in the Canadian context.

Core Knowledge

One of the prerequisites to the recognition of early childhood education and care as a profession is identifying the **core knowledge** that defines the early childhood profession. A characteristic of any profession is a specialized body of knowledge and skills shared by all its members. This core of common knowledge is defined by answering these two questions (Willer, 1994):

1. "Is this knowledge or skill required of every early childhood professional, regardless of level or setting or professional role?" (p. 13). That is, does every early childhood teacher have to know and be able to do this in order to be effective?
2. "Does the sum of this body of knowledge and competencies uniquely distinguish the early childhood professional from all other professionals?" (p. 13).

That is, although other professionals may share some of this knowledge, is most of it different from that required by other professionals?

Core knowledge in early childhood education and care, or what an early childhood practitioner needs to be able to do, was described in Chapter 4. This description was compiled from a number of initiatives that have identified the key skills and competencies—or core knowledge—for early childhood practitioners working in early childhood development settings. It is adapted to reflect the core knowledge needed to fulfill the occupational requirements in the Theory into Practice box on page 202.

The core knowledge of the early childhood workforce combines both theoretical and practical knowledge. Elaine Ferguson from Child Care Connections in Nova Scotia has written extensively on the need to recognize and value both theory and practice elements of core knowledge. She recognizes the "importance of a [theoretical] foundation in early childhood education and care and the practice skills necessary to integrate the theory into developmentally appropriate practice" (Ferguson, 1997, p. 2). Ferguson (1995) cautions us to remember this balance and to not devalue or forget the less-visible skills of practical caring as we move forward with more precise definitions of our core knowledge.

In the United States, a similar body of core knowledge has been identified by comparing common elements in the guidelines for professional teacher preparation programs at the associate and baccalaureate levels (NAEYC, 1991; 1995) and in the Child Development Associate Professional Preparation Program (Phillips, 1991). These common components suggest that all early childhood practitioners should be able to do the following:

- demonstrate an understanding of *child development* and apply this knowledge in practice;
- *observe and assess children's behaviour* in planning and individualizing teaching practices and curriculum;
- establish and maintain a *safe and healthy environment* for children;
- *plan and implement a developmentally appropriate curriculum* that advances all areas of children's learning and development, including social, emotional, intellectual, and physical competence;
- establish supportive relationships with children and implement developmentally appropriate techniques of *guidance and group management;*
- establish and maintain positive and productive *relationships with families;*
- support the development and learning of individual children, recognizing that children are best understood in the context of *family, culture, and society;* and
- demonstrate an understanding of the early childhood profession and make a commitment to *professionalism.* (Willer, 1994, p. 13).

Phillips (1994) identifies five themes that address content that every professional in the field should know:

1. children develop in context;
2. strategies for working with children are constructed each day;
3. effective practice requires a comprehensive set of skills;
4. early childhood professionals know they belong to a profession; and
5. even skilled professionals have limitations.

2000
1950
1900
1850
1800
1750
1700
1650

1980
United States has the highest percentage of children living in poverty of all western nations, a figure that continues to rise throughout the 1980s and 90s.

Toronto Board of Education releases its *Comprehensive Child Care Report.*

1980s
Nobody's Perfect, a parenting program, is developed through public health groups in Atlantic Canada and is widely adopted across Canada.

THEORY INTO PRACTICE

The Theory and Practice of Early Childhood Education and Care

An early childhood practitioner will
- ensure that the physical environment and daily practices of caregiving promote the *health, safety, and well-being* of children in their care;
- establish a *working partnership with families* that supports their responsibilities to their children;
- develop and maintain a responsive relationship with each child and the group of children to *guide children's behaviour and manage group dynamics;*
- plan and provide daily learning opportunities, routines, and activities that promote *positive child development;*
- *observe and assess* children's activity and behaviour;
- act in a manner consistent with the *principles of fairness, equity, and diversity* to support the development and learning of individual children within the context of family, culture, and society; and
- work in *partnership with other community members* to support the well-being of families.

Source: Adapted from Beach et al. (1998), p. 4.

In recent years the shift from a focus on centre-based preschool to one that is inclusive of all children from infancy through age twelve in home- and centre-based programs and related early childhood services has changed the core of knowledge that early childhood practitioners need. Early childhood practitioners gain core knowledge from professional preparation programs, experiences in early childhood education and care settings, and ongoing professional development.

Everything that is learned is used, as professionals increasingly refine their skills and broaden their knowledge. This ongoing discussion about what early childhood professionals need to know will enrich and include, rather than stagnate and exclude.

Codes of Ethics

We noted earlier in the chapter that a profession has a code of ethics, which is a "statement of principle that governs moral behaviour and ethical decisions" (Beach et al., 1998 p. 129). In Canada several professional organizations have developed code of ethics statements for early childhood practitioners, to be used as a guide in making day-to-day decisions.

Early childhood practitioners are frequently faced with ethical dilemmas in their work with young children and their families.
- What would you do when a parent demanded to know who bit her child?
- What would you do when a co-worker complained to you about another co-worker's treatment of a child?

- What would you do if a neighbour told you she heard bad things about the last centre you worked in?
- What would your responsibility be when another teacher told you that symptoms made her suspect child abuse, but that she was afraid to report it?

Often, early childhood practitioners have to take action in situations in which all the facts are not known, or there is no single course of action that is clearly right or wrong. It is sometimes difficult to decide what an ethical response might be. Making ethical decisions and taking ethical actions in early childhood education and care may require being able to see beyond short-term consequences to consider long-range consequences.

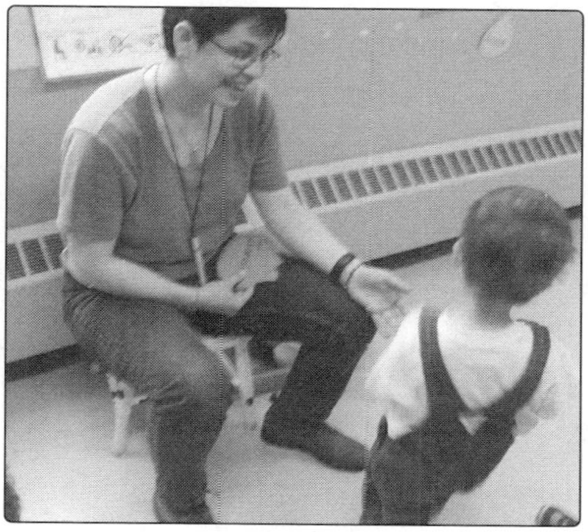

Early childhood practitioners face ethical decisions each day (courtesy of Yes I Can! Nursery School).

When you work in early childhood education and care settings, you will be asked to make many decisions about appropriate behaviour. Some of these decisions will require more than your accumulated knowledge of child development or educational practice. Some will pose genuine moral dilemmas, where you have to weigh your actions carefully in considering the parties involved.

The issues raised above, and others like them, are answered only by considering professional ethics, or the system of morals that defines a profession's proper work practices. A code of ethics is "a set of statements that helps us to deal with the temptations inherent in our occupations ... helps us to act in terms of that which we believe to be right rather than what is expedient—especially when doing what we believe is right carries risks" (Katz, 1991, p. 3). Such risks could be losing a job or alienating others with whom one must work. A code expresses the profession's belief about correct, rather than expedient, behaviour; about good, rather than merely practical, actions; and about what professionals must never do or condone, for the good and protection of those they serve.

The value in having an explicit code is that members have a document that helps them go past their intuition and individual beliefs to focus on core professional values; "it is not so much what I care about but rather what the good early childhood practitioner should care about" (Kipnis, 1987, p. 28). The code can remind early childhood practitioners of their priorities and responsibilities, and it can provide solid guidance and professional support for the decisions and behaviour of an individual early childhood practitioner.

There is no single code of ethics to guide early childhood practitioners across Canada. Instead, provincial/territorial professional organizations adopt their own codes. Early Childhood Educators of British Columbia (ECEBC) developed a code of ethics statement in 1992, which guides early childhood practitioners in their daily professional practice. Early childhood practitioners in that organization adopted the ECEBC Code of Ethics after a considerable period of discussion and consultation with members, government representatives, and college and university faculty. It has been a model for the development of code of ethics statements in other provincial organizations.

In the United States, the NAEYC developed its code of ethical conduct gradually, after gaining insights from many professionals and practitioners. The NAEYC board appointed an ethics commission to explore and clarify the early childhood profession's understanding of its ethics. The commission surveyed members to learn concerns, held workshops to identify and explore issues, and followed up with another survey of members to help formulate principles of ethical action. After further refining a draft code, the NAEYC's governing board approved the final document in July 1989. (The National Education Association has had a code of ethics since 1929 for teachers in school systems; however, there is nothing in the current code about parent-teacher relationships, so the NAEYC code is likely more relevant to your practice in early education.) The preamble states that the focus of the guidelines is on daily practice with children and their families in programs and classrooms, although many of the provisions also apply to specialists who do not work directly with children.

Standards of Practice

Standards of practice indicate the skills and abilities needed to perform tasks effectively. Included are standards of performance; input standards, or the skills and knowledge an individual brings to an occupation; and process standards, or the tasks required by an occupation (Employment & Immigration Canada, 1993).

> In child care, supporting and fostering child well-being and development (performance standards) requires the provision of certain experiences (process standards). The extent to which experiences are appropriate and effective depends, in part, on the skills, knowledge, and abilities of the child care provider (input standards) (Doherty, 1998, p. 1).

MAKING IT HAPPEN

Code of Ethics, Early Childhood Educators of British Columbia

Early childhood practitioners:

- promote the health and well-being of children
- use developmentally appropriate practices when working with all children
- demonstrate caring for children in all aspects of their practice
- work in partnership with parents, supporting them in meeting their responsibilities to their children
- work in partnership with colleagues and other service providers in the community to support the well-being of families
- work in ways to enhance human dignity and adopt an anti-bias approach
- pursue, on an ongoing basis, the knowledge, skills and self-awareness needed to be professionally competent
- demonstrate integrity in all of their professional relationships

Source: ECEBC (1992). Reprinted with permission.

Standards of practice are useful to a profession in establishing its own benchmarks of quality beyond the basics of regulatory requirements and contributing to the overall quality of early childhood education and care settings (Griffin, 1994). Teaching and nursing professional organizations establish and monitor standards of practice for their practitioners (Beach et al., 1998). Standards of practice can be used along with the body of core knowledge to guide professional preparation and development, to establish criteria for certifying early childhood practitioners, and to provide a basis for job descriptions and performance evaluation procedures. A coherent standard of practice statement that has broad support and endorsement can be a useful tool in informational advocacy activities aimed at raising public understanding of value of early childhood practitioners' work.

To date in Canada, the professional organizations in early childhood education and care have not adapted standards of practice. However, the Canadian Child Care Federation developed standards of practice for early childhood practitioners and administrators as part of the development of the National Framework for Quality Assurance (Doherty, 2000). The standards include input and process standards, which are drawn from previous efforts to identify the elements of competent knowledge, skills, and abilities.

THEORY INTO PRACTICE

Standards of Practice for Early Childhood Settings

The CCCF developed standards of practice for early childhood practitioners in family child care homes, child care centres, or in family resource programs.

Occupational Standards for Practitioners in Early Childhood Settings

Section A: Caring for Children

- Protect and promote the psychological and physical safety, health, and well-being of each child being cared for
- Develop and maintain a warm, caring and responsive relationship with every child and the group of children
- Plan and provide daily experiences that support and promote each child's physical, emotional, social, communication, cognitive, and creative skills
- Use a variety of observation techniques to identify children's skills, abilities, interests, and needs and to evaluate the daily experience provided to the children

Section B: Partnerships with Children's Families

- Establish and maintain an open, cooperative partnership with each child's family that supports the family in meeting its responsibilities for the child
- Strengthen the adults in the family in their roles as parents, nurturers, and providers and empower them to act on their own behalf

(cont'd)

Standards of Practice for Early Childhood Settings (cont'd)

Section C: Partnerships with Colleagues and Other Community Services

- Establish and maintain a collaborative working relationship with supervisors, colleagues, assistants and alternatives
- Establish and maintain cooperative working relationships with other community services

Section D: Professionalism

- Adhere to ethical standards in all aspects of practice
- Balance one's personal and work roles
- Reflect on one's own practice and identify areas where obtaining additional knowledge or changing approaches is indicated
- Establish, implement, review, and revise an achievable, realistic professional development plan

Section E: Operation of a Family Child Care Home

- Operate the family child care home as a small business consistent with legislative requirements.
- Manage the family child care home's financial resources so that they are used effectively.
- Interact positively and effectively with a broad range of stakeholders including, but not limited to, families, colleagues, assistants, alternates, practitioner's family members, other community services and regulatory and/or agency staff and officials.
- Hire, develop and maintain knowledgeable, sensitive, and motivated colleagues, assistants and alternates.
- Ensure a safe, healthy, well-organized environment and purposeful daily program that meets the needs of the children being cared for and their families.

Source: Doherty. (2000). Reprinted with permission.

Credentialing

Credentials are the evidence that an individual has both the knowledge and skills needed to practice in a specific occupation. **Credentialing** is the process of evaluating knowledge and skills. Doherty (1997) defines three methods of credentialing in the North American early childhood education and care: equivalency validation, certification or licensing of early childhood education graduates, and competency-based assessment. We will define each of these processes and consider their application in Canada.

- Equivalency validation credentials verify that an early childhood practitioner who does not meet a particular jurisdiction's qualification requirements has comparable or equivalent education and experience. All provinces and territories that have educational qualification requirements for early childhood practitioners in early childhood education and care settings have some established procedures for assessing qualifications and credentials from other jurisdictions.
- Credentials of graduates from ECE certificate, diploma, and degree programs may be issued by the government, as a license or classification, or they may be

issued by a professional organization, as voluntary recognition or certification. For example, in British Columbia, early childhood practitioners are required to have a government license to practice to be recognized as qualified staff in a regulated child care centre or preschool. To be eligible for the license to practice, early childhood practitioners must have an ECE academic qualification and documented work experience. Volunteer certification is provided by a few professional organizations and is discussed in more detail later in this chapter.

- Competency-based assessment "testifies that a person whose formal educational qualifications related to child care are less than those required for entry into the field has the practical competencies required to provide competent child care" (Doherty, in press, p. 1). The only competency-based assessment process (outside of prior-learning assessment in colleges and universities) is in Manitoba and is conducted by the provincial government.

Voluntary Certification

Voluntary certification recognizes and endorses the educational qualifications and, sometimes, the performance levels of early childhood practitioners, regardless of their work setting. It is a mechanism to monitor standards of practice and to promote both the quality of early childhood education and care services and the professionalism of the early childhood workforce.

Three provincial professional organizations offer voluntary certification to their members: the Association of Early Childhood Educators of Ontario (AECEO), the Early Childhood Professional Association of Alberta (ECPAA), and the Certification Council of Early Childhood Education of Nova Scotia (CCECENS).

- The AECEO requires eligible applicants to have an ECE diploma or equivalent, to be members of the organization, and to be working in an early childhood education and care setting for at least twenty hours a week. Applicants submit a written personal profile of their philosophy of early childhood education and take an examination within a year of their original application. The examination is based on the expected learning outcomes for students graduating from Ontario's community college ECE diploma programs. Once certification is granted to an early childhood practitioner, she is a professional member of the organization and maintains her certification credentials as long as she remains a member of the AECEO (Trainer, 1998).
- The ECPAA certification process requires that applicants have educational qualifications, experience, and a letter from their current employer describing their work performance. Neither observation nor examinations are required.
- The CCECENS certification process stipulates that eligible applicants must have an ECE education qualification and at least two years' experience working in an early childhood education and care setting with young children. Applicants are observed at their workplace by a facilitator (who is an early childhood practitioner with CCECENS certification) to assess their performance against fifteen standards of practice, which have been developed by the CCECENS. The process can be repeated several times until the applicant meets the criteria. At this point a validator reviews written evaluations, observes the applicant, and makes a recommendation to the CCECENS, where the final decision is made. A certified

1981
Laurier Lapierre and Ada Scherman write *To Herald a Young Child.*

The Ontario Coalition for Better Day Care (later renamed the Ontario Coalition for Better Child Care) is established.

1982
Mary Ainsworth continues the discussion of attachment in infancy, in *Early Caregiving and Later Patterns of Attachment.*

There are 125 000 day care spaces in Canada.

The Toronto Board of Education decides to facilitate the development of parent-operated centres in vacant classroom space.

The second National Child Care Conference is held in Winnipeg. Child care activists and early educators pass resolutions calling for a national day care act, national standards, and support the goal of universally accessible day care for all Canadians.

early childhood practitioner maintains her certification by paying an annual fee and by providing proof of her participation in professional development.

The voluntary certification processes in Canada aim to improve quality and increase professionalism. However, there are some limitations. Only a small proportion of early childhood practitioners in each of these provinces is certified. Although in some early childhood education and care programs, certification is recognized and valued, there is little overall recognition or motivation to become certified, as it is not required. In both Alberta and Ontario the ECE educational qualification establishes eligibility and, therefore, provides the standards of practice to be assessed. In Nova Scotia the process has introduced an assessment of an early childhood practitioner's practice based on criteria established by other practitioners. Unfortunately, the automatic renewal in certification in the three provinces fails to provide a mechanism to ensure that early childhood practitioners' knowledge remains current (Doherty, 1997).

Self-Governance

The self-governance of a profession refers to its authority to establish bodies to regulate its members and their practice through certification, standards of practice, discipline, and requirements for professional education. Self-governance recognizes bodies outside government or organizations that represent the economic or professional development interest of practitioners. The recognition is usually embedded in legislation.

Provincial/territorial professional organizations have taken on tasks—such as voluntary certification, development of codes of ethics, consultation with governments on issues related to equivalency validation, and appropriate course content for professional education programs—which are often associated with self-governance. There is only one self-governing body in Canada, the Association of Early Childhood Educators, Newfoundland.

The Certification Council of Early Childhood Education, Nova Scotia, the Association of Early Childhood Educators Ontario, and the Early Childhood Educators of British Columbia have proposed legislated or mandated self-regulatory bodies, which would regulate early childhood practitioners within each of the provincial jurisdictions. To date, the proposals have not been adapted.

Professional Development

Not all early childhood professional education and development is part of postsecondary education programs or part of professional preparation (Beach, 1999). In fact, most of the professional preparation we have considered in this chapter will equip you only with the entry-level skills and knowledge of an early childhood practitioner. Most early childhood practitioners take part in other activities, often offered by professional organizations, which promote ongoing learning, professional development, and professional practice in working with young children and their families.

The range of development opportunities that promote skills and knowledge includes workshops, seminars, conferences, publications, and networking. These resources are offered by early childhood education and care organizations, government departments, community groups, family resource programs, family child care agencies, and child care organizations.

Publications

Most national and provincial/territorial organizations produce a regular newsletter for members. There are also a number of journals prepared for the early childhood workforce that report on research studies, policy developments, and innovations in early childhood education and care programs.

- *Interaction* is a bilingual journal published bimonthly by the Canadian Child Care Federation. It includes articles on current research and public policy in early childhood education and care and related fields. It also profiles individuals and organizations from across the country and provides a comprehensive listing of upcoming events.
- *Young Child* is a peer-reviewed journal from the Canadian Association of Young Children that reports on applied research findings and innovative models of practice from early childhood education and care programs.
- *Canadian Journal of Research in Early Childhood Education* is a bilingual journal that features Canadian research projects related to early childhood education and care. Published twice yearly by Concordia University in Montreal, its mandate is to inform early childhood practitioners of findings from past and current Canadian research. It also includes reviews of early childhood education texts and notes upcoming conferences and seminars.

Professionalization Dilemmas

As the early childhood workforce becomes more like a profession in Canada, it is faced with a number of dilemmas:

- As the workforce becomes more professionalized, a hierarchy develops, and the direct work with young children is the least valued and rewarded.
- The drive for increased professionalization is often isolated and separate from other challenges facing early childhood practitioners.
- Currently, only a small percentage of qualified early childhood practitioners and an even smaller percentage of all caregivers are members of professional organizations.
- The numbers of early childhood practitioners who have educational or equivalent qualifications are increasing in early childhood education and care settings, but most individuals working with young children do not have them. The professionalization of this workforce may exclude those many experienced caregivers and limit the participation of caregivers from particular ethnocultural backgrounds.
- The diversity of work environments in early childhood education and care programs makes it difficult to find common ground to build professional practices for early childhood practitioners.

2000
1950
1900
1850
1800
1750
1700
1650

1983
The Canadian Day Care Advocacy Association (later the Canadian Child Care Advocacy Association) is established.

Howard Gardner publishes *Frames of Mind*, describing his theory of multiple intelligences.

- Many of these issues are raised and discussed in the course of the child care sector study. The final report notes that the early childhood workforce "has the opportunity to define a new form of professionalism which includes all caregivers who want to provide quality experiences for young children and their families, and recognizes various levels of credentials, experience and employment setting" (Beach et al., 1998, p. 128).

THEORY INTO PRACTICE

Coming of Age

Within the last couple of generations in Canada, providing child care services for pay began to emerge as a legitimate occupation, with recognized standards, training and public recognition. This process is still very incomplete and most of the shared concerns and future tasks articulated in this report [Child Care Sector Study] are related to the intense desire and need by child care workers to define themselves as a legitimate, respected and reasonably compensated occupation and to get the public acceptance and legislative recognition they deserve.

Child care as an occupation (or series of related occupations) is marked by its origins. For most of this century, in North America at least, the task of raising children has been one that women were expected to do by staying at home when children were young. As women moved into the labour force, they typically made arrangements through the family and extended family or with neighbours to provide child care, generally in exchange for non-monetary types of rewards (as part of intra-family obligations, or mutual help arrangements with neighbours, etc.).

Then, it would have seemed foolish to suggest that those providing child care needed special training; women were assumed to learn these talents at their mothers' knees. Child care as a distinct occupation requiring formal education and ongoing training, regulations and standards, has taken a long time to emerge from this history. The attitudes of many towards child care and its workers is rooted here. Current lack of self-organization and divisions between different types of caregivers have much to do with this history.

There are many types of child care and of caregivers. This report [Child Care Sector Study] and the general set of occupations with which it deals include early childhood practitioners in licensed centre care, in nursery schools, and in licensed family day care homes. It includes nannies and informal caregivers who organize themselves into associations and who provide care as a continuing business or occupation. It does not include the very heterogeneous group of caregivers who provide care informally within their own neighbourhoods and may provide care for only a short period.

For all of the caregivers with whom this Report is concerned, a central objective is to convince parents, the general public and governments that there are, at least on average, dramatic differences in reliability and positive child development effects of the care they provide and that are provided in a more ad-hoc way by this part of the informal sector. Care in this less organized part of the informal sector (partly because of its favourable tax treatment, partly because of its poorer facilities, partly because its workers do not have good labour market alternatives) is typically cheaper than for other child care services. One way of interpreting the central shared objective of child care workers as we have described it comes down to convincing parents, the general public and governments that the service they provide is well worth its extra cost.

Workers across the child care sector have a strong interest in convincing parents, the public and governments that better quality child care matters—that it has very important differential effects on child development, socially, intellectually, emotionally, etc. Similarly, child care workers have a collective interest in

Coming of Age (cont'd)

anything that educates parents about child care quality and that helps them to know how to judge the different aspects of quality. Unless parents appreciate the distinctive contribution of good child care, there will be no basis to resolve issues of lack of respect, appreciation, poor remuneration, etc. that plague the sector. There are many ways to contribute to this: supporting studies of the effects of quality child care on children's development by add-ons to the National Longitudinal Survey of Children and Youth, accreditation of centres and homes (seal of quality), public awards and honours drawing attention to good quality child care, etc. This list is only suggestive; there are lots of other ways of studying, evaluating and generally raising the profile of quality issues in child care.

Child care is leaving its adolescence and coming of age; there are difficult issues of direction to resolve at this crucial time. It is the end of a rather long period of growth of all parts of the sector as the demand for non-parental paid child care has grown largely for mothers' labour force reasons. Although there may be continuing growth, it will not be so rapid and will depend very much on government initiatives, like the ones recently announced in Quebec (of course, it is unclear whether they will have the budget to fulfill this commitment).

The predominant governmental trends are still in the opposite direction: move to the CHST from CAP, failure to live up to the Red Book commitments, reduction of support for the licensed sector, encouragement of use of informal care, opening subsidies to all segments of the informal sector, etc. What government does matters a great deal to the future of child care and child care workers.

The child care sector is beset with a series of environmental issues such as: perception of a lack of recognition of the value of the work done, concern about low pay and poor benefits, public perception that child care is, generally, not a lifelong occupation but a temporary way station, concern that there is no career ladder or reward for improving your skills in the sector. These issues appear to be linked to the lack of clarity in the sector and amongst parents and the public about what the main purpose of child care is.

From the point of view of many parents and many child care providers, it appears that facilitating labour force participation is the key objective of child care. The implication is that child development occurs at home, or later in the child's life at school, and that a warm, caring, custodial child care arrangement is quite satisfactory. The further implication is that good child care is cheap child care, that the convenience of the parents is the paramount concern, and that the caregiver need only be someone the parent feels comfortable with (similar background, education level, etc.).

The alternative perspective is that the key purpose of child care is child development. In this view, trained child care workers should be seen as distinctly different from untrained ones, there should be pressure to increase the length and intensity of child care training, there should be tools which facilitate the choice by parents of good, rather than not-so-good child care centres, and centres should be competing to hire the best quality staff.

The point of the above discussion is that it makes sense for child care workers of different types to join in working to clarify in the public mind how child care should be seen. This choice determines one's attitude to most issues of public policy towards the sector and towards the compensation and regulation of the workers in it.

The dilemma facing child care workers is a difficult one. Child care workers work in an industry/sector which is extremely labour-intensive and in which, compared to other labour-intensive sectors, there is no obvious way in which computerization, or intelligent rethinking of the service provided can lower the amount of labour required. This means that every dollar of wages or benefits received by workers translates directly

(cont'd)

Coming of Age (cont'd)

into higher cost child care. And, every improvement of the staff-child ratio, or the ratio of trained to untrained staff, directly increases the cost of child care. And, this is not a cheap service either. Child care is a big-ticket item.

In a country like Canada, in which the price of labour is relatively high, good child care is seen by many as a luxury item. Many parents settle for not-very-good child care (you can see this by looking at the average expenditures of families on child care; some spend a lot, but many spend not very much).

Child care has, then, a choice (a difficult choice) of self-definition. Either it goes for what parents are able and willing to pay for, which is not-very-good child care. In this case, the downward pressure on wages and benefits is inevitable, but the private demand for the service will be strong. Or, it goes for a strong child development, quality-enhancement, self-definition. This may be the right choice in the long run, but it requires both increased private and public dollars.

This means convincing parents and opinion leaders of the key importance of good child care in early years. It means continual public education programs and encouragement of research into what a difference quality child care makes. It might (or might not) mean accreditation programs to publicly distinguish good quality child care. It might (or might not) mean moves towards professionalization of the occupation (not in a narrow, craft unionist sense). It might (or might not) mean increased contact with kindergarten teachers to emphasize the similarity of the child development roles that these workers play. It might (or might not) mean encouraging investigation of how much difference a B.A. (E.C.E.) versus a 2 year diploma, versus a one-year certificate makes. And so on.

Consultation with the sector suggests that the second route is the one that child care workers should take and that various structures and processes should be established which could facilitate this.

It is within the above perspectives that we have to judge the current type of training that ECEs receive across the country. Is the level and type of education adequate and more than adequate? Are ECEs well prepared to provide a level and type of care that will be recognized as distinct and developmental? Do ECE schools do a good enough job trying to raise standards in the occupation (all parts)? Do they provide home care training? Do they effectively restrict entry into the occupation in a desirable way (improving quality, enhancing standards)? Do colleges require some level of academic performance in high school as a minimum condition for entry? Are workers well enough trained in communication skills to represent the occupation well to parents and others? Should there be room on the career ladder for university educated child care workers with in-depth knowledge about child development, education, child care programming, etc.?

We will need to reach conclusions on ways that child care workers can organize themselves to work collectively to get recognition for the skills and importance of their occupation. Is there a need for a sector council to work over time on common projects and interests related to quality promotion, encouragement of explicit standards of performance and training in different parts of the sector, of lobbying government to recognize the centrality of quality issues in relation to child development? Would moves towards professionalization hurt or help the occupation to come of age? What role should various child care associations/organizations play? What positive role do unions have in standardizing the pay and benefits of child care workers and encouraging a longer term view on the part of employers in the child care field? What do current and potential relationships with kindergarten teachers have to do with this?

Source: Gordon Cleveland (unpublished) prepared for the Child Care Sector Study (Beach, Bertrand & Cleveland, 1998), reprinted with permission.

Early Childhood Education and Care Organizations

The early childhood workforce is coming of age (courtesy of George Brown College).

There are early childhood education and care organizations and associations at the national, provincial/territorial, and local levels that support the early childhood workforce and advocate for its recognition. In addition to their involvement in advocacy and activities related to professionalism, these groups carry out a range of professional education and development activities.

The Child Care Sector Study (Beach et al., 1998) found that organizations face several challenges in their ability to meet the needs of the early childhood workforce:

- Most organizations operate with little or no public funding and rely on considerable volunteer labour to carry out their activities.
- The ability of early childhood practitioners and early childhood education and care programs to sustain organizations through membership fees is limited, reflecting, in part, the low compensation that early childhood practitioners receive.
- Membership in early childhood education and care organizations is voluntary, and most early childhood practitioners do not belong to any organization.

Canadian Child Care Federation (CCCF)

The CCCF is a national organization committed to improving the quality of early childhood care and education services across Canada. Its membership includes thirteen affiliated provincial/territorial organizations, with a total of more than 9000 members across Canada. The CCCF provides leadership to the early childhood workforce on a number of fronts, including development of national principles for quality early childhood education and care programs and guidelines for professional education programs. It also provides research, publications, conferences and workshops, and information services. The CCCF sponsors projects to develop a national framework for quality assurance and guidelines for family child care training, and is also working with the Association of Canadian Community Colleges to consider an accreditation process for ECE certificate and diploma programs in Canada.

Provincial/Territorial Early Childhood Education and Care Organizations

There is at least one organization in every province and territory (except for the Northwest Territories and Nunavut) that represents the early childhood workforce. These groups have grown from the needs of early childhood practitioners and do not have government-defined mandates.

THEORY INTO PRACTICE

Early Childhood Education and Care Organizations in Canada

Early childhood practitioners have access to professional, advocacy, and resource organizations across Canada. The following list is an overview of the organizations and resource groups and their primary purpose. You can find links to their Web sites at www.ece.nelson.com. There are also local groups that provide support to early childhood practitioners.

National

Canadian Child Care Federation
Works to improve quality of child care services across Canada, build an infrastructure for the early childhood education and care community, and provide extensive information and resource services. Includes thirteen provincial and territorial affiliate members. Provides a complete listing of all organizations, professional education and development programs, and related resource groups.

Child Care Advocacy Association of Canada
Advocates for comprehensive, universally accessible, high-quality, and nonprofit child care programs.

Canadian Association of Family Resource Programs
Provides consultation and support for program development, publications, and professional development to a national network of family resource programs.

Canadian Association for Young Children
Promotes professional development and communication among early childhood practitioners in child care programs and school settings.

SpeciaLink
Promotes inclusion of children with special needs in child care and other community programs and provides research, resources and information, networking, and training to early childhood practitioners and related services.

Childcare Resource and Research Unit
Provides public education, consultation, and publications; organizes and disseminates information and resources; conducts child care–related research projects; and provides a circulating library and database of resources.

Child and Family Canada
On-line network of organizations related to child and family issues in Canada.

Newfoundland

Association of Early Childhood Educators, Newfoundland and Labrador
Provides awareness of quality child care and advocates for early childhood practitioners.

Prince Edward Island

Early Childhood Development Association, P.E.I.
Promotes knowledge of early childhood education, promotes professionalism, and provides support to membership.

Early Childhood Education and Care Organizations in Canada (cont'd)

Nova Scotia

Child Care Advocacy Association of Nova Scotia
Represents caregivers in nonprofit child care centres and family child care.

Nova Scotia Family Day Care Association
Provides networking opportunities for caregivers in regulated child care.

Certification Council of Early Childhood Education of Nova Scotia
Offers voluntary certification process for early childhood practitioners since 1988.

Child Care Professional Development Association of Nova Scotia
Provides licensure of child care practitioners.

Child Care Connection—Nova Scotia
Connects child care professionals to resources, promotes certification and administration, operates a resource centre library, and sponsors professional development events.

New Brunswick

Early Childhood Coalition Petite Enfance
Promotes and supports quality of life for children in New Brunswick.

Quebec

Quebec Association for Preschool Development
Promotes and supports child care administration.

Concentration Inter-régionale des garderies du Québec
Advocates for quality, nonprofit child care.

Association of Early Childhood Practitioners, Québec
Represents early childhood practitioners and promotes professionalism.
Provides voluntary certification process for early childhood practitioners.

Association des services de garde en milieu scolaire du Québec
Promotes increased access to school-age child care.

Ontario

Association of Day Care Operators of Ontario
Represents and promotes private operators.

Association of Early Childhood Educators, Ontario (affiliate member of CCCF)
Represents early childhood practitioners and promotes professionalism.
Provides voluntary certification process for early childhood practitioners.

Home Child Care Association of Ontario
Promotes and supports quality home child care, with primary focus on regulated home child care.

Ontario Network of Home Child Care Provider Groups
Promotes and advocates for family child care providers.

(cont'd)

Early Childhood Education and Care Organizations in Canada (cont'd)

Ontario Coalition for Better Child Care
Advocates and promotes quality nonprofit child care and represents nonprofit child care programs.

Ontario Association for School Age Care
Supports and promotes quality school-age child care programs.

Ontario Association of Family Resource Programs
Represents and provides support to provincial family resource programs.

Manitoba

Manitoba Child Care Association (CCCF affiliate)
Advocates for a quality child care system, represents nonprofit child care programs and regulated family child care, provides services to membership, and advances child care as a profession.

Saskatchewan

Saskatchewan Child Care Association (CCCF affiliate)
Supports caregivers in centre-based care and family child care.

Alberta

Alberta Family Day Care Association
Provides support to regulated family child care.

ECE Professional Association of Alberta (CCCF affiliate)
Supports ECE staff and promotes professional development. Provides voluntary certification process for early childhood practitioners.

British Columbia

Early Childhood Educators of British Columbia (CCCF affiliate)
Promotes and supports early childhood practitioners.

Western Canadian Family Child Care Association
Promotes, supports, and advocates for quality family child care.

British Columbia School Age Association
Promotes and supports school-age child care and represents school-age child care staff.

Coalition of Child Care Advocates of BC
Promotes and advocates for quality child care.

Westcoast Child Care Resource Centre
Provides child care resources, information and referral services, and professional development with and through affiliate early childhood education and care organizations.

Yukon

Yukon Child Care Association
Promotes centre-based and family child care.

MAKING IT HAPPEN

Provincial Early Childhood Education and Care Organizations

The Association of Early Childhood Educators, Ontario (AECEO)

The AECEO is a professional organization concerned with the quality of care and education for young children. It provides leadership and vision as the unified voice of early education in Ontario.

The AECEO was established in 1959 as the Nursery Education Association of Ontario (NEAO). It evolved from the increasing demand for child care after World War II, the absence of government regulations, and the lack of formal training in child care studies.

In the early years, the AECEO worked tirelessly to foster the development of training programs. Cooperating with the Institute of Child Studies at the University of Toronto, Ryerson Polytechnical Institute (now Ryerson Polytechnical University), and the Ontario Agricultural College (now University of Guelph), the AECEO began some of the first courses in child studies. Later, under AECEO sponsorship, the extension departments of six Ontario universities began offering evening and summer training programs for those already working in the field. Parent resource centres were also set up in some localities.

In the 1960s AECEO was a major help in the development of the two-year diploma program for the new community colleges. The diploma course is now the standard for qualification as a professional early childhood practitioner in Ontario.

The AECEO established a certification process in 1964 to encourage individual achievement in formal child care training and practical experience.

In 1969 the organization changed its name to the Association for Early Childhood Education, Ontario (AECEO). In 1980 the AECEO held its first annual Week of the Child promotion, following the UN International Year of the Child in 1979. Every October a broad series of public events and professional development activities are sponsored across Ontario during the Week of the Child.

In 1993 the Association changed its name to the Association of Early Childhood Educators Ontario to better reflect its current mission, which is "to be the leader in promoting the professional development and recognition of early childhood practitioners, on behalf of the children of Ontario."

Today the AECEO has more than 2500 members distributed across 24 branches that represent the entire province. It is governed by a board of directors with representation from the membership. Its strategic goals are

- to continue to strengthen the organization and base of the operations;
- to pursue legislative recognition for ECEs; and
- to enhance public education and awareness of the importance of quality early childhood education.

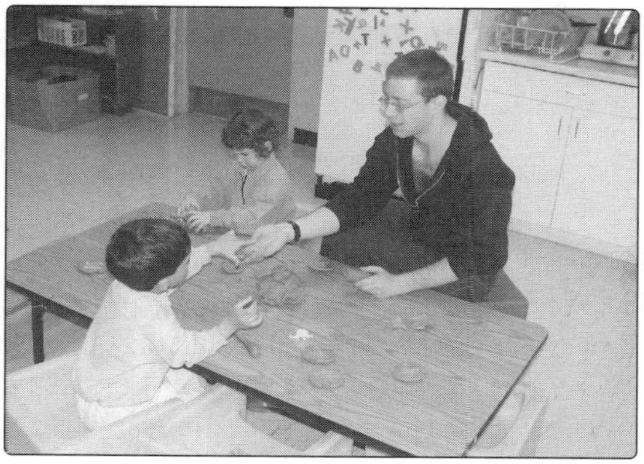

Early childhood practitioners often seek out opportunities to share experiences and learn from their colleagues (courtesy of Play and Learn).

Sources: Thomas (2000) and Szereszewski (1994).

SUMMARY

As you have read about the work of professional organizations, it should be obvious that there are many resources and vehicles for early childhood practitioners to achieve professionalization, both as individuals and with colleagues.

Joining a professional organization as a student or new teacher is critical for professional awareness and growth.

REVIEW QUESTIONS

1. Describe several of the characteristics of a profession, and discuss how early childhood care and education matches the standard.
2. Name two major early childhood professional organizations, discussing the services of each.
3. Discuss the ECEBC code of ethics, its component parts, and its uses for early childhood practitioners.
4. Discuss the importance of the position statement on developmentally appropriate practice.

STUDY ACTIVITIES

1. Visit your school library to learn what journals from professional early childhood organizations are available. Read several articles from a representative issue, and write brief reports that indicate the kinds of knowledge you obtain. Examine the issues for regular features that would be helpful to early childhood teachers in their practice and in their learning about policy and wider issues of the profession.
2. Using your community resource guide and contacts with early childhood teachers, learn what professional organizations are available and used by teachers in your area. If possible, attend a local meeting. Talk with local teachers who are members of the organization to learn how they have benefited from their membership. If there are no local groups, call or write the nearest affiliate to get a listing of their current activities and to get on their mailing list, if possible.
3. Find out if there is a Week of the Young Child activity held in your community. If there is, find out how you and other students can become involved. If there is not, what could you do to initiate one?
4. With classmates, use the ECEBC code of ethics to discuss the following issues that could occur in a classroom.
   a. A parent asks an infant caregiver to please not feed the baby in the late afternoon, so that she can feed her at home and put her to sleep. The baby cries each day until her mother comes.
   b. A co-worker has discussed a family at your child care centre with her boyfriend's family, who ask you about the situation.

c. A four year old is angry when his mother leaves him each morning, and he treats other children aggressively. The assistant teacher is bothered by the lead teacher's response, which is to put him in time-out for long periods.

d. An early childhood practitioner is told by her supervisor that a child who is absent has contagious diarrhea, so teachers should wash hands carefully. The early childhood practitioner is surprised that a notice is not posted to inform parents.

e. An early childhood practitioner working with three year olds is unaware when she is first employed that the number of children in her room exceeds the adult-child ratio standards of the provincial/territorial day care regulations. When the licensing officer comes to inspect, the director tells him that the cook also works regularly in the program.

f. A mother asks you not to let her four-year-old son nap in the afternoon, since he then wants to stay up late at night and she has to get up early to go to work. The child seems to need his nap to play happily in the afternoon.

5. Check out early childhood organizations in your province or territory at www.ece.nelson.com. Are there any student membership rates? What services are available?

KEY TERMS

code of ethics: Statements of a profession that govern moral behaviour and ethical decisions. The NAEYC adopted and published its early childhood code of ethics in 1989.

core knowledge: Basic knowledge a professional group acknowledges to be needed by all its members.

credentialing: Evaluating professional qualifications from recognized educational institutions or professional organizations.

professional: Practitioner who has met the standards of knowledge and performance required by a profession.

SUGGESTED READINGS

Buck, Linda. (1987, January). Directors: How to sell accreditation to staff, board, and parents. *Young Children,* 42 (2), 46–49.

Canadian Child Care Federation. (2000). *Tools for practitioners in child care settings: Standards of practice, code of ethics, guide to self-reflection.* Ottawa.

Ferguson, E.E. (1995). *Child care ... Becoming visible.* Halifax: Child Care Connection-Nova Scotia.

Ferguson, L. (1991). The child care crises: Realities of women's caring. In C. Baines, P. Evans, & S. Newysmith (Eds.), *Women's caring, feminist perspectives on social welfare.* Toronto: McClelland & Stewart, 21–36.

Radomski, Mary Ann. (1986, July). Professionalization of early childhood practitioners: How far have we progressed? *Young Children,* 41 (5), 20–23.

CHAPTER NINE
Advocacy

OBJECTIVES

After studying this chapter, students will be able to

- discuss the relationship between quality experiences for children, compensation and economic conditions for the early childhood workforce, and accessibility to programs for families;
- identify personal and collective approaches to advocating for early child development programs and its workforce;
- recognize the role of advocacy organizations and other organizations in addressing the issues of quality, compensation, and accessibility; and
- discuss several personal actions that will contribute to successful growth as an early childhood practitioner and advocate.

Advocacy is the act of defending or stating a cause for yourself or another. The role of advocacy is played out within early childhood education and care settings and in the world at large. Personal advocacy involves our day-to-day actions that promote children's healthy development, support families, and recognize our own value. Public policy advocacy is the collective effort to address three interconnected issues within early childhood education and care programs and the early childhood workforce in Canada: quality, compensation, and accessibility. Informational advocacy seeks to raise public awareness of the importance of children's early years and the potential of early childhood education and care programs to contribute to healthy children, strong families, and cohesive communities.

In this expanding early childhood workforce, responses to the needs of children, families, and early childhood practitioners are propelling changes that affect us all. Some of these trends will strengthen the workforce itself, making it more likely that you will see enormous changes during your career. As you enter the early childhood workforce, it is important that you consider the complexity of these issues and decide how you can advocate for yourself and for the children and families with whom you work.

Before you read any further, note in your journal the ideas that you believe might be included in this discussion of advocacy issues. Think about the ideas you have encountered already in your own work experiences and in the course of your early childhood education studies. Reflect on the challenges raised in the first and second sections of this book, where you examined early child development programs and the early childhood workforce, and in Chapter 7, where you considered their historical development.

As you continue your ECE studies and embark on your journey as an early childhood practitioner, there are guidelines for personal daily practice. The most effective advocate for early childhood education and care programs is a dedicated, reflective practitioner. Your knowledge, skills, and abilities shape the daily lives of young children and families. You also influence the public (local and global) perceptions about how we should move the early child development agenda forward in the twenty-first century.

Personal Advocacy

Early childhood practitioners are advocates for young children (courtesy of George Brown College).

Every early childhood practitioner can be a personal advocate for herself and for the children and families with whom she works, as well as for other early childhood practitioners, children, and families. Personal advocacy is based on the way in which you carry out your daily practice and communication with others. It can be as simple as gently correcting a friend or neighbour who suggests that what you do is "just babysitting—something anyone could do" or taking the initiative to ensure a child you are working with has access to needed specialized services. It may mean going out into your community to find out what types of early child development programs are really needed and wanted by families and then trying to match them with local resources.

Personal advocacy includes being a watchdog and a whistle-blower. As you learn about child development and appropriate practices, you are more likely to resist pressure to accept less-than-optimal conditions in early childhood settings. You will advocate for conditions that support the delivery of high-quality care and education. This means being aware of situations in your local communities that may have an adverse affect on children's health, safety, or well-being. For instance, there may be physical dangers in playgrounds or schoolyards or a need for safer crossing points on busy streets. Keeping yourself informed and becoming involved in making changes to improve children's environments are part of personal advocacy.

Children cannot vote on issues that affect them, but you can. You should be registered to vote where you live and exercise your right to vote during elections. To influence how you vote, you'll need to know which politicians represent you and keep an eye on how they vote on issues that affect children and early childhood education and care programs.

Public Policy Advocacy

In early education, public policy advocacy is the act of defending or stating the cause of early childhood education and care programs and the workforce. Kagan suggests public advocacy in child care and early education has three main rationales:

1. to safeguard what has been achieved;
2. to increase the capacity to provide services; and
3. to make important changes (Kagan, 1988).

1984
The Royal Commission on Equity in Employment recommends a national child care act to fund and set standards for high-quality child care.

In the current climate of deregulation, decentralization, and deficit-cutting, these will continue to be powerful motivations for early childhood practitioners to be involved in public policy advocacy.

Public policy advocacy for early child development programs and the early childhood workforce is based on the view that children and families should have access to quality services and that early childhood practitioners should be recognized as a critical component of these services.

Understanding the Links: Quality, Compensation, and Accessibility

This section examines the links between **quality, compensation, and accessibility**. The future of the early childhood workforce and programs depends on these three intertwined issues. There is a clear relationship between the quality of care and education children receive and the compensation and working conditions of early childhood practitioners. There is also a clear relationship between children's and families' access to quality programs and the costs, which mainly go toward covering early childhood practitioners' wages.

This section then considers various strategies to address these issues and to advocate for quality programs for children, equitable compensation for early childhood practitioners, and accessibility for families. These issues are important to all early childhood professionals. You will learn about how early childhood practitioners personally advocate for public policies, work together on campaigns, and participate in raising public awareness and understanding of the underlying issues. As you continue with your professional development, you will be involved in some of the processes that work toward solutions, resolutions, and further actions. There are many exciting developments in our immediate and distant futures.

Quality and compensation. To explore the first link between quality of early childhood education and programs and compensation, we will rely on Canadian and U.S. research. There is only one Canadian large-scale, on-site study of the quality of centre-based and home-based early childhood education and care settings. But we can compare the findings of small Canadian studies with those of U.S. studies. We can assume that the well-established link between quality and compensation established in the U.S. studies is applicable in the Canadian context.

In Canada, the recent research project, *You Bet I Care!* included studies of the quality of child care centres (Goelman, Doherty, Lero, LaGrange, & Tougas, 2000) and regulated family child care homes (Doherty, Lero, Goelman, Tougas, & LaGrange, 2000). The media release for the quality study stated that "while safe environments with supportive adults are the norm for child care in Canada, fewer than 1 in 3 preschoolers and 1 in 4 infants are in programs that stimulate the child's social, language and thinking skills, say two comprehensive studies on child care quality released today in Toronto and Vancouver" (University of Guelph Centre for Families, Work, and Well-Being, 2000). The centre-based study looked at the relationships between centre quality and centre characteristics (such as size of centre and auspice—nonprofit or commercial); program staff wages and working conditions; and program staff characteristics and attitudes (Goelman et al., 2000). The study found that the majority of programs provided safe environments with caring adults. But less than half (44.3 percent) of the preschool rooms and just over a quarter (28.7 percent) of the infant/toddler rooms were providing environments that were stimulating and supportive of children's optimal early development.

The study of regulated family child care homes considered the relationship between quality and the child care provider's characteristics and attitudes; child care provider's income levels and working conditions; and the child care provider's use of resource programs and networking and professional development opportunities (Doherty et al., 2000). The results showed just over one-third of the family

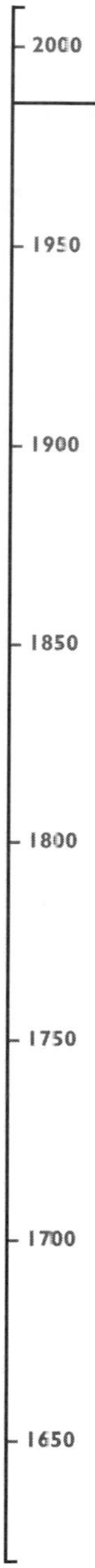

1985
The report of the government's Task Force on Child Care recommends a national system of child care in Canada.

NAEYC establishes a voluntary accreditation system for centres.

FIGURE 9.1
Quality of Regulated Child Care in Canada

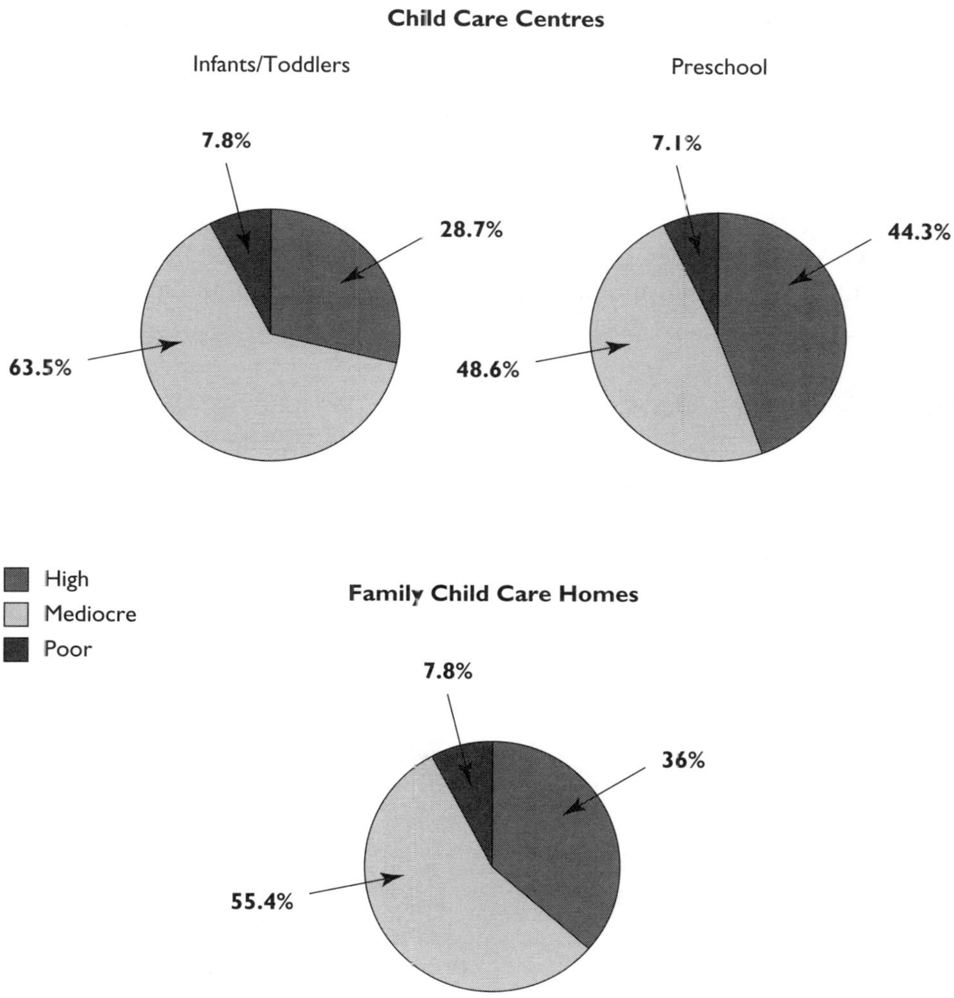

Sources: Goelman et al. (2000); Doherty et al. (2000).

1986
The report of the federal government's Special Committee on Child Care recommends increased tax credits for families, continuation of child care subsidies to low-income families, and small operating grants to child care programs.

child care homes provided environments that would stimulate children's development.

Both the centre- and home-based studies of quality found that the provider's or staff's level of income and level of education predicted the quality of the environment. That is, higher-quality environments were more likely when there were higher income and educational levels. Also higher levels of satisfaction were found more often in the higher-quality settings. Both studies found that compensation directly influenced the level of quality.

Many early childhood practitioners, like most Americans and Canadians, were jolted by newspaper and television news feature stories in early 1995. "Most Day Care Poor to Just OK" announced *USA Today* (2 June 1995); "Child Care Is Worse Than Believed, with Safety Jeopardized, Study Suggests" proclaimed the *Wall Street Journal* (2 June 1995); "Child Care Attacked," summed up Connie Chung on the CBS evening news. Behind these worrisome headlines is a complicated report, *Cost, Quality, and Child Outcomes* (Cost, Quality, and Outcomes Study Team, 1995), which once again linked the quality of the child care in the four states studied (400 randomly selected centres in California, Colorado, Connecticut, and North Carolina) to specific variables, most of which are related to economic factors. The study found that, though there were wide variations among states and among individual programs within states, the care at the majority of the centres studied was poor to mediocre when rated according to standards for health and safety, of relationships characterized by warmth and support of adults, and of attention to children's learning needs (see Figure 9.2). In fact, 40 percent of infant and toddler care was rated at less-than-minimal quality. Another finding of the study indicates

FIGURE 9.2
Child Care Quality in U.S. Centres

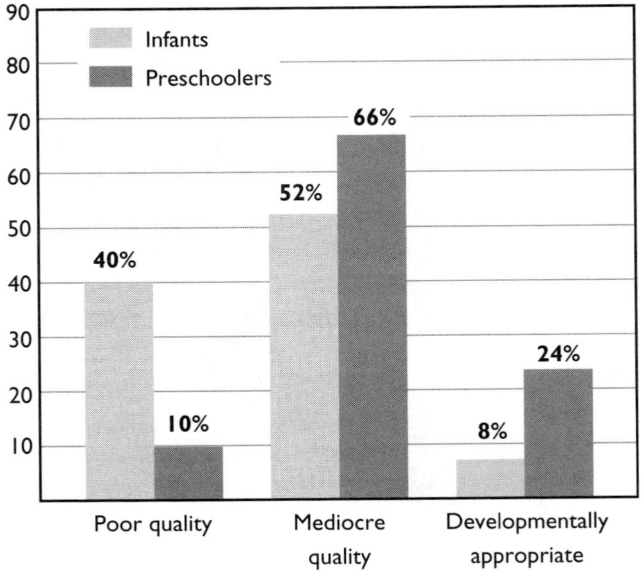

A major U.S. study on child care quality indicates that the majority of children are in centres that offer poor or mediocre quality in specific areas.

that children's later cognitive and social development are positively related to the quality of their child care experiences, no matter what the variables in maternal education, ethnicity, or child gender. Therefore, it seems imperative that quality child care be part of every child's optimal early experiences. Yet the finding that so few children have this quality child care is certainly shocking in a society that voices concern for and recognition of the importance of early childhood experiences. It is even more disturbing that the study points out that quality is linked to specific variables that are significantly related to economic factors.

In 1999, a follow-up study was released by The Cost, Quality and Outcomes Study Team, after the children from the original study were followed into Grade Two. The study found that that preschool child care quality had an effect on Grade Two outcomes. "Findings reveal that children who have more positive child care experiences during the preschool years have better outcomes through the elementary school years" (Cost, Quality, and Outcomes Study Team, 1999, p 9).

The Canadian Child Care Sector Study (Beach, Bertrand, & Cleveland, 1998) reviewed Canadian research studies, consulted with caregivers in all types of early childhood education and care settings from across Canada, and interviewed government officials, representatives from early childhood education/child care organizations, and experts in the field. The study reports that the research, consultations, and interviews identified concerns and questions about the quality of child care in regulated and unregulated settings.

> Research from other countries suggests many child care settings in Canada are not providing optimal environments for young children. There is a considerable body of U.S. research which finds high turnover rates, poor compensation, and a lack of related caregiver education qualifications are associated with poor quality child care [reported in Doherty, 1996]. These problems exist in the Canadian context and it is reasonable to assume that they have a similar influence on the quality of care (Beach et al., 1998, p. 35).

Before the *Cost, Quality, and Child Outcomes* studies, previous U.S. research had already linked quality with economic factors and the education of early childhood practitioners. In 1989 the National Child Care Staffing Study reported some significant findings relating quality in child care to variables in the adult work environment. Major findings of this study include the following:

- The education of child care teaching staff and the arrangement of their work environments are essential determinants of the quality of programs for children. As you might expect after reading the earlier discussions in this text, early childhood practitioners provided more responsive and appropriate care for children if they had had formal training in early childhood education at the college level, earned higher wages and better benefits, and worked in centres whose budgets used a higher percentage of their funds for early childhood practitioner salaries and benefits.
- The most important predictor of the quality of care children receive, among all the variables related to adult work environments, is staff wages. Better-quality centres had higher wages for early childhood practitioners and, consequently, lower rates of teaching staff turnover.
- Child care teaching staff earn very low wages, especially considering the fact that they have higher levels of formal education than the average North

1987
NAEYC publishes its position statement on developmentally appropriate practice, revised from a preliminary statement in 1984.

The Canadian Child Day Care Federation (later the Canadian Child Care Federation) is established.

1989
The Canadian government passes a resolution to eliminate poverty among Canadian children by the year 2000.

There are 943 000 poor children in Canada.

American worker. At the time of the study, early childhood practitioners were earning less than one-half of what women with comparable education levels were earning in the general work force, and less than one-third of men with comparable education. Even worse, the salaries were decreasing; in the previous decade, wages had decreased more than 20 percent when adjusted for inflation. Employment benefits were minimal: fewer than half of the early childhood practitioners in the study received medical coverage, and only a small percentage had a retirement plan. Other than sick leave and paid holidays for some, the only other benefit for most staff was reduced child care fees for their own children.

- The turnover rate of teaching staff in child centres had nearly tripled from the late 1970s to the late 1980s; by the late 1980s, the annual turnover rate for early childhood practitioners was about 41 percent. To underscore the fact that the turnover rate was related to the salary level, the study found that the turnover rate for early childhood practitioners earning the lowest wages was about twice that of early childhood practitioners with the highest wages.

The National Child Care Staffing Study, published in 1989, recommended that

- child care early childhood practitioner salaries be raised as a means of recruiting and retaining a qualified professional work force;
- opportunities for formal education and training be provided for early childhood practitioners to enhance their ability to offer developmentally appropriate learning experiences for young children;
- state and federal regulations regarding adult-child ratios, staff education, and compensation be adopted as a means of raising the quality in the United States' child care programs;
- standards for the adult work environment be developed within the child care industry to minimize differences in quality between various types of programs; and
- public education about the crucial importance of well-trained and adequately compensated early childhood practitioners be promoted, to gain community support for the full cost of quality care (see Whitebrook, Howes, & Phillips, 1989).

That these recommendations have not yet succeeded is evident from the findings of an update of the study (Whitebook, Phillips, & Howes, 1993). Four years later, less than 20 percent of the centres included in the original study offered full health benefits for all staff. In 1993 the average child care centre early childhood practitioner was earning, at the most, $15 400, even though about 75 percent of them had at least some college-level training or a B.A. degree. At the same time, the average salary for women with college degrees or some college training was $33 300, and $51 804 for men. The lack of increase in salary over the four-year period affected the stability of staff and centres; indeed, only 30 percent of the early childhood practitioners and assistants interviewed in the original study were still employed in the same setting, and of the 227 centres in the original study, 34 had closed (Bellm et al., 1994).

Early childhood practitioners' habit of caring for others extends even to this form of sacrificing their own incomes. But awareness has been growing that this sacrifice hurts those whom teachers have been trying to protect in the long run—

children and families—through the staff turnover that results from early childhood practitioners' dissatisfaction.

Canadian studies report a connection between compensation levels and staff turnover (the number of staff leaving the centre in that year, divided by the total number of staff employed). In *You Bet I Care!* (Doherty et al., 2000), the annual average turnover was 21.7 percent in 1998. The most common reason for leaving a centre voluntarily was dissatisfaction with pay. In an earlier study, *Caring for a Living* (CCCF & CDCAA, 1992), the total turnover rate averaged 26 percent across child care centres. An analysis of the *Caring for a Living* data (Beach et al., 1998) finds that low wages significantly increase the probability of turnover. Another compensation factor that affects turnover is pensions. In centres where pension plans (either partially or fully paid by employers) are available, staff were much less likely to indicate that they planned to leave.

High-quality infant child care is linked to low staff turnover (courtesy of George Brown College).

Clearly there is a connection between low wages and staff turnover. In too many communities, early childhood practitioners discover that they cannot pay all their bills on the salaries they earn as early childhood practitioners, and discover they need to take on an additional job to meet their financial responsibilities. Working with young children all day long is already physically exhausting and emotionally and cognitively challenging; to have to then go on to a different job puts heavy burdens on early childhood practitioners, to say nothing of strains on their personal lives. And there is more to it than that: the low salaries imply a lack of societal support and respect for the importance of what early childhood practitioners do, which ultimately eats away at self-esteem and a sense of professional worth.

One early childhood practitioner who chose other career directions after five years of working with four year olds said, "I got tired of always being broke days before my next paycheque. And hoping against hope that I wouldn't get sick enough to have to go to a doctor, let alone take time off my job, since sick days weren't one of our centre's benefits. And most of all I got tired of seeing the friends I had gone to college with be able to go on trips and stuff that I knew I'd never be able to do if I stayed in that job. I miss the work and the kids, but I decided if changes were coming, they weren't coming soon enough to suit me." It is not hard to imagine the feelings of frustration, hurt, and disappointment that lie behind such words, and behind such a final act as leaving the early childhood profession. *Cost, Quality, and Child Outcomes in Child Care Centres* documents that quality suffers with these high rates of early childhood practitioner turnover. Several characteristics of quality are directly related to the issue of compensation. The study found that high-quality centres shared these characteristics: small groups of children were assigned to each adult; college-educated early childhood practitioners worked competently with children, basing their actions on their knowledge of child development; experienced administrators managed the programs; staff were stable, resulting from the programs' ability to recruit and retain good staff; and mechanisms existed to meet high standards, such as strong licensing requirements and voluntary accreditation.

Training is important. But it is quite unreasonable to expect a high level of professional preparation without the matched levels of compensation. When

1990s
There are 320 288 day care spaces in Canada.

The average annual salary for an early childhood practitioner working in a child care centre is $11 639; the average salary of a teacher in a kindergarten program is $32 501.

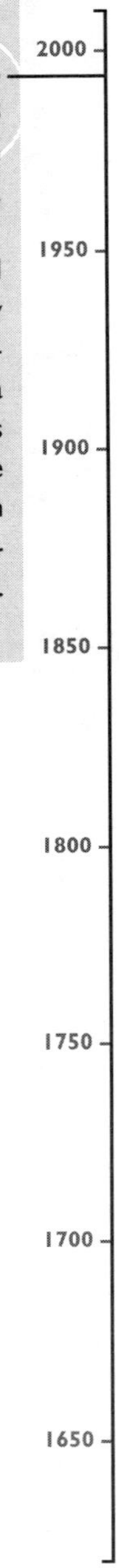

trained early childhood practitioners cannot be recruited, administrators who are desperate to hire staff to at least maintain the adult-child ratios may lower their standards for early childhood practitioners, with a resulting drop in the quality of experiences for children. In centres without well-trained staff, children's development suffers, as is clearly shown in all the studies.

Even when staff have appropriate training, if they are inadequately compensated, children and families still suffer because of the high rates of staff turnover. When children are left in the care of adults other than their parents, it is critical that they be able to form trusting, warm relationships with those caregivers. Learning and stimulation, as we have said, occurs within the context of relationships, with teaching embedded in interaction. Learning to feel safe and secure with adults is an important developmental task for young children. It is impossible to accomplish this task when a young child's question is, "Mommy, who's going to be my teacher today?" (Whitebook and Granger, 1989). The disruptions in style of care and in routines are disturbing to children's security, especially that of younger children.

Just consider all the changes children must adapt to when an early childhood practitioner leaves. The daily schedule may be different; there are likely new rules and expectations; the new early childhood practitioner will not know, at least for some time, each child's needs, likes, and style; and the songs she sings or the things she says are unfamiliar to the young child. The security of the daily rituals, and the person who was associated with them are gone, shaking children's sense of safe belonging. And for parents who are already struggling with turning the care of their children over to others, having to learn to trust still another adult takes time, energy, and a feeling that it may not be worthwhile to invest much in building yet another relationship. This is an attitude that will inhibit the comfortable communication that benefits children, parents, and early childhood practitioners. The centre as a whole is disrupted when staff members leave, with those who remain having to take on additional responsibilities, including orienting another new early childhood practitioner, while wondering how much longer they can remain or grow in this context of continual disruption. The dilemma, to paraphrase Marcy Whitebook, is that many early childhood practitioners cannot afford to stay, while the children cannot afford to have them leave (Whitebook, in Whitebook and Granger, 1989).

The issues related to inadequate compensation are pervasive and completely counterproductive to overall quality, professional growth, and development of individuals. As we have seen, there is a vicious cycle at work here: there are many factors related to compensation that push early childhood practitioners out of the field, which results in replacing them with less-qualified staff, therefore diminishing the quality of programs for children. When the public perceives this poor quality—and reads headlines such as those we saw earlier, which reported the recent study without seeing the complexity and interrelatedness of factors that result in quality—their response is to devalue the work of early childhood practitioners. Then we're right back to the problem of low status, one of those factors that relate to compensation and to a decreased sense of professionalism in the first place!

Compensation and accessibility. The compensation levels of early childhood practitioners are also linked to the accessibility of early childhood education and care programs to families. The issue of affordability (which determines accessibility) for families is directly tied to the issue of early childhood practitioners' compensation. In nonprofit child care programs, by far the largest percentage of the budget that drives parent fees is the salaries and benefits for the professional staff. Another aspect of child care budgets that affects compensation costs is the adult-child ratio, or the number of children that are cared for by one adult. Obviously, lower adult-child ratios are more costly, as well as more effective for quality programs.

Let's consider the phrase **full cost of quality.** Although parents find their budgets strained by the cost of their child care, they—and policymakers—do not realize that child care costs more than is apparent from the fees that parents pay. The full cost includes hidden costs, such as goods and services that may be donated, occupancy of the facility, donations, and, most importantly for this discussion, **foregone annual wages** and benefits—that is, the difference in wages between the actual wage received as an early childhood practitioner and the amount the individual could have earned in another occupation. As you can see in Figure 9.3, foregone wages amount to nearly 20 percent of the full cost of quality. By accepting low wages, child care professionals are, in fact, subsidizing the child care costs of the families in their communities.

1991
The first Worthy Wage Day in the United States, organized by the Child Care Employee Project, is held on April 9.

The federal government introduces a ten-week parental leave benefit, in addition to the existing fifteen-week maternity leave benefit, as part of the Unemployment Insurance program.

FIGURE 9.3
Full Cost of Quality

Average full cost = $2.83/child hour
Average hidden cost = 72¢/child hour (represented by cut slice)

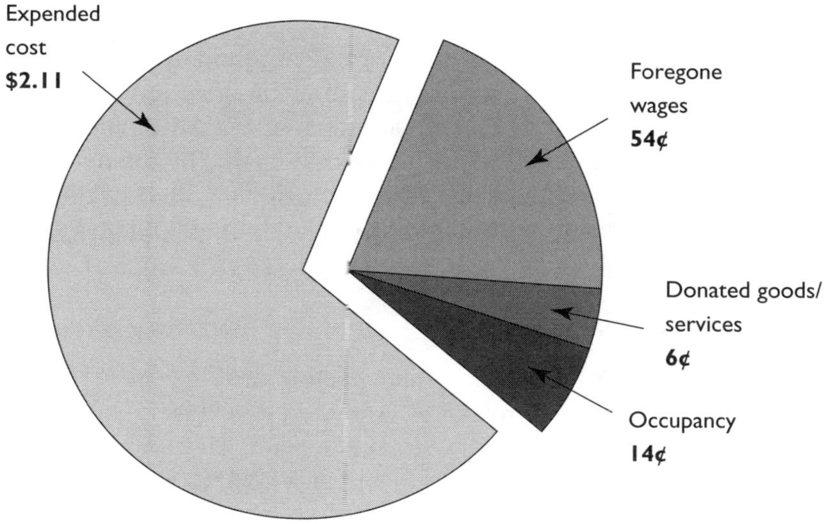

Foregone wages (those amounts early childhood practitioners would have earned in other occupations) account for nearly 20 percent of the cost of quality.

1992

Caring for a Living, a national report on child care centre staff working conditions and wages, is released.

The Live-In Caregiver Program is introduced by the federal government.

Quality and accessibility. The third link is between the quality of an early childhood education and care program and its accessibility to children and their families. As quality of early childhood programs increases, their availability and affordability often decrease. Families base their child care decisions on what they can afford and on what is available in a given geographic area; the level of quality that they can afford depends on the choices available.

The result? Parents who cannot afford the fees of high-quality child care programs will choose lower-quality and less-expensive services, or they may not use any service at all. If high-quality child care programs that meet individual children's and families' needs do not exist within a reasonable travel distance, again, parents will choose to not use child care or to use a lower-quality form of child care.

Doherty-Derkowski reports that studies in both Canada and the United States "indicate that children from less stressed families with higher socioeconomic status are more likely to be enrolled in high quality early childhood programs than are their peers from stressed and/or low socioeconomic status homes" (1995, p. 175). Stressed families are likely less able to seek out child care programs that may be within the availability circle of less-stressed families. Families of low socioeconomic status clearly are less able to pay for quality child care.

This focus on the issue of quality may bring parents into the battle for fair compensation as co-supporters, in ways that have not yet happened with other initiatives. For some time now, it has seemed to many early childhood practitioners that they were going to have to "face the harsh reality of the economic marketplace. No one else is going to solve our problem for us. We must become more and more assertive on our own behalf. It is child care employees themselves who hold the key to improving compensation" (Morin, 1989, p. 18).

The expanded understanding that we now have is that the link between increasing compensation and increasing parent fees can, and must, be broken. Consumers simply cannot bear the burden of an appropriate income for all early childhood early childhood practitioners. Serious commitments of public and private funds will be essential, and recent efforts show progress in this direction. By removing the link between early childhood practitioner salaries and parent fees, both early childhood practitioners and parents, unhindered by the need to protect personal vested interests, will be able to focus on the idea that quality for early childhood practitioners means corresponding quality for children and programs.

Public Policy Options

The cycle must be broken if there are to be viable futures for all of us, including you, who are just entering the profession. And the cycle must be broken because the healthy future of children and families also depends on it. Actually, the headlines *should* read, as Marcy Whitebook (1995) said, "What's Good for Child Care Early Educators Is Good for Our Country's Children." In the past, early childhood practitioners have been reluctant to lobby for their own interests, thinking it made them appear less professional and more self-centred, rather than focused on serving others and concerned about the increased costs of care for parents. But the studies discussed earlier make it clear that it is necessary to lobby for changes in policies and conditions that affect quality for children and themselves, simultaneously. In addition, because parents have a clear interest in obtaining the highest possible quality of care for their children, they should understand that supporting early childhood practitioners' efforts to establish higher standards in regulations and finding access to extra resources beyond parent fees will benefit their own children.

The quality, compensation, accessibility triangle is at the crux of the challenge facing early childhood practitioners. The Introduction identified the range of public policies that shape early child development programs in Canada today. These policies are caught in four interrelated contradictions:

1. Government policies at the federal and provincial/territorial level see the benefits of early education for all children, but the costs seem prohibitive. Governments move to provide early childhood development opportunities to the targeted groups of children and families who are likely to benefit the most.
2. The federal versus provincial/territorial jurisdiction of child care continues to thwart either government's attempts and resolve to address early child development policy.
3. Canada is ambivalent about whether early childhood education and care is a public or private responsibility. The ambivalence about women's equality, preferred family structures, and the rights and obligations of parents all come into play. Are early childhood education and care programs a public right to Canada's youngest citizens, or is it the responsibility of their parents to care for and educate children until they reach school age?
4. For some, the primary purpose of early childhood education and care programs is to promote children's optimal development, whereas others view programs as services that enable parents to participate in the labour force or prepare for labour force participation.

Policymakers and government officials often seem captive to these dilemmas and are unable to address the quality, compensation, accessibility dilemma. Child care advocates must open their minds to put forward a range of policy options to meet the needs of children, their families, and early childhood practitioners. Policy options that fail to improve all three problems of quality, compensation, and accessibility will not be real improvements. Increasing the amount of child care at prices affordable to parents, without ensuring reasonable wages for competent early childhood practitioners, is not a solution. Increasing wages and benefits for early childhood practitioners by increasing the cost of programs to families is not a solution either. Nor is improving requirements for early childhood practitioner qualifications and reducing adult-child ratios, while passing on the increased costs to parents.

Below is a shopping list of the public policy options often put forward to address early childhood education and care issues. Many were introduced in Chapter 1 when you reviewed the public policy that now shapes early childhood education and care programs in Canada. Consider how each measures up to the task of addressing the quality, compensation, accessibility dilemma. Policies that cannot measure up in addressing all three issues are not solutions and may become part of the problem. As you consider each policy, ask the following questions:

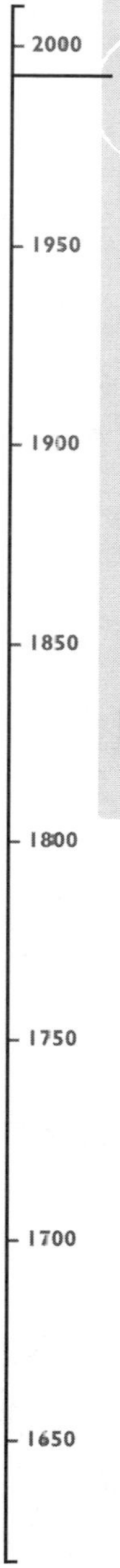

1993
The Perry Preschool Project publishes its data on the now 27 year olds who were once in its preschool programs, showing the continued benefits in social adjustments and proving economic effectiveness of preschool education.

The Manitoba Child Care Association initiates the Worthy Wage Campaign.

- Does the policy support improved quality by increasing the knowledge and abilities of early childhood practitioners and reducing turnover?
- Does the policy support improved compensation for early childhood practitioners?
- Does the policy ensure that higher-quality early childhood education and care programs, with better compensated early childhood practitioners, will be more available to families and more affordable?

Integration of children/family services. In Canada, provincial/territorial public policy initiatives and programs for children and their families tend to be divided across health, education, and social service ministries. In most jurisdictions, governments are moving toward a more integrated approach, which either moves responsibilities for child and/or family services into one ministry or establishes a children's secretariat to coordinate related policy and services. This direction offers the potential to bring together various early childhood education and care services, such as kindergarten programs in schools that are part of education; early intervention programs, which are operated by health ministries; and regulated child care services in social service ministries.

Expand early childhood education and care programs located in the schools. Many out-of-school programs, child care centres, nursery schools, and family resource programs are located in school buildings. Early childhood practitioners often share information, resources, and expertise with educators in the school system and may receive low-cost or rent-free space.

Integrate early childhood education and care programs into the education system. Kindergarten programs are offered as part of the school system and could be expanded to full-day programs. The school system could also expand full-day programs to younger children, as well as operate out-of-school programs for school-age children. The Quebec, Nova Scotia, and New Brunswick governments have extended full-day programs for kindergarten.

Work-related centres can meet the needs of young children and their families (courtesy of George Brown College).

Entrepreneurship. The private sector could be encouraged to operate more early childhood education and care programs as businesses, based on market demand.

Work-related child care. Workplace child care centres, child care information and referral services, and support to employees who use neighbourhood early childhood education and care programs are all considered work-related child care. In Canada, a survey of all child care centres located on or near the work site reported that they account for less than 3 percent of licensed child care spaces (Beach, Friendly, & Schmidt, 1993). The survey findings noted that most work-related child care centres charge comparable fees to other programs and receive few funds from employers.

National early child development system. Many advocates in Canada recommend that the federal government establish a national policy with broad principles and cost-share funding with provincial/territorial governments that pulls together all forms of early childhood education and care programs, family support programs, and early years community initiatives.

Direct or block grants to early childhood education and care. Funding for programs may be targeted to specific operating costs, such as salaries or in-service training. Direct grants can also be used for the overall operating costs, reducing or even eliminating fees to parents.

Child care tax deductions/benefits. Income tax deductions and benefits direct public funds to parents who are paying fees for early childhood education and care programs. The Child Care Tax Deduction now in effect allows parents to deduct up to $7000 for their child care fees for children under seven years. The deduction must be claimed by the parent with the lowest income, and a receipt from a family child care provider or an early childhood education and care program must be provided. The resulting reduction in income tax is greater for higher-income earners. An income tax benefit could be targeted to lower-income families.

Complementary work and family policies. Early childhood education and care programs cannot address all of the supports families need to raise their children and earn a living. Families need work policies (such as flexible hours and family responsibility leave) that help parents balance work and family life.

THEORY INTO PRACTICE

This excerpt from the executive summary of the Child Care Sector Study summarizes the policy challenges to both the child care sector and its workforce and presents the study's key recommendations.

Moving Forward

Our Child Care Workforce is framed around strong agreement on the urgent need to improve the quality, accessibility, affordability and accountability of child care in Canada. A human resource strategy for caregivers is integral to reaching these goals. It is a key component of an overall strategy which includes:

- supporting public policy, legislation and funding to enhance child care quality and make it accessible to all children;
- advocating for fair compensation and good working conditions for caregivers;
- continuing to build a professional framework to support the child care workforce;
- increasing the accessibility and relevance of education and training in child development and early childhood education; and,
- encouraging the pursuit of a pragmatic research agenda to support public policy development in the child care sector.

Source: Beach (1998), p. 9.

Taking Action

There is no simple solution or quick policy fix for early childhood education and care and its workforce. Despite the obstacles, the prospects have never been brighter to finally make the early childhood workforce a well-recognized and well-rewarded profession (Bellm et al., 1994, p. 168). Early childhood practitioners are powerful advocates who can take action. The first step is identifying policy options to meet the objectives of quality, compensation, and accessibility. The next step is to take action.

Worthy wage campaigns. Worthy wage campaigns draw attention to early childhood practitioners' inadequate compensation and to the value of the work itself. Often, not even parents and family members are aware of the wages and benefits early childhood practitioners actually receive. Campaigns inform the public of the wide discrepancies between the responsibilities and skills required and the compensation levels. Often a worthy wage campaign begins with a survey of wages and benefits. The campaigns use innovative actions to publicize the information and to celebrate the early childhood workforce. The purpose is to mobilize support for improvements to compensation levels, which will also support quality but not reduce accessibility by raising the fees that parents must pay.

You can clearly see the importance of the compensation issue to the profession, to children, families, and communities, and to you. Progress is being made, albeit slowly. Nevertheless, these efforts will not succeed unless all early childhood practitioners make it their business to understand the issues and the solutions and to recruit others, both parents and early childhood practitioners, to join in the

THEORY INTO PRACTICE

The Worthy Wage Campaign in the United States

A group of child care early childhood practitioners in the San Francisco Bay area, who were concerned about the low pay and status of their work, established an organization called the Child Care Employee Project in 1977. This was the force behind the National Child Care Staffing Studies of 1988 and 1993 that have so clearly linked the issues of compensation and quality. Now called the National Centre for the Early Childhood Work Force, this is the most powerful voice advocating for fair and decent employment conditions for early childhood professionals. It is under this organization that the Worthy Wage Campaign efforts are organized. This is a grassroots attempt to mobilize child care professionals, parents, and others to fight for solutions to the issues of low compensation and resultant turnover.

In recent years, the efforts of the campaign have focused on various local and national activities on Worthy Wage Day, held during the Week of the Young Child in April. It attempts to draw community attention to the facts of compensation and advocate for improving conditions. For example, in April 1995, for the fourth annual Worthy Wage event, hundreds of citizens nationwide were invited to participate in a "Job Shadow." They worked alongside an early childhood practitioner in a classroom or family child care home, and then received a cheque that was equal to the early childhood practitioner's real wage for that period. In such ways, the Worthy Wage Campaign is trying to support community efforts to improve working conditions for early childhood practitioners.

effort. Make it your concern, now as a student and later as an early childhood practitioner, to find out about Worthy Wage campaign efforts in your community.

Position papers and presentations. Local, provincial/territorial, and federal governments need to hear from early childhood practitioners. Task force committees, commissions, and public hearings present opportunities to communicate your policy recommendations and to respond to government policy initiatives and budget announcements. This is one way to influence government policy and to draw attention to the challenges facing early childhood education and care programs.

The following steps outline how to prepare a position paper, which outlines your perspective and recommendations on a particular issue or policy (Bertrand, 1990):

1. Identify who you are representing—perhaps yourself, a group of early childhood practitioners, a particular early child development program, an early childhood organization, or a coalition of organizations and programs.
2. Briefly state your position.
3. Define the issue. Clearly state the problem or concerns you are addressing in the position paper. It may be useful, here, to review key points in a government initiative or to relate a series of events in chronological order to help identify the central problem or concern.
4. Support the position. Find common ground with the government's policy and state your differences. Provide specific arguments or reasons to back up your position. Include examples of support and agreement from the community or from other government officials.
5. Conclude with strength. Restate your position and include an optimistic view about the positive outcome that is likely if your recommendations are followed.

Building coalitions. Coalitions build the base of support for public policy changes at the local, provincial/territorial, and national level. They often include child care programs, parents, early childhood practitioners, community associations, child welfare organizations, trade unions, teachers' federations, child care resource

MAKING IT HAPPEN

Ontario Coalition for Better Child Care

The Ontario Coalition for Better Child Care (OCBCC) brought together child care activists, early childhood practitioners, trade unions, teachers' federations, women's groups, social service organizations, and parents in 1981 to form a coalition of interests and organizations at the provincial level.

Since its beginning, the OCBCC has been a powerful force in the campaign for more affordable, accessible, high-quality child care in that province. It has also spearheaded a number of initiatives to improve the compensation of early childhood practitioners, including provincial wage grants and pay equity legislation and grants. The OCBCC continues to play an important role in Ontario's and Canada's struggle for a system of early childhood care and education.

1994
Fisher Price Toy Company sells $800 million worth of toys.

Loris Malaguzzi, founder of the schools in Reggio Emilia, Italy, dies, while the philosophy of his schools continues to influence programs around the world.

2000
1950
1900
1850
1800
1750
1700
1650

groups, human service agencies, women's groups, and early childhood practitioner students.

Media. Television, radio, newspapers, and the Internet are the most effective, powerful tools in maintaining a sustained advocacy campaign. Appearing daily to a broad cross-section of the population, the media provide an opportunity to reach many people outside committed supporters and confirmed opponents. The media are an important part of a strategy to bring about significant change.

Contact members of the media in person, on the telephone, or through press releases. It is helpful to cultivate contacts with particular media persons who seem to have an interest in the issue and will put forward a fair position.

For a major issue or event, a press conference can be assembled. You might organize a press conference when you want to respond to a particularly critical situation or when you are initiating a major campaign, such as a local Worthy Wage Campaign.

Unionization. Unionization of early childhood practitioners and other staff members in early childhood education and care settings can raise wages, improve working conditions, and contribute to better-quality care and education. Although early childhood practitioners have been reluctant to form unions to lobby on their own behalf, unionization is a clear strategy to improve compensation and working conditions, as well as to improve the quality of the early childhood education and care programs.

To date, only a small proportion of early childhood practitioners are members of trade unions in Canada. As the Child Care Sector Study points out

> the employment structures and funding in the sector present barriers for the organization of the child care workforce. In centre-based programs, it is difficult to organize relatively small caregiver groups into collective bargaining units. Demands for increased compensation are immediately faced with the ability of parents to afford increased fees. Caregivers employed in child care centres which are operated within larger institutions (such as municipalities in Ontario or community colleges) may be included as a small proportion of a broad bargaining unit and it may be difficult to address issues specific to the child care workforce (Beach et al., 1998, p. 125–26).

Informational Advocacy

Because of their education and experience, early childhood practitioners can lead the way in advocating for children and families. Those who have knowledge of child development have an obligation to share their expertise with others to help them make informed decisions. Often in the name of accountability, fiscal economy, and efficiency, policies are made that can be harmful to children's development.

Informational advocacy, most often in the form of public education, is a powerful force early childhood practitioners can use to counter these policy directions. Many of the activities discussed in the earlier section on public advocacy will also

MEET THE EARLY CHILDHOOD WORKFORCE

Jamie Kass

Jamie Kass is the National Child Care Co-ordinator for the Canadian Union of Postal Workers. She coordinates a $2 million fund that creates flexible, quality child care spaces for postal workers.

Jamie became interested in working with young children while she was studying at university and taking child development courses. In the summer of 1975, she worked in a school-age program and found the work so meaningful that she was hooked. In 1976, Jamie started to work full-time in a community-based child care centre. During her three years at the centre she was a preschool teacher, a cook, and a coordinator. At the same time, Jamie pursued studies in the ECE diploma program at Algonquin College.

In 1978, child care staff faced provincial cutbacks that threatened both their salaries and the survival of child care programs. Jamie was a leader in bringing child care workers together to discuss how to fight for better wages and working conditions. The next step was looking at the feasibility of joining a union. After much debate and consideration of the options, Jamie and others opted to join the Canadian Union of Public Employees (CUPE). It was a national union with research and communication resources. The new local was CUPE 2204, and child care workers from five centres joined.

Jamie continued to work within the union movement to advance the wages and working conditions of the child care workforce and the quality of child care programs. In her role today at CUPW, Jamie is enthusiastic about how the fund is able to extend child care options for shift workers and workers with children with special needs. She is able to work directly with local child care communities to create more flexible child care services. Jamie plays a key role in bridging the understanding of the child care sector and the union movement.

Jamie sees the challenges and opportunities for people entering the early childhood workforce. It is difficult to find the balance of enjoying the work with young children and their families and making a commitment to make it better. In order to improve the quality of child care for young children, the conditions of work must be improved. She calls on new members of the workforce to become advocates for themselves and more child care. Jamie points out that political or public advocacy are important to improving quality of today's child care in Canada. By raising these kinds of issues within local neighbourhoods and communities, there are real opportunities to make a difference for the children, families, and the child care workforce.

"There are rich opportunities and potential when you work in this field. Part of the potential is in the direct work with children and part is in joining the child care advocacy movement."

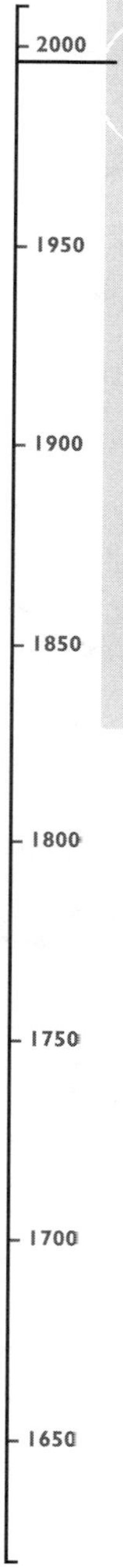

1995

There are 1 472 000 poor children in Canada.

In Canada, approximately three million children under the age of twelve have parents who work in the labour force or are pursuing further education and training. There are 474 969 regulated child care spaces.

serve to inform and educate the general public. As well, early childhood practitioners can build awareness about what ECEC is and why it is important to young children by

- taking part in information sessions at conventions, shopping malls, or entertainment events;

- early childhood environments using photographs, videos, multimedia presentations, three-dimensional models, and children's own art work; and
- seeking out and recognizing individual early childhood practitioners whose work with young children is an inspiration to others.

Essentials for Early Childhood Practitioners

Having decided that working with young children is the career for you, you will now continue in a planned program preparation at a college or university. This course of study has been designed to offer you both the knowledge base that you will need to provide developmentally appropriate responses, environments, and learning activities for the children you will work with, and the opportunities to develop your skills in supervised practica. The path through your course work has been designed quite clearly for you, to offer the most appropriate sequence of learning experiences.

But how you travel that path is up to you. As an advocate for young children and their families, for quality early childhood education and care programs and for the early childhood workforce, you must start with a commitment to being a competent, learning early childhood practitioner.

Young children rely on reflective practitioners who will advocate for their best interests (courtesy of George Brown College).

It is important to understand that the same principles that we recognize as optimal for children's learning apply equally well to adults. We know that children learn best when they take initiative, make choices among interesting possibilities, act on materials, and interact with people. So, too, adults construct their increasingly sophisticated new ideas by interaction with the concepts and experiences they encounter. Whether your past learning environments have encouraged you to play this active role of a learner or not, allow yourself now to assume an active role. You will become an advocate for yourself, the young children and families whom you will work with, and for the early childhood workforce.

If you are to be a successful early childhood practitioner, you must invent your own way of teaching; you will use the knowledge and experience you gain, and you will reflect on your own meanings and experiences. You will come out of your professional preparation program ready to assume your own identity as an early childhood practitioner only if you have done more than passively meet the assigned requirements, by creating your own knowledge and increasing your self-understanding.

Begin inventing your own way of teaching now, by framing some of your own questions and using some of the additional resources suggested at the end of each chapter to delve further into the material. Use every encounter and opportunity to create your expanded understanding of what it means to be an early childhood practitioner, and how you will do it.

Attitudes and Dispositions

The attitudes and dispositions that you now begin to form will guide you as you become an effective professional early childhood practitioner. In general, you will need to have a positive attitude toward yourself, the children you work with as students, their parents, your supervising teachers and other administrators, and your colleagues as a student and, later, as an employee. This positive attitude allows you to see the potential in any situation or person, rather than narrowly focusing on limits and restrictions. Early childhood practitioners who nurture this perspective find less frustration and greater pleasure in their working situations.

In acknowledging that there is no one right way to teach and that there are always unknowns, the disposition of remaining open to new ideas and possibilities develops. It is important that you realize that completing a professional preparation program is not the end of something, but only another step along the path. So many students seem to feel that graduation means that the learning part of their career is over, and they are now moving into the practice aspect. It is vital that you understand that continual learning comes from the practice, if early childhood practitioners are to remain able to grow and adapt to new demands. Learning becomes a passion for open, growing teachers. "The exceptional teachers I know are passionate about learning ... They see connecting points everywhere" (Perrone, 1991, p. 117).

Along with this disposition to remain open to insights and learning goes the disposition of risk-taking. You must dare to try new things, to risk the mistakes and failure that sometimes accompany new directions, and to learn from what you have risked. These are characteristics that you will want to cultivate. Students are often unnecessarily hard on themselves, expecting the polished performances that they see when visiting the classrooms of mentor teachers. Recognize now that much of your learning as a student will come in evaluating experiences afterward and figuring out what you will do differently the next time. Mistakes and less-than-positive experiences are a necessary part of this learning.

Closely related to risk-taking is the disposition of restlessness: the inability to accept and drift with the status quo. For your healthy professional future, it is important that you avoid being pulled into any rut and maintain a perspective of restless searching for what step, idea, or experience comes next—for you and for your early childhood program.

Another disposition to develop is what Van Arsdell (1994) describes as "seeing themselves less as managers of learning and more as witnesses to learning" (p. 94). Being able to see the excitement in others' active learning, and not to feel one has to control the directions of that learning, allows teachers to retain true excitement about children's learning and growth.

Perhaps the most important disposition is the capacity to be passionate about what you do. To retain the edge of excitement and freshness, early childhood practitioners develop strong convictions about what is appropriate for children's healthy growth. That underlying passion propels teachers with determination that doesn't falter, even in the face of frustrating circumstances. Care about what you do, and keep caring. These attitudes and dispositions will be indispensable as you travel your professional path.

Collegiality and Connection

Your growth and learning as a professional will have to be in the context of dialogue with others in the early childhood workforce. For those of you planning to

1996

The Canada Health and Social Transfer (CHST) block grant, which combines funding for social services (including child care subsidies), social assistance, health care, and postsecondary education, replaces the Canada Assistance Plan.

The National Crime Prevention Council identifies early child care and education programs as essential components of an integrated approach to reduce the long-term incidence of crime.

The National Forum on Health calls for a broad, integrated child and family strategy.

1997
The federal and provincial/territorial governments agree to develop a National Children's Agenda, which is intended to be a comprehensive strategy to improve the well-being of Canada's children.

The Quebec government announces major child care and education reforms, which will expand the provision of early childhood education and care programs in that province.

do family child care, this will be harder than for the others, but the principle remains the same. It is within those professional relationships that you will find the support, the challenges, the ideas that you will need. When you find yourself within a community of others who care about common issues, as well as for each other's welfare, you will be encouraged in the difficult periods and fortified in the good times of teaching. Where are you going to find these connections?

First, deliberately seek employment in a setting that values collegiality. You will be able to tell if this is so by asking questions about staff meetings and staff development opportunities, by looking at the staff bulletin board, reading job descriptions, and talking with teachers. Some programs make collegiality and **team teaching** a vital part of their day-to-day life, and it is evident in the relationships of the co-workers.

When newly employed, convey an attitude of openness to others by asking for their assistance and opinions about issues that will occur. New early childhood practitioners are often so anxious to prove themselves and their competence to others that they put up invisible barriers that prevent them from getting the support and assistance that they really need. Seek out the experienced early childhood practitioners whose programs you admire, and let them know you would value their ideas and help in designing your practice. Find out if there are formal mentoring programs in your program; many communities have designed such pairings to support new early childhood practitioners in their first year or two of teaching. If there is such a system, volunteer to become involved. Not only will this relationship foster your developmental growth as an early childhood practitioner but also will benefit the experienced colleague. If such systems do not exist, read about and promote discussion of mentoring systems.

Make sure that you become involved in whatever local early childhood organization is available to you. Here you will find a ready-made group of individuals who have similar interests and commitment to children's issues. You will find members with a cross-section of experience, job description, and training; within such diversity you will find others to support you. Take the initiative to seek out what you need. If you gain more by talking with experienced early childhood practitioners, do so. Or if it is more helpful to talk with other novice early childhood practitioners who are undergoing the same kinds of adjustments to working with young children that you are, create these opportunities.

There are several ways to form early childhood practitioner support groups. If a number of graduates from your program are working in the same area, you may be able to form the core of a group for regular meetings to discuss common concerns. Or the professional association may offer such options. Both new and experienced early childhood practitioners benefit from the exchange of questions and ideas in such a group. Sometimes groups read and discuss a common book; others simply raise issues of personal or program concern. Sometimes such groups work on community early childhood issues, finding a common voice together, but for beginning teachers, their main benefit is support for professional growth. Whatever form it takes, teachers who expect to grow and learn, through good times and bad, will discover that they must find ways of **networking,** or making connections with colleagues.

Reflective Teaching

Throughout this text, you have been encouraged to discover the importance of reflection on your relationship to teaching and caring for young children **reflective teaching** and your life. As you move into early childhood settings, it will continue to be important for you to use your daily experience as material for you to consider, muse, and meditate upon, and contemplate and speculate upon. "The basic and comprehensive question during reflection is, 'What am I doing and why?'" (Valverde in Cruickshank, 1987, p. 3).

Many early childhood practitioners find that recording events and questions to reflect upon in a personal journal allows them to gain different perspectives for approaching their work. Finding a few minutes when children are napping or at the end of the day may seem too difficult for new early childhood practitioners, who are overwhelmed with program plans, material preparations, and physical environment maintenance. But it is likely one of the most important things you can do to help keep your focus on your professional development. Through using a reflective journal, early childhood practitioners record their "experiences within and outside of the classroom that bear directly on classroom life" (Cruickshank, 1987, p. 10). This record of thoughts, actions, beliefs, and attitudes will help teachers learn much from the children and the classroom. When ideas and experiences are recorded, they provide a frame of reference for early childhood practitioners to refer back to and see growth, or see patterns of questions that will identify needs for additional research. Make it your resolution to begin a journal on the first day of your new job, and to record in it regularly. Keeping a journal forces an early childhood practitioner to assume the posture of reflection. The early childhood practitioner asks: "How did I come to do it this way? What might I do differently?" Early childhood practitioner growth and development lie in the answers.

Fine-Tuning Skills and Deepening Knowledge Base

As has been stated, completing a program of professional preparation at a college or university is a beginning. During the early years of your practice, you will discover that there are many classroom skills that you will need to fine-tune. There are many things that you learned in theory, and likely had limited opportunities to practice, in your early education courses and practicums.

However, it is only through regular and frequent use of these skills in the context of your own early childhood education program that you will become comfortable, confident, and proficient in their use, and that you will discover what additional knowledge you need. Instead of purely concentrating on survival during your first months and years in the early childhood setting, you will benefit from consciously identifying skills that you need to develop, and setting small goals for each on a regular basis. For example, you will likely need to develop your skills of observation, organization of your record-keeping, communication, and decisionmaking.

A beginning early childhood practitioner might set the goal of recording an observation on every child at least once a week. Recognizing that the early childhood practitioner preparation program was just a beginning, good early childhood practitioners continue to develop their own goals and skills.

2000 · 1950 · 1900 · 1850 · 1800 · 1750 · 1700 · 1650

1998
The federal and provincial/territorial governments introduce the National Child Benefits System.

The child care sector report *Our Child Care Workforce: From Recognition to Remuneration* is released.

1999
Developmental Health and the Wealth of Nations, by Daniel Keating and Clyde Hertzman, is published.

2001
The federal/provincial/territorial Early Child Development Agreement is signed.

From our discussions of the developments of the profession in this chapter, and the issues facing the profession in Chapter 6, you know that your knowledge must grow in these two areas in order to remain current.

Self-Evaluation

One of the most difficult things for many new early childhood practitioners is to develop the skill of self-evaluation (Duff, Brown, & Van Scoy, 1995). Perhaps because so many life experiences depend on others' telling us how well (or not) we are doing, most beginning early childhood practitioners have had little experience in evaluating their own performance. Any well-run program will provide you with evaluation from your supervisor, but usually such feedback is so occasional that teachers must, instead, rely on their own evaluation of their performance to determine their strengths and to set their own goals for improvement. Your personal reflections and your journal can be helpful in starting this process. You will likely discover a number of evaluation tools in the course of your studies and as you visit various centres. The questions at the end of each chapter in Carol Hillman's *Teaching Four-Year-Olds* (1988) act as useful informal self-evaluation guides on the various aspects of an early childhood practitioner's practice. Don't hesitate to evaluate yourself honestly and fairly; your professional growth depends on it.

Personal Philosophy of Education

To evaluate how close you are keeping to your personal pathway, it is important that you have a guideline of your personal philosophy of early childhood education. As your last act in this preliminary examination of early child development, organize your thoughts on what you believe about teaching and learning into a written statement. No doubt this document will grow and be adapted as you proceed through subsequent course work in your professional preparation. Plan to review it at the end of each semester or year. And take it with you, making it something you refer to and reflect on as you complete each year, evaluation session, or some regular period in your professional life. As you evaluate your day-to-day practice in the light of your vision and philosophy, you may be able to keep your ideals and realities aligned.

Rhythms of Working with Young Children

As is true of so many things in life, there is a rhythm, pacing, and pattern to working with young children, ranging from the excitement and exhilaration of the first experiences, to the plateaus of sameness and day-to-day duties, to the depths of frustration. Recall the stages of early childhood practitioner development that we discussed in Chapter 5, and give yourself the benefit of time to grow, to develop, to become. Be aware that the first year of working in an early childhood setting is generally recognized as difficult for most new early childhood practitioners. Don't be afraid to admit your dissatisfactions and disappointments and to seek the help and support you need during this difficult time. As you will be patient with your children's progress, so, too, be patient with your own. You won't be the same early childhood practitioner in the first month that you will be in your twelfth and then in your twentieth. Know that your decision was right and good, find the supports you need, and watch your own growth and development as an early childhood practitioner. Don't let discouragement take you from the field. Stay, for all the reasons that brought you here.

New skills, new insights, new growth—the daily life of an early childhood practitioner! (courtesy of George Brown College)

The early childhood workforce is a challenging profession with many wonderful aspects. It provides a way to stay young at heart, to maintain a lifetime of active learning, to be a special part of the world of the present *and* the future, while having opportunities to delve into the past. It is, in every respect, a profession of hope (Perrone, 1991, p. 131).

A story is told of two men visiting an ocean. They saw, much to their amazement, that a storm had tossed thousands, perhaps, millions, of starfish up on the shore. As far as the eye could see, piles of starfish lay there out of the water in the sunshine; it was obvious that very soon, they would begin to die. One of the men stooped down and began to pick up starfish and throw them back into the water. "What are you doing?" cried his friend. "Don't you see that it is hopeless? There are millions of starfish here, and they are all going to die!" Picking up another and throwing it back into the waves, the first man replied, "Not that one." Keep your eyes firmly fixed on the lives you can touch.

SUMMARY

Many issues are part of dialogue and study in early child development programs today. The role of advocacy continues to be important in addressing the challenges of compensation, quality, and accessibility. Early childhood practitioners can take on personal, public, and informational advocacy roles and learn to be an effective voice in raising both the awareness of and support for quality early education and child care opportunities for children and their families in Canada.

The first step in being an advocate is to be the best possible early childhood practitioner you can be. You are just beginning the journey.

REVIEW QUESTIONS

1. Discuss the relationships between early childhood practitioner compensation, quality, and accessibility of programs for children.
2. Identify the Worthy Wage Campaign. How does it influence increasing professionalism?
3. Identify and discuss personal and collective approaches to advocacy in ECEC.

STUDY ACTIVITIES

1. Visit the Web site www.ece.nelson.com and note the current issues and challenges facing the early childhood workforce in your region.
2. Do some informal research on the relationship between factors associated with quality in your community.
   a. Determine the adult-child ratios in each child care facility. Compare the salaries and benefits paid to early childhood professionals. Try to compare positions that might be similar, such as lead teachers or assistant teachers. Learn if there is a different wage paid to professionals with training or advanced degrees.
   b. Through conversations, try to learn how long early childhood practitioners have been in their positions at the centre. You may want to cooperate on gathering this information with other students and combine your data. Draw up a table to allow you to juxtapose this information. Do your findings support the correlations found in the studies referred to in this chapter?
3. Discover whether there is a Worthy Wage Campaign in your community. If there is, find out how you can become involved.

KEY TERMS

advocacy: Defending or stating the cause of another. In early education, supporting the ideas and issues of the profession.
foregone annual wages: Additional amounts that practitioners, given their education, could have earned annually in another occupation.
full cost of quality: An initiative promoted by NAEYC, to help parents and policymakers understand the hidden subsidy provided in child care by the inferior wages of practitioners.
quality, compensation, and accessibility: An initiative sponsored by NAEYC to draw attention to correlations between staff compensation and the factors that produce quality in programs.
networking: Making connections with others in the profession for mutual support and information, and for professional development and advancement.
reflective teaching: Process of thinking back over teaching experiences to form questions, set goals, and grow as a practitioner.
team teaching: Status and responsibility of early childhood practitioners in an early childhood setting are equal, rather than hierarchical.

SUGGESTED READINGS

Canadian Child Care Advocacy Association. (1993). *Children: Our hope, your future.* Ottawa: CDCAA.
Bloom, Paula Jorde. (1993, March). But I'm worth more than that! *Young Children,* 48 (3), 65–68.
———. (1993, May). But I'm worth more than that! Implementing a comprehensive compensation system. Part 2. *Young Children,* 48 (4), 67–72.

Cleveland, G., & Krashinsky, M. (1998). *The benefits and costs of good child care: The economic rationale for public investment in young children.* Toronto: Childcare Resource & Research Unit.

Doherty, G., & Dunster, L. (2000) "Guide to self-reflections for practitioners in child care settings" In *Partners in quality: Tools for practitioners in child care settings.* A project of the Canadian Child Care Federation.

Doherty, G., Lero, D., Goelman, H., Tougas, J., & LaGrange, A. (2000). *You Bet I Care! Caring and learning environments: Quality in regulated family child care across Canada.* Guelph: Centre for Families, Work and Well-Being, University of Guelph, Ontario.

Goelman, H., Doherty, G., Lero, D., LaGrange, A, & Tougas, J. (2000). *You Bet I Care! Caring and learning environments: Quality in child care centres across Canada.* Guelph: Centre for Families, Work and Well-Being, University of Guelph, Ontario.

Modigliani, Kathy. (1988, March). Twelve reasons for the low wages in child care. *Young Children,* 43 (3), 14–15.

NAEYC. (1993, January). The effects of group size, ratios, and staff training on child care quality. *Young Children,* 48 (2), 65–67.

———. (1990, November). NAEYC position statement on guidelines for compensation of early childhood professionals. *Young Children,* 46 (1), 30–32.

Ontario Coalition for Better Child Care. (1997). *A guide to child care in Ontario.* Toronto: OCBCC.

Russell, Sue. (1993, July). Linking education and compensation: A wholistic model. *Young Children,* 48 (5), 64–68.

APPENDIX
The Canadian Child Care Federation National Statement on Quality Child Care

Preamble

The National Statement on Quality Child Care was developed with the assistance of many, many individuals from across the country. It is a tangible demonstration of the child care community's ability and commitment to enhance the quality of child care in Canada. We wish to thank all those individuals who took the time to forward their valuable comments.

> This statement is targeted to a broad, national application in recognition of the diverse service realities. Studies indicate that there are several contextual factors which impact on quality child care: auspice, funding, licensing and enforcement, and sponsorship. As a living document, the quality statement is a mechanism to increase public understanding and support for quality child care as essential to the well-being of Canada's children.*
>
> —*Quality Committee*

*Approved by the Board of Directors in October, 1992. Ratified at the A.G.M., May 1993. Reprinted with permission of the Canadian Child Care Federation.

National Statement on Quality Child Care: Centre-Based

Quality child care is provided by trained caregivers who are child centred and supportive of families in Canada, their values, and needs. It is developmentally and culturally appropriate for each child in its service. Quality care encompasses all types of care provision available to all families in Canada, regardless of choice of service, age or developmental needs of children, or residence of family. Care is the responsibility of parents who have access to facilities provided by the community and society as a whole. Quality is ensured by building, maintaining, and adopting national principles, which reflect regional and cultural variables. Child care providers integrate their work and constantly adapt their skills—always thinking of the development of the whole child while meeting the needs of the group. At all times, the best interests of children and their families are of primary concern.

Principles of quality care used in this statement reflect the responsibilities of care providers, programs, parents, educational institutions, the community, and government. They are divided into seven significant areas, each with a statement of general philosophy. These philosophical principles are defined in more detail in three categories, people, practice, and place.

Specific indicators of quality care are based upon literature review, and the energies and beliefs of Canadians who have shared their perspectives during the compilation and integration of our statement.

Suitability and Training of Care Providers

Quality child care employs graduates in early childhood education and care as primary care providers.

The People

- have experience and formal post-secondary accredited early childhood education and care. Their training minimally includes the study of child development and developmentally appropriate practice for the early years and includes supervised practicum experiences to ensure transfer of knowledge
- have a developed code of ethics which speaks to confidentiality, ongoing professional development, and advocacy for all children and families
- are competent and committed early childhood professionals who seek out continued opportunities for professional growth through ongoing in-service training
- are active participants in local, regional, provincial, and national organizations and associations
- may be intuitive and experienced individuals without formal training who have demonstrated equivalent or superior competencies as graduates of traditional training approaches, through assessment by formal post-secondary programs and/or provincial or territorial accredited bodies
- when working with infants, school-age, special needs or as centre administrators have completed specialized training at the post-diploma or degree level where available
- are guaranteed working conditions benefits such as sick leave, professional in-service time, breaks and optional packages such as dental and disability coverage, etc.
- are warm, nurturing, and flexible care providers who genuinely enjoy their work with children and families
- model respect for and care of the natural environment
- receive remuneration appropriate to their level of professionalism and experience

The Practice

- offers clearly stated job descriptions and regular performance appraisals in consultation with supervisors and, on the basis of these evaluations, forms goals to improve the effectiveness of care providers in the program
- recognizes obvious career commitment of staff in their care of children and level of academic achievement, and increases their responsibilities and remuneration accordingly

- recognizes that fair and equitable salaries for care providers reduces staff turnover and ensures consistent, quality care
- accepts and trains practicum students to facilitate their professional development
- ensures administrators comply with legislated financial and record keeping requirements to ensure the stability of their programs, and/or uses sound business principles and practices as prepared for in specialized training
- supports and directs the on-site supervisors of practicum students, through regular contact with the staff of the training institutions regarding expectations
- provides a positive emotional environment for work through team building, open communication, and collaboration
- maximizes the varying strengths and interests of staff members and their complementary backgrounds
- encourages support among colleagues both within and across disciplines that begins early in training and continues throughout their professional lives

Child Development and the Learning Environment

Quality child care responds to children's needs by offering continuous opportunities for learning and nurturance. The goals of the service are determined by the needs of children and the shared philosophies of parents and care providers. A reliable framework of routines which support the physical, social, intellectual, and emotional needs of the children facilitates their learning. All practices that take place are based on sound child development theories and current research.

The People

- plan and facilitate meaningful experiences based on knowledge of children's development
- share information with parents regarding child rearing practices, child development, educational materials and community resources
- demonstrate a knowledge of and appreciation for a variety of cultures
- observe and record the development of individual children within a small group
- recognize, in children, developmental delays and/or physical or mental challenges and refer parents to appropriate agencies, subsequently collaborating with parents and agency in delivery of services suitable for the child
- encourage children to explore their environment and communicate with others
- ensure that children receive individualized attention, as well as small and large group time
- model appropriate social behaviours illustrating respect, concern, and consideration for others in thought, word, and action
- involve children in planning, discussion, and evaluation of activities and projects
- respect children's choices and unique ways of working with materials appropriate to their stage of development
- assist children in learning to identify and express feelings
- appreciate and communicate that being engaged in the creative process is more important than the product created

- recognize that individual unstructured daily routines are appropriate for infants
- recognize, accept, value, and acknowledge each child's unique creative expression
- model creative thinking and expression
- encourage children in problem-solving as well as identification and expression of feelings
- hold infants in arms as much as possible
- engage in positive social behaviour
- encourage children to offer and accept assistance from one another
- share information with parents about children's social development thereby together setting realistic expectations for children
- are proficient in the use of positive guidance, and strive for continuity between staff's handling of children
- guide individuals and groups, ensuring a creative and secure environment

The Practice

- fosters a knowledge and appreciation of a variety of cultures through the respectful introduction of art, music, foods, clothing, literature, and customs
- reflects the pluralistic nature of Canadian society
- involves children in learning experiences within the community
- bases daily activities on the continuous observation, guidance, and assessment of individual children and their interaction with others
- encourages children to actively engage in a variety of developmentally appropriate experiences which will:
  - foster each child's positive self-concept
  - respect cultural diversity of themselves and others
  - enhance social skills
  - nurture communication and language development
  - stimulate creative expression
  - extend each child's capacity for thinking, reasoning, questioning, and experimentation
  - provide sound health, safety, and nutritional practices
  - develop physical competence and co-ordination
  - foster each child's development of self-control
  - foster in children a respect for the natural environment and encourage environmentally sound principles
- recognizes the importance of sensory experiences in early development
- uses caregiving as an opportunity to deepen relationships with infants and supports their development through sensitivity to the child while performing routine tasks
- recognizes and attends to the individual rhythms of each infant
- attends to the affective needs of those present
- provides a predictable daily routine to instill a sense of security, which flexibly allows for individual preferences and independent choices. Each day will incorporate a balance of:
  - indoor and outdoor play

    - quiet and active times
    - large group and small group activities as well as individual activity
    - activities using both large and small muscles
    - child initiated activities as well as adult initiated activities
- provides learning opportunities through:
    - the many types of play
    - familiar day-to-day routines
    - opportunities for social interaction with peers and adults
    - opportunities which are challenging both physically and intellectually
    - transition times
- ensures that the daily plan is designed to encourage positive behaviours
- respects the child's first language and encourages it as much as possible
- displays children's creations
- provides opportunities for spontaneous play each day
- encourages children to assert their rights in socially acceptable ways
- nurtures children's understanding and respect of the rights of others
- develops and facilitates policies of guiding children's behaviour

The Place

- provides sufficient space for a variety of play and learning experiences
- is aesthetically pleasing and arranged to meet the needs of children and adults
- is equipped with sufficient numbers and varieties of equipment and materials which encourage and stimulate learning
- is arranged so that children have regular access to toys, equipment, and materials they use daily
- is designed and equipped to accommodate those with disabilities
- is arranged so there are spaces for individual play, small group and/or large group activity
- is equipped with individual storage space for the personal belongings of each child and adult
- presents indoor space with both hard and soft surfaces suitable for a variety of activities
- has furniture and fixtures of a size and location suitable for the children in attendance
- has separate space for play and for rest for infants and toddlers
- provides adjacent outdoor play space which offers a variety of surfaces and natural elements of sand, grass, bushes, trees, earth for digging, planting, etc.
- provides materials appropriate to the range of developmental levels of the children
- provides outdoor toys and equipment available in sufficient numbers for children in attendance
- has a variety of toys and equipment including items for: active and quiet play, construction, fine and gross motor activity, sensory and creative activity
- has a variety of books and toys which support dramatic play, and promote cognitive development, language, and multiculturalism

Group Size and Ratios

Quality child care recognizes that small group size contributes to the quality of interaction among the children and care providers, which determines the frequency of individualized attention children receive. Small groups (either multi-age or same age) foster the development of independence, cooperation, and mutual respect in children.

The People

- interact frequently with children
- observe and attend to the group dynamics
- modify the adult:child ratio to the ages and special needs of the children

The Practice

- plans daily activities to accommodate small groups
- allows for occasional special large group activities during which adult:child ratios are maintained
- recognizes that consistency of care providers is necessary for secure attachments particularly for infants and toddlers
- implements group size and ratios according to the chart on page [266 in this appendix].

The Place

- provides separate, self-contained play space for each group of children
- is arranged in relation to small group and individual activity

Adult Relationships

Quality child care maintains an open, friendly, and informative relationship with each child's family and encourages their involvement. The interaction among all persons in a child care setting reflects mutual respect, trust, and cooperation. The family and the care providers become partners who communicate openly for the mutual benefit of the children and themselves.

The People

- arrange for orientation of new families
- ask for day-to-day information about children from parents
- keep parents informed about children's daily experiences
- discuss children's progress frequently with parents
- provide information to parents regarding resources available within the community
- communicate with other agencies involved with individual children
- share a common philosophy and develop common goals with parents
- maintain a positive working environment through mutual support and frank, respectful resolution of disagreements
- work as a team

The Practice

- supports parents' desires to be involved in programs to the degree that meets individual family needs, and respects and welcomes their involvement at that level through home visits, regular newsletters and opportunities for family involvement, etc.
- suggests and/or provides ideas for parents to share with children at home
- recognizes and respects cultural and social backgrounds of individual families
- observes strict confidentiality regarding children and families

The Place

- has a parent resource component
- provides privacy for parents to meet, uninterrupted, with care providers
- has some adult size furniture for parents' comfort while interacting with children

Health and Nutrition

Quality child care practices sound health principles and promotes these among children, their families and care providers.

The People

- share with parents information regarding sound health practices and community health resources
- model the use of sound health practices
- conduct activities in a positive and tension free manner
- attend to children's physical needs and nurture children's self sufficiency in these areas according to each child's capability
- attend to children's emotional needs and nurture each child's understanding and expression of these needs according to her/his capability
- communicate regularly and as needed with parents regarding the child's health and well-being
- join the children in a relaxed mealtime routine, making eating pleasant for each child

The Practice

- establishes and adheres to procedures with regard to:
  - regular hand washing
  - personal hygiene
  - toileting and diapering
  - isolation of children who become ill during the day
  - food preparation and storage
  - laundering, storage, and use of personal items
  - general cleaning and housekeeping
- excludes children and care providers suffering from communicable diseases
- maintains up-to-date health records for children and care providers
- provides information regarding children's allergies and/or medical conditions to care providers

- provides for documented parental authorization for the administration of medications
- provides opportunities, on a daily basis, for children which encourage the development of: physical, dental, emotional, and nutritional health practices
- facilitates children's learning in the above areas through contact with community health care workers
- meets children's nutritional requirements through meals which accommodate cultural backgrounds and follow the Canada Food Guide
- supplements the nutritional requirements of high risk children
- ensures meals and snacks are served to children in small groups
- meets infant/toddler needs for individualized feeding schedules
- provides care providers with breaks, allowing them to be away from children in a designated staff room area, while maintaining adult:child ratios

The Place

- has natural and artificial light, is well ventilated, and is maintained at a comfortable temperature
- accommodates varying community realities such as lack of running water
- is designed and furnished so as to facilitate regular cleaning and disinfecting
- is fitted with noise absorbing materials
- provides at least 4 m^2 of useable play space indoors per child
- provides at least 13.8 m^2 of useable play space outdoors per child
- is equipped with a sufficient number of appropriately sized tables and chairs for meals and snacks
- has a space designated specifically for staff use
- has a separate sleeping area and accommodates non-nappers sensitively
- is equipped with a utility sink for use in general cleaning and sanitizing
- is equipped for proper storage and preparation of food and for clean up and disposal of garbage
- is equipped with a diapering and handwashing area which is separate from other areas
- has a separate area to isolate a child who has become ill where there is provision for rest, nutrition, personal hygiene, and privacy until parents arrive
- is physically designed to accommodate those with special needs

Safety

Quality child care provides for the safety of each child, and promotes safe practices among children, their families, and care providers.

The People

- are aware of and prepared for potentially hazardous situations
- are competent in the use of safety equipment
- regularly examine facilities, equipment, and toys (in and out of doors) to ensure that all are sound and in suitable condition for use
- provide supervision both indoors and outdoors which is appropriate to the children's needs
- are trained in the practice of standard first aid and infant/child CPR

- use safe pedestrian, auto, and bus practices when travelling with children
- share safety information with children and parents
- provide for the emotional safety of the child

The Practice

- maintains first aid and emergency supplies
- establishes emergency procedures which are posted and practised regularly by care providers and children
- maintains a current easily accessible list of phone numbers of parents, guardians, alternates, and emergency services
- records and regularly updates specific information regarding each child, i.e.: physician's name and phone number, child's health care number, allergy information, etc.
- documents and notifies parents of the occurrence and treatment of any illness or injury
- establishes procedures for reporting of suspected child abuse/neglect to the appropriate authority, and the follow-up support for the family

The Place

- is sanitary and free of hazards
- is equipped with operable safety devices
- is equipped with first aid supplies that are accessible indoors and outdoors, and on excursions for the numbers of adults and children normally present
- provides locked storage space for tools, cleaning products, medicines, etc.
- defines limits both indoors and outdoors in the interest of safety
- accommodates the physical environment of children with special needs including ramps, handrails, and physically accessible buildings
- is furnished with toys, equipment, and play materials, which are safe for the development level and needs of the children in attendance

Partnerships

Quality child care serves the best interests of children and families in a partnership of parents, professionally trained care providers, all levels of governments, training institutions, and provincial and territorial and national organizations who carry out complementary responsibilities.

Practice

- recognizes the role of the parent as the primary caregiver, and the role of the trained care provider as one which is supportive and enhancing to child and family
- utilizes parent boards, advisory committees, cooperatives, etc., to maintain a philosophy consistent with the needs of the families served
- ensures parents/boards have access to ongoing board development and training to ensure effective management of the program
- has a developed program philosophy and posted policies for behaviour management, safety/emergency procedures, and the reporting of suspected child abuse

- shares resources among parents and care providers within a supportive and complementary environment
- adheres to all applicable provincial/territorial legislation with commitment to exceed legislation where possible
- ensures that ongoing and annual program reviews/evaluations are conducted, both internal and external, to ensure delivery of quality service
- initiates and maintains positive liaison/networking with community resources such as schools, religious institutions, Indian band offices, community boards and agencies, and professional associations
- provides for an administrative office for use by the director, where confidential records of children and staff are kept

Training Institutions

- are accredited by the Department of Post-Secondary Education in each province/territory
- employ faculty who have completed recognized advanced professional training in early childhood education and adult education and have direct experience in the field
- develop and deliver curriculum based upon the principles of this National Statement
- place practicum students in centres which have been identified as appropriate settings, employ trained staff, and have compatible philosophies
- ensure that on-site supervisors of practicum students receive support and direction

Provincial/Territorial Associations

- take a pro-active role in ensuring all partners understand, adopt, and implement the principles of this quality statement
- have support programs which provide opportunities for information sharing, networking, and professional development

Government

- encourages a variety of flexible delivery models to meet the diverse needs of families
- recognizes current research and social policy issues as the foundation on which licensing standards are built
- employs individuals within the licensing body who have completed recognized advanced professional training and experience related specifically to the field
- provides information to enable parents to make meaningful choices regarding quality care
- ensures that quality care is accessible and affordable
- coordinates various departments and levels at the municipal, regional, provincial/territorial, and national levels to ensure quality child care

BIBLIOGRAPHY/SUGGESTED READINGS

Bailey, Gordon D. (1989). *Children's programmes: The new neighbourhood* (Appendix A: Day care and sleep). Unpublished paper.

Canadian Child Care Federation, Training Committee. (1991). *Issues in post-secondary education for quality early childhood care and education: A discussion paper.* Paper presented at Children: The Heart of the Matter. National Child Care Conference, Charlottetown, PEI, June 5–8.

Day Care: Competency Based Assessment: Policies and Procedures. (1986). Winnipeg, Manitoba: Manitoba Community Services.

Early-childhood teacher-education: Traditions and trends: An executive summary of colloquium proceedings. (1988, November), Editorial. *Young Children,* 44 (1): 53–57.

Esbensen, S. (1986). *The early childhood playground: An outdoor classroom.* Ypsilanti, MI: High/Scope Press.

Factors related to quality in child care: A review of the literature. (1990). Toronto, ON: Ontario Ministry of Community and Social Services.

Factors related to quality in child care: Annotated bibliography. (1990). Toronto, ON: Ontario Ministry of Community and Social Services.

Fenicel, E.S., & Eggbeer, Linda. (1990). Preparing practitioners to work with infants, toddlers and their families: Issues and recommendations for the professions. Washington, DC: Centre for Clinical Infant Programs.

Manitoba Child Care Task Force: Report. (1989). Winnipeg, Manitoba: The Task Force.

Promoting nutritional health during the preschool years. (1989). Ottawa: National Institute of Nutrition; Network of the Federal-Provincial-Territorial Group on Nutrition.

Savage, Brenda. (1991). Presentation to B.C. Task Force on Child Care on behalf of the Pacific Child Care Society. Unpublished paper.

Weinstein, Carol Simon, & David, Thomas G. (1987). *Spaces for children: The built environment and child development.* New York: Plenum Press.

Whitebook, Marcy, et al. (1989, November). Research report: Who cares? Child care teachers and the quality of care in America. *Young Children,* 45 (1), 41–45.

Yeates, McKenna, & Warberg, Chandler. (1990). *Administering early childhood settings: A Canadian perspective.* Columbus, OH: Merrill Pub. Co.

National Statement on Quality Child Care: Family-Based

Family day care is child care offered in the home of the care provider. It is not the conversion of a home to a day care setting but the sharing of the home with the children in care. Family day care is a flexible model of care. It is community based and can be shaped to meet many diverse family needs, both in urban and rural areas. It can accommodate an individual child or small multiple age groupings from infancy to teens. Providers contract directly with parents regarding the care of children or work under the auspices of a supervised agency.

Family day care provides the largest proportion of child care in the country yet is perhaps the least visible of the child care options. A variety of models exist including regulated and unregulated, supervised and unsupervised care:

1. licensed family day care agencies
2. licensed family day care providers
3. registration of family day care providers

4. resource supported regulated family day care providers
5. resource supported unregulated family day care providers
6. unregulated family day care providers

Indicators of quality family day care listed in the following sections reflect those items or aspects which can be measured, but the judgement about what quality family day care really is cannot be determined solely by using this as a checklist. Because this model of care occurs within a family environment, the settings where care takes place are infinitely variable and often unique; hence the need to use flexibility when judging the quality. The variance in socioeconomic, cultural, and geographic settings must be considered and quality child care can be that care which reflects the best standards of that neighbourhood. Quality child care is ultimately the combination of good personal relationships between the provider and children and their parents, compliance with standards, and a willingness to learn and grow.

Licensing alone is not a measure of quality family day care; however it is in place to establish and ensure that standards are met. In addition to licensing, the value of providing support to providers must be stressed. This resourcing role is vital in light of providers' isolation and the changing role of the family. Licensing and resourcing also offer a degree of visibility and professionalism to the service. In the end, it is the willingness and ability of providers, parents, agencies, associations, and government to meet set standards which will determine whether there is good quality family day care for Canadian children.

Role of Parents in Day Care

As parents are the primary caretakers of children, they play a key role in the day care arrangements for their children. It is important that the parent and provider work together to ensure that the child's day care experience is a positive one. To facilitate this, parents using family day care:

- strive for open and honest communication and take an interest in the care
- supply complete and up-to-date information
- ensure that their child is brought to the family day care well rested, fed, healthy and clean
- ensure that the home care provider has sufficient and appropriate diapers, clothing, toys, and supplies
- respect the provider's home, family, property, privacy, culture, and religion
- respect other children in care
- ensure that the confidentiality of the family day care home is maintained

Suitability and Training of Care Providers

A family day care provider is an adult who is able to transmit a genuine feeling of fondness, caring, and nurturing to each child in care. A family day care provider is also sensitive to the ethnic, cultural, and religious background of children in care.

In terms of learned skills for working with young children, it is important that a provider has an understanding of the developmental needs of children. This information can be acquired through formal child related courses of study and through a variety of life experiences.

A conscientious provider will take advantage of training opportunities. Providers can access a variety of mediums for growth such as books, correspondence courses, specific provider courses, conferences or workshops, training videos or, where available, on a one-to-one agency support basis. What is important is that training models be flexible and offer a variety of opportunities and curriculum to a population with very diverse backgrounds.

Coordination of the Home and the Family Day Care Service

The family day care provider maintains an open, friendly, and informative relationship with each child's family and encourages their involvement. The provider and parent become partners in child rearing and communicate openly and directly to ensure that the day care arrangement works for the child.

The Home

- has an area which can accommodate the meeting or occasional visit by parents to discuss the day-to-day care of their child

The Family Day Care Provider

- develops a partnership with parents who together work out common goals for the child
- asks parents for day-to-day information about their child
- communicates with parents about their child's daily experience
- negotiates a mutually convenient time with parents, when children are not present, to discuss their child's progress or any concern which may arise
- seeks information from the community about available resources and shares that information with parents
- cooperates with other organizations, agencies, or professionals that may be involved with the children in care
- recognizes and respects the cultural and social background of individual children and families
- observes strict confidentiality regarding children and families

Child Development and the Learning Environment

The family day care provider responds to a child's developmental needs by offering continuing opportunities for nurturing and learning. The goals of the care should reflect the shared philosophies of the parents and provider. Practices which are followed are based on discussion with parents, sound child development principles, and current research.

Routines which are followed in the home should support the physical, social, intellectual and emotional needs, and creative instincts of the child in order to facilitate learning. This includes the life skills that a child learns by observing appropriate role models.

The Home

- is arranged, or can be adapted, so that the children have easy access to toys, equipment, and materials
- is equipped with sufficient play materials, books, and craft supplies for indoor and outdoor use which encourage and foster learning. This includes items for active and quiet play, fine and gross motor activity, creative and cognitive, social and imaginative play.
- contains storage space for each child's personal belongings
- has furniture which is child size or suitable for children's use

The Family Day Care Provider

- offers a flexible, yet predictable, daily routine which promotes the development of self-esteem in children. The optimal atmosphere is one where children feel secure, stimulated, and valued. This may be accomplished by:
  - assisting children in learning to identify and express their feelings appropriately
  - recognizing, accepting, and acknowledging each child's creative expression
  - providing appropriate role models which promote respect and consideration in others
  - respecting different cultures and languages and incorporating different cultures into the daily routine and activities
  - offering a variety of appropriate activities or creating play opportunities which include: indoor and outdoor, quiet and active, individual and group, fine and gross motor, structured or spontaneous, and imaginative play
  - using materials in the home creatively to provide stimulating play and learning situations
- involves children, as developmentally appropriate, in community based activities: schools, libraries, museums, play groups, and other centres of interest
- makes appropriate use of television
- shares with parents information concerning community resources which may be of interest
- acquires additional skills through training opportunities to benefit children in care, and for personal growth
- uses daily household tasks to promote learning

Interaction/Relationships

The interaction among all persons involved in a family day care setting should reflect mutual respect, trust, and cooperation.

The Home

- provides comfortable areas where children can interact as well as areas where children can enjoy privacy
- provides a personal storage space for each child's belongings
- clearly identifies areas which may be reserved for the provider's family only

The Family Day Care Provider

- does not permit or practise any form of physical punishment, verbal or emotional abuse or deny physical necessities to any child in her/his care
- uses a positive approach which values the worth of a child, and is clear and consistent when dealing with negative behaviour
- encourages children to identify and express feelings and to assert their rights in age appropriate and socially acceptable ways
- is aware of the effect that events and changes in their home can have on children in care
- strives for consistency in behaviour management and guidance
- explains limits and expectations to children in terms they understand
- encourages children to respect the rights of others
- encourages children to offer and accept assistance from one another
- provides a role model which is socially acceptable
- acts in concert with parents in the setting of realistic expectations and goals for children's social development
- uses appropriate community resources, for example, play groups, drop-in centres, toy libraries

Health and Nutrition

The family day care provider engages in sound health practices and promotes these among the children cared for and their families. This is achieved by familiarity with guidelines on health and nutrition, and by following appropriate routines within the home.

The Home

- is clean, adequately lighted, ventilated, and maintained at a comfortable temperature
- allows for indoor space sufficient for the number, ages, and developmental level of the children cared for
- has access to outdoor play space and equipment suitable for the number, ages and developmental level of the children cared for
- equipment is selected to meet the developmental needs of the children cared for, and is easy to clean and disinfect

The Family Day Care Provider

- is in good general health as are other members of the household
- is aware of possible stress inducers, such as isolation and fatigue, and has strategies to reduce them
- establishes and practices recommended sanitary procedures with regards to:
  - general cleaning and housekeeping
  - toileting and diapering
  - personal hygiene
  - food preparation and dishwashing
  - storage and laundering of personal items
  - toys and equipment

- animals
- care of children who are ill
- well water, where applicable

- follows sound health practices and, in consultation with parents, establishes daily routines which promote good physical, mental, dental, and nutritional health in children
- maintains current information pertaining to the health records for the children in care
- ensures that children eat nutritional meals and snacks which are age, developmentally, and culturally appropriate
- establishes a relaxed mealtime routine for children which includes being with the children during meals
- ensures that children have developmentally appropriate sleep or rest periods daily
- excludes children suffering from communicable diseases during the infectious period
- obtains written permission from parents or physicians for the administration of any drugs or medication
- ensures that appropriate hygiene and handling procedures for pets are followed and that dogs and cats are inoculated against rabies
- consults with parents over health concerns and with health professionals if required

Safety

The family day care provider ensures that the home is a safe environment, promotes safe practices among children, and acts on behalf of a child's safety.

The Home

- permits easy exit in the event of necessary evacuation
- is clean, in good general repair, and free from dangers
- is equipped with functioning safety devices, including smoke detectors, fire extinguishers, gates, electrical outlet covers
- is adequately equipped with readily available first aid supplies
- contains storage space which is inaccessible to children for tools, firearms, hazardous and combustible products, drugs and medicines, and other dangerous items
- contains, or has access to, outdoor play space which is maintained in a safe condition

The Family Day Care Provider

- keeps the home (indoors and outdoors) safe for children
- is able to use safety devices
- establishes emergency procedures that are rehearsed regularly in order to ensure that children understand and will follow them

- has readily available telephone numbers of the various emergency services: police, fire, hospitals, poison control
- maintains accessible and current records and emergency information on the children in care including: telephone numbers of parents, physician's name and telephone number, child's medical health number, allergy information, and emergency person's name and telephone number
- obtains a signed release for medical emergencies
- obtains permission and instructions from parents for procedures requiring specific consideration, restrictions, or permission: medical treatment, allergies, special diets, rest, outdoor play, outings
- seeks safety information in the community and shares this information with parents
- documents and notifies parents of the occurrence and treatment of any illness, injury, or accident
- maintains up-to-date first aid certificate including infant CPR
- recognizes and knows how to report suspected child abuse or neglect
- clearly defines "off limit" areas for both indoor and outdoor play
- provides a level of supervision, both indoors and outdoors, which is appropriate to the child's developmental level
- uses safe pedestrian, automobile, and bus practices when travelling with children
- ensures that animals are kept in a safe environment and that children who handle animals are well supervised
- obtains appropriate liability insurance for the home and vehicle

Group Sizes and Ratios

Group size contributes to the quality of interaction among children and providers. Small ratios allow opportunities for more individual attention and through small family-like groups, children learn how to share, cooperate, and develop a sense of independence and compassion.

The Home

- offers adequate play and private space, for example, areas for meals, rest, personal hygiene, etc.

The Family Day Care Provider

- ensures that the number and age of children permits any necessary evacuation
- organizes space which can be used for individual play or activities for small groups
- provides sufficient space for all children to take part simultaneously in group activities
- ensures that the capacity in the home conforms to provincial/territorial ratios and, at the same time, allows children to receive care best suited to their needs
- ensures that adequate supervision for all indoor and outdoor activities or outings can be provided

Management of a Family Day Care Home

Effective management of a family day care home requires the careful balancing of household and care related duties. Quality family day care occurs when children's needs are being met.

The Family Day Care Provider

- is committed to providing a quality service to children and families and is clear to parents about limitations and expectations involving that care
- seeks out and takes advantage of professional development opportunities to enrich and improve her/his understanding and handling of children and families
- maintains necessary confidential records concerning children in care
- obtains and maintains liability insurance for the home and vehicle
- maintains complete and accurate business records, including the provision of receipts

Partnerships

In addition to the importance of the direct partnership between parents and providers in the provision of quality family day care, it must be stressed that other bodies also play a significant role in ensuring this end. Thus agencies, support services, local, provincial and territorial associations, and governments must work together to carry out their specific responsibilities related to the children's care.

Government Responsibilities

Similar government responsibilities exist for family day care as for centre-based care. These broad government responsibilities are:

- encourage a variety of flexible delivery models to meet the diverse needs of families
- recognize current research and social policy issues as the foundation upon which standards are built
- employ individuals who have completed recognized professional training related to family day care
- ensure that flexible models of training opportunities are available and that there are incentives provided for professional training
- provide accessible information to enable parents to make meaningful choices regarding family day care
- provide funding for resources and support services which can be accessed by providers
- ensure that quality care is accessible and affordable
- ensure that there are ample subsidies for those in need

Government, by direct licensing of providers, by licensing or contracting with agencies, or by supporting a variety of resources which are accessible to providers, contributes to the quality of family day care.

Role of Agencies

Agencies are licensed, or contracted with, to supervise the family day care arrangement. The mandate of agency staff varies, but includes some or all of the following functions:

- screens and approves family day care homes for the placement of children
- provides professional support and information to providers and parents in all of the significant areas outlined
- ensures that legislative requirements, which encompass all the significant areas, are being met (provincial, territorial, local, or band)
- matches families and providers
- assesses level of care being offered
- supplements supplies and equipment to providers
- encourages ongoing communication between the parent and provider and acts as an intermediary if needed
- assists in referral of individual children or families to appropriate services in the community where necessary
- provides care for families eligible for government subsidies
- takes a leading role in planning for, or providing, training opportunities and other resources in the community
- maintains accountable financial and case records

Role of the Licensing Authority Where Family Day Care Providers Are Directly Licensed

- issues a licence to operate in accordance with provincial, territorial, band, or local regulations (which encompass all the significant areas)
- monitors licensed homes on an ongoing basis to ensure that the quality of care meets legislated standards
- provides resources and support to licensed family day care providers
- takes a leading role in promoting the development of training opportunities and resources in the community

Role of Support Services

The need for support services as they pertain to family day care is a growing one. Community resources can offer an important contribution to quality family day care. Such supports provide networking opportunities and emotional support for providers in their day-to-day functioning, promote professional and personal growth for providers, play opportunities for children, and support and information for parents. Listed below are a variety of valuable supports. Some are offered hand-in-hand under the auspices of a family day care agency, a day care centre, a community organization, or a child care association; others are "stand alone" services set up in response to a community need:

- provider associations
- play groups
- resource centres or mobile resource units
- toy and equipment lending libraries

- telephone support lines
- provider registries
- newsletters
- training and professional development sessions
- parent groups

BIBLIOGRAPHY

Corsini, David A., Wisensale, Steven, & Caruso, Grace-Ann. (1988, September). Family day care: System issues and regulatory models. *Young Children* 43 (6), 17–23.

Deller, J. (1988). *Summary report of family day care internationally: A literature review.* Ontario: Ministry of Community and Social Services.

National statement on quality child care, second draft. (1990, November). Ottawa: Canadian Child Care Federation.

Private-home day care manual. (1983). Ontario: Ministry of Community and Social Services. [Loose-leaf Update 1989.]

The Canadian Child Care Federation develops a process of regular review of the National Statement on Quality Child Care so that it may remain a living statement. The review process will ensure the Quality Statement is updated as research information and knowledge about child development advances.

Group Size

Age of Children[a]	*6*	*8*	*10*	*12*	*14*	*16*	*18*	*20*	*22*	*24*	*26*	*28*	*30*
Infants (Birth–12 mos)	1:3 [b]												
Toddlers (12–24 mos)	1:3	1:4											
Two-year-olds		1:4	1:5	1:6									
Three-year-olds			1:5	1:6	1:7								
Four- & five-year-olds						1:8	1:9						
Five- & six-year-olds						1:8	1:9						
Six- to nine-year-olds								1:10	1:11	1:12			
Ten- to twelve-year-olds										1:12	1:13	1:14	1:15

[a] Ratios are determined with the adult having full-time responsibility for the care of children. If the other administrative duties or support services must be carried out, then an additional adult is required to maintain the designated ratios.

[b] Where multi-age grouping exists, the adult:child ratio and group size requirements shall be based on the age of the majority of children in the group. When infants are included, ratios and group size for infants must be maintained.

Adapted from: *Guide to Accreditation,* Washington, DC: National Academy of Early Childhood Programs, a division of National Association for the Education of Young Children, 1985.

References

Ade. W. (1982, March). Professionalization and its implications for the field of early childhood education. *Young Children, 37* (3), 25–32.

AECEO. (1997). Meet the affiliates. The Association of Early Childhood Educators, Ontario (AECEO). *Interaction* (Spring), 3.

Allen, Jeannie. (1991). Caregiver's Corner, *Young Children,* 46(6), p. 18.

Almy, M. (1988). The early childhood educator revisited. In Bernard Spodek, Saracho, Olivia N. & Donald L. Peters (Eds.). *Professionalism and the early childhood practitioner.* New York: Teachers College Press.

Ambrose, Richard P. (1993). Personal narratives and professional development. *Childhood Education,* 69 (5), 274–276.

Antler, Joyce. (1987). *Lucy Sprague Mitchell: The making of a modern woman.* New Haven, CT: Yale University Press.

Association of Canadian Community Colleges (ACCC). (1997). *Canadian training inventory.* www.accc.ca/english/programs/ecce/institutions.cfm

Ayers, William. (1993). *To teach: The journey of a teacher.* New York: Teachers College Press.

———. (1989). *The good preschool teacher.* New York: Teachers College Press.

Baker, Amy C. (1992, July). A puzzle, a picnic, and a vision: Family day care at its best. *Young Children,* 47 (5), 36–38.

Balaban, Nancy. (1992, July). The role of the child care professional in caring for infants, toddlers, and their families. *Young Children,* 47 (5), 66–71.

Basset, M. (1998). *The professional nanny.* Albany, NY: Delmar Publishers

Beach, J. & Bertrand, J. (2000). *More than the sum of the parts: An early child development system for Canada.* Occasional Paper No. 12. Toronto: Childcare Resource and Research Unit, University of Toronto.

Beach J., Friendly, M., & Schmidt, L. (1993). *Work-related child care in context: A study of work-related child care in Canada. Occasional Paper No. 3.* Toronto: Childcare Research & Resource Unit, University of Toronto.

Beach, Jane. (1998). *Our child care workforce: From recognition to remuneration: Executive Summary.* Ottawa: Child Care Human Resources Steering Committee.

Beach, Jane, Bertrand, Jane, & Cleveland, Gordon. (1998). *Our child care workforce: From recognition to remuneration.* Ottawa: Child Care Human Resources Steering Committee.

Beatty, Barbara. (1995). *Preschool education in America: The culture of young children from the colonial era to the present.* New Haven, CT: Yale University Press.

Beers, C. David. (1993). Telling our stories: The CDA process in Native American Head Start. In Elizabeth Jones (Ed.), *Growing teachers: Partnerships in staff development.* Washington, DC: NAEYC.

Behrmann, Michael M., & Lahm, Elizabeth A. (1994). Computer applications in early childhood special education. In *Young children: Active learners in a technological age.* Washington, DC: NAEYC.

Bellm, Dan, Gnezda, Terry, Whitebook, Marcy, & Breunig, Gretchen Stahr. (1994). Policy initiatives to enhance child care staff compensation. In Julienne Johnson & Janet B. McCracken (Eds.), *The early childhood career lattice: Perspectives on professional development.* Washington, DC: NAEYC.

Benner, Alison. (In press). *Quality child care and community development: What is the connection?* Ottawa: Canadian Child Care Federation.

Berk, Laura & Winsler, Adam. (1995). Scaffolding children's learning: Vygotsky and early childhood education. Washington, DC: National Association for the Education of Young Children.

Bernhard, J., Pollard, J., Eggers-Pierola, C., & Morin, A. (2000). Infants and toddlers in Canadian multi-age child care settings: Age, ability and linguistic inclusion. In *Research Connections Canada* 4, pp. 79–154. Ottawa: Canadian Child Care Federation.

Bertrand, J. (2001). *Summary of research findings on children's developmental health.* Ottawa: Canadian Institute of Child Health & Canadian Child Care Federation.

———. (1990). *Child care management guide.* Toronto: Ontario Coalition for Better Child Care.

Bloom, Benjamin. (1964). *Stability and change in human behaviour.* New York: Wiley.

Bloom, Paula Jorde. (1993, March). Full cost of quality report. But I'm worth more than that! *Young Children,* 48 (3), 65–68.

Bowman, Barbara T., & Beyer, Elizabeth R. (1994). Thoughts on technology and early childhood education. In *Young children: Active learners in a technological age.* Washington, DC: NAEYC.

Bradshaw, Douglas (1997). Front Line Profile, *IDEAS,* Volume 4 (2) (December 1997), 19.

Bredekamp, S. (1995, January). What do early childhood professionals need to know and be able to do? *Young Children,* 50 (2), 67–69.

———. (1992, January). Composing a profession. *Young Children* 47 (2), 52–54.

———. (1991). *Accreditation criteria and procedures of the National Academy of Early Childhood Programs.* (Rev. ed.). Washington, DC: National Association for the Education of Young Children.

———. (1990). An overview of NAEYC's criteria for high quality in early childhood programs. In Barbara Willer (Ed.), *Reaching the full cost of quality in early childhood programs.* Washington, DC: NAEYC.

———. (1990, September). Extra-year programs: A response to Brewer and Uphoff. *Young Children,* 45 (6), 20–21.

———. (1987). *Developmentally appropriate practice in early childhood programs serving children from birth through age 8.* Washington, DC: NAEYC.

———. (Ed.). (1987). *Developmentally appropriate practice in early childhood programs serving children from birth through age 8.* Washington, DC: NAEYC.

Bredekamp, Sue, & Rosegrant, Teresa. (1994). Learning and teaching with technology. In *Young children: Active learners in a technological age.* Washington, DC: NAEYC.

Bredekamp, Sue, & Willer, Barbara. (1993, March). Professionalizing the field of early childhood education: Pros and cons. *Young Children,* 48 (3), 82–84.

Brewer, Jo Ann. (1990, September). Transitional programs: Boon or bane? *Young Children,* 45 (6), 15–18.

Broere, A. Supervisor, Rural Roots Children's Centre front line profile. (1995, December). *IDEAS: The Journal of Emotional Well-Being in Child Care,* 1 (3): 15–16.

Brooke, Gretchen E. (1994, September). My personal journey toward professionalism. *Young Children,* 49 (6), 69–71.

Bruner, Jerome. (1991). Play, thought, and language. In N. Lauter-Klatell (Ed.), *Readings in Child Development.* Mountain View, CA: Mayfield Publishing Co.

Burts, D. C., Charlesworth, R., & Fleege, P. O. (1992). Observed activities and stress behaviors in developmentally appropriate and inappropriate kindergarten classrooms. *Early Childhood Research Quarterly,* 7, 297–318.

Caldwell, Bettye M. (1990). Educator: A new professional identity. *Dimensions,* 16, 3–6.

———. (1986). Professional child care: A supplement to parental care. In Bettye Caldwell & N. Gunzenhouser (Eds.), *Group care for young children: Considerations for child care and health professionals, public policy makers and parents.* Skillman, NJ: Johnson and Johnson Baby Products.

———. (1985). *What is quality child care?* Washington, DC: NAEYC.

Caldwell, Bettye M., & Hilliard, III, Asa G. (1985). *What is quality child care?* Washington, DC: NAEYC

Canadian Child Care Federation. (2000). *Tools for practitioners in child care settings.* Ottawa: CCCF.

———.(1996). Facts on family day care in Canada. *Interaction* (Summer), 11–12.

———. (1995). *Towards excellence in ECE training programs: A self-assessment guide.* Ottawa: author.

———. (1994). *National guidelines for training in early childhood care and education.* Ottawa: author.

———. (1993). *National child care policy statement.* Ottawa: author.

———. (1991). National Statement on Quality Child Care. Ottawa.

Canadian Day Care Advocacy Association & Canadian Child Day Care Federation. (1992). *Caring for a living: Final report: A study on wages and working conditions in Canadian child care.* Ottawa: author.

Canadian Education Association. (1993). *Admission to faculties of education in Canada: What you need to know. A Report from the Canadian Education Association.* Toronto: author.

Canadian Intergovernmental Conference Secretariat. (2000). *First ministers' communique on early child development.* Ottawa: First Ministers' Meeting.

Carew, (1980). *Observation study of caregiver and children in day care homes.* Paper presented at the Society for Research in Child Development meeting. San Francisco, California.

Carson, Linda. (1978). Notes from a municipal day care worker. In K. G. Ross (Ed.), *Good day care: Fighting for it, getting it, keeping it.* Toronto: The Women's Press.

Carter, M., & Curtis, D. (1994). *Training teachers: A harvest of theory and practice.* St. Paul, MN: Redleaf Press.

Caruso, J. (1977, November). Phases in student teaching. *Young Children,* 32 (1), 57–63.

Cassidy, D. J., & Lancaster, C. (1993, September). The grassroots curriculum: A dialogue between children and teachers. *Young Children,* 48 (6), 47–51.

Chandler, Phyllis A. (1994). *A place for me: Including children with special needs in early care and education settings.* Washington, DC: NAEYC.

Childcare Resource and Research Unit. (2000). *Early childhood care and education in Canada: Provinces and territories.* Toronto: Childcare Resource and Research Unit, University of Toronto.

———. (1997). *Childcare in Canada: Provinces and territories 1995*. Toronto: Childcare Resource and Research Unit, University of Toronto.

———. (1988). *Information Sheet*. Toronto: Centre for Urban and Community Studies, University of Toronto.

Cicerelli, V., et al. (1969). *The impact of Head Start. An evaluation of the effects of head start on children's cognitive and affective development*. Westinghouse Learning Corporation and Ohio University. Washington DC: Government Printing Office.

Clarke-Stewart, K. (1987). In search of consistencies in child care research. In D. Phillips (Ed.). *Quality in child care: What does the research tell us?* 105–120. Washington, DC: National Association for the Education of Young Children.

Clements, Douglas. (1994). The uniqueness of the computer as a learning tool: Insights from research and practice. In *Young children: Active learners in a technological age*. Washington, DC: NAEYC.

Combs, Arthur W., Blume, Robert A., Newman, Arthur J., & Wass, Hannelore L. (1974). *The professional education of teachers: A humanistic approach to teacher education*. (2nd ed.). Boston: Allyn and Bacon, Inc.

Corbett, Barbara. (1989). *A century of kindergarten education in Ontario*. Mississauga: The Froebel Foundation.

Corsini, David A., Wisensale, Steven, & Caruso, Grace-Ann. (1988, September). Family day care: System issues and regulatory models. *Young Children*, 43 (6), 17–23.

Cost, Quality, and Outcomes Study Team. (1995, May). Cost, quality, and child outcomes in child care centres: Key findings and recommendations. *Young Children*, 50 (4), 40–50.

Cruickshank, Donald R. (1987). *Reflective teaching: The preparation of students of teaching*. Reston, VA: Association of Teacher Educators.

Culkin, Mary L., Helbrun, Suzanne W., & Morris, John R. (1990). Current price versus full cost: An economic perspective. In Barbara Willer (Ed.), *Reaching the full cost of quality in early childhood programs*. Washington, DC: NAEYC.

Cyander, Max, & Mustard, Fraser. (1997). Brain development, competence and coping skills. In *Entropy: Founders' Network Report in Support of the Canadian Institute for Advanced Research*, 1 (1), 5–6.

Davidson, Jane, & Wright, June L. (1994). The potential of the microcomputer in the early childhood classroom. In *Young children: Active learners in a technological age*. Washington, DC: NAEYC.

DelCampo, Robert L., & DelCampo, Diana S. (1995). *Taking sides: Clashing views on controversial issues in childhood and society*. Guilford, CT: The Dushkin Publishing Group, Inc.

Derman-Sparks, Louise. (1989). *Anti-bias curriculum: Tools for empowering young children*. Washington, DC: NAEYC.

Diamond, Karen E., Hestenes, Linda L., & O'Connor, Caryn E. (1994, January). Integrating young children with disabilities in preschool: Problems and promise. *Young Children*, 49 (2), 68–73.

Doherty, G. (2000). Issues in Canadian child care: what does the research tell us? In *Research Connections Canada* 5, pp. 5–106.

———. (1997). *Zero to six: The basis for school readiness*. Research paper, Applied Resources Branch. Ottawa: Human Resources Development Canada.

Doherty, G., & Stuart, B. (1996). *A profile of quality in Canadian child care centres.* Guelph: University of Guelph.

Doherty, G., Lero, D., Goelman, H., LaGrange, A., & Tougas, J. (2000). *You bet I care! A Canada-wide study on: Wages, working conditions and practices in child care centres.* Guelph: Centre for Families, Work, and Well-being, University of Guelph, Ontario.

Doherty, G., Lero, D., Goelman, H., Tougas, J., & LaGrange, A. (2000). *You Bet I Care! Caring and learning environments: Quality in regulated family child care across Canada.* Guelph: Centre for Families, Work and Well-Being, University of Guelph, Ontario.

Doherty, Gillian. (1997). Credentialing as a strategy for promoting quality in child care. Draft. (July). Ottawa: Canadian Child Care Federation.

———. (1996). *The great child care debate: The long-term effects of non-parental child care.* Occasional Paper No. 7. Toronto: University of Toronto, Child Care Resource and Research Unit.

Doherty-Derkowski, Gillian. (1995). *Quality matters: Excellence in early childhood programs.* Reading, MA: Addison-Wesley Publishers Ltd.

Dollase, Richard H. (1992). *Voices of beginning teachers: Visions and realities.* New York: Teachers College Press.

Doxey, Isabel. (Ed.). (1990). *Child care and education: Canadian dimensions.* Toronto: Nelson Canada.

Duff, R. Eleanor, Brown, Mac H., & Van Scoy, Irma J. (1995). Reflection and self-evaluation keys to professional development. *Young Children,* 50 (4), 81–88.

Early Childhood Educator of British Columbia (ECEBC). (1992). *Code of ethics.* Vancouver: ECEBC.

Edwards, Carolyn. (1994). Partner, nurturer, and guide: The roles of the Reggio teacher in action. In Carolyn Edwards, Lella Gandini, & George Forman (Eds.), *The hundred languages of children: The Reggio Emilia approach to early childhood education.* Norwood, NJ: Ablex Publishing Corporation.

Elkind, D. (1988). *Miseducation: Preschoolers at risk.* New York: Alfred A. Knopf.

———. (1983, January). Montessori education: Abiding contributions and contemporary challenges. *Young Children,* 38 (2), 3–10.

Employment and Immigration Canada. (1993). *Occupational standards and certification. Canadian practices.* Ottawa: Minister of Supply and Services. Publication No. LM-271-03-93E.

Epstein, Ann S. (1993). *Training for quality: Improving early childhood programs through systematic inservice training.* Ypsilanti, MI: High/Scope Educational Research Foundation.

Exchange top 100. (1994). The nation's largest early childhood organizations. *Child Care Information Exchange,* (November–December) 72.

Feeney, Stephanie. (1990, May). Update on the NAEYC Code of Ethical Conduct. *Young Children,* 45 (4), 20–21.

Feeney, Stephanie, Christensen, Doris, & Moravcik, Eva. (1996). *Who am I in the lives of children: An introduction to teaching young children.* (5th ed.). Columbus, OH: Merrill Publishing Co.

Ferguson, E. (1998). *Caring in practice: Child care substitute youth internship program. A Synopsis.* Halifax: Child Care Connection-NS.

Ferguson, Elaine E. (1997). *Child care administrator credentialing: A work in progress.* Halifax: Child Care Connection-NS.

———. (1995). *Child Care ... Becoming Visible.* Halifax: Child Care Connection-NS.

Fewster, Gerry. (1990). *Being in child care: A journey into self.* New York: The Haworth Press.

Freire, P. (1970). *Pedagogy of the oppressed.* New York: The Seabury Press.

Friendly, M. (2000). *Child care and Canadian federalism in the 1990s: Canary in a coal mine.* Toronto: Childcare Resource and Research Unit, University of Toronto.

Friendly, Martha. (1994). *Child care policy in Canada: Putting the pieces together.* Don Mills, ON: Addison-Wesley Publishers, Inc.

Frontline Profile. (1997, December). *IDEAS: The Journal of Emotional Well-Being in Child Care, 4* (3), p. 19.

Gaffield, C. (1982). Schooling, the economy, and rural society in nineteenth-century Ontario. In J. Parr (Ed.), *Childhood and family in Canadian history.* Toronto: McClelland & Stewart.

Galant, Kim, & Hanline, Mary Frances. (1993). Parental attitudes toward mainstreaming young children with disabilities. *Childhood Education,* 69 (5), 293–297.

Galinsky, E., Howes, C., Kontos, S., & Shinn, M. (1994). The study of children in family child care and relative care—key findings and recommendations. *Young Children* 50 (1), 58–61.

Gandini, Lella. (1994). Educational and caring spaces. In Carolyn Edwards, Lella Gandini, & George Forman (Eds.), *The hundred languages of children: The Reggio Emilia approach to early childhood education.* Norwood, NJ: Ablex Publishing Corporation.

———. (1993, November). Fundamentals of the Reggio Emilia approach to early childhood education. *Young Children,* 49 (1), 4–8.

Gersten, Russell, & Keating, Thomas. (1990). Long-term benefits from direct instruction. In Mary A. Jensen & Zelda W. Chevalier (Eds.), *Issues and advocacy in early education.* Boston: Allyn and Bacon.

Gestwicki, Carol. (1996). *Home, school, community relations.* (3rd ed.). Albany, NY: Delmar Publishers.

———. (1995). *Developmentally appropriate practice: Curriculum and development in early education.* Albany, NY: Delmar Publishers Inc.

Giles T.E., & Proudfoot, A.J. (1994). *Educational administration in Canada.* (5th ed.) Calgary, Alberta: Detselig Enterprises Ltd.

Goelman, H., Doherty, G., Lero, D., LaGrange, A, & Tougas, J. (2000). *You Bet I Care! Caring and learning environments: Quality in child care centres across Canada.* Guelph: Centre for Families, Work and Well-Being, University of Guelph, Ontario.

Goelman, H., & Pence, A. (1987). Effects of child care, family and individual characteristics on children's language development: The Victoria day care research project. In D. Phillips (Ed.), *Quality child care: What does research tell us?* Research monograph, Vol. 1. Washington, DC: National Association for the Education of Young Children.

Gordon, Ann, & Browne, Kathryn Williams. (1996). *Beginnings and beyond.* (4th ed.). Albany, NY: Delmar Publishers Inc.

Goss Gilroy (1999). *The family child care provider survey.* Ottawa: Canadian Child Care Federation.

Goulet, M. (2001). *Handout #5: Curriculum development.* Toronto: George Brown College.

Granucci, Pamela Lesiak. (1990, March). Kindergarten teachers: Working through our identity crisis. *Young Children,* 45 (3), 6–11.

Greenberg, Polly. (1990, January). Why not academic preschool? (Part 1). *Young Children,* 45 (2), 70–80.

———. (1987, July). Lucy Sprague Mitchell: A major missing link between early childhood education in the 1980s and progressive education in the 1890s–1930s. *Young Children,* 42 (5), 70–84.

Griffin, S. (1994). *Professionalism: The link to quality care.* Ottawa: Canadian Child Care Federation.

Guy, K. (1997). *Our promise to children.* Ottawa: Canadian Institute of Child Health.

Harms, Thelma, & Clifford, Richard. (1980). *Early childhood rating scale.* New York: Teacher's College Press.

Health Canada, Human Resources Development Canada, and Indian & Northern Affairs Canada (2001). *Federal/Provincial/Territorial Early childhood development agreement: Report on government of Canada activities and expenditures 2000–2001.* Ottawa: Minister of Public Works and Government Services Canada.

Helburn, S., Culkin, M., Morris, J., Morcan, N., Howes, C., Phillipsen, L., Bryant, D., Clifford, R., Cryer, D., Peisner-Feinberg, E., Burchinal, M., Kaga, S., & Rusticic, J. (1995). *Cost, quality & child outcomes in child care centres.* University of Colorado, University of California, University of North Carolina, and Yale University, Bloomington, IN: Phi Delta Kappa Educational Foundation.

Hewlett, Sylvia Ann. (1991). *When the bough breaks: The cost of neglecting our children.* New York: HarperCollins Publishers.

Hill, Patty Smith. (1987, July). The function of the kindergarten. *Young Children,* 42 (5), 12–19.

Hillman, Carol. (1988). *Teaching four-year-olds: A personal journey.* Bloomington, IN: Phi Delta Kappa Educational Foundation.

Hohmann, C., Carmody, B., & McCabe-Branz, C. (1995). *High/Scope buyer's guide to children's software.* (11th ed.). Ypsilanti, MI: High/Scope Press.

Honig, Alice S. (1994, March). Mental health for babies: What do theory and research teach us? *Young Children* 48 (3), 69–76.

Howe, N., Jacobs, E., & Fiorentino, L. (2000). The curriculum. In L. Prochner & N. Howe (Eds.), *Early childhood care and education in Canada.* Vancouver: UBC Press.

Human Resources Development Canada (HRDC). (1995). *Child Care Initiatives Fund evaluation.* Ottawa: Government of Canada.

Human Resources Development Canada. (1994). *Child care and development: A supplementary paper.* Ottawa: Government of Canada.

Hunt, J. McVicker. (1961). *Intelligence and experience.* New York: Ronald.

Hymes, James, Jr. (1978–79). *Living history interviews. Books 1–3.* Carmel, CA: Hacienda Press.

———. (1974). *Teaching the child under six.* (2nd ed.). Columbus, OH: Charles E. Merrill Publishing Co.

Hyson, Marion C. (1982, January). Playing with kids all day: Job stress in early childhood education. *Young Children,* 37 (2), 25–31.

Hyson, Marion C., & Eyman, Alice P. (1986, September). Approaches to computer literacy in early childhood teacher education. *Young Children,* 41 (6), 54–59.

Irwin, S., Lero, D., & Brophy, K. (2000). *A matter of urgency: Including children with special needs in child care in Canada.* Wreck Cove, NS: Breton Books.

Irwin, Sharon. (1995). *Charting new waters.* Wreck Cove, NS: Breton Books.

Isenberg, J., & Quisenberry, N. (1988). Play: A necessity for all children. *Childhood Education,* 64 (3), 138–145.

Jalongo, Mary Renck. (1991, September). Stephanie: One teacher's story. *Young Children,* 46 (6), 62–64.

Janmohamed, Z. (2001). Unpublished paper. Toronto: George Brown College, 20–27.

Janus, M., and Offord, D. (2000). Readiness to learn at school. *Isuma,* 1 (2), 71–75.

Joffe, C. (1977). *Friendly intruders.* Berkeley, CA: University of California Press.

Johnson, K., Lero, D., & Rooney, J. (2001). *Work-Life compendium 2001: 150 Canadian statistics on work, family & well-being* Guelph: Centre for Families, Work and Well-Being, University of Guelph.

Jones, E. (Ed.). (1993). *Growing teachers: Partnerships in staff development.* Washington, DC: NAEYC.

Jones, E. (1991, May). Do ECE people really agree? Or are we just agreeable? *Young Children,* 46 (4), 59–61.

Jones, E. & Nimo, John. (1994). *Emergent curriculum.* Washington, DC: NAEYC.

Jones, Elizabeth, and Reynolds, G. (1992). *The play's the thing: Teachers' roles in children's play.* New York: Teachers College Press.

Kagan, Sharon L. (1994, Fall). Families and children: Who is responsible? *Childhood Education,* 71 (1), 4–8.

———. (1988, May). Dealing with our ambivalence about advocacy. *Child Care Information Exchange,* (61), 31–34.

Kagan, Sharon Lynn, & Newton, James W. (1989, November). For-profit and non-profit child care: Similarities and differences. *Young Children,* 45 (1), 4–10.

Kaiser, J., & Rasminsky, S. (1991). *The good day care book.* Toronto: Little, Brown & Company (Canada) Ltd.

Kamii, C. (Ed.). (1990). *Achievement testing in the early grades: Games grown-ups play.* Washington, DC: NAEYC.

Katz, L. (1993). *Five perspectives on quality in early childhood programs.* Perspectives from ERIC/EECE. [Monograph series, 1]. Urbana, IL: ERIC Clearing House on Elementary and Early Childhood Education.

———. (1995). *Talks with teachers of young children.* Norwood, NJ: Ablex Publishing Corporation.

———. (1995). What is basic for young children? In *Talks with teachers: A collection.* Norwood, NJ: Ablex Publishing Corp.

———. (1995). The nature of professions: Where is early childhood education? In *Talks with teachers of young children.* Norwood, NJ: Ablex Publishing Corp.

———. (1991). Ethical issues in working with young children. In Lillian Katz & Evangeline Ward (Eds.), *Ethical behaviour in early childhood education.* (Expanded ed.). Washington, DC: NAEYC.

———. (1984, July). The professional early childhood teacher. *Young Children, 39* (5), 3–10.
———. (1980). Mothering and teaching—some significant distinctions. In L. Katz (Ed.), *Current topics in early childhood education, 3,* Norwood, NJ: Ablex Publishing Corporation.
Katz, Lilian G., Evangelou, Demetra, & Hartman, Jeanette Allison. (1991). *The case for mixed-age grouping in early education.* Washington, DC: NAEYC.
Keating, D. & Hertzman, C. (1999). *Developmental health and the wealth of nations.* New York: Guilford Press.
Kellerman, Maureen. (1995). *The 1995 report on family resource programs across Canada.* Ottawa: Canadian Association of Family Resource Programs.
Kidder, T. (1989). *Among schoolchildren.* Boston, MA: Houghton Mifflin Co.
King, A., & Peart, M. J. (1990). *The good school.* Toronto: Ontario Teacher's Federation.
Kipnis, Kenneth. (1987, May). How to discuss professional ethics. *Young Children,* 42 (40), 26–30.
Klein, Tovah, Bittel, Calley, & Molnar, Janice. (1993, September). No place to call home: Supporting the needs of homeless children in the early childhood classroom. *Young Children,* 48 (6), 22–31.
Kochendorfer, Leonard. (1994). *Becoming a reflective teacher.* Washington, DC: National Education Association.
Kyle. I., & Kellerman, M. (1998). *Case studies of Canadian family resource programs: Supporting families, children, and communities.* Ottawa: Canadian Association of Family Resource Programs.
Lazer, I., & Darlington, R. (1982). Lasting effects of early education: A report from the consortium for longitudinal studies. *Monographs of the Society for Research in Child Development* 47, 2–3.
Lazar, I., Hubbell, V. R., Murray, H., Rosthe, M., & Royce, J. (1997). *Summary Report. The persistence of preschool effects: A long-term follow-up of fourteen infant and preschool experiments.* Washington, DC: U.S. Dept. of Health, Education, and Welfare. OHDS 78-30129.
Lortie, Dan. (1975). *School teacher: A sociological study.* Chicago: University of Chicago Press.
Lyon, Patricia, & Canning, Mary. (1995). *The Atlantic day care study.* St. John's, NF: Memorial University of Newfoundland.
Malaguzzi, Loris. (1994). History, ideas, and basic philosophy: An interview with Lella Gandini. In Carolyn Edwards, Lella Gandini, & George Forman (Eds.), *The hundred languages of children. The Reggio Emilia approach to early childhood education.* Norwood, NJ: Ablex Publishing Corporation.
———. (1993, November). For an education based on relationships. *Young Children,* 49 (1), 9–12.
Mathien, Julie. (2001). *Children, families and institutions in late 18th century and 20th century Ontario.* Unpublished master's thesis, Ontario Institute for Studies in Education, University of Toronto.
———. (1990). *School programs for young children in Ontario and Toronto.* (Unpublished manuscript).
McCain, M. & Mustard, J.F. (1999). *Early Years Study.* Toronto: Ontario Children's Secretariat.

McLean, S. Vianne. (1993). Learning from teachers' stories. *Childhood Education, 69* (5), 265–268.

Meisels, S., & Shonkoff, J. (1990). Preface. In *Handbook of early interventions,* S. Meisels and J. Shonkoff (Eds.). New York: Cambridge University Press.

Melhuish, E. C., Mooney, A., Martin, S., & Lloyd, E. (1990a). Type of childcare at 18 months: I. Differences in interactional experience. *Journal of Child Psychology and Psychiatry,* 31 (6), 849–59.

Melhuish, E. C., Mooney, A., Martin, S., & Lloyd, E. (1990b). Type of childcare at 18 months—II. Relations with cognitive and language development. *Journal of Child Psychology and Psychiatry,* 31 (6), 861–70.

Millichamp, Dorothy. (1974). Epilogue to *The adult and the nursery school child,* by Margaret Fletcher. Toronto: University of Toronto Press.

Mitchell, Lucy Sprague. (1953). *Two lives.* New York: Simon and Schuster.

Modligiani, Kathy. (1988, March). Twelve reasons for the low wages in child care. *Young Children,* 43 (3), 14–15.

Morgan, Gwen, Azer, Sheri L., Costley, Joan B., Genser, Andrea, Goodman, Irene R., Lombardi, Joan, & McGimsey, Bettina. (1993). *Making a career of it. The state of the state's report on career development in early care and education.* Boston MA: The Centre for Career Development in Early Care and Education at Wheelock College.

Morris, J., & Murphy, C. (1996). *Meet the affiliate.* The Association of Early Childhood.

Moss, P. (2000). Workforce issues in early childhood education and care. Prepared for consultative meeting on International Developments in Early Childhood Education and Care, May 11–12, 2000. New York: Columbia University.

Moustakas, Clark. (1966). *The authentic teacher.* Cambridge, MA: Howard A. Doyle Publishing Co.

Moyer, J., Egerston H., & Isenberg, J. The child-centred kindergarten. *Childhood Education,* 64, 235–242.

NAEYC. (1995). *Guidelines for preparation for early childhood professionals associate baccalaureate, and advanced levels. Position statement.* Washington, DC: NAEYC.

———. (1995, May). Cost, quality, and child outcomes in child care centres: Key findings and recommendations. *Young Children,* 50 (4), 40–44.

———. (1994). *Position statements.* Washington, DC: NAEYC.

NAEYC position statement: A conceptual framework for early childhood professional development. (1994, March). *Young Children,* 49 (3), 68–77.

NAEYC. (1993, January). The effects of group size, ratios, and staff training on child care quality. *Young Children,* 48 (2), 65–67.

———. (1993, September/October). What parents want. *Child Care Information Exchange,* 93, 63–67.

———. (1991a, November). Early childhood teacher certification: A position statement of the association of teacher educators and the National Association for the Education of Young Children. *Young Children,* 47 (1), 16–21.

———. (1991b, September). NAEYC to launch new professional development initiative. *Young Children,* 46 (6), 37–39.

———. (1990, November). Position statement on school readiness. *Young Children,* 46 (1), 21–23.

NASBE. (1988). *Right from the start.* Alexandria, VA: Author.

National Academy. (1990, January). Reaccreditation: A snapshot of growth and change in high-quality early childhood programs. *Young Children,* 45 (2), 58–61.

New, Rebecca. (1994). Cultural variations on developmentally appropriate practice: Challenges to theory and practice. In Carolyn Edwards, Lella Gandini, & George Forman (Eds.) *The hundred languages of children: The Reggio Emilia approach to early childhood education.* Norwood, NJ: Ablex Publishing Corporation.

Nicholson, Simon. (1974). How not to cheat children: The theory of loose parts. *Alternate Learning Environments.* In G. Coates (Ed.), Stroudsberg, PA: Dowden, Hutchinson and Ross.

Norris, N. Catherine. (1994, September). The NAEYC accreditation process isn't really so scary—Why not try it? *Young Children,* 49 (6), 72–74.

Olmsted, Patricia P. (1992, Fall). Where did our diversity come from? *High/Scope Resource,* 4–9.

Ontario Coalition for Better Child Care. (1997). *Child Care in Ontario.* Toronto: OCBCC.

Osborn, D. Keith. (1991). *Early childhood education in historical perspective.* (3rd ed.). Athens, GA: Education Associates, Div. of The Daye Press, Inc.

Paley, Vivian Gussin. (1995). *Kwanzaa and me.* Cambridge, MA: Harvard University Press.

———. (1990). *The boy who would be a helicopter.* Cambridge, MA: Harvard University Press.

Pence, Alan. (1990). The child care profession in Canada. In Isabel Doxey (Ed.), *Child care and Canada.* Toronto: Nelson Canada.

Perrone, V. (1991). *A letter to teachers: Reflections of schooling and the art of teaching.* San Francisco: Jossey-Bass Publishers.

Perry, B. (1996). Neurodevelopmental adaptations to violence: How children survive the intragenerational vortex of violence. In *Violence and childhood trauma: Understanding and responding to the effects of violence on young children.* Cleveland: Gund Foundation.

Peters, R. (2001). *Developing capacity and competence in the Better Beginnings, Better Futures communities: Report summary.* Kingston: Research Coordination Unit, Queen's University.

Phillips, C. (1991). At the core: What every early childhood professional should know. In Julienne Johnson & Janet B. McCracken (Eds.), *The early childhood career lattice: Perspectives on professional development.* Washington, DC: NAEYC.

———. (Ed.) (1991). *Essentials for child development associates working with young children.* Washington, DC: Council for Early Childhood Professional Recognition.

———. (1990, March). The child development associate program: Entering a new era. *Young Children,* 45 (3), 24–27.

Phillips, Deborah A. (Ed.). (1987). *Quality in child care: What does research tell us?* Washington, DC: NAEYC.

Platt, Elizabeth Balliett. (1991). *Scenes from day care: How teachers teach and what children learn.* New York: Teachers College Press.

Poelle, Lisa. (1992). I'll visit your class, you visit mine: Experienced teachers as mentors. In Elizabeth Jones (Ed.), *Growing teachers: Partnerships in staff development.* Washington, DC: NAEYC.

Pollard, J. (1996). *Student and sponsor-educator relationships during early childhood education practicum.* Unpublished master's thesis, the School of Child and Youth Care, University of Victoria.

Pratt, Caroline. (1970). *I learn from children.* New York: Harper and Row. (Original work published 1948).

Prochner, L. (2000). A history of early education and child care in Canada, 1820–1966. In L. Prochner & N. Howe (Eds.), *Early childhood care and education in Canada.* Vancouver: UBC Press.

———. (1996). Quality in care in historical perspective. *Early Childhood Research Quarterly,* 11, 5–17.

Raphael, Ray. (1985). *The teacher's voice: A sense of who we are.* Portsmouth, NH: Heinemann.

Rasmussen, J. (2000). Personal Communication.

Raymond, Jocelyn. (1991). *The nursery years.* Toronto: University of Toronto Press.

Riley, Sue Spayth. (1979). *How to generate values in young children.* Los Angeles: The New South Company.

Robinson, B. (1988, September). Vanishing breed: Men in child care programs. *Young Children,* 43 (6), 54–57.

Roemer, Joan. (1989). *Two to four from nine to five.* New York: Harper and Row.

Rogers, Dwight L. (1990, Fall). Are questions the answer? Enhancing young children's learning. *Dimensions,* 19 (1), 3–5.

Rubenstein, J., & Howes, C. (1979). Caregiving and infant behaviour in day care and in homes. *Developmental Psychology,* 15 (1), 1–24.

Russell, Sue, & Derrick, Teresa M. (1993). *Careers in early childhood.* Chapel Hill, NC: Day Care Services, Inc.

Ryan, Kevin, (Ed.). (1970). *Don't smile until Christmas: Accounts of the first year of teaching.* Chicago: University of Chicago Press.

Saracho, Olivia N. (1988). Cognitive style and early childhood practice. In Bernard Spodek, Olivia N. Saracho, & Donald L. Petus (Eds.), *Professionalism and the early childhood practitioner.* New York: Teachers College Press.

Saracho, Olivia N. & Donald L. Peters (Eds.). (1975). *The early childhood educator at work.* New York: McGraw Hill.

Saracho, Olivia, & Donald L. Peters (Eds.). *Professionalism and the early childhood practitioner.* New York: Teachers College Press.

Sarason, S. B. (1972). *The creation of settings and the future societies.* San Francisco: Jossey-Bass.

Schulz, Patricia. (1978). Day care in Canada: 1850–1962. In *The good day care book.* Toronto: The Women's Press.

Schweinhart, L., & Weikart, D. (1993). Changed lives, significant benefits: The High/Scope Perry Preschool Project to date. *High/Scope Resource* (Summer), 10–14.

Schweinhart, L., Weikart, D., & Larner, M. (1986). Consequences of three preschool curriculum models through age 15. *Early Childhood Research Quarterly,* 1, 15–45.

Schweinhart, Lawrence J., & Weikart, David P. (1986, January). What do we know so far? A review of the Head Start Synthesis Project. *Young Children,* 41 (2), 49–55.

Seifert, Kelvin. (1988). Men in early childhood education. In Bernard Spodek, Olivia N. Saracho, & Donald L. Peters (Eds.), *Professionalism and the early childhood practitioner.* New York: Teachers College Press.

Shonkoff, J., & Meisels, S. (Eds.). (2000). *Handbook of early childhood intervention.* (2nd ed.). Cambridge: Cambridge University Press.

Shore, R. (1997). *Rethinking the brain.* New York: Families and Work Institute.

Silin, Jonathan G. (1985, March). Authority as knowledge: A problem of professionalization. *Young Children,* 40 (3), 41–46.

Singleton, C. (1997). *The development and implementation of a reflective mentoring program for early childhood educators.* Unpublished master's thesis, Memorial University of Newfoundland.

Smith, Marilyn. (1987, March). NAEYC at 60: Visions for the year 2000. *Young Children,* 42 (3), 33–39.

South East Grey Community Outreach. (1999). *Survive and thrive.* South East Grey County.

SpeciaLink. (1996). Provincial and Territorial Early Intervention Directory SpeciaLink Supplement.

Spodek, Bernard. (1994). The knowledge base for baccalaureate early childhood teacher education programs. In Julienne Johnson & Janet B. McCracken (Eds.), *The early childhood career lattice: Perspectives on professional development.* Washington, DC: NAEYC.

Spodek, Bernard. (1988). Professionalizing the tasks ahead. In Bernard Spodek, Olivia N. Saracho, & Donald L. Petus (Eds.), *Professionalism and the early childhood practitioner.* New York: Teachers College Press.

Spodek, Bernard, Saracho, Olivia N., & Petus, Donald L. (1998). Professionalism, semi-professionalism, and craftsmanship. In Bernard Spodek, Olivia N. Saracho, & Donald L. Petus (Eds.), *Professionalism and the early childhood practitioner.* New York: Teachers College Press.

———. (Eds.). (1988). *Professionalism and the early childhood practitioner.* New York: Teachers College Press.

Stanton, Jessie. (1990). The ideal nursery school teacher. Reprinted in *Young Children,* 45 (4), 19. Used with permission of New York State AEYC. Originally published in New York Nursery Education News (Winter 1954).

Statistics Canada. (1992). *National child care study: Introductory report.* Cat. No. 89–526. Ottawa: Author.

Surbeck, Elaine. (1994, Summer). Journal writing with preservice teachers. *Childhood Education,* 70 (4), 232–35.

Sutherland, N. (1976). *Children in English-Canadian society: Framing the 20th century consensus.* Toronto: University of Toronto Press.

Sylva, K. (1994). School influences on children's development. *Journal of Child Psychology and Psychiatry,* 35 (1), 135–70.

Thomas, B. (1998). *Family Literacy in Canada.* Welland, ON: Soleil Publishing.

Trawick-Smith, Jeffrey, & Lambert, Laura. (1995, March). The unique challenges of the family child care provider: Implications for professional development. *Young Children,* 50 (3), 25–32.

Tremblay, R., & Craig, W. (1995). Developmental crime prevention. In M. Tonry & D. P. Farrington (Eds.), *Building a safer society: Strategic approaches to crime prevention.* Chicago: University of Chicago Press, 151–239.

Tzelepis, A., Giblin, P. T., & Agronow, S. J. (1983). Effects of adult caregiver's behaviours on the activities, social interactions, and investments of nascent preschool day care groups. *Journal of Applied Developmental Psychology*, 4, 201–216.

United Nations. (1989). *Convention on the rights of the child.* Ottawa: Human Rights Program Department of Canadian Heritage.

Uphoff, James. (1990, September). Extra-year programs: An argument for transitional programs during transitional times. *Young Children*, 45 (6), 19–20.

Van Arsdell, Marian. (1994). Preparing early childhood teachers for careers in learning. In *New perspectives in early childhood teacher education: Bringing practitioners into the debate.* New York: Teachers College Press.

van Manen, Max. (1991). *The tact of teaching: The meaning of pedagogical thoughtfulness.* Albany, NY: State University of New York Press.

Vander Ven, K. (1994). Professional development: a contextual model. In J. Johnson, & J. McCracken (Eds.), *The early childhood career lattice*. Washington, DC: National Association for the Education of Young Children and National Institute for Early Childhood Professional Development.

———. (1988). Pathways to professional effectiveness for early childhood educators. In Bernard Spodek, Olivia N. Saracho, & Donald L. Peters (Eds.), *Professionalism and the early childhood practitioner.* New York: Teachers College Press.

————. (1986). You've come a long way baby: The evolution and significance of caregiving. In Karen Vander Ven & Ethel Tittnich (Eds.), *Competent caregivers-competent children: Training and education for child care practice.* New York: The Haworth Press.

Wasserman, Selma. (1990). *Serious players in the primary classroom.* New York: Teachers College Press.

Whitebook, M., Howes, C., & Phillips, D. (1989). *Who cares? Child care teachers and the quality of care in America.* Final report of the National Child Care Staffing Study. Oakland CA: Child Care Employee Project.

Whitebook, M., Phillips, D., & Howes, C. (1993). *The national child care staffing study revisited: Four years in the life of center-based child care.* Oakland, CA: Child Care Employee Project.

Whitebook, Marcy. (1995, May). What's good for child care teachers is good for our country's children. *Young Children*, 50 (4), 49–50.

Whitebook, Marcy, & Granger, Robert C. (1989, May). Mommy, who's going to be my teacher today? Assessing teacher turnover. *Young Children*, 44 (4), 11–14.

Whitebook, Marcy, Hnatiuk, Patty, & Bellm, Dan. (1994). *Mentoring in early care and education: Refining an emerging career path.* Washington, DC: National Center for the Early Childhood Work Force.

Wiler, B. (1988, May). Nourishing your enthusiasm: Staying excited about your job. *Child Care Information Exchange* (61), 9–10.

Willer, B. (Ed.). (1990). *Reaching the full cost of quality in early childhood programs.* Washington, DC: NAEYC.

Willer, B. (Ed.). (1994). A conceptual framework for early childhood professional development. NAEYC position statement. In Julienne Johnson & Janet B. McCracken (Eds.), *The early childhood career lattice: Perspectives on professional development.* Washington, DC: NAEYC.

Willer, Barbara, & Bredekamp, Sue. (1993, May). A new paradigm of early childhood professional development. *Young Children,* 48 (4), 63–66.

———. (1990, July). Redefining readiness: An essential requisite for educational reform. *Young Children,* 45 (5), 22–24.

Willms, J. D. (2000). Three hypotheses about community effects relevant to the contribution of human and social capital to sustaining economic growth and well-being. Report prepared for the Organization for Economic Cooperation and Development and Human Resources Development Canada. Fredericton, NB: Canadian Research Institute for Social Policy, University of New Brunswick.

Wolery, Mark, & Wilbers, Jan S. (Eds.). *Including children with special needs in early childhood programs.* Washington, DC: NAEYC.

Wollman, P. Greenberg. (1994). *Behind the playdough curtain: A year in my life as a preschool teacher.* New York: Charles Scribner's Sons.

Yonemura, M. V. (1986). *A teacher at work: Professional development and the early childhood educator.* New York: Teachers College Press.

Youcha, Geraldine. (1995). *Minding the children: Child care in America from colonial times to the present.* New York: Scribner.

Young, N. (1994). *Caring for play: The school and child care connection.* Toronto: Exploring Environments.

Zigler, E. (1990). Forward. In *Handbook of early interventions,* S. Meisels and J. Shonkoff (Eds.). New York: Cambridge University Press.

Zigler, Edward F., & Lang, Mary E. (1991). *Child care choices: Balancing the needs of children, families, and society.* New York: The Free Press.

Index

A

Aboriginal Headstart Program, 18–19
Absorbent mind, 35, 58
Accreditation, 91, 97, 213, 223
ACEI, 69, 98
Action Day Care, 188–89, 200
Adequacy, feelings of, 122–23, 126
Administrative style, 83
Adult–child
- programs, 43
- ratios, 11, 30, 78–79
- relationships, 72–73, 108–9

Advocacy, 220–38, 244
- accessibility, 229–30
- coalitions, 235
- compensation, 222–29
- informational, 236–38
- media involvement, 236
- personal, 231
- position papers, 235
- public policy, 221–38
- public policy options, 231–35
- quality, 222–29, 230, 234–35
- unionization, 236
- worthy wage campaigns, 234–35

Alberta Family Day Care Association, 216
Almy, Milly, 124, 195
Antibias curriculum, 51–54, 58
Assessment, 80, 97
Association des services de garde en milieu scolaire du Québec, 215
Association of Childhood Education International (ACEI), 69
Association of Day Care Operators of Ontario, 215
Association of Early Childhood Educators, Newfoundland and Labrador (AECENL), 214
Association of Early Childhood Educators, Ontario (AECEO), 186, 215, 217
Association of Early Childhood Educators, Quebec, 215
At risk, 18
Attachment, 147, 154–55
Authors/editors, 147
Autonomy, 123

B

Bad Guys Don't Have Birthdays (Paley), 75
Behavioural problems, xxii
Behavioural theory, 45, 168, 191
Better Beginnings, Better Futures, 19, 111
Birch, Margaret, 194
Blatz, William, 182, 188
Block grants, 233
Bloom, Benjamin, 169
Blow, Susan, 49, 70
Bott, E. A., 181
Boy Who Would Be a Helicopter, The (Paley), 75
Brain development, xvii, xviii–xix, 49
British Columbia School Age Association, 216
Burnout, 154, 161

C

Canada Health and Social Transfer (CHST), 86
Canadian Association for Young Children, 214
Canadian Association of Family Resource Programs, 214
Canadian Child Care Federation (CCCF), 16, 30, 69, 75, 77, 86, 94–96, 119, 134, 214, 225
- standards of practice, 90–91
- statement of quality, 67, 247–66

Canadian provinces/territories. *See* Provinces/territories
Career lattice, 147–48, 161
Career opportunities, 141–48
Caregiving, 107, 115
Caring for a Living study, 227, 230
Certificate/diploma programs. *See* Postsecondary education
Certification Council of Early Childhood Education of Nova Scotia, 215
Certification, voluntary, 89, 207
Child abuse, xxi
Child–adult ratios. *See* Adult–child
Child and Family Canada, 214
Child care
- administrators, 83
- centres, 8–12, 30, 67, 142
- commercial, 9
- funding, 11, 85, 86
- nonprofit, 9
- *See also* Early childhood education and care (ECEC); Nonparental care

Child Care Advocacy Association of Canada, 119, 189, 209, 214
Child Care Advocacy Association of Nova Scotia, 215
Child Care Connections—NS, 215
Child Care Professional Development Association of Nova Scotia, 215
Child care sector study, 64–65, 119, 133, 149, 213, 228–29, 236, 241
Child development specialist, 143
Child study movement, 181–83, 191
Childcare Resource and Research Unit, 214
Childhood and Society (Erikson), 151
Coalition of Child Care Advocates of BC, 216
Coalitions, 235
Code of ethics, 197, 202–4, 219
Cognitive development, 79, 98, 168, 186
College/university programs. *See* Postsecondary education
Comenius, John, 6
Commercial child care centres, 9, 31
Common School Act, 27, 173
Common Sense Book of Baby and Child Care, The (Spock), 149
Community Action Program for Children (CAP-C), 18, 27
Compensation, 84, 85, 149–50, 222–29
Compensatory programs, 18–19, 31, 187–88, 229
Concentration Inter-régionale des garderies du Québec, 215
Consultants, 145
Content of Children's Minds (Hall), 53, 181
Cooperative learning, 43, 58
Core knowledge, 200–202, 219
Cost, Quality, and Child Outcomes in Child Care Centres study, 227, 228–29
Crèches, 170–72
Credentialing, 129, 206, 219
Curriculum, 33–57, 76
- academic learning, 44
- antibias curriculum, 51–54
- developmentally appropriate practice (DAP), 34–35, 56
- emergent curriculum, 39–40
- emergent literacy, 47–48
- emergent numeracy, 48
- High/Scope, 35
- inclusion, 49–51
- issues/approaches, 41–57
- models, 34–41
- Montessori schools, 35
- problem-based play, 41
- quality, 69–81
- readiness, 44–48
- Reggio Emilia, 39–41
- special-needs children, 49–51
- technology, 55–57

D

Day care centres. *See* Early childhood education and care (ECEC)
Demographics, xx
Derman-Sparks, Louise, 52–53
Developmental health, 168, 191, 242
Developmentally appropriate practice (DAP), 34–35, 56, 58, 91
Dewey, John, 70, 72, 180–81
Didactic Montessori materials, 35
Dingle, Della, 182
Diploma programs. *See* Postsecondary education
Direct guidance, 109, 115
Disability, 49–51
Discovery play, 41
Dispositions, 122–24, 125
DISTAR, 172
Diversity, 70, 77, 98, 127
Doherty, Gillian, 44, 65, 76, 204
Dominion–Provincial Wartime Day Nurseries Agreement, 143, 184
Dotsch, Daisy, 185

E

Early academics, 44, 45, 58
Early child development programs. *See* Early childhood education and care (ECEC)
Early Childhood Coalition Petite Enfance, 215
Early Childhood Development Association, P.E.I., 214
Early childhood education and care (ECEC), 3–27
- child care centres, 8–12, 67, 142, 149
- early childhood intervention, 9, 17–22
- early years communities, 26
- family child care, 9, 13–15, 66, 142
- family support programs, 23–25
- in-home child care, 9, 16–17, 142
- kindergarten, 9, 22–23, 66, 174–76
- leisure and recreation programs, 9, 22, 143

nursery schools, 9, 12–13, 179–80
Early childhood education and care organizations, 213–17
Canadian Child Care Federation, 16, 30, 67, 69, 75, 77, 86, 90–91, 94–96, 119, 134, 214, 225, 247–66
Child Care Advocacy Association of Canada, 119, 189, 209
National Association of the Education for Young Children, 35, 59, 67, 69, 86, 91
provincial/territorial organizations, 213–17
Early Childhood Education, 5. *See also* Postsecondary education
Early childhood educators, 129, 130–35. *See also* Early childhood practitioners
Early Childhood Educators of British Columbia, 203, 204, 216
Early Childhood Environment Rating Scale (ECERS), 92
Early childhood intervention, 17–18
compensatory programs, 17, 18, 31
Early Intervention (EI) programs, 18, 21, 31
inclusive ECEC programs, 17, 49–51
Early childhood settings. *See* Early childhood education and care (ECEC)
Early childhood practitioners, 3, 103–37
careers, 7, 141–47
certification, 129
challenges, 148–60
compensation, 84, 149–50
definition, 104
diversity, 70, 127
education and training (*see* Postsecondary education)
employment contracts, 150
guidelines for success (*see* Guidelines for success)
health and safety, 151
men, 104–7
motivations, 117–21
qualities/dispositions, 122, 238–39
roles, 5, 107–12
skills/competencies, 120, 239–43
stress and isolation, 151–56
quality, 67–69, 75–81
postsecondary education, 130–35 (*see also* Postsecondary education)
stages of development, 124–26
See also Early childhood workforce
Early childhood workforce, xviii, 103–60
careers, 141–47
diversity, 70, 104
men, 104–7
professionalism, 193–218
work environment, 83–84, 141–60
working conditions, 84, 148–57
See also Early childhood practitioners
Early Development Instrument (EDI), 49
Early Intervention (EI) programs, 144
Early Years Study, The (McCain & Mustard), xvii, 164
Early years, xvii–xxiii, 27
ECEC curriculum. *See* Curriculum
ECEC organizations. *See* Early childhood education and care (ECEC)
ECE Professional Association of Alberta, 216
Emergent curriculum, 39–40, 58
Emergent literacy, 47–48, 58
Emergent numeracy, 48, 58
Emile (Rousseau), 167
Emotional development, xxii, 79, 88
Empathy, 123
Employment contract, 150
Equivalency, 129–30
ERIC clearinghouse, 179
Erikson, Eric, 151, 167
Evolution of early childhood care. *See* Historical roots

F

Family child care, 9, 12–16, 66, 142
Family child care certificate programs, 136
Family literacy, 24, 26, 27, 31
Family resource programs, 24–26
Family support programs, 23–27
Family specialists, 143
Feelings of adequacy, 122–23
Ferguson, Elaine, 137
Fine motor skills, 79, 98
Foregone annual wages, 229, 244
Frames of Mind (Gardner), 209
Frank, Lawrence, 118
Freud, Sigmund, 119, 167
Froebel, Friedrich, 23, 26, 174–76
Full cost of quality, 229, 244
Full-day programs, 23, 31
Funding, 11–12, 84–86

G

Gardner, Howard, 168, 209
Gender bias, 54

Gesell, Arnold, 113, 149
Gross motor skills, 79, 98
Guidance, 109
Guidelines for success, 238–42

H

Hall, G. Stanley, 53, 180, 181
Head Start, 18, 20–21, 81, 170, 187
Health and safety, 86, 151
High/Scope, 35–39, 188
Hill, Patty Smith, 58, 112, 181
Hincks, Clare, 181
Historical roots
  baby boom generation, 184–86
  child care, 170, 184, 188–90
  children's playgrounds, 177
  child study movement, 180–82, 199
  crèches, 170–71
  day nurseries, 184, 145
  Froebel kindergartens, 23, 174–76
  How, Hester, 171
  infant schools, 63, 169
  kindergarten, 35, 49, 53, 174–76
  McMillan sisters, 180
  Montessori, Maria, 179–80
  nursery schools, 179, 181, 184
  philosophical views, 167
  progressive education, 180–81
  public schools, 172–74
  reason for studying, 166–67
  recent developments, 168–69, 188–90
  Timeline, 168
Home visiting, 27
How, Hester, 63, 171, 175
Hughes, Ada Maream, 175
Hughes, James L., 174, 175
Hunt, J. McVicker, 158

I

Inclusion, 17, 48–57, 59
Indirect guidance, 109, 115
Individualized family plan, 144
Individualized program plan, 144
Infant development programs, 19
Infant schools, 25, 169
Informational advocacy, 236–38
Inherited, definition of, 120, 140
In-home child care, 16
In-service training and education. *See* Professional development
Instructors, 146
Integrated curriculum, 80, 198
Intelligence and Experience (Hunt), 158
Interaction, 209
Interest centres, 82
Interrelationships, 65, 98

J

Jobs. *See* Career opportunities
Johnson, Harriet, 91, 108
Journal, 140
Junior kindergarten, 23, 146

K

Key experiences, 36–37, 59
Kindergartener, 175, 191
Kindergartens, xxii, 7, 22–23, 45, 49, 53, 66–67, 174–76
Kohlberg, Lawrence, 168, 198

L

Language and Thought of the Child, The (Piaget), 120
Lapierre, Laurier, 207
Libraries, 22
Local initiatives program (LIP) grants/child care programs, 186–87
Live-in Caregiver Program, 17, 230
Live-in caregivers, 16–17
Living and Learning report, 179
Locke, John, 7, 167
Longitudinal studies, 46, 119, 140, 167, 200

M

Mainstreaming, 49
Malaguzzi, Loris, 39, 73, 236
Male early childhood practitioners, 104–7
Manitoba Child Care Association, 216
Maream, Ada, 175
Marsh Report on Social Security, 146
McMillan, Margaret and Rachel, 78, 83, 180, 181
Mentor, 137, 140
Mentoring, 136–38
Merchandiser, career as, 147
Mitchell, Lucy Sprague, 91, 94
Mixed-age groupings, 42–43, 59
Mollie Is Three (Paley), 75
Moncton Headstart Inc., 20–21, 188
Montessori, Maria, 35, 77, 90, 179–80, 181
Montessori Method, The (Montessori), 179
Montessori schools, 35, 59, 83, 156
Moral development, 79, 98
Moral Stages and Moralization (Kohlberg), 198

Mothercraft infant caregiver training program, 124
Motivations for choosing early childhood education, 117–21
Multicultural, 52, 59
Multicultural issues in child care, 53–54
Multiple intelligences, 168, 191
Museums, 22

N

NAEYC position statement, 91
Nannies, 16
National Association of the Education for Young Children (NAEYC), 35, 59, 67, 69, 86, 91
National Guidelines for Training in Early Childhood Care and Education (CCCF), 134–35
Networking, 240, 244
Neurons. *See* Brain development
Nonparental care, xx, 7
Nonprofit, 9, 11, 26
Nursery Education Association of Ontario (NEAO), 151, 186
Nursery School, The (M. McMillan), 108
Nursery schools, xxii, 12–13, 179–80, 191
Nurturing, 108, 115

O

Ontario Association for School Age Care, 216
Ontario Association of Family Resource Programs, 216
Ontario Coalition for Better Child Care (OCBCC), 207, 216, 235
Ontario Network of Home Child Care Provider Groups, 21, 215
Open-ended, definition of, 98
Ottawa Day Nursery, 83
Our Child Care Workforce study. *See* Child care sector study
Owen, Robert, 16, 169

P

Parent cooperative, 185, 191
Parenting, 6, 26
Partners in Quality project, 94–96
Peabody, Elizabeth, 49, 175–76
Perry Preschool Project, 167, 187, 231
Personal advocacy, 221
Personal philosophy of education, 242
Pestalozzi, Johann, 17
Philanthropy, 170
Philosophical views, 170
Philosophy, 33, 47, 59
Physical development, 79
Physical environment, 80–82, 98
Piaget, Jean, 36, 112, 120, 167, 168, 186
Plan-do-review, 36, 59
Play, Dreams, and Imitation in Children (Piaget), 167
Playgrounds, 76, 177–78
Position statement, 67, 98. *See also* NAEYC position statement
Postsecondary education, 127–35, 197
    ECE certificate and diploma programs, 77–79, 130–35, 177
    equivalency, 129
    family child care certificate program, 136
    postcertificate/postdiploma programs, 136
    provincial/territorial requirements, 87–89
    Red River Community College, 131–32
    regulation, 129
    stages of student early childhood practitioner development, 125–26
    teacher education, 136
    university degree programs, 78
Poverty, xxi
Practical life skills, 35
Pratt, Caroline, 84, 85
Pregnancy, 6
Preschool programs (preschools), 12
Preservice training and education. *See* Postsecondary education
Primary caregivers, 72, 98
Professional, definition of, 219
Professional development, 208–9
Professional education
    mentoring, 136–38
    stages of professional development, 124–26
    stages of student early childhood practitioner development, 125–26
    See also Postsecondary education
Professionalism, 184–217
    code of ethics, 197, 202–4, 219
    core knowledge, 200–202, 219
    credentialing, 206–7, 219
    dilemmas, 209–12
    professional development, 208–9
    self-governance, 208
    standards of practice, 197, 204
    voluntary certification, 207–8

Professionalization dilemmas, 209–12
Professional organizations. *See* Early childhood education and care organizations
Progressive education, 191
Provinces/territories
  child care, 10
  early intervention programs, 20
  family child care, 10, 14–15, 16
  family resource programs, 26
  governments, xxii, 4
  kindergartens, 23
  professional organizations, 214–16
  public funding, 86
  regulated ECEC services, 10–11
  staffing requirements, 87–89
  wages (compensation), 150
Psychological Care of Infant and Child (Watson), 122
Psychosocial theory, 191
Publications, 209
Public funding, 86, 189
Public health, 52, 123, 178–9
Public policy advocacy, 221–33
Public policy options, 230–33
Public schools, 172–74

Q

Quality, 62–96
  accessibility, 230
  accreditation, 91
  administrative style, 82
  bottom-up (children's) perspective, 63
  CCCF national statement, 247–66
  compensation, 84, 222–28
  context components, 68, 182–92
  dynamic components, 68, 69–75
  evaluation tools, 92
  framework components, 68, 75–82
  funding, 82–84
  indifference, 1
  inside (staff) perspective, 63
  integration of learning experiences, 79–80
  NAEYC position statement, 67
  number of children, 78–79
  outside (community/society) perspective, 63
  outside–inside (parents) perspective, 63
  physical environment, 80–82
  poor quality, characteristics of, 92–94
  reciprocity, 73–75
  regulation, 86–89, 223
  respect, 69–72
  responsiveness, 71–75
  staff–child ratios, 78–79
  top–down (expert) perspective, 63
  underlying developmental understanding, 76–78
  voluntary standards, 89–90
  work environment, 82–83
Quality, compensation, and accessibility, 223–31, 244
Quebec Association for Preschool Development, 215

R

Readiness, xix, 44–48, 59
Reciprocity, 73–75, 98
Recreation leaders, 22
Red River Community College, 131–32
Reflective teaching, 241, 244
Reggio Emilia, 39–41, 59
Regulation, 11, 86–89, 129, 144–45
Report on a System of Elementary Public Instruction for Upper Canada (Ryerson), 172–73
Resource teacher, 142
Respect, 69–72, 98, 156
Responsiveness, 71–75, 98
Rockefeller, Laura Spelman, 181
Roles of early childhood practitioners, 107–11
Rote learning, 180, 191
Rousseau, Jean Jacques, 167
Rural Roots Children's Centre, 13
Ryerson, Egerton, 172–73, 174

S

Salle d'asile Saint Joseph, 43
Saskatchewan Child Care Association, 216
Scherman, Ada, 207
Schurz, Margarethe, 175
Self-control, 46
Self-discovery, 122
Self-esteem, 46, 70
Self-evaluation, 242
Self-governance, 208
Self-identity, 70
Self-regulation, 70, 98
Senior kindergarten, 23
Sense of the big picture, 124
Sensitive period, 35, 59
Separation, 53, 161

Sequential steps, 35
Sesame Street, 180, 187–88
Skinner, B. F., 168
Social development, 79, 98
Social services, 32
Special education, 49, 59
Special needs, 32, 49
SpeciaLink, 214
Spitz, Rene, 147
Spock, Benjamin, 149
St. George's School for Child Study, 118, 120, 182–83
Stability and Change in Human Characteristics (Bloom), 169
Staff–child ratios. *See* Adult–child
Staff turnover, 98
Standardized tests, 59
Standards of practice, 197, 204–5
Stress, 151–52, 161

T

Teacher education programs, 136
Teachers. *See* Early childhood practitioners
Team teaching, 240, 244
Technology, xix, 54, 59
Territories. *See* Provinces/territories
Theories, 76–78, 98, 167–69
Timeline, 165
To Herald a Young Child (Lapierre & Scherman), 207
Towards Excellence in ECCE Training Programs: A Self-Assessment Guide, 134
Transitions, 82, 99

U

Understanding the Early Years project, 28
Unionization, 236
Unit blocks, 85
University degree programs, 78, 130, 135–36

V

Vancouver Infants Centre, 83
Vygotsky, Lev, 168

W

Wages. *See* Compensation
War-time day nurseries, 143, 145, 184
Watson, John, 128
Weikart, David, 167
Westcoast Child Care Resource Centre, 216
Western Canadian Family Child Care Association, 216
Whole child, 79, 99
Work environment, 82–84
Worthy wage campaigns, 229, 231, 234–35
Wright, M. J., 188

Y

You Bet I Care! study, 68
Young Child journal, 209
Yukon Child Care Association, 216

Notes

Notes

Notes

Notes

Notes

Notes

Notes